MEMBRANE PHYSIOPATHOLOGY

DEVELOPMENTS IN CARDIOVASCULAR MEDICINE

121. S. Sideman, R. Beyar and A. G. Kleber (eds.): Cardiac Electrophysiology, Circulation, and Transport. Proceedings of the 7th Henry Goldberg Workshop (Berne, Switzerland, 1990). 1991.ISBN 0-7923-1145-0.
122. D. M. Bers: Excitation-Contraction Coupling and Cardiac Contractile Force. 1991. ISBN 0-7923-1186-8.
123. A.-M. Salmasi and A. N. Nicolaides (eds.): Occult Atherosclerotic Disease. Diagnosis, Assessment and Management. 1991. ISBN 0-7923-1188-4.
124. J. A. E. Spaan: Coronary Blood Flow. Mechanics, Distribution, and Control. 1991. ISBN 0-7923-1210-4.
125. R. W. Stout (ed.): Diabetes and Atherosclerosis. 1991. ISBN 0-7923-1310-0.
126. A. G. Herman (ed.): Antithrombotics. Pathophysiological Rationale for Pharmacological Interventions. 1991. ISBN 0-7923-1413-1.
127. N. H. J. Pijls: Maximal Myocardial Perfusion as a Measure of the Functional Significance of Coronary Arteriogram. From a Pathoanatomic to a Pathophysiologic Interpretation of the Coronary Arteriogram. 1991. ISBN 0-7923-1430-1.
128. J. H. C. Reiber and E. E. v. d. Wall (eds.): Cardiovascular Nuclear Medicine and MRI. Quantitation and Clinical Applications. 1992. ISBN 0-7923-1467-0.
129. E. Andries, P. Brugada and R. Stroobrandt (eds.): How to Face "the Faces" of Cardiac Pacing. 1992. ISBN 0-7923-1528-6.
130. M. Nagano, S. Mochizuki and N. S. Dhalla (eds.): Cardiovascular Disease in Diabetes. 1992. ISBN 0-7923-1554-5.
131. P. W. Serruys, B. H. Strauss and S. B. King III (eds.): Restenosis after Intervention with New Mechanical Devices. 1992. ISBN 0-7923-1555-3.
132. P. J. Walter (ed.): Quality of Life after Open Heart Surgery. 1992. ISBN 0-7923-1580-4.
133. E. E. van der Wall, H. Sochor, A. Righetti and M. G. Niemeyer (eds.): What is new in Cardiac Imaging? SPECT, PET and MRI. 1992. ISBN 0-7923-1615-0.
134. P. Hanrath, R. Uebis and W. Krebs (eds.); Cardiovascular Imaging by Ultrasound. 1992. ISBN 0-7923-1755-6.
135. F. H. Messerli (ed.): Cardiovascular Disease in the Elderly, 3rd ed. 1992. ISBN 0-7923-1859-5.
136. J. Hess and G. R. Sutherland (eds.); Congenital Heart Disease in Adolescents and Adults. 1992. ISBN 0-7923-1862-5.
137. J. H. C. Reiber and P. W. Serruys (eds.): Advances in Quantitative Coronary Arteriography. 1993. ISBN 0-7923-1863-3.
138. A.-M. Salmasi and A. S. Iskandrian (eds.): Cardiac Output and Regional Flow in Health and Disease. 1993. ISBN 0-7923-1911-7.
139. J. H. Kingma, N. M. van Hemel and K. I. Lie (eds.): Atrial Fibrillation, a Treatable Disease? 1992. ISBN 0-7923-2008-5.
140. B. Ostadal, N. S. Dhalla (eds.): Heart Function in Health and Disease. 1993. ISBN 0-7923-2052-2.
141. D. Noble and Y.E. Earm (eds.): *Ionic Channels and Effect of Taurine on the Heart.* Proceedings of an International Symposium (Seoul, Korea, 1992). 1993. ISBN 0-7923-2199-5.
142. H.M. Piper and C.J. Preusse (eds.): *Ischemia-reperfusion in Cardiac Surgery.* 1993. ISBN 0-7923-2241-X.
143. J. Roelandt, E.J. Gussenhoven and N. Bom (eds.): *Intravascular Ultrasound.* 1993. ISBN 0-7923-2301-7.
144. M.E. Safar and M.F. O'Rourke (eds.): *The Arterial System in Hypertension.* 1993. ISBN 0-7923-2343-2.
145. P. W. Serruys, D.P. Foley and P.J. de Feyter (eds.): *Quantitative Coronary Angiography in Clinical Practice.* 1993. ISBN 0-7923-2368-8.
146. J. Candell-Riera and D. Ortega-Alcalde (eds.): *Nuclear Cardiology in Everyday Practice.* 1993. ISBN 0-7923-2374-2.
147. P. Cummins (ed.): *Growth Factors and the Cardiovascular System.* 1993. ISBN 7923-2401-3.
148. K. Przyklenk, R.A. Kloner and D.M. Yellon (eds.): *Ischemic Preconditioning: The Concept of Endogenous Cardioprotection.* 1993. ISBN 0-7923-2410-2.

MEMBRANE PHYSIOPATHOLOGY

Edited by
GHASSAN BKAILY

SPRINGER-SCIENCE+BUSINESS MEDIA, LLC

Library of Congress Cataloging-in-Publication Data

Membrane physiopathology / edited by Ghassan Bkaily.
p. cm. -- (Developments in cardiovascular medicine ; DICM 159)
Includes index.
ISBN 978-0-7923-3062-2 ISBN 978-1-4615-2616-2 (eBook)
DOI 10.1007/978-1-4615-2616-2
1. Membrane disorders--Pathophysiology. I. Bkaily, Ghassan.
II: Series: Developments in cardiovascular medicine ; v. 159.
[DNLM: 1. Cell Membrane--physiology. 2. Cardiovascular Diseases--physiopathology. 3. Cardiovascular System--cytology.
4. Gastrointestinal Diseases--physiopathology. QH 601 M53255 1994]
RB113.M46 1994
616.07--dc20
DNLM/DLC 94-33097
for Library of Congress CIP

Printed on acid-free paper.

DEDICATION

This book is dedicated to the memory of my father,
Mohamed Rustom El-Bkaily and my mother Karia Zayed-El-Bkaily,
whose constant encouragement and help
were essential to the completion of this work.
I shall forever be in debt to them.

ACKNOWLEDGEMENT

The author is grateful to Christiane Ducharme
for her excellent assistance
in the preparation of this book.

CONTENTS

LIST OF CONTRIBUTORS

Elias Abdulnour
Department of Surgery
Faculty of Medicine
University of Sherbrooke
Sherbrooke, Quebec, Canada

Nasir Afzal
Division of Cardiovascular Sciences
St. Boniface General Hospital Research Center
University of Manitoba, Faculty of Medicine
351 Tache Avenue
Winnipeg, Manitoba, Canada

Habib M. Alloush
Department of Physiology
Faculty of Medicine
American University of Beirut
Beirut, Lebanon

Jean-Luc Ardilouze
Department of Medicine, Endocrinology Service
Faculty of Medicine
University of Sherbrooke
Sherbrooke, Quebec, Canada

Jean-François Beaulieu
Groupe de recherche en biologie du développement
Département d'anatomie et de biologie cellulaire
Faculté de médecine
Université de Sherbrooke
Sherbrooke, Québec, Canada

Richard Béliveau
Groupe de recherche en biothérapeutique moléculaire
Université du Québec
Montréal, Québec, Canada

Sylvie Bernier
Department of Pharmacology
Faculty of Medicine
University of Sherbrooke
Sherbrooke, Quebec, Canada

David Bielefeld
Department of Physiology and Biophysics
College of Medicine
University of Cincinnati
Cincinnati, Ohio 45267-5636 USA

Anwar B. Bikhazi
Department of Physiology
Faculty of Medicine
American University of Beirut
Beirut, Lebanon

Ghassan Bkaily
Department of Physiology and Biophysics
Faculty of Medicine
University of Sherbrooke
Sherbrooke, Quebec, Canada

Dominique Boivin
Groupe de recherche en biothérapeutique moléculaire
Université du Québec
Montréal, Québec, Canada

Shufen Cai
Department of Physiology and Biophysics
Faculty of Medicine
University of Sherbrooke
Sherbrooke, Quebec, Canada

Mouna Chakir
Department of Pharmacology
Faculty of Medicine
University of Sherbrooke
Sherbrooke, Quebec, Canada

Audrey Claing
Department of Pharmacology
Faculty of Medicine
University of Sherbrooke
Sherbrooke, Quebec, Canada

Fabien deLorenzi
Department of Physiology and Biophysics
Faculty of Medicine
University of Sherbrooke
Sherbrooke, Quebec, Canada

Naranjan S. Dhalla
Division of Cardiovascular Sciences
St. Boniface General Hospital Research Center
University of Manitoba, Faculty of Medicine
351 Tache Avenue
Winnipeg, Manitoba, Canada

Pedro D'Orleans-Juste
Department of Pharmacology
Faculty of Medicine
University of Sherbrooke
Sherbrooke, Quebec, Canada

Jianmin Duan
Department of Pharmacology and Toxicology
Medical Sciences Building
University of Western Ontario
London, Ontario, Canada

Demetri Economos
Department of Physiology and Biophysics
Faculty of Medicine
University of Sherbrooke
Sherbrooke, Quebec, Canada

Muna El-Kasti
Department of Physiology
Faculty of Medicine
American University of Beirut
Beirut, Lebanon

Najla A. Fakruddin
Department of Physiology
Faculty of Medicine
American University of Beirut
Beirut, Lebanon

Pallab K. Ganguly
Division of Cardiovascular Sciences
St. Boniface General Hospital Research Centre
University of Manitoba, Faculty of Medicine
351 Tache Avenue
Winnipeg, Manitoba, Canada

Line Garneau
Research Group on Membrane Transport
Department of Physiology
University of Montreal
Montreal, Quebec, Canada

Denis Gingras
Groupe de recherche en biothérapeutique moléculaire
Université du Québec
Montréal, Québec, Canada

Gaétan Guillemette
Department of Pharmacology
Faculty of Medicine
University of Sherbrooke
Sherbrooke, Quebec, Canada

George E. Haddad
Department of Physiology and Biophysics
Faculty of Medicine
University of Sherbrooke
Sherbrooke, Quebec, Canada

Yoshihito Inoue
Department of Physiology and Biophysics
College of Medicine
University of Cincinnati
Cincinnati, Ohio, USA

Danielle Jacques
Department of Physiology and Biophysics
Faculty of Medicine
University of Sherbrooke
Sherbrooke, Quebec, Canada

Gaétan Jasmin
Department of Pathology
Faculty of Medicine
University of Montreal
2900 Edouard MontPetit
Montreal, Quebec, Canada

Michael Koutsilieris
Endocrine-Oncology Clinic Research Center
Laval University Hospital Center
and Hormonal Bioregulation Laboratory
Laval University
Quebec, Canada

Kanzhi Liu
Division of Cardiovascular Sciences
St. Boniface General Hospital Research Centre
University of Manitoba, Faculty of Medicine
351 Tache Avenue
Winnipeg, Manitoba, Canada

John Lorenz
Department of Physiology and Biophysics
College of Medicine
University of Cincinnati
Cincinnati, Ohio 45267-5636 USA

Daniel Ménard
Groupe de recherche en biologie du développement
Département d'anatomie et de biologie cellulaire
Faculté de médecine
Université de Sherbrooke
Sherbrooke, Québec, Canada

Margaret P. Moffat
Department of Pharmacology and Toxicology
Medical Sciences Building
University of Western Ontario
London, Ontario, Canada

Makoto Nagano
Department of Internal Medicine
Aoto Hospital - Jikei University
Tokyo, Japan

Radha Naik
Department of Physiology and Biophysics
Faculty of Medicine
University of Sherbrooke
Sherbrooke, Quebec, Canada

Yusuke Ohya
Department of Physiology and Biophysics
College of Medicine
University of Cincinnati
Cincinnati, Ohio, USA

Grant N. Pierce
Division of Cardiovascular Sciences
St. Boniface General Hospital Research Centre
University of Manitoba, Faculty of Medicine
351 Tache Avenue
Winnipeg, Manitoba, Canada

Gérard E. Plante
Departments of Medicine,
Physiology and Pharmacology
University of Sherbrooke
Sherbrooke, Quebec, Canada

Libuse Proschek
Department of Pathology
Faculty of Medicine
University of Montreal
2900 Edouard MontPetit
Montreal, Quebec, Canada

Jihong Qu
Department of Physiology and Biophysics
Faculty of Medicine
University of Sherbrooke
Sherbrooke, Quebec, Canada

Darren E. Richard
Department of Pharmacology
Faculty of Medicine
University of Sherbrooke
Sherbrooke, Quebec, Canada

Domenico Regoli
Department of Pharmacology
Faculty of Medicine
University of Sherbrooke
Sherbrooke, Quebec, Canada

Peter A. Rogers
Endocrine-Oncology Clinic Research Center
Laval University Hospital Center
and Hormonal Bioregulation Laboratory
Laval University
Quebec, Canada

Elena Ruiz-Petrich
Department of Physiology and Biophysics
Faculty of Medicine
University of Sherbrooke
Sherbrooke, Quebec, Canada

Heinz Rupp
Physiologishes Institute II
Universitat Tubingen
Tubingen, Germany

Animesh Sahai
Division of Cardiovascular Sciences
St. Boniface General Hospital Research Center
Faculty of Medicine, University of Manitoba
351 Tache avenue
Winnipeg, Manitoba, Canada

Rémy Sauvé
Research Group on Membrane Transport
Department of Physiology
University of Montreal
Montreal, Quebec, Canada

Otto F. Schanne
Department of Physiology and Biophysics
Faculty of Medicine
University of Sherbrooke
Sherbrooke, Quebec, Canada

Keiichi Shimmamura
Department of Physiology and Biophysics
College of Medicine
University of Cincinnati
Cincinnati, Ohio, USA

Adrian Sculptoreanu
Department of Pharmacology
University of Washington
Seattle, Washington, USA

Nick Sperelakis
Department of Physiology and Biophysics
College of Medicine
University of Cincinnati
Cincinnati, Ohio, USA

Nobuakira Takeda
Department of Internal Medicine
Aoto Hospital - Jikei University
Tokyo, Japan

Sabine Télémaque
Department of Pharmacology
Faculty of Medicine
University of Sherbrooke
Sherbrooke, Quebec, Canada

Fadia H. Uthman
Department of Physiology
Faculty of Medicine
American University of Beirut
Beirut, Lebanon

Rui Wang
Research Group on Membrane Transport
Department of Physiology
University of Montreal
Montreal, Quebec, Canada

Shimin Wang
Department of Physiology and Biophysics
Faculty of Medicine
University of Sherbrooke
Sherbrooke, Quebec, Canada

Christopher A. Ward
Department of Pharmacology and Toxicology
Medical Sciences Building
University of Western Ontario
London, Ontario, Canada

Zhiling Xiong
Department of Physiology and Biophysics
College of Medicine
University of Cincinnati
Cincinnati, Ohio 45267-5636 USA

PREFACE

The need for publishing a comprehensive review of a number of different membrane pathologies of muscle and non-muscle cells in illnesses ranging from diabetes to heart disease and cancer lies on to the fact that there are several books dealing with the properties of normal cell membranes, although there are very few books focussing on the abnormal membrane behavior. Since the membrane is the critical outer barrier of a cell, this membrane could be the first structure to be affected in some diseases. Research is advancing at the cellular level at a very rapid rate. We can now address questions such as: "How and by what is the mechanism underlying membrane ion channel and receptor dysfunction leading to abnormal cell function?" and "What substances cause dysfunction in specific ion channels or receptors?". Such questions bring together the microscopic world of the cell with the macroscopic manifestation of disease. We believe that a book such as this one would help researchers, physicians, and students to better understand the relationship between cell membrane dysfunction and abnormal function of the cell and tissue. This book is intended for practicing clinicians and academic researchers, as well as resident physicians, medical students and graduate students. Hopefully, such treatise will help to fill an important gap between basic science and clinical science. We are greatly indebted to all the distinguished and highly-qualified researchers from university and industrial milieus who contributed to this book. Finally, we would like to thank the publishers for their confidence and cooperation in making this book available for the medical sciences.

Ghassan Bkaily

INTRODUCTION

This book is a timely reminder of the importance of fundamental aspects of pathology at the cell membrane which plays the role of a selective filtre and of a mirror that reflects the functional image of every active nuclei, cytoplasmic entity and catalyser.

Thus the membranes of the sarcolemma, intracelular organelles as well as nuclei may frequently play a primary role in the onset and development of disease. Clinical aspects of a pathology alone may result in a sterile and unproductive future in this field. Thus, knowing the fundamental aspect of membrane pathophysiology will allow clinical colleagues to integrate the elementary components of pathology with the overall aspect of disease and create a cross-fertilization between basic and clinical sciences.

The major difficulty in preparing this book was the selection of the diseases to be covered. We chose to focus on membrane pathophysiology of the cardiovascular system as well as kidney and cell differentiation without forgetting the most important aspect of the membrane functioning such as receptors, ionic channels and Ca^{2+} metabolism.

Ghassan Bkaily

CHAPTER 1

HAMSTER CARDIOMYOPATHY: NEW INSIGHTS IN THE PATHOGENESIS OF THIS HEREDITARY DISEASE

Gaétan Jasmin and Libuse Proschek,
University of Montreal, Montreal, Canada

CLINICAL COURSE AND GROSS PATHOLOGY IN UM-X7.1 CARDIOMYOPATHIC HAMSTER

Since 20 years ago, a new cardiomyopathic (CM) Syrian hamster colony, the UM-X7.1, has been established and raised under optimal environmental conditions (1). With a careful selection of breeders, it has been possible to preserve the original trait from the BIO 14.6 progenitors (2) and to arrive at a uniform and predictable expression of the disease. The average survival time of this new line is 250 days. Hence, it has been possible to carry out pathogenic studies in a reliable manner. Apart from a progressive blueing of the eyes due to venous congestion, the clinical pathological signs of this polymyopathy are very subtle. One may detect some impaired locomotor activity, becoming more evident with forced muscular exercise (3). Abnormal electrocardiogram patterns, as we will see in a moment, become significant after 150 days (4) when the heart tends to hypertrophy with concomitant dilatation of the right chambers and development of progressive changes of circulatory failure. Premature death may result from ventricular arrhythmia; otherwise, more than 50% of the remaining animals exhibit variable degree of cardiac failure. Signs of passive congestion are more obvious at autopsy when examining the lung or the liver. The hepatic tissue is often sclerotic with liquid effusion in the peritoneum and in the thoracic cavity, and ultimately one may observe a giant anasarca. By then, the entire myocardium becomes flabby due to dilatation of the heart cavities which often contain a mural thrombus (Fig. 1).

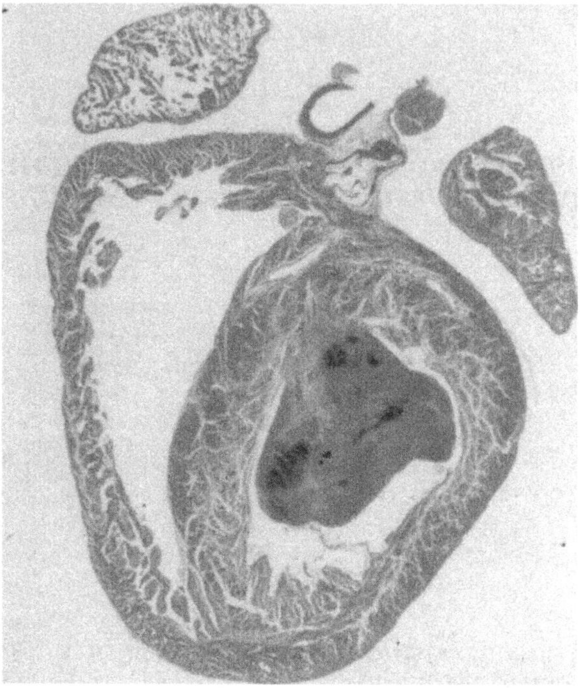

Figure 1. *Transection of a myopathic heart during the late stage of heart failure. Note the presence of a mural thrombus in the right ventricle (from Jasmin et al. (4)).*

Histopathology of the Cardiac Damage

In light microscopy, myocardial lesions are readily detectable at 45 days. The necrotizing process is focal and diffuse and predominates in the left myocardium. It is characterized by sarcoplasmic coagulative changes which become granular and often calcified or by mere dissolution of the myofibrils. Hence, the cardiocytes may be dissolved in the absence of cellular infiltration leaving sarcolemmal remnants and occasional, few nuclei.

Histogenesis of the myocardial degeneration has shown that the progression of the disease follows a phasic pattern; necrotization culminates at 80 days and is followed by a scaring process which lasts until 150 days. The calcified fragments of cardiocytes are then sequestered within the fibrotic tissue; by that time, the heart muscle has become hypertrophic especially in the septal area. Interestingly, the myolytic changes tend to aggravate with occurrence and severity of heart failure suggesting an activation of degradative enzyme(s) activity (5).

Though the cardiac symptoms predominate in the hamster myopathic disease, let us recall that the pathologic changes equally involve skeletal muscles with similar degenerative changes running parallel with the cardiomyopathy. By 15 days of age, necrotic lesions are detectable by macroscopic examination of intercostal muscles. They appear as white streaks and correspond to coagulative necrosis with less calcification than in myocardium; they last until 120 days of age. Thereafter, regeneration occurs with little scar tissue formation and almost no interference with the animal mobility, a basic difference with progression of human muscular dystrophy. With subsidence of the necrotizing process, the elevated serum creatine kinase tends to diminish, but never to normal values.

Terminal Stage of Heart Failure

Over 150 days, early manifestations of heart failure with premature death are occasionally seen, but these become more conspicuous when subcutaneous edema become tangible by 220 days of age: ECG abnormalities are also quite relevant as well as the progressive lost of myocardial contractility along with the depletion of energy reserves. These pathologic changes have been ascribed to a drop in adrenergic tone (6, 7), membrane alterations, myolysis and a progressive shift of α into β myosin heavy chain isoform (8, 9). A chronic verapamil treatment succeeded in preserving the energy reserve in as much as the contractility of the myocardium, thereby prolonging the life span of treated cardiomyopathic hamsters (8, 10-13).

In Summary. Cardiomyopathy in hamsters develops in a phasic pattern with well defined and predictable stages: 1- a necrotic phase with disseminated lesions in the myocardium elapsing between 30 and 120 days; 2- a healing phase characterized by a scarring process with ventricular hypertrophy and dilatation until 200 days of age and 3- a terminal phase with typical changes of heart failure until death occurs between 250-300 days.

ECG RECORDINGS

To study electrophysiological parameters of myopathic hearts by means of a non-invasive method, we undertook evaluating the progression of the myopathic changes through ECG recordings.

ECG parameters were investigated in 50 normal and 165 myopathic hamsters (aged between 30-230 days) to further characterize the necrotizing and scarring process in the myocardium and the severity of heart failure. ECG readings, obtained by the computer averaging technique, were carried out under anesthesia using Na-pentobarbital (45-50 mg/kg i.p.) 10-15 min. prior to the recording (4). After 5 min. of equilibration, within the range of 100 beats, the following variables were then computed: RR interval (ms), PR interval (ms), QRS width (ms), QT interval (from

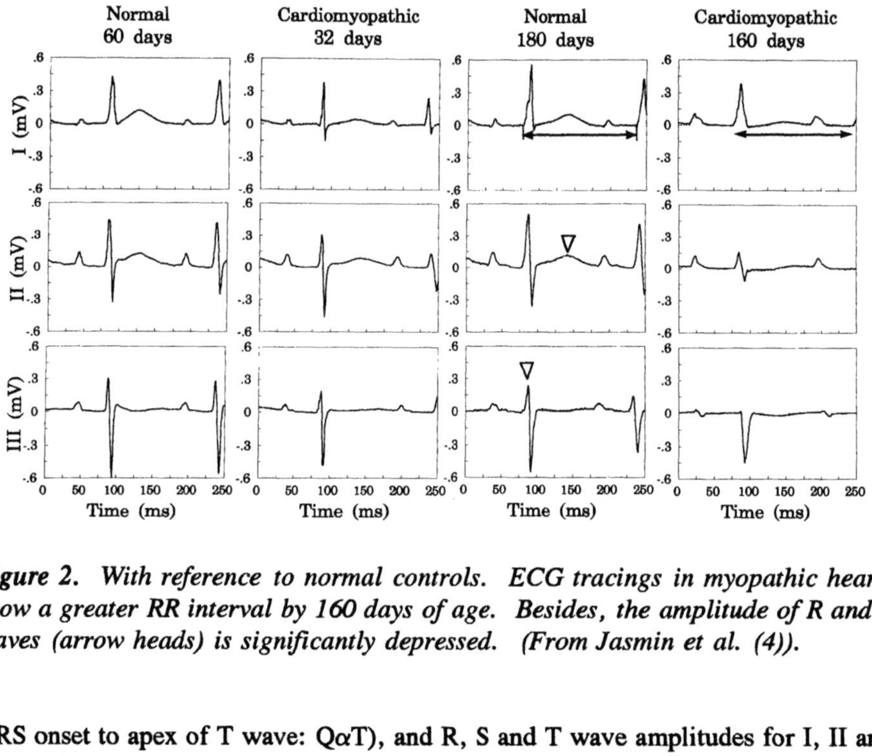

Figure 2. With reference to normal controls. ECG tracings in myopathic hearts show a greater RR interval by 160 days of age. Besides, the amplitude of R and T waves (arrow heads) is significantly depressed. (From Jasmin et al. (4)).

QRS onset to apex of T wave: QαT), and R, S and T wave amplitudes for I, II and III standard leads. QαT was normalized for RR variations ($Q\alpha Tc = Q\alpha T/\mathrm{sqrt}(RR)$). To assess the reproducibility of the tracings, 10 normal controls were used on different days and their ECGs recorded; no appreciable changes were noticed.

Representative ECG computerized tracings of younger and older adult normal and CM hamsters are shown in Fig. 2. The ECGs of the normal animals show taller and deeper waves and shorter intervals than those of CM hamsters; the earliest and most consistent changes observed in myopathic hearts were: a) prolongation of the intervals; b) progressive flattening, even inversion of T-wave and c) a decrease of the amplitude of R-wave, mostly evident in lead III. A progressive decrease of heart rate with age was observed in CM and, to a lesser extent, in normal hamsters as shown in Fig. 3 *left*. (R-R interval is inversely related to heart rate). A marked early prolongation in the duration of P-R (Fig. 3 *right*) and QTc (Fig. 4 *left*) intervals during the development of the disease was striking. Unexpectedly, the P-R intervals were even significantly longer in CM females (independently of a previous pregnant state) than those in their age matched male counterparts (Fig. 4 *right*). Such sex difference did not exist in normal controls. Surprisingly the duration of QRS complex was similar in CM and normal hamsters (not shown). Except for a consistently low amplitude of R wave (Fig. 5 *left*), the flattening or even inversion of T-wave were prominent and regular features in CM animals (Fig. 5 *right*). Most likely, this is attributable to increase in edema, in as much to the thoracic size and to membrane depolarization abnormalities in CM hearts respectively.

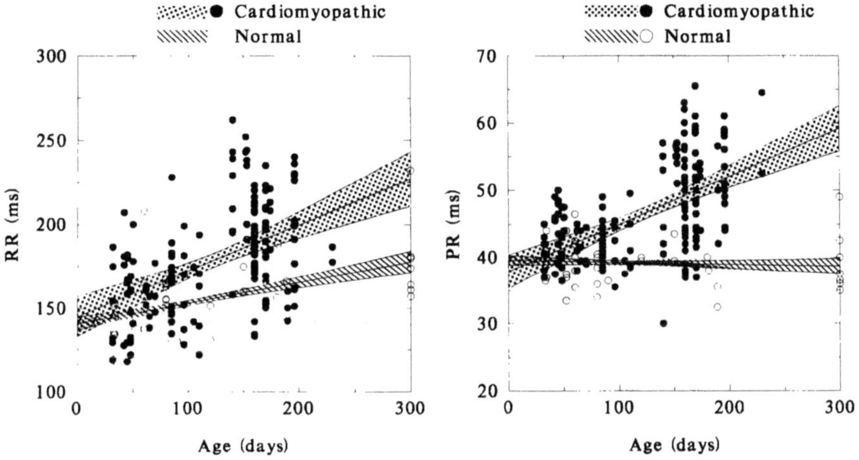

Figure 3. *A prolonged (left) RR and (right) PR interval is readily seen in CM hamsters by 50 days of age. (From Jasmin et al. (4)).*

In Summary. Through these ECG studies, we have learned: 1- that the overall electrical activity in myopathic hearts is reduced; 2- that the heart rate becomes slower with progression of the disease; 3- that the PR interval becomes elongated at an early stage of the disease, suggesting a delay in atrio-ventricular conduction that may lead to a complete A-V block; 4- that the cardiomyopathy is more severe in females than in males; 5- that CM hamsters are prone to develop arrhythmia as evidenced by a prolongation of the QT interval.

The Concept of a Defective Membrane with Disturbance in Calcium Metabolism

It has often been claimed that a rise in calcium level within muscle cells is the earliest manifestation of the membrane genetic defect (14-18). High intracellular calcium concentration would inactivate calcium slow channels and disrupt the homeostasis. Interestingly, only liposoluble calcium channels blockers such as verapamil or diltiazem can restore inner cell electrolyte equilibrium (19) and at the

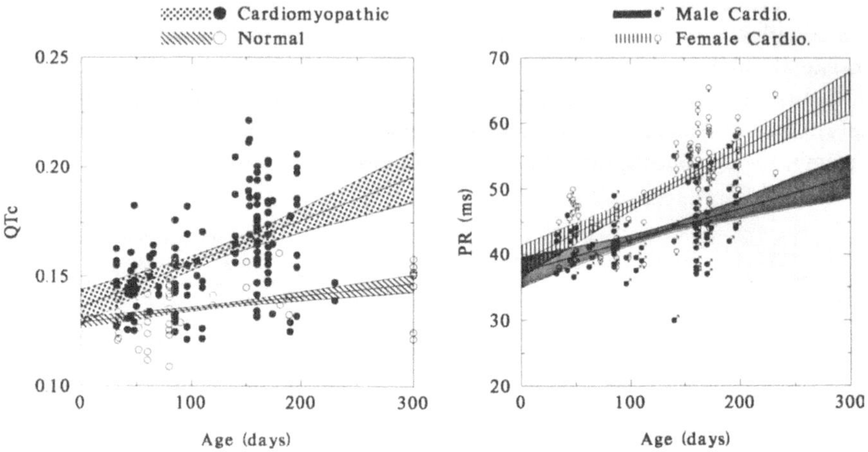

***Figure 4.** The elongation of QTc (left) at 30 days of age is one of the earliest ECG findings in myopathic hamsters. On the right side, the difference in PR intervals between males and females. (From Jasmin et al. (4)).*

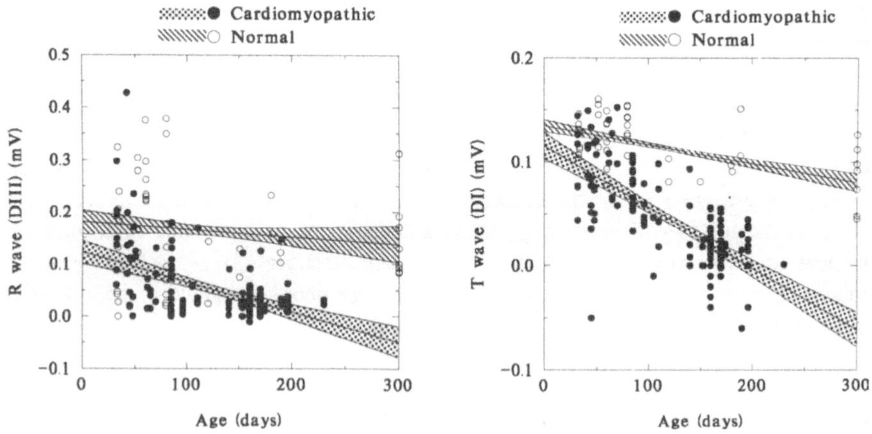

***Figure 5.** Except for a lowering of R wave amplitude (left), the flattening of the T wave is most relevant of a defective depolarization in the myopathic heart (right). (From Jasmin et al. (4)).*

same time prevent the necrotizing process. Cardioprotection, however, can be achieved by dissimilar, even oppositely acting drugs such as isoproterenol (Fig. 6) and propanolol which interact with adrenergic receptors or even by treatment with prostaglandin E_2 (20), a potent cytoprotector, as well as propylthiouracil, interfering with T_4 synthesis (21). Seemingly, these agents would exert their beneficial effects on membrane receptors, thereby improving the transduction process with activation of cAMP and the maintenance of cellular calcium homeostasis, enabling the calcium-induced, calcium-release process (22). Within the same line of reasoning, we have

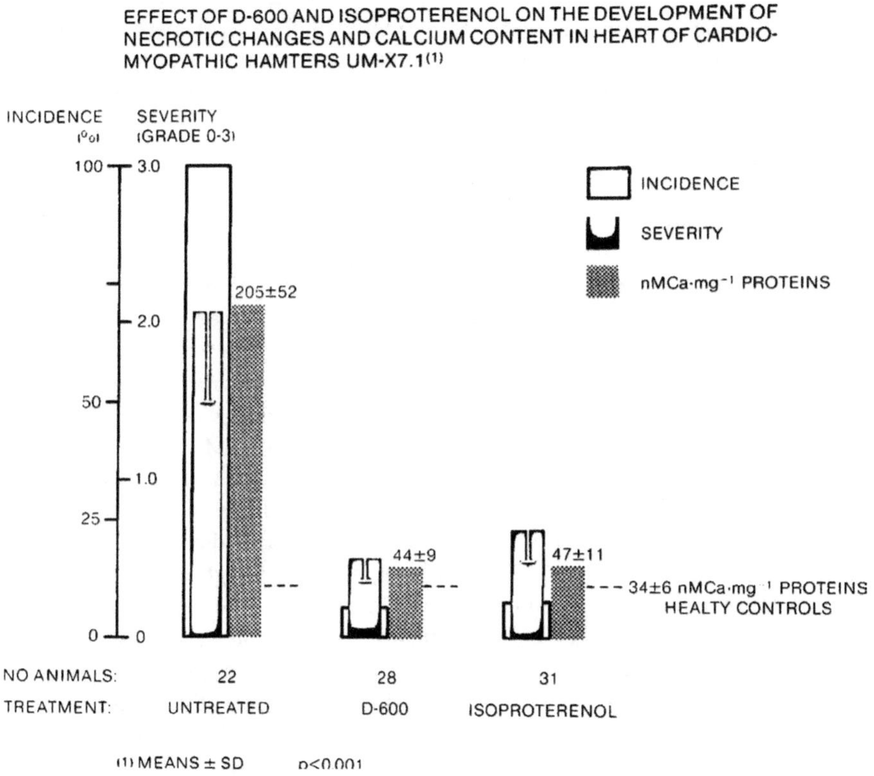

Figure 6. Twenty-eight day-old CM hamsters were treated during 4 weeks. D-600 (gallopamil) at a dose of 2.6 mg/kg/day and isoproterenol at a dose of 3 mg/kg/day were injected sc twice daily.

found that such therapeutic measures prevent the decrease in mitochondrial respiration rate and improve partly uncoupled oxidative phosphorylation (11). Hence, the functional integrity of cardiocytes closely relates to that of mitochondria (23). Moreover, a long term treatment with verapamil or with gallopamil succeeded in preserving energy reserves long enough to delay the development of heart failure (13, 24). Thus, an adequate intracellular calcium metabolism with sufficiently high level of energy phosphates plays a major role in the excitation-contraction coupling process.

In Summary. In the past years, numerous investigations on structural and functional changes in CM hamsters have contributed to emphasizing a primary intrinsic membrane defect (25-28), presumably of genetic origin. Very recent findings indicate the relative decrease of some dystrophin-associated proteins in the sarcolemma of hamster skeletal and cardiac muscle (29), which may play a role in the development of necrotic changes. Moreover, the same glycoprotein surface membrane complex has been found to be deficient in patients having severe childhood autosomal recessive muscular dystrophy (30).

Recent Investigations with Soluble Cations

There exists a good deal of evidence derived from voltage clamp studies suggesting that several cations can influence, positively or negatively, membrane ionic currents (31). In the following experiments, we have selected four cations previously shown to act preferentially on calcium L-type calcium channels. To our knowledge, *in vivo* studies have never been carried out in CM hamsters.

These *in vivo* experiments have been carried out in 58 CM hamsters aged 29 $\pm$ 1 days. They were divided into 5 groups and treated with a Ca-gluconate, La-nitrate, Co-acetate and Cd-chloride for 28 days. The drugs were solubilized and injected twice daily or given orally in drinking water as indicated in the table of results. At autopsy, the hearts were removed, blotted, weighed and examined by naked eye for an overall estimation of necrotic changes; some hearts were fixed and processed for routine histology using haematoxylin-phloxine-saffron as a stain.

Microassessment of cardiac necrosis was based on a previously described arbitrary scale of 0 to 3 (20). Using a double-blind procedure, the degree of damage was rated 1, 2 or 3 (with half marks when necessary) to designate slight, moderate or severe necrosis, respectively. The highest score on tissue sections corresponds to 50% damage of the entire ventricle. Below 1.4, we consider that a drug exerts a protective effect. The statistical significance was determined following the Student t-est; $p > 0.05$ was considered as insignificant.

Observations

Heart and skeletal muscle (presently the diaphragm) lesions were studied after 4 weeks of treatment with the various cations. Treatment with Ca^{++}, La^{+++} and Co^{++} salts exerted a favourable effect upon the pathologic changes in the heart, but skeletal muscle lesions remained unchanged except for a slight reduction resulting from Co^{++} salt treatment. The salts may be ranked in the following order of efficacy: Ca^{++}, La^{+++} and Co^{++} (Table 1). More work is needed to elucidate their *in vivo* mode of action. It may be suggested that Ca-gluconate enhanced myocardial contractility and overall energetics due to improved mitochondrial respiration since calcium is translocated by the calcium-anion symporter (gluconate) or by activation of sarcolemma ionic pumps including Na^{++}/Ca^{++} exchanger. La^{+++} or Co^{++} may act directly on the cell membrane mainly through blockage of L-type calcium channels.

TABLE 1
EFFECT OF CATIONS ON THE DEVELOPMENT OF HEART AND SKELETAL MUSCLE LESIONS IN CM HAMSTERS UM-X7.1 LINE*

Treatment (no)	Daily dose (mg/kg/2 inj) 28 days	Body weight gain (g) (Initial = 51 ± 2	Myocardial necrotic lesions Severity Grade (0-3)	Myocardial necrotic lesions Incidence %	Hemidiaphragm necrotic lesions Severity Grade (0-3)	Hemidiaphragm necrotic lesions Incidence %
Untreated (28)	Vehicles	40 ± 2	2.62 ± 0.06	100	1.86 ± 0.26	100
Ca-gluconate aqueous (8)	800 i.p.	30 ± 2[a]	0.50 ± 0.00[b]	38	1.87 ± 0.10	100
La-nitrate paraffin emulsion (8)	80 s.c.	21 ± 1[b]	0.75 ± 0.09[b]	25	1.88 ± 0.26	100
Co-acetate aqueous (8)	200 oral	33 ± 1	1.06 ± 0.20[b]	100	1.56 ± 0.21	100
Cd-chloride aqueous (6)	10 oral	39 ± 2	2.25 ± 0.11	100	1.91 ± 0.15	100

*Means ± SE; a, b significantly different, $p<0.05$ and 0.001 respectively.

In Summary. The necrotic cardiac lesions in myopathic hamsters are significantly reduced in incidence and severity by oral or parenteral administration of Ca^{++}, La^{+++} or Co^{++} salts. Since the treatment proves to be efficient only against myocardial necroses, it is assumed that the solubilized salts act as inorganic calcium channel blockers *in vivo*.

Acknowledgements

We are grateful to Michel Vermeulen for his expertise in ECG recordings, to Claudette Gemme for assistance in the editorial work and the McDonald Stewart Foundation for financial support of this research.

REFERENCES

1. Jasmin G and Bajusz E (1973) Ann Anat Pathol 18:49-66
2. Bajusz E (1969) Am Heart J 77:689-696
3. Homburger F, Nixon CW, Eppenberger M and Baker JR (1966) Ann NY Acad Sci 138:14-27
4. Jasmin G, Proschek L, Vermeulen M, Li Fen and Cardinal R (in press) In: Negano M, Takeda N, Dhalla NS (eds) The Cardiomyopathic Heart. Raven Press, New York
5. Fitchett DH, Scott, J, Stephens HR and Peters TJ (1979) Cardiovasc Res 13:260-268
6. Sole MJ, Lo CM, Laird CW, Sonnenblick EH and Wurtman RJ (1975) Circ Res 37:855-862
7. Jasmin G and Proschek L (1982) In: Chazov E, Saks V and Rona G (eds) Advances in Myocardiology. Vol. 4, Plenum Press, New York pp.45-53
8. Jasmin G, Proschek L, Dechesne C and Léger J (1988) Proc Soc Exp Biol Med 188:142-148
9. Malhotra A, Karell M and Scheuer J (1985) J Mol Cell Cardiol 17:95-107
10. Jasmin G, Solymoss CB and Proschek L (1979) Ann NY Acad Sci 317:338-348
11. Jasmin G and Proschek L (1980) In: Fleckenstein A and Roskamm H (eds) Calcium Antagonismus. Springer-Verlag Berlin, New York pp. 144-150
12. Wikman-Coffelt, Sievers R, Parmley WW and Jasmin G (1986) Heart Circ Physiol H22-H28
13. Jasmin G and Proschek L (1989) In: Bender F, Meesmann (eds) Treatment with Gallopamil. Steinkopft Verlag, Darmstadt, pp. 13-23
14. Dhalla NS, Singh A, Lee SL, Anand MB, Bernatsky AM and Jasmin G (1975) Clin Sci Mol Med 49:359-368
15. Slack BE, Boegman RJ, Downie JW and Jasmin G (1980) J Mol Cell Cardiol 12:179-185

16. Harrow JAC, Singh N, Jasmin G and Dhalla NS (1975) Can J Biochem 53:1122-1127
17. Wrogemann K and Nylen EG (1978) J Mol Cell Cardiol 10:185-195
18. Jasmin G and Proschek C (1984) Can J Physiol Pharmacol 62:891-898
19. Jasmin G and Preoschek L (1984) In: Sperelakis N, Caufield JB (eds) Calcium Antagonists, Mechanism of Action of Cardiac Muscle and Vascular Smooth Muscle. Martinus Nijhoff, Boston pp. 229-239
20. Jasmin G and Proschek L (1987) In: Kawai C and Abelmaun H (eds) Pathogenesis of Myocarditis and Cardiomyopathy. University of Tokyo Press, Tokyo pp. 79-89
21. Jasmin G, Proschek L, Brisson G and Dhalla NS (1987) In: Dhalla NS, Singal PK, Beamish RE (eds). Pathology of Heart Disease. Martinus Nijhoff, Boston, pp. 311-322
22. Fabiato A (1983) Am J Physiol 245:C1-2/10
23. Proschek L and Jasmin G (1983) Muscle and Nerve 5:26-32
24. Capasso JM, Sonnenblick EH and Anversa P (1990) Circ Res 67:1381-1393
25. Makino N, Jasmin G, Beamish RE and Dhalla NS (1985) Biochem Biophys Res Commun 133:491-497
26. Panagia V, Lee SL, Singh A, Pierce G, Jasmin G and Dhalla NS (1986) Can J Cardiol 2:236-247
27. Berry B, Poulsen R, Yunge L, Brunewald P, Fitchett D, de Chastonay C, Gabbiani g and Hüttner I (1983) J Mol Cell Cardiol 15:503-513
28. Caufield JB (1966) Ann NY Acad Sci 138:151-152
29. Roberds SL, Ervasti JM, Anderson , Ohlendieck K, Kahl SD, Zoloto D and Campbell KP (in press) J Biol Chem
30. Matsumura K, Tome FMS, ollin H, Azibi K, Chaouch M, Kaplan JC, Fardeau M and Campbell KP (1992) Nature 359:320-322
31. Lakshminarayanaiah N (1991) In: Hurwitz L, Partridge LD and Leach JK (eds) Calcium Channels: Their Properties, Functions, regulation and Clinical Relevance. CRC Press Inc. Boca Raton, Flo, USA

CHAPTER 2

MEMBRANE PHYSIOPATHOLOGY IN HEREDITARY CARDIOMYOPATHY

Ghassan Bkaily and Danielle Jacques

Department of Physiology and Biophysics,
Faculty of Medicine, University of Sherbrooke,
Sherbrooke, Quebec, Canada

CARDIOMYOPATHY

Hereditary cardiomyopathy in the hamster provides unique possibilities for studying the pathology and clinical course of primary congestive cardiomyopathies. The autosomal recessive disorder is readily transmissible with 100% incidence in the offspring, although the defective gene has not yet been identified. The cardiomyopathy develops in a characteristic, well-defined, predictable stage (see Chapter 1). Verapamil (an L-type Ca^{2+} blocker) and isoproterenol (an indirect L-type Ca^{2+} stimulator) were highly efficient in preventing the development of necrotic cardiac changes in the cardiomyopathic hamster (1 and references therein). Necrotic changes in cardiomyopathic hamsters become fully expressed at the critical age of 55 days. However, no biochemical or morphological changes were found in new born cardiomyopathic hamster. At this stage, early-embryonic tetradotoxin (TTX) and Mn^{2+}- insensitive slow sodium channels were detected in the newborn cardiomyopathic hamster (1). This slow Na^+ channel has the same kinetics as the one reported in 3-day-old chick embryonic heart cells (2-4). A similar TTX- and Mn^{2+}- insensitive slow Na^+ channel was found in 10-19-week-old human fetal heart cells. This early fetal slow Na^+ channel was similar in kinetics and pharmacology to the slow Na^+ channels in newborn cardiomyopathic heart cells. This type of channel was sensitive to agents that were shown to prevent the development of cardiac necrosis in the cardiomyopathic hamster.

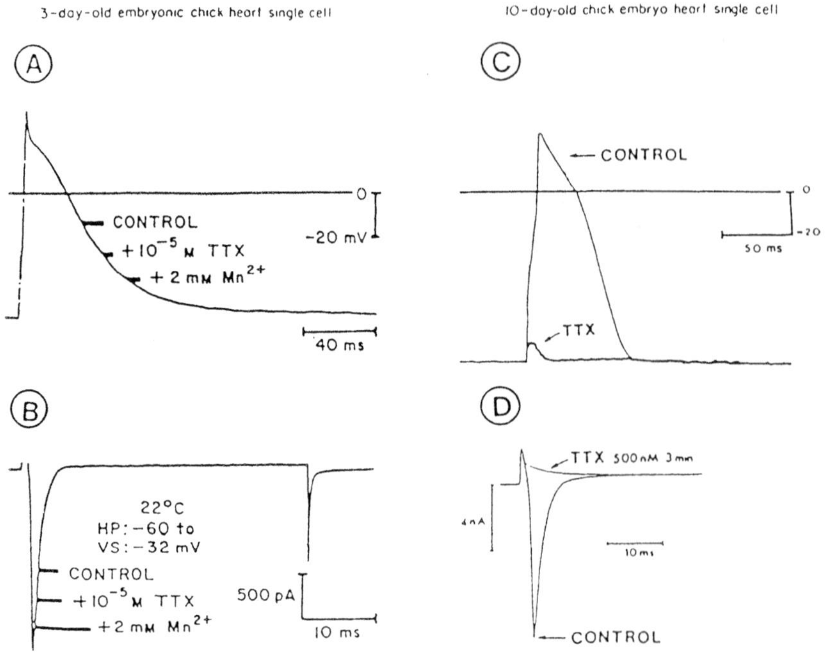

Figure 1. Characteristics of the action potentials (APs) and the inward currents of single embryonic chick hearts at different stages of development (A-D). Action potentials (A and C) and inward currents (B and D) recorded from single cells by switching between current clamp and voltage clamp modes. (A, B) The APs and the inward slow current of single 3-day-old embryonic chick ventricular cells in culture were insensitive to 10^{-5}M TTX and 2 mM Mn^{2+}. (C, D) Blockade of the APs (C) and inward current by 500 nM TTX in single heart cells from the 10-day-old embryonic chick. The frequency of stimulation was 0.02 Hz. HP = holding potential and VS = voltage step. Current traces in panels (C) and (D) were taken from two different single cells. (Modified from Bkaily et al. (2)).

TTX- and Mn^{2+} insensitive slow Na^+ channels in early fetal development of the heart

During embryonic development of the chick heart, sequential changes in the cardiac action potential (AP) takes place and the ionic inward currents generating the rising phase of the AP vary according to the developmental stage (5-8). Three-day-old embryonic hearts have slow-rising APs (10-30 V/s) dependent on Na^+ ions and independent of Ca^{2+} ions (Figures 1A and B). These slow Na^+-dependent APs are unaffected by TTX and Mn^{2+} (Figure 1 A and B). They are blocked by verapamil and D600, (L-type Ca^{2+} blockers), whereas mesudipine and bepridil (L-type Ca^{2+} channel blockers) are less effective (8). The fast Na^+ channel in embryonic chick heart which is sensitive to TTX makes its initial appearance on about day 4, and increases in density until about day 18. After day 8, the APs are completely abolished by TTX (Fig. 1C). This indicates that the AP generating channels consist predominantly of fast Na^+ channels (Fig. 1D), most of the slow Na^+ channels having been lost (functionally) so that insufficient numbers remain to support regenerative excitation (6).

In 10 weeks of age human fetal heart, a large number of ventricular heart cells show only the presence of the slow Na^+ channels while some other cells from the same heart, show both the slow and fast Na^+ channels. Figures 2A and E show examples of the absence of the fast, TTX- sensitive Na^+ inward current in 10-week old fetal human heart cells. However, in heart from the 20 week old human fetus, most of the ventricular heart cells show a fast Na^+ current. Very few cells show both fast and slow Na^+ currents. At 22 weeks of age, all fetal ventricular cells show only fast Na^+ channels.

The biophysical characteristics of the slow Na^+ channel in both 3-day-old embryonic chick heart cells (2-4) and 10-20 week old human fetal heart cells (7) are similar. The kinetics of the slow Na^+ current in chick and human heart cells at early fetal development is almost similar to that of TTX-resistant (9-11) and TTX-insensitive (12-13) Na^+ currents. Molecular cloning of TTX- resistant rat heart Na^+ channel isoform (14) as well as the primary structure, functional expression and molecular cloning of human heart TTX-insensitive slow Na^+ channel was recently reported (15, 16). Cardiac cell line (MCM1) originated from a transgenic mouse shows a TTX-insensitive Na^+ current (17) that seems to be similar in origin to the TTX-insensitive sodium current in fetal heart cells and hamster cardiomyopathic heart cells.

The slow Na^+ channel is permeable to Li^+ and Ba^{2+} (4). This type of channel is insensitive to inorganic L-type calcium channel blockers such as Ni^{2+}, Cd^{2+} and La^{3+} (Table 1). However, it is sensitive to a few L-type Ca^{2+} antagonists such as verapamil and isradipine (Table 1). This channel is also sensitive to some toxins, mainly melitin (2) and apamin (4).

Presence of fetal slow Na^+ channel in ventricular cells of the cardiomyopathic hamster

As was mentioned in Chapter 1, several reports in the literature suggest that calcium overload in cardiomyopathic hamster heart cells could be due to a defect in the functioning of the L-type Ca^{2+} channel. This suggestion was based on the fact that verapamil had a beneficial effect on this model of cardiomyopathy. However, a closer look at the literature shows that many L-type Ca^{2+} antagonists are not as effective as verapamil. The most striking result against the implication of the L-type channel (in the preventive effect of verapamil in hereditary cardiomyopathy of the

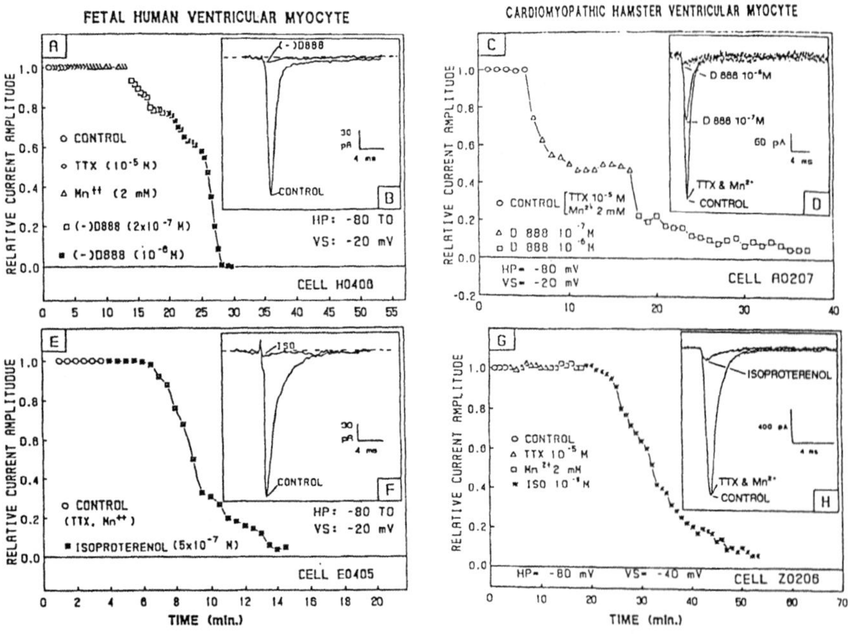

Figure 2. Blockade of the fast transient TTX- and Mn^{2+}-insensitive Na^+ current by (-)D888 and isoproterenol (ISO) in single heart cells from a 10-week-old human fetus (A,E) and newborn cardiomyopathic hamster (C,G). (A,C) Time-course of inhibition of the peak TTX- and Mn^{2+} -insensitive Na^+ current by (-) D888 in single heart cells of both human fetus, and newborn cardiomyopathic hamster respectively. (B,D) Current traces in the absence and presence of (-) D888. (E,G) Time-course of inhibition of the peak TTX- and Mn^{2+} -insensitive slow Na^+ current by isoproterenol, in single cells of both 10-week-old human fetus and newborn cardiomyopathic hamster respectively. (F,H) Current traces in the absence and presence of isoproterenol. (Bkaily et al., unpublished results).

hamster) is that isoproterenol which is known to stimulate this type of channel was found to mimic the effect of verapamil. Since verapamil or its derivatives, as well as isoproterenol both blocked the slow Na^+ channel in fetal ventricular cells, it is possible that this channel continues to be expressed and functions in heart cells of the young cardiomyopathic hamster. The coexistence of this channel with the fast Na^+ channel could contribute to an increase of intracellular Na^+ which may force the Na^+-Ca^{2+} exchanger to pump Na^+ ions out, thus creating a calcium influx through this exchanger. Such a phenomenom would allow Ca^{2+} overload in the heart cells.

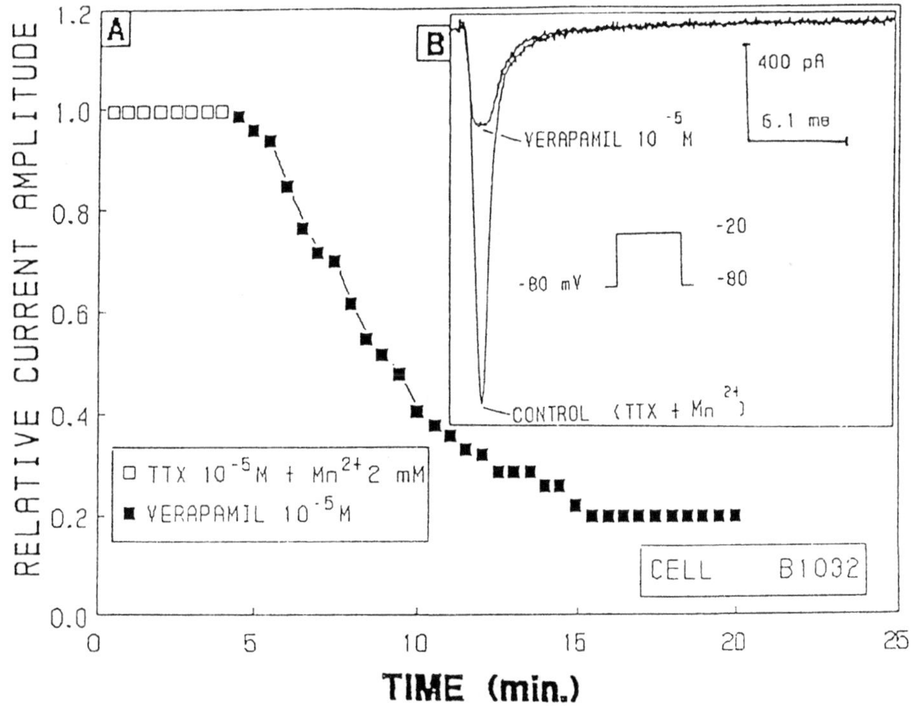

Figure 3. Blockade of TTX and Mn^{2+} insensitive slow Na^+ current in 3-day-old cardiomyopatic hamster single heart cells by verapamil. (A) Time course inhibition of the peak slow Na^+ current by verapamil ($10^{-5}M$). (B) Current traces in the absence and presence of verapamil. (Bkaily et al., unpublished results).

A close look at ventricular heart cells of the newborn normal hamster shows only the presence of the fast Na^+ current (1). However, ventricular cells of newborn (1 to 2-days old) cardiomyopathic hamster show mainly the presence of the slow Na^+ current. As can be seen in Figure 2 (C and D) in fetal human ventricular cells, as in heart cells from the one day old cardiomyopathic hamster, there is a TTX and Mn^{2+}- insensitive, slow inward Na^+ current. This slow Na^+ current in both fetal human and newborn cardiomyopathic hamster ventricular cells was sensitive to the verapamil derivative, D888, and isoproterenol. This slow Na^+ channel, continues to be functional during the development of the cardiomyopathic hamster. Figure 3 shows an example of ventricular cells of a 3-day-old animal. As can be seen in this figure, verapamil, which is known to prevent Ca^{2+} overload in the ventricular cells of the cardiomyopathic hamster, largely decreased the amplitude of the slow Na^+ current.

The kinetics of the slow Na^+ channel in both fetal human ventricular heart cells and cardiomyopathic cells is similar. Figure 4 shows that the slow Na^+ channel is fully available for voltage activation at a membrane holding potential near -94 mV and that 50% of the channels are inactivated at a membrane potential near -55 mV. At -42 mV, all the channels are inactivated.

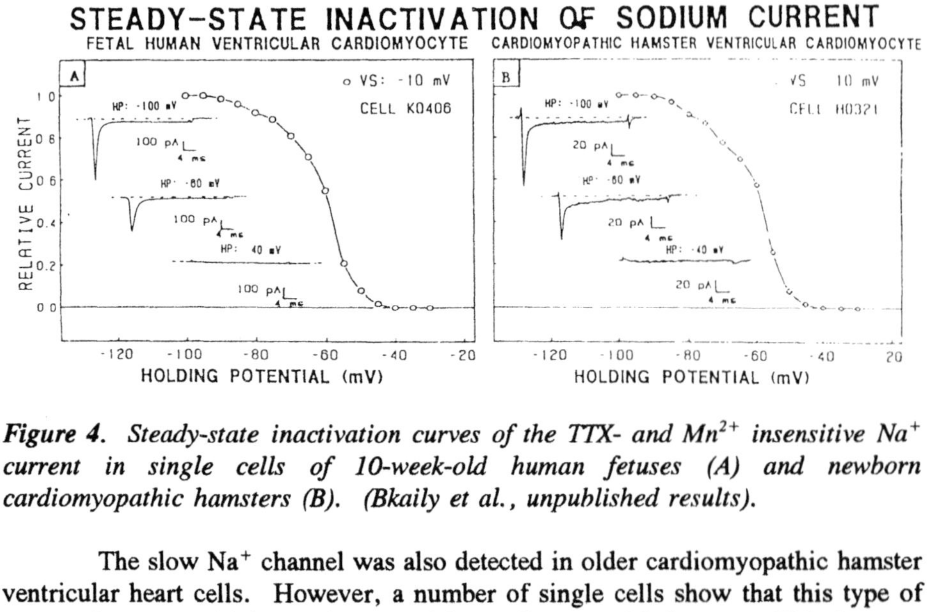

Figure 4. Steady-state inactivation curves of the TTX- and Mn^{2+} insensitive Na^+ current in single cells of 10-week-old human fetuses (A) and newborn cardiomyopathic hamsters (B). (Bkaily et al., unpublished results).

The slow Na^+ channel was also detected in older cardiomyopathic hamster ventricular heart cells. However, a number of single cells show that this type of current decreases with increasing age of the cardiomyopathic hamster. This decrease in the number of ventricular heart cells with functional slow Na^+ channels may correlate with the development of intracellular Ca^{2+} overload and focal myocardial necrosis in this pathological model.

Increase of the window T-type Ca^{2+} current in heart cells of the cardiomyopathic hamster

Both T and L-type Ca^{2+} currents were found in single heart cells of the one-day- old cardiomyopathic hamster. Most the cardiomyopathic ventricular single cells showed both types of calcium currents (I_{Ca}). When compared to single heart cells of the normal hamster, the cardiomyopathic single cells express, to a large extent, a continuation of both types of Ca^{2+} currents. Also, the number of single heart cells that showed only the T-type I_{Ca} was larger in cardiomyopathic than normal hamster heart.

However, the number of single cells that showed only the L-type I_{Ca} largely decreased in cardiomyopathy when compared to normal hamster heart. There was no difference in the shape, apparent threshold, peak potential and reversal potential of the current voltage relationship (I/V) of the T-type I_{Ca} in cardiomyopathic and normal newborn hamsters. No statistical differences were noted between the shape of the I/V curve of the L-type I_{Ca} in normal and cardiomyopathic newborn hamsters.

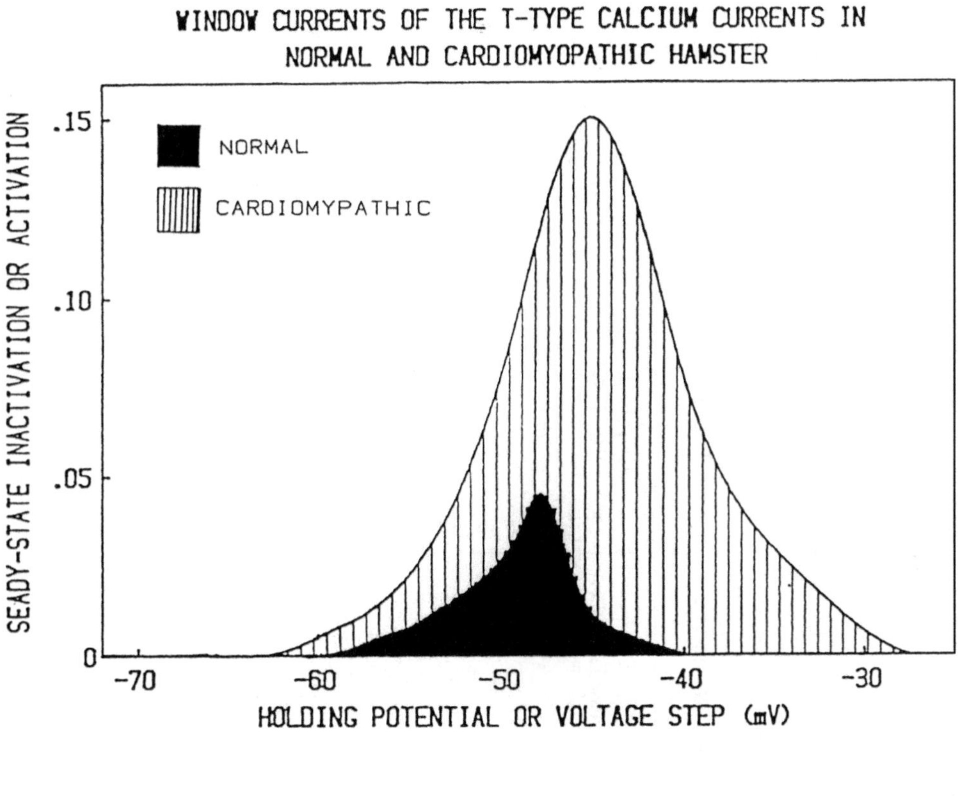

Figure 5. The window currents of the T-type I_{Ca} in normal (■) and cardiomyopathic (▥) hamster ventricular myocytes. The ratio of the areas under the curves (I_{Card}/I_{Norm}) is about 10. (Bkaily et al., unpublished results).

Table 1

PHARMACOLOGY OF THE TTX AND Mn^{2+} INSENSITIVE Na^{+} CURRENT IN THE HEARTS OF THE 3 DAY OLD EMBRYONIC CHICK, THE 10-11 WEEK OLD HUMAN FETUS, AND THE NEWBORN AND 8-10 DAY OLD CARDIOMYOPATHIC HAMSTER

10-11-WEEK-OLD HUMAN FETUS	3-DAY-OLD CHICK EMBRYOS	SUBSTANCES			NEW BORN CARDIOMYOPATHIC HAMSTER
INHIBITION (+) NO INHIBITION (-)	INHIBITION (+) NO INHIBITION (-)				INHIBITION (+) NO EFFECT (-)
++	++	10^{-5}M	Nifedipine	10^{-5}M	++
-	-	10^{-3}M	Cadmium	10^{-3}M	-
-	-	10^{-4}M	Nickel	10^{-4}M	-
-	-	10^{-4}M	Lanthanum	10^{-4}M	N.D
-	-	10^{-3}M	Cobalt	10^{-3}M	-
+++	+++	10^{-4}M	Azelastine	10^{-4}M	+++
+++	+++	10^{-6}M	D-888	10^{-6}M	+++
N.D	++	10^{-5}M	Verapamil	10^{-5}M	+++
N.D	++	10^{-5}M	D-600	10^{-5}M	N.D
+++	+++	10^{-10}M	Apamin	10^{-6}M	++
N.D	+++	10^{-8}M	Melittin	10^{-8}M	+
++	++	10^{-3}M	8-Br-cAMP	10^{-3}M	N.D
++	+++	10^{-3}M	8-Br-cGMP	10^{-3}M	N.D
+++	+++	10^{-8}M	Isoproterenol	10^{-8}M	+++
+++	+++	10^{-8}M	Propanolol	10^{-8}M	+++
++	+++	10^{-8}M	PN 200-110	10^{-6}M	+++
N.D	N.D	10^{-4}M	Clentiazem	10^{-4}M	+++

+ , inhibition; - , no effect; **N.D.** , not determined

There were no significant differences between the values of the L-type I_{Ca} inactivations obtained from single cells of normal and newborn cardiomyopathic hamsters. This suggests that the kinetics of the L-type I_{Ca} did not undergo a significant change in the heart of the newborn cardiomyopathic hamster.

Unlike the voltage dependence of the L-type I_{Ca}, the voltage dependence of the inactivation of the T-type I_{Ca} in ventricular cells of the newborn cardiomyopathic hamster is significantly different from that of normal newborn hamster cardiomyocytes. As figure 5 shows, the cardiomyopathic $V_{0.5}$ is shifted from -61.2 $\pm$ 0.3 mV in normal hamster to -54.7 $\pm$ 0.3 mV. Also, the V_0 of the cardiomyopathic T-type inactivation curve (-30 mV $\pm$ 6 mV) was also displaced to a more positive potential relative to the V_0 of the T-type I_{Ca} of normal hamster (-44 $\pm$ 1mV). The shift of the T-type I_{Ca} inactivation curve in the cardiomyopathic hamster may increase the window T-type Ca^{2+} current. This seems to be the case. Figure 5 shows that the activation curve of the T-type I_{Ca} did not change in both normal and cardiomyopathic heart cells (respectively $V_{0.5}$ is -36.3 $\pm$ 0.9 mV and -37.5 $\pm$ 0.3). Thus the shift of the inactivation of the T-type I_{Ca} to the right is largely responsible for the increase of the T-type window current. Finally, the increase in the number of single cells expressing a detectable T-type I_{Ca} as well as an increase of its window current may contribute in part to the gradual loading of the cardiac cells with Ca^{2+} during the development of cardiomyopathy. No detectable changes in the window current of L-type I_{Ca} were noticed.

CONCLUSION

It is apparent from the data presented in this chapter that the biophysical and pharmacological properties of the slow Na^+ channel in early fetal heart cells and in the newborn cardiomyopathic hamster are similar. As seen in Table 1, compounds that were reported to have a beneficial effect on the heart of the cardiomyopathic hamster (see Chapter 1) are blockers of the slow Na^+ channel. Also, the fact that this channel exists before the appearance of any other detectable changes in the heart cells highly suggests that the first pathological signal in the heart cells of the cardiomyopathic hamster takes place at the membrane level.

Also, another defect at the membrane level seems to take place early in this heart pathology. This defect seems to be at the T-type calcium channel level. Both, the presence of slow Na^+ channels and an increase of the window T-type Ca^{2+} current will contribute highly to Ca^{2+} overload in some single ventricular cells. In addition to other effects, Ca^{2+} accumulation can induce the activation of several Ca^{2+}-dependent mechanisms such as phosphalipases and proteases which promote damage to the membrane (18, 19). Also, Ca^{2+} overload could be responsible for the increased preproenkephain A mRNA synthesis which greatly contributes to the pathogenesis of the disease (20, 21). Calcium overload could also be indirectly responsible for the reported defect in the ryanodine-sensitive sarcoplasmic reticulum release channel of the cardiomyopathic hamster. Finally, a progressive postnatal intracellular overload may take place and could be largely due to the presence of the slow Na^+ channel and the increased window T-type Ca^{2+} current in ventricular cells. These early membrane defects preceed development of the disease and play a key role in the pathogenesis of hereditary cardiomyopathy of the hamster.

Acknowledgements

This chapter is based on work that was supported by a grant from the MRCC (MT 16496). Dr G. Bkaily is a Merck Frosst-FRSQ Professor and Miss D. Jacques is a Ph.D. fellow of the CHF. These studies were performed in my laboratory by Miss D. Jacques, Mr A. Sculptoreanu and Miss N. Perron.

REFERENCES

1. Jasmin G, Pasternac A, Bkaily G and Proschek L (1991) In: Calcium Channels: Their Properties, Functions, Regulation and Clinical Relevance (Hurwitz L, Partridge D, Leach J eds) Telford Press, New Jersey pp 295-307
2. Bkaily G, Jacques D, Yamamoto T et al (1988) Can J Physiol Pharmacol 66: 1017-1022
3. Bkaily G, Peyrow M, Yamamoto T et al (1988) Mol Cell Biochem 80: 59-72
4. Bkaily G, Jacques D, Sculptoreanu A et al (1991) J Mol Cell Cardiol 23: 25-39
5. Sperelakis N (1980) In: Slow Inward Current and Cardiac Arrhythmias (Zipes DP, Bailey JC, Elharrar V eds) Martinus Nijhoff Publisher, Boston pp. 221-262
6. Sperelakis N and Bkaily G (1985) In: Pathophysiology of Cardiovascular Injury (Stone HL, Weglicki WB) Martinus Nijhoff Publisher, Boston pp 109-144
7. Bkaily G (1992) In: In Vitro Methods in Toxicology (Jolles G, Cordier A eds) Academic Press, London pp 289-334
8. Kojima M and Sperelakis N (1983) Eur J Pharmacol 94: 9-18
9. Ikeda S and Schofield GG (1987) J Physiol (Lond) 389: 255-270
10. Roy ML and Narakash T (1992) J Neurosci 12: 2104-2111
11. Schlichter R, Bader CR and Bernheim L (1991) J Physiol (Lond) 442: 127-145
12. Anderson PAV (1987) J Exp Biol 133: 231-248
13. Ogata N and Tatebayashi H (1992) Devel Brain Res 65: 93-100
14. Rogart RB, Cribbs LL, Muglia LK et al (1989) Proc Natl Acad Sci USA 86: 8170-8174
15. Gellens ME, George Jr AL, Chen L et al (1992) Proc Natl Acad Sci USA 89: 554-558
16. George Jr AL, Knittle TJ, Tamkun MM (1992) Proc Natl Acad Sci USA 89: 4893-4897
17. Sculptoreanu A, Morton M, Gartside CL et al (1992) Am J Physiol 262: C724-C730

18. Titus EO (1983) Adv Exp Med Biol 161: 509-518
19. Kondo T (1987) Cardiovasc Res 21: 248-254
20. Proschek L and Jasmin G (1982) Muscle and Nerve 5: 26-32
21. Ouellette M and Brakier-Gingras L (1988) Biochem Biophys Res Comm 155: 449-454

CHAPTER 3

RESTRUCTURING OF SARCOPLASMIC RETICULAR MEMBRANE DURING THE DEVELOPMENT OF HEART DISEASE

Naranjan S. Dhalla, Nasir Afzal,
Heinz Rupp, Nobuakira Takeda and Makoto Nagano

Division of Cardiovascular Sciences,
St. Boniface General Hospital
Research Centre, Winnipeg, Canada
Physiologishes Institute II,
Universitat Tubingen, Tubingen, Germany
and Department of Internal Medicine,
Aoto Hospital, Jikei University, Tokyo, Japan

INTRODUCTION

The sarcoplasmic reticulum (SR), a highly specialized network of membranes, is believed to participate in cardiac contraction and relaxation by raising and lowering the cytoplasmic levels of Ca^{2+}. This membrane system is considered to represent a rapidly exchangeable calcium pool which plays an important role in heart function and metabolism (1-3). The main functions of SR are known to be Ca^{2+} sequestration, Ca^{2+} storage and Ca^{2+} release. The wave of excitation travelling along the transverse tubule is known to release Ca^{2+} from the sarcoplasmic reticular stores for initiating cardiac contraction. On the basis of its ability of accumulate Ca^{2+} by energy-dependent mechanisms, SR lowers the intracellular concentration of Ca^{2+} and initiates the relaxation phase of cardiac muscle. The sequestrated Ca^{2+} is stored in SR for release in the next contraction cycle. Thus any alteration in the function of SR can affect the cardiac contraction-relaxation cycle (1-2). Since SR has been shown to play a major part in the regulation of intracellular Ca^{2+} in the cardiac cell, it is the objective of this article to review some of the pertinent literature indicating the participation of different components of SR in regulating the intracellular Ca^{2+}

in myocardium in health and disease. Furthermore, it is intended to focus discussionon the role of SR in the process of excitation-contraction coupling. An effort is made to identify the mechanisms for restructuring of SR membrane and its components during the development of heart disease. The sequence of events indicating the participation of SR in raising and lowering the intracellular concentration of Ca^{2+} in cardiac cell is shown in Figure 1.

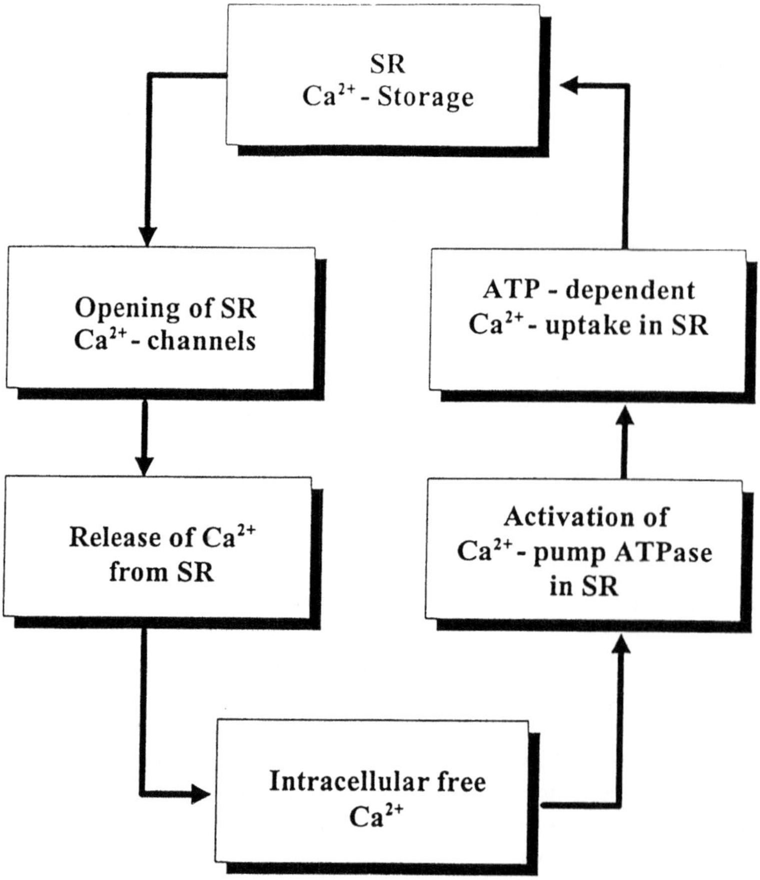

Figure 1. Sequence of events depicting the role of sarcoplasmic reticulum (SR) in raising and lowering the intracellular concentration of free Ca^{2+} in cardiac cell.

SR STRUCTURE AND COMPOSITION

A role for SR in cellular homeostasis of Ca^{2+} in muscle was first indicated about 30 years ago by experiments of Hasselbach and Makinose (4) in which a preparation of muscle microsomes was shown to remove a significant amount of Ca^{2+} from the medium in a process that was dependent on ATP. Now it is evident that a definitive spectrum for the relative contribution of SR towards Ca^{2+} transient exists with respect to the type of muscle and the species. For example, rat myocardium receives the largest contribution from SR and rabbit the least (5). Since SR regulates intracellular Ca^{2+} and the contractile state of the muscle, it is not unexpected that muscles with different contractile properties contain correspondingly varying amounts of the SR membrane. Although quantitative estimates of SR content (volume/surface area) are correlated well with the physiological profile of the muscle, thestereological data reveal considerable difference not only in total SR complement in different muscle types, but also in distinct regions of the SR in each sarcomere. Regardless of species, atrial tissue SR is nearly double the volume of that in ventricle and regardless of muscle type, the volume of terminal cisternae is related inversely to the isometric contraction time. Depending on the type of muscle, SR constitutes only 0.5 to 10 % of the cellular volume, yet it plays a major role in the regulation of muscle contractility. Such quantitative data are important when considering ultrastructural changes in certain muscle pathologies (6). For example, it was found that in the left ventricle of renal hypertensive rats, the surface area and volume of SR is increased in comparison to the control value and that the volume of SR in the epicardium is larger than that in the endocardium (6).

The SR membrane system has been a subject of numerous in vivo and in vitro studies using structural, physiological and biochemical techniques. The literature is full of morphological terms for various subregions or distinct features of SR; however, in general terms, the total SR complement may be broadly divided into two morphological categories: junctional SR and nonjunctional SR. The SR tubules, which are closely opposed and attached to external membranes, the sarcolemma and transverse tubules (T-tubules), are known as junctional SR (7). In cardiac muscle this region of SR is called terminal cisternae and this anatomical apposition forms a dyad at the level of each Z band (8). The major content of this part of SR is a calcium binding glycoprotein called calsequestrin (9). Bridging the 12 to 14 nm gap between the junctional SR and T tubule membranes in cardiac muscle are regularly spaced, densely staining projections, termed as SR feet or junctional feet (10). The function of the SR feet seems to be the mechanical adhesion of the junctional SR to the transverse tubule membrane. The longitudinal reticulum, which faces the myofibrils, is known as free SR because it does not participate in the formation of junctions with other membrane systems (11). Compared to the degree of specialization in junctional SR, the nonjunctional SR is considerably simpler, although not without morphological differentiation. The major function of the longitudinal reticulum appears to be the sequestration of Ca^{2+} during the relaxation phase (12); the Ca^{2+} pump is uniformly distributed in this part of the SR membrane. Unlike the junctional

SR, the longitudinal network of SR does not contain calsequestrin (6).

In view of the unavailability of suitable techniques for assessing the function of SR in vivo, various investigators have employed isolated fragments of SR (microsomal fraction) for studying its Ca^{2+} accumulating properties under in vitro conditions. The SR vesicles can be easily isolated and purified in considerable quantities from cardiac muscle by differential centrifugation of muscle homogenates. The major nonmembranous contaminant, actomyosin, can be removed by extracting the preparation of SR vesicles with 0.6 M KCl (13). The homogenization and isolation do not give rise to randomized inversion of the SR vesicles and therefore they retain the proper orientation for Ca^{2+} transport (14). These membrane vesicles are composed of several proteins and lipids; various proteins are situated within the basic matrix of a lipid bilayer (15). The SR vesicles are identified by their ability to accumulate Ca^{2+} and hydrolyze ATP due to the presence of Ca^{2+}-stimulated Mg^{2+}-dependent ATPase (Ca^{2+}-pump ATPase). The ATP-dependent Ca^{2+} accumulation by these organelles in the presence of permeant ions such as oxalate and phosphate is termed as Ca^{2+} uptake in which process Ca^{2+}-stimulated ATPase serves as a calcium pump.

Ebashi (16) was first to show that isolated SR vesicles accumulated Ca^{2+} with high affinity in the presence of ATP whereas Hasselbach (17) was first to describe Ca^{2+} accumulation as an active transport process against Ca^{2+} gradient which derives its energy from the hydrolysis of ATP through an ATPase enzyme incorporated in the SR membrane. Purified cardiac SR vesicles have been shown to consist of a heterogenous population of vesicles with respect to structure and function (18). It has been found that SR vesicles with different buoyant densities, termed light SR and heavy SR differed with respect to protein composition and electron-dense contents. The biochemical and morphological studies have shown that heavy SR vesicles are derived from the terminal cisternae of the SR whereas the light SR vesicles are derived from the longitudinal SR (19). The heavy junctional SR vesicles have been characterized by the presence of high calsequestrin content as well as their sensitivity to ryanodine (18).

The SR membranes from canine heart have been studied extensively and several protein constituents (Table 1) have been identified and characterized (20). These include Ca^{2+}-transport ATPase (21,22), phospholamban (22,23); calsequestrin (24,25) and ryanodine-sensitive Ca^{2+} release channels (26,27). Although the presence of two other glycoproteins has been observed in the SR membrane, their function remains undefined (25). The Ca^{2+}-transport ATPase of cardiac SR, like phospholamban, is the product of a gene expression in cardiac and slow-twitch skeletal muscle (21,22) but this is not the case in fast-twitch skeletal muscle. Calsequestrin is the product of two separate genes, one predominates in fast and slow-twitch skeletal muscle and the other predominates in cardiac muscle (24). Less is known regarding the genetic aspect of other SR proteins, 53 and 130 KD glycoproteins. The genetic of ryanodine-sensitive Ca^{2+} channels, whose cardiac and skeletal muscle forms are similar in apparent molecular weight and chemical properties (26,27), remains undefined. Significant differences in the ryanodine-

Table 1. Sarcoplasmic reticular proteins in the myocardium.

Protein	Molecular weight	Putative function
Ca^{2+}-pump ATPase	105 KD	Ca^{2+}-uptake
Phospholamban	27 - 28 KD	Stimulation of Ca^{2+}-uptake
Calsequestrin	53 - 55 KD	Storage of Ca^{2+}
Ca^{2+}-release channels	400 KD	Ca^{2+}-release
Unidentified protein	130 KD	Intraluminal Ca^{2+} binding

binding properties of cardiac and skeletal muscle SR have been noted thus raising the possibility that the cardiac and skeletal muscle proteins are the products of separate genes or these are alternate-splice or post-translationally mediated products of the same gene. These differences might just as plausibly result from differences in the molecular environments of the ryanodine receptors in cardiac and skeletal muscle sarcoplasmic reticulum. It is possible that changes in the synthesis or degradation of the SR proteins may lead to restructuring of the SR membrane and this may lead to the development of heart dysfunction.

SR CALSEQUESTRIN

Campbell et al (25) were first to identify and purify calsequestrin, the second major protein constituent of SR, from the canine cardiac muscle. Calsequestrin is a high capacity, moderate affinity, Ca^{2+} binding protein which is localized in the lumen of junctional SR in cardiac and skeletal muscles (28,29). This acidic protein, which binds 300 nmol Ca^{2+} per milligram protein is probably responsible for Ca^{2+} storage within SR (20). The apparent molecular weight of cardiac calsequestrin is 45 KD when measured in neutral SDS gels and 55 KD when measured in alkaline SDS gels. Cardiac calsequestrin stains blue with Stains-All and has been identified as a component of ryanodine-sensitive cardiac SR system (25). Recent anatomical evidence from rat myocardium indicates that Ca^{2+} accumulated across the lumen of SR is then bound to calsequestrin in the junctional SR (30). It has been suggested that the junctional SR is a potential source of the releasable Ca^{2+}. Although the exact role of calsequestrin in cardiac muscle is not clear, results from a study using skeletal muscle suggest that Ca^{2+}-dependent conformational changes in calsequestrin affect the junctional face of the membrane-bound proteins including the 400 KD Ca^{2+}-release channel and that calsequestrin may then regulate Ca^{2+} release into the cytoplasm (31). It is known that cardiac and skeletal forms of calsequestrin bind similar amounts of Ca^{2+}, even though structural differences exist between the cardiac and skeletal types (32,33), presumably by acting as a charged surface protein rather than by presenting multiple discrete Ca^{2+} binding sites (32). In recent years, the major SR proteins controlling Ca^{2+} release, reuptake and storage have been isolated and their primary structures have been determined (34). Calcium exchanges between the cytosol and SR are influenced by calsequestrin, which acts as a calcium buffer (35). Two isoforms of calsequestrin, a cardiac isoform and a fast-twitch muscle isoform have been identified (36,37). In a study involving the expression of calsequestrin mRNA, it was found that the expression of calsequestrin did not change in hyperthyroid and hypothyroid ventricular muscles (38) indicating that the storage capacity of the SR was unaltered in these pathological states. It has also been found that there is no difference in the level of cardiac calsequestrin mRNA between the patients with end-stage heart failure and their normal controls (39). However, further studies using several other experimental models of heart disease are required before any meaningful conclusion can be made regarding the status of calsequestrin changes under pathological conditions.

SR Ca^{2+}-RELEASE CHANNELS

Excitation of the surface membrane of mammalian cardiac muscle releases Ca^{2+} from the SR into the cytoplasm which then activates the myofilaments and initiates contraction. During the past several years, three general hypotheses have been proposed to explain the nature of excitation-contraction coupling (40) (a). Charge movement coupled Ca^{2+} release, in which voltage dependent movement of fixed charges in sarcolemma mechanically gates the release of Ca^{2+} from SR. (b). Ca^{2+} induced Ca^{2+} release, in which Ca^{2+} influx across sarcolemma triggers the release of Ca^{2+} from SR and (c). Inositol 1,4,5-triphosphate (IP3) induced Ca^{2+} the release, in which IP3 acts as a diffusible second messenger to trigger release of Ca^{2+} from SR. It should be noted that the cisternae of SR in the proximity of transverse tubules, which inflect periodically from sarcolemma at the level of the Z lines, is anchored to tubular system by a fatlike projection (11) forming the so-called dyadic junction in heart muscle. The depolarization of the cell membrane is believed to be transmitted through the transverse tubules to the interior of the muscle cell, thus causing efflux of Ca^{2+} from the neighbouring cisternae of SR. Evidently, a portion of the SR membrane acts as a sensor for the excitation wave spreading from the plasma membrane enveloping the cell to the transverse tubules. It is considered that the fatlike projections that connect the transverse tubules with the reticular cisternae are physically involved in the transmission of the excitation to the interior of the cell but the nature of the signal (whether mechanical, electrical, or chemical that will release Ca^{2+} from SR) is not known (3). Charge movement in cardiac muscle has many similarities to that in skeletal muscle thus raising the possibility that Ca^{2+} release may be coupled to charge movements. However, several lines of evidence indicate that charge movement coupled Ca^{2+} release may not occur in cardiac muscle (40).

The hypothesis of Ca^{2+} induced Ca^{2+} release was originally proposed for the mechanism of excitation-contraction coupling in skeletal muscle (41,42) and later it was extended to the cardiac muscle (43-45). According to this theory, the rapid influx of Ca^{2+} ions across sarcolemma triggers the release of Ca^{2+} from SR via Ca^{2+}-release channels, thereby providing a large quantity of Ca^{2+} required for initiating contraction. Experimental evidence from mechanically skinned cardiac cells (43-46), isolated SR (47-50) and intact cardiac myocytes (51) show that Ca^{2+} satisfies the first requirement of an excitation-contraction coupling messenger, in that it can release Ca^{2+} from the SR. Although the key role of sarcolemmal Na^+-Ca^{2+} exchange is to extrude Ca^{2+} from the cytoplasm, this system might also serve to drive Ca^{2+} into the cell. Leblanc and Hume (52) showed that Ca^{2+} release in guinea pig ventricular myocytes, which is small in amplitude, can occur in the virtual absence of an inward Ca^{2+} current. Since this release was associated with activation of tetrodotoxin sensitive Na^+ channels and required extracellular Ca^{2+}, it was concluded that Ca^{2+} release was triggered by the influx of Ca^{2+} via Na^+-Ca^{2+} exchange in response to Na^+ influx through tetrodotoxin sensitive Na^+ channels. It should be pointed out that Sham et al (53) disputed these findings which they believe are due to incomplete block of the Ca^{2+} channels or incomplete voltage control. Nonetheless, the SR protein that mediates Ca^{2+} release has been isolated and purified (26,54,55) and is generally referred to as the SR Ca^{2+} release channel or ryanodine receptor. Other experiments have shown that in the presence of micromolar concentrations of Ca^{2+}, the SR Ca^{2+} channel is activated allowing Ca^{2+} release into the cytosol (26,54-56). Studies on intact cardiac cells provided evidence that Ca^{2+} release is specific for Ca^{2+} influx but not for Ba^{2+} or Na^+ (57). Although detailed mechanisms of physiological SR Ca^{2+} release remain to be determined, rapid mixing-vesicle ion flux and planar lipid bilayer-single channel measurements have suggested that cardiac SR Ca^{2+} release is mediated by a high conductance Ca^{2+}-channel that is activated by micromolar Ca^{2+} and millimolar ATP and inhibited by millimolar Mg^{2+} (58,59). The positive inotropic effect of caffeine, which is associated with a prolonged Ca^{2+} transient is known to be due to its interaction with the Ca^{2+}-release channels by increasing the open time and sensitivity of the channels to activation by Ca^{2+} (60,61). The presence of a similarly ligand-gated Ca^{2+} conducting channel has been previously observed in junctional derived heavy SR fraction from the skeletal muscle (58,59).

The Ca^{2+} release behaviour of heavy cardiac and skeletal SR vesicles is significantly affected by the neutral plant alkaloid, ryanodine, which binds specifically to the cardiac and skeletal Ca^{2+}-release channels (49). At nanomolar concentrations, ryanodine locks the channel into a long-lived open subconductance state, while at concentrations above 10 μM, ryanodine completely closes the channel (62). Using $[^3H]$-ryanodine as a Ca^{2+}-release channel-specific ligand, the ryanodine receptor from skeletal muscle has been purified by sequential column chromatography and by density gradient centrifugation (26). Lai et al (27) showed that the cardiac ryanodine receptor, like the skeletal receptor, migrated through sucrose gradients as a large complex and displayed a major polypeptide band of molecular weight of 400 KD on the SDS-polyacrylamide gel. Incorporation of the purified receptor into planar lipid bilayers induced single Ca^{2+}-channel currents with conductance and gating kinetics similar to those of native cardiac Ca^{2+}-release channels. It has been suggested that the purified cardiac ryanodine receptor complex contains the Ca^{2+} conducting pore of the cardiac Ca^{2+}-release channel, although retention of the full range of regulatory sites of the native channel has not been demonstrated (27).

A further clue to the precise subcellular origin of ryanodine sensitive and ryanodine insensitive cardiac SR components is given by measuring the responses of the isolated SR subfractions to the inorganic, polycationic staining agent, ruthenium red (63). It has been demonstrated that ruthenium red increases Ca^{2+} uptake by ryanodine sensitive heavy SR vesicles but has no significant effect on Ca^{2+} uptake by ryanodine insensitive light SR vesicles isolated from the heart (63,64). The effects of ruthenium red on Ca^{2+} uptake by cardiac SR vesicles support the idea that ryanodine sensitive cardiac SR vesicles originate from junctional SR in myocardium and that ryanodine insensitive cardiac SR vesicles originate from free SR (65). Ruthenium red stimulates net Ca^{2+} uptake by junctional cardiac SR vesicles at much lower concentrations than does ryanodine. Half-maximal stimulation by ruthenium red occurs at a concentration near 10^{-7} M whereas several hundred fold higher concentration is required for a similar effect by ryanodine (66,67). The combination of ryanodine and ruthenium red added to junctional cardiac SR vesicles gives no further stimulation of Ca^{2+} uptake than does either agent alone, suggesting that both agents increase Ca^{2+} uptake by a similar mechanism (65-67). As in the case of ryanodine, no increase in Ca^{2+}-ATPase activity is associated with ruthenium red stimulation of Ca^{2+} uptake by junctional cardiac SR vesicles. This observation suggests that ruthenium red increases net Ca^{2+} uptake into junctional cardiac SR vesicles by blocking Ca^{2+} efflux from the vesicles in a manner similar to that which has been proposed to explain the effect of ryanodine on net Ca^{2+} transport in the SR vesicles (65-67).

The role of IP_3 as a second messenger is well established in many cell types where it functions by releasing Ca^{2+} from the intracellular stores (68). Some studies have shown that in these cells IP_3 acts by directly opening Ca^{2+}-release channels (69,70). The properties of this IP_3 gated Ca^{2+}-release channel (IP_3 receptor) seem to be quite different from the Ca^{2+} gated Ca^{2+}-release channel (ryanodine receptors) (69). Although hormone stimulated IP_3 synthesis has been clearly demonstrated in cardiac myocytes, the functional role of IP_3 in excitation-contraction coupling is as yet unclear and controversial (71). However, the limited information regarding IP_3 induced Ca^{2+} release in cardiac cells suggests that IP_3 should not be considered as a major physiological trigger mechanism for Ca^{2+} release. Recent data from skinned cardiac cells indicate that IP_3 only plays a pathological role in cardiac arrythmogenesis by enhancing spontaneous Ca^{2+} oscillations (72). The mechanism by which IP_3 enhances Ca^{2+} oscillations or induces Ca^{2+} release is not yet clear. It may either directly gate Ca^{2+}-release channels (73) or modify the gating of ryanodine receptor (74).

Although some information concerning calcium release from SR in regional schemia in dogs hearts is available, SR release activity has not been examined in a detailed manner ir other types of heart failure. In a rat model of global ischemia, it was found that a decrease in Ca^{2+} uptake caused by ischemia is not due to any defect in Ca^{2+} pumping capability but is due to an increased efflux through the ryanodine sensitive Ca^{2+}-release channel of cardiac SR (75). When this activity was measured in the presence of ryanodine no significant difference was observed in Ca^{2+} uptake properties of SR in ischemic hearts as compared to control. In this regard it should be noted that low concentrations of ryanodine were found to prevent ventricular arrhythmias during acute ischemia and reperfusion in the rat heart (76). In another study, junctional SR membrane vesicles were isolated from chronically failing human hearts, incorporated into an artificial planar phospholipid bilayer and the activity of a single Ca^{2+}-release channel was investigated under voltage-clamp conditions. The properties of these channels were similar to those previously reported from normal animal tissue and this provided evidence that the function of individual Ca^{2+} release channels is not altered in failing hearts (77).

The primary structure of the Ca^{2+}-release channel has been determined by Otsu et al (78); two distinct isoforms of the ryanodine receptor have been identified in cardiac and skeletal muscle. In one study regarding ryanodine receptor mRNA, the expression of mRNA coding the ryanodine receptor was upregulated in hyperthyroidism and downregulated in hypothyroidism; these results were in parallel with mRNA levels for Ca^{2+}-pump ATPase (39). Efforts have also been made to explore the molecular basis for the systolic and diastolic dysfunction, which are characteristics of end-stage heart failure in humans. The mRNA levels for ryanodine receptors were found to be decreased in ischemic cardiomyopathy but there was no change in idiopathic dilated cardiomyopathy as compared to control (79). These findings indicate that changes in SR Ca^{2+}-release channels depend on the type of heart disease.

SR Ca^{2+}-PUMP ATPase

The SR membrane is comprised of about one third lipids and two thirds proteins. The Ca^{2+}-stimulated ATPase, accounting for about two thirds of the total proteins of the SR membrane, carries out the enzymatic function of Ca^{2+} transport (80). The localization of Ca^{2+}-pump ATPase in skeletal and cardiac muscle has been determined by Jorgensen et al (81). Indirect immunofluorescent staining has shown that the Ca^{2+}-pump ATPase is concentrated in the longitudinal SR and in nonjunctional regions of the terminal cisternae. It has a highly asymmetric structure with respect to the membrane in which a major portion (60 %) of its mass protrudes into the cytoplasm as a pear-shaped head with a narrow stalk, while the rest is buried within the membrane. The Ca^{2+}-pump ATPase consists of a single polypeptide that is able to couple the hydrolysis of one molecule of ATP to the active transport of two Ca^{2+} ions across the SR membrane (82). The ATP-binding site is located in the cytoplasm portion of the enzyme, whereas the Ca^{2+} binding site is formed from its transmembrane segments. Because these two functional sites are presumably separated from each other, the coupling of ATP hydrolysis and Ca^{2+} movement probably occurs indirectly. In addition to one active site and two high affinity Ca^{2+} binding sites, this enzyme also contains one phosphorylation site per ATPase molecule (82). The purified ATPase possesses all of the properties attributed to the Ca^{2+}-pump system and when reconstituted into phospholipid bilayer, it transports Ca^{2+} against a concentration gradient (83). MacLennan (9) was first to successfully purify the Ca^{2+}pump -ATPase with high enzymatic activity from rabbit skeletal muscle SR.

The structure of the Ca^{2+}-pump ATPase molecule has been extensively studied and information has been obtained about its primary structure and various functional sites (15,82). The ATPase has an apparent molecular weight of 100 to 120 KD based on SDS gel electrophoresis (23,84) and can be cleaved at specific sites by limited digestion of intact SR vesicles with trypsin (84,85). The first cleavage produces two fragments of molecular weight 50 to 60 KD (fragment A) and 45 to 56 KD (fragment B). The second cleavage of the ATPase produces further subfragments which have molecular weight of 30 (A_1) and 20 (A_2) KD (84,85). Localization of several major functional sites within the ATPase has been determined by using limited digestion (84,86). The site for phosphoenzyme formation has been shown to reside in the intact ATPase and in the A fragment or A_1 subfragment after limited digestion (84). The high affinity Ca^{2+} binding sites of ATPase, directly involved in the transport mechanism, are defined as Ca^{2+} transport sites and have been localized to the A fragment or A_2 subfragment of the ATPase (87). It appears that A_1 subfragment contains the site involved in ATP hydrolysis and the A_2 subfragment contains the high affinity Ca^{2+} binding sites involved in Ca^{2+} transport (20).

Binding of two Ca^{2+} ions to the high affinity Ca^{2+} site from the cytoplasmic side of the SR membrane induces phosphorylation of the enzyme by ATP to form the ADP-sensitive phosphoenzyme (E1P). Ca^{2+} ions in the E1P are occluded in that they cannot be exchanged with Ca^{2+} ions in the aqueous medium on either side of the SR membrane. E1P is then converted spontaneously to the ADP-insensitive phosphoenzyme (E2P) and at the same time, Ca^{2+} ions are exposed on the luminal surface and released. E2P is subsequently hydrolyzed to yield a dephosphoenzyme, which then starts another round of the reaction cycle. The Ca^{2+}-stimulated ATPase utilizes a Mg^{2+}-ATP complex as a physiological substrate. It is well known that Mg^{2+} is required for the rapid turnover of this enzyme; Mg^{2+} derived from the Mg^{2+}-ATP complex remains bound to the ATPase at least until E2P is hydrolyzed and that the binding of this Mg^{2+} is responsible for the induction of a high rate of enzyme turnover. Binding of Ca^{2+} with unphosphorylated Ca^{2+}-stimulated ATPase, as analyzed in equilibrium measurements, revealed that the stoichiometry for high affinity Ca^{2+} binding is 2 mole per mole of the phosphorylated site (80,82).

SR PHOSPHOLAMBAN

The ß-adrenergic actions of catecholamines represent a key control mechanism that regulates the metabolic, electrical, and mechanical performance of the myocardium. Adenosine 3',5', monophosphate (cAMP) serves as the intracellular second messenger for these regulatory processes in which ß-adrenergic activation of the sarcolemmal adenylyl cyclase causes intracellular cAMP levels to rise (88). This second messenger appears to exert most of its cellular effects by activating cAMP dependent protein kinase (cAMP-PK). This enzyme catalyzes phosphorylation of a number of proteins which alter several aspects of the excitation-contraction process in the myocardium including increases in the flux of ions across the sarcolemma and SR (88). It should be noted that cardiac SR can form two classes of phosphoprotein which differ in both molecular weight and chemical characteristics (89). The Ca^{2+} pump ATPase of the cardiac SR, which serves as an energy transducer for the translocation of Ca^{2+} across the membrane, forms a phosphoprotein intermediate by incorporating the terminal phosphate of ATP into a hydroxylamine-labile acylphosphate (89). The cardiac SR is able to form a second type of phosphoprotein when reacted with cAMP-PK (90). In this phosphoprotein, the terminal phosphate of ATP is incorporated mainly into a serine residue of a 22 KD protein that exhibits the stability characteristics of a phosphoester (90). This phosphorylatable protein, which represents one of the sites at which cAMP-PK exerts its action, appears to function as a modulator of the Ca^{2+}-stimulated ATPase and Ca^{2+}-uptake by cardiac SR. The formation of a phosphoester phosphoprotein when cardiac SR is incubated with cAMP in the absence and presence of exogenous cAMP-PK is independent of Ca^{2+} up to 0.1 mM but is markedly dependent on cAMP between 0.1 and 10 μM (90). Other lower molecular weight proteins are also phosphorylated by cAMP-PK and the chemical characteristics of these smaller proteins are distinct from those of phospholamban (91).

Phospholamban phosphorylation occurs in cardiac SR preparations which are virtually freed from sarcolemmal vesicles by means of density gradient centrifugation. Phospholamban is intimately associated with the SR membrane and appears to be an acidic proteolipid (91). A portion of this molecule is exposed at the surface of SR vesicles as evidenced by the observation that it can be phosphorylated by exogenous cAMP-PK (91). The phospholamban content of the cardiac SR is estimated to be about 4-6 % of the total protein (88). During Ca^{2+} transport into the SR, the ATPase undergoes a complex series of reactions in which several phosphorylated intermediates (EP) are sequentially formed and degraded (89). Steady state EP levels are not altered by phospholamban phosphorylation at saturating concentrations of Ca^{2+} and ATP but the rate-limiting step of EP decomposition is markedly enhanced (89). Major changes in the Ca^{2+} affinity occur in two of the rate determining steps during the turnover of the ATPase that are enhanced by phospholamban phosphorylation suggesting that phopholamban may exert its action by regulating the Ca^{2+}-mediated conformational changes of the ATPase. This interpretation is supported by the findings of Kirchberger et al (90) that phospholamban phosphorylation alters the sensitivity of the SR Ca^{2+} pump.

Phospholamban is a product of a single gene. It is expressed in SR of cardiac, slow-twitch skeletal and smooth muscle where the SERCA 2 type Ca^{2+} pump is expressed. Phospholamban consists of five monomers with molecular weight of 60 KD. The monomer is an amphipathic polypeptide with C-terminal transmembrane and N-terminal cytoplasmic domain. The cytoplasmic domain has phosphorylation sites catalyzed by cAMP and calmodulin-dependent PK. One line of evidence indicates that phospholamban acts as a suppressor of Ca^{2+} pump ATPase and that phosphorylation suppresses its inhibiting effects. Recent studies with synthetic peptides of phospholamban revealed that phospholamban inhibits the Ca^{2+} pump ATPase at two sites in different manners. These include the cytoplasmic domains for Vmax and the intramembrane domain for Kca; phosphorylation of phospholamban suppressed these two inhibiting effects (92). Thus, phospholamban may act as a regulator of Ca^{2+} pump ATPase in cardiac SR by a direct protein-protein interaction.

It has been reported that phosphate could be incorporated into phospholamban in the presence of Ca^{2+} and calmodulin and formation of this phosphoprotein with stability characteristics of a phosphoester was dependent on calmodulin but independent of cAMP-PK (91). It is postulated that phospholamban could be phosphorylated by an endogenous PK associated with the SR which is activated by calmodulin and Ca^{2+} (91). In this regard preincubation of cardiac SR with calmodulin was found to enhance the rate of calcium uptake and that calmodulin dependent phosphorylation of phospholamban was accompanied by an enhanced rate of Ca^{2+} transport. At physiological concentrations of Ca^{2+} (0.1-10 μM), stimulation of Ca^{2+} transport by either cAMP-PK or calmodulin-PK systems appears to occur independently. It has been reported that these effects on Ca^{2+} transport were additive which is in accord with the finding that phosphorylation by two kinases occurs independently (88). Phosphorylation of cardiac SR by either cAMP or Ca^{2+}-calmodulin-dependent mechanisms results in approximately two fold increase in the steady-state rate of oxalate or phosphate-facilitated Ca^{2+} transport measured at optimal physiological Ca^{2+} concentration (88). The cAMP mediated acceleration of Ca^{2+} uptake by cardiac SR may explain two of the major mechanical effects of catecholamines on heart muscle, abbreviation of systole and increased contractility (93,94). Thus, the cAMP mediated increase in the rate of Ca^{2+} accumulation by SR could explain the faster rate of relaxation because of the increased rate at which Ca^{2+} would be removed from troponin in catecholamine-stimulated myocardium. Enhanced accumulation of SR Ca^{2+} could increase the amount of Ca^{2+} stored within the cell as some of the Ca^{2+} that would otherwise be lost during diastole (95). This could increase the amount of Ca^{2+} available for delivery to the contractile proteins in subsequent contraction thereby producing the positive inotropic effect due to catecholamines. The abbreviation of systole seen in the catecholamine-stimulated heart might be explained by an action of phosphorylated phospholamban to accelerate spontaneous transition from Ca^{2+} uptake to Ca^{2+} release in the SR (88).

MODIFICATION OF SR FUNCTION

The SR membrane might be looked upon as a specialized lipid-protein environment which is sensitive to alterations in either constituent or in their ratio. There is considerable evidence for a highly specific association of certain phospholipids with the Ca^{2+}-stimulated ATPase molecule and any alteration in this lipid-protein interaction may have dramatic effects on Ca^{2+} transport and ATPase activity (18,96). Isolated SR contains 90 to 110 phospholipid molecules per ATPase molecule together with minor amounts of neutral lipids and cholesterol. These phospholipids comprise mainly of phosphatidylcholine as the major component, together with phosphatidylethanolamine, phosphatidylserine and phosphatidylinositol. Phosphatidylethanolamine has been reported to be distributed primarily in the outer (cytoplasmic) monolayer of the membrane while phosphatidylcholine is evenly distributed and phosphatidylserine is primarily found in the inner monolayer. For studying the lipid-protein interaction, it is important that procedures are established by which it is possible to substitute the normal lipid environment with any other lipid species. The most utilized procedures employ addition of exogenous lipid to SR or purified ATPase with cholate or deoxycholate followed by sucrose density centrifugation to isolate lipid-substituted vesicles or lipid-protein complexes from the detergent containing samples. From these types of studies, it has been learned that the hydrophobic region of the bilayer probably is the decisive factor for the expression of enzyme activity (97).

There are a number of functionally important amino acid residues which regulate the ATPase enzyme activity in its denatured or native conformation stage (98). Most important of these are cysteine, lysine and histidine residues. Out of 24-25 cysteine residues, 17-19 SH groups are titrated with DTNB in denatured enzyme whereas 13-15 SH groups are reactive in the native conformation. More recently it was found that the number of crucial thiols is two or more and blocking of different thiols inhibits different elementary steps of the enzyme reaction (99). Chemical modification of lysine residues of the ATPase with fluorescamine results in the inhibition of Ca^{2+}-stimulated ATPase activity by blocking EP formation or EP decomposition (100). One or more histidine residues may also be involved in the function of Ca^{2+}-stimulated ATPase since photoinactivation of SR membrane in the presence of Rose Bengal or Methylene Blue results in a relative modification of histidine residues accompanied by inactivation of both Ca^{2+} uptake and ATPase activities (101).

Monovalent cations exert regulatory influences on the Ca^{2+}-stimulated ATPase of SR. In detailed studies, carried by Jones and Watanabe (102), the potency of various monovalent cations like K^+, Na^+, NH_4^+, Cs^+, and Li^+ for stimulating Ca^{2+}-stimulated ATPase activity was determined and K^+ was found to be the monovalent cation most effective. It was suggested that K^+ stimulates the enzyme activity primarily by increasing the rate of decomposition of phosphoprotein. The increase in activity induced by K^+ is modulated by ATP concentration present in the incubation medium; both K^+ and ATP exert allosteric control over the rate-limiting step of the reaction (102).

It is postulated that the effects of drugs, fatty acids and ethanol on the SR membrane are mediated either directly by interaction with the Ca^{2+}-pump protein of the SR membrane or indirectly by an alteration of the physico-chemical properties of the bulk lipid phase and lipid-protein interface of the SR lipid bilayer (103). Local anesthetics such as lidocaine and tetracaine, have been shown to inhibit Ca^{2+} transport and ATPase activity of both cardiac and skeletal SR (103,104). General anesthetics such as halothane and enfluane also affect the functional properties of the SR with many of these effects exhibiting a complex concentration-dependence similar to local anesthetics (103). Calcium channel blockers and ß-adrenergic receptors blocking agents appear to act on the SR nonspecifically rather than through mechanisms related to their calcium channel or ß-adrenergic blocking actions. At concentrations $< 10^{-5}$ M where Ca^{2+}-antagonists act as calcium channel blockers in whole heart, verapamil and D-600 were found to have no effect on active calcium transport and passive Ca^{2+} release of cardiac SR (105) whereas inhibition of Ca^{2+} transport was observed at concentrations in the range of 10^{-5} to 10^{-3} M (106). ß-adrenergic receptor blocking drugs such as propranolol and timolol have been shown by several investigators to inhibit calcium transport and total ATPase activity of both cardiac and skeletal SR (107). Fatty acids and fatty acid derivatives such as palmitylcarnitine have complex effects on SR membrane ATPase systems ranging from complete inhibition to two-fold increase in ATPase activity and other coupled biological reactions (103). Alcohols, especially, ethanol have been shown to affect the functional and presumably structural states of several biological membranes including SR. Partial solubilization of the lipid bilayer by ethanol may be the cause of increased ATPase activity whereas inhibition of calcium transport may be caused by a direct interaction with the calcium pump protein (108,109). Thus changes in membrane phospholipid or membrane proteins in SR by a wide variety of chemical agents and drugs can be seen to alter the function of SR.

SR Ca^{2+}-TRANSPORT IN HEART FAILURE

Varying degrees of defects in SR Ca^{2+} transport have been identified in different experimental models of heart failure as well as in failing human hearts during the past 25 years (1). One of the first experiments designed to assess SR function was carried out in microsomes prepared from spontaneously failing canine heart-lung preparations which showed depressed Ca^{2+} uptake and Ca^{2+}-stimulated ATPase activities (102). Depressed activity of the sarcoplasmic reticulum Ca^{2+}-pump ATPase could lead to a reduced amount of Ca^{2+} available for contraction and ultimately could alter the function of the cardiac pump by slowing the normal relaxation phase (2,110). Observations of a depressed rate of SR Ca^{2+} uptake have been made in failing heart muscle obtained from humans (111) as well as in a number of experimental models employing rabbits (112) and hamsters (113,114). Alterations in SR Ca^{2+} uptake were obvious with the development of left ventricular failure in rabbits with hypertrophied hearts secondary to aortic insufficiency (115). Depressed calcium uptake activities of the SR fraction were noted without any change in the Ca^{2+}-stimulated ATPase activity indicating uncoupling of Ca^{2+} pump in cardiomyopathic hamsters (113) and bacterial infected cardiomyopathy in rabbits (116). The SR vesicles isolated from left ventricles of human hearts with idiopathic dilated cardiomyopathy did not show any change in Ca^{2+} uptake properties and accordingly it was suggested that abnormal Ca^{2+} handling in the myocardium of these patients was not a result of any intrinsic alteration of SR Ca^{2+} uptake system (117).

Attempts to unravel the rational basis for the pathogenesis of heart failure are always accelerated by new ideas and the application of new technologies. Molecular biology has provided new insight into the mechanisms responsible for the alterations of contraction and relaxation observed during cardiac hypertrophy and heart failure. Determination of the expression of genes encoding various SR proteins are under investigation in various models of heart failure. Recent reports have indicated that alterations observed in SR function in heart failure are at the transcriptional level as hypertrophy of rat and rabbit hearts was associated with decreased myocardial level of SR Ca^{2+}-stimulated ATPase mRNA and protein concentration (118,119). Altered SR Ca^{2+}-stimulated ATPase gene expression as compared to controls has been reported in a group of patients with idiopathic dilated cardiomyopathy, coronary artery disease with myocardial infarction, or valvular insufficiency (120). Accordingly, it was concluded that a decrease in the SR ATPase mRNA in the myocardium plays an important role in the alterations of Ca^{2+} movements and cardiac relaxation reported during end-stage heart failure (120). Molecular mechanisms of hypertrophied failing heart and abnormalities in the diastolic properties and contractility were studied and it was reported that the mRNA levels of Ca^{2+}-pump ATPase were decreased significantly by pressure overload and were correlated with corresponding ATPase protein levels (121).

Many membrane proteins are difficult to purify and once purified are difficult to reconstitute into their native environments. In this regard, monoclonal antibodies may be useful probes for the evaluation of the function and the mechanism of action of such membrane proteins (122). Antibodies are exquisitely selective proteins that can bind to a single target among millions of irrelevant sites. Monoclonal and polyclonal antibodies to the major SR proteins of rabbit skeletal and canine cardiac muscle were used to identify and characterize the corresponding components of human cardiac SR (123). In this study, immunoblots of SR from the normal left ventricles and excised failing left ventricles of patients with idiopathic dilated cardiomyopathy were compared in search of qualitative differences in the protein patterns of the failing hearts. No such differences were found with respect to the Ca^{2+}-pump ATPase, 53 KD glycoprotein, the ryanodine-sensitive Ca^{2+}-channel, calsequestrin or phospholamban (123). Since phospholamban plays a regulatory role in Ca^{2+}-pump ATPase activity, a defect in the phospholamban protein may lead to abnormal ATPase activity. This possibility was studied with the use of phospholamban-specific monoclonal antibodies, where binding of this antibody to isolated SR from normal and failing hearts was taken as simulation of the phospholamban phosphorylation by cAMP PK (124). Stimulation of SR Ca^{2+}-pump ATPase was found to be similar among failing and normal hearts and these investigators suggested that the pathogenesis of heart failure in idiopathic dilated cardiomyopathy is independent of intrinsic alteration of this mechanism. Furthermore, oxalate-supported Ca^{2+}-uptake by cardiac SR was examined in microsomes prepared from left ventricle free wall with idiopathic dilated cardiomyopathy and no significant difference in Ca^{2+}-uptake was found as compared to control. On the basis of these observations it was concluded that abnormalities in the regulation of Ca^{2+} in failing human myocardium cannot be ascribed to changes in the level or function of the Ca^{2+}-transport ATPase or phospholamban in the SR (125). However, it is pointed out that these conclusions do not include the possibility that the changes in molecular events leading to SR dysfunction may depend upon the stage and type of heart failure.

It is important to consider the particular etiology of heart failure when studies on the function of subcellular organelles are carried out in humans because changes in intracellular Ca^{2+} handling are not identical in every type or stage of heart failure (1). Furthermore, consideration of specific abnormalities of cellular membrane function in the pathogenesis of any given type of heart failure is necessary (1). With regard to animal models of heart failure, intracellular Ca^{2+} overload is believed to exist in experimental models of genetic cardiomyopathy and isoproterenol induced cardiomyopathy (1,2), hypertrophy due to pressure overload (126), diabetic cardiomyopathy (127,128), and acute ischemia-reperfusion injury (129). In another study intracellular Ca^{2+} transients were measured with aequorin from patients with idiopathic dilated cardiomyopathy, ischemic heart disease, valvular insufficiency and hypertrophic cardiomyopathy and were found to be markedly diminished and prolonged in comparison to the control (130). During different pathological conditions, the sensitivity of myocardium to extracellular Ca^{2+} has also been shown to be changed (126,131-133). In one study, the response of infarcted heart to extracellular Ca^{2+} was monitored and it was observed that the inotropic response to a decrease in extracellular Ca^{2+} was greater in infarcted hearts than in sham controls (134). Although changes in myofibrillar sensitivity to Ca^{2+} may be the cause of abnormalities in cardiac function in patients, the sensitivity of the myofibrils to Ca^{2+} is known to be unaffected by different structural alterations of the thick filaments including a shift in myosin isozymes which occur during heart failure (135).

Relaxation abnormalities are often more important than contraction abnormalities in the pathogenesis of the clinical syndrome of heart failure (136). Experimental studies have shown that impaired relaxation is among the initial abnormality which appears in failing hearts. In fact, diastolic heart failure is characterized by increased resistance to diastolic filling of cardiac ventricles (116). Impaired Ca^{2+} handling has been reported to occur in a variety of conditions like hypertrophic cardiomyopathy, myocardial ischemia and congestive heart failure, characterized by diastolic dysfunction and is considered to play a unifying role as a basic mechanism for the occurrence of diastolic heart failure (110). Ca^{2+} transients of muscles from failing hearts have been reported to be markedly prolonged and these muscles have shown a diminished capacity to restore a low resting Ca^{2+} levels during diastole; this is consistent with reduced rate of sequestration of Ca^{2+} (126,127). Accordingly, it can be argued that impaired uptake of Ca^{2+} by SR results in a rise in resting tension, leading to diastolic heart failure. The exact role of SR in terms of Ca^{2+} uptake and Ca^{2+} release at different stages of congestive heart failure has not yet been cleary defined. Recently alterations in SR Ca^{2+} transport in hemodynamically assessed preparations were observed in congestive heart failure secondary to myocardial infarction as a result of coronary ligation (138). Interestingly, while Ca^{2+}-uptake by SR from the left ventricle was decreased, the right ventricle showed augmented Ca^{2+}-uptake activity at early and moderate stages of heart failure (138). Thus it appears that there occurs a restructuring of the SR membrane which may determine its function during the development of heart disease. Defective SR function has also been reported to occur in diabetic cardiomyopathy (139) whereas augmented SR function was observed in hyperthyroid hearts (140).

CONCLUDING REMARKS

By virtue of its ability to accumulate, store and release Ca^{2+}, SR is now known to be intimately involved in raising and lowering the intracellular concentration of free Ca^{2+} and thus in heart function on a beat to beat basis. The accumulation of Ca^{2+} in SR has been demonstrated to be due to the presence of Ca^{2+}-pump ATPase in the SR membrane and this event is primarily regulated by the presence of phospholamban which is phosphorylated by cAMP-PK or calmodulin-PK. On the other hand, Ca^{2+} storage and Ca^{2+} release are due to the presence of calsequesterin and Ca^{2+}-channels in SR membrane, respectively. In addition, different membrane phospholipids are known to determine the SR Ca^{2+}-pump activity. Thus any changes in one or more protein and phospholipid components of SR membrane can be seen to affect the SR function and subsequently both cardiac contraction and relaxation properties. Several studies by employing different experimental models of heart failure as well as failing human hearts have revealed varying degrees of abnormalities in SR function. Such alterations seem to occur either due to defective molecular mechanisms in terms of gene expression and/or changes in the phospholipid composition of the SR membrane. It is suggested that there occurs a restructuring of the SR membrane and this aspect plays a crucial role in the pathogenesis of cardiac dysfunction depending upon the type and stage of heart disease.

Acknowledgements

The research reported in this article is supported by the Medical Research Council Group in Experimental Cardiology. Dr. N. Afzal was a postdoctoral fellow of the Medical Research Council of Canada.

REFERENCES

1. Dhalla NS, Das PK and Sharma GP (1978) J Mol Cell Cardiol 10:363-385
2. Dhalla NS, Pierce GN, Panagia V, Singal PK and Beamish RE (1982) Basic Res Cardiol 77:117-139
3. Carafoli E (1987) Annu Rev Biochem 56:395-433
4. Hasselbach W and Makinose M (1961) Biochem Z 333:518-528
5. Fabiato, A (1982) Fed Proc 41:2238-2244
6. Anversa P, Loud A, Giacomelli F and Wiener J (1978) Lab Invest 38:597-609
7. Johnson E and Sommer JR (1967) J Cell Biol 33:103-129
8. Scales DJ (1981) J Mol Cell Cardiol 13:373-380
9. Maclennan DH and Wong PTS (1971) Proc Natl Acad Sci USA 68:1231-1235
10. Franzini-Armstrong C and Nunzi G (1983) J Muscle Res Cell Motil 4:233-252
11. Franzini-Armstrong C (1980) Fed Proc Fed Am Soc Exp Biol 39:2403-2409
12. Somlyo AV, Gonzalez-Serratos H, Shuman H, McClellan G and Somlyo AP (1981) J Cell Biol 90:577-594
13. Martonosi A (1968) J Biol Chem 243:71-81
14. Hasselbach W (1978) Biochem Biophys Acta 515:23-53
15. Tada M, Yamamoto T and Tonomura Y (1978) Physiol Rev 58:1-79
16. Ebashi S (1958) Arch Biochem Biophys 76:410-418
17. Hasselbach W (1964) Prog Biophys Mol Biol 14:169-185
18. Jones LR and Cala SE (1981) J Biol Chem 256:11809-11818
19. Meissner G (1975) Biochim Biophys Acta 389:51-68
20. Campbell KP (1986) In: Entman ML, Winkle WBV (eds) Sarcoplasmic Reticulum in Muscle Physiology, CRC Press, Florida, Vol 1, pp 65-99
21. Brandl CJ, Green NM, Korczak B and Maclennan DH (1986) Cell 44:597-607
22. Brandl CJ, Deleon S, Martin DR and Maclennan DH (1987) J Biol Chem 262:3767-3774
23. Tada M, Kirchberger MA, Repke DI and Katz AM (1974) J Biol Chem 249:6174-6180
24. Fliegel LE, Leberer E, Green N and Maclennan DH (1989) FEBS Lett 242:297-300
25. Campbell KP, Maclennan DH, Jorgensen AO and Mintzer MC (1983) J Biol Chem 258:1197-1204
26. Inui M, Saito A and Fleischer S (1987) J Biol Chem 262:15637-15642
27. Lai FA, Anderson K, Rousseau E, Liu Q-Y and Meissner G (1988) Biochem Biophys Res Commun 151:441-449
28. Jorgensen AO, Shen AC-Y and Campbell KP (1985) J Cell Biol 101:257-268

29. Jorgensen AO, Shen AC-Y and Campbell KP and Maclennan DH (1983) J Cell Biol 97:1573-1581
30. Jorgensen AO, Broderick R, Somlyo AP and Somlyo AV (1988) Circ Res 63:1060-1069
31. Ikemoto N, Ronjat M, Meszaros LG and Koshita M (1989) Biochem 28:6764-6771
32. Scott BT, Simmerman HKB, Collins JH, Nadal-Ginard B and Jones LR (1988) J Biol Chem 263:8958-8964
33. Mitchell RD, Simmerman HKB and Jones LR (1988) J Biol Chem 264:1376-1381
34. Lytton J and Maclennan DH (1992) In: Fozzard HA, Haber E, Jennings RB, Katz AM, Morgan HE (eds) The Heart and Cardiovascular System, Raven Press Publishers, New York, Vol 2
35. Maclennan DH, Campbell KP and Reithmeier RAF (1983) In: Cheung WY (ed) Calsequestrin, Calcium and Cell Function, Academic Press, New York, Vol 4, pp 151-173
36. Fliegel L, Ohnishi M, Carpenter MC, Khanna VK, Reithmeier RAF and Maclennan DH (1987) Proc Natl Acad Sci USA 84:1167-1171
37. Zarain-Herzberg A, Fliegel L and Maclennan DH (1988) J Biol Chem 263:4807-4812
38. Arai M, Otsu K, Maclennan DH, Alpert NR and Periasamy M (1991) Circ Res 69:266-276
39. Takahashi T, Allen PD, Lacro RV, Marks AR, Dennis AR, Schoen FJ, Grossman W, Marsh JD and Izumo S (1992) J Clin Invest 90:927-935
40. Callewaert G (1992) Cardiovas Res 26:923-932
41. Ford LE and Podolsky RJ (1970) Science 167:58-59
42. Endo M, Tanaka M and Ogawa Y (1970) Nature 228:34-36
43. Fabiato A (1983) Am J Physiol 245:C1-C14
44. Fabiato A (1985) J Gen Physiol 85:247-289
45. Fabiato A (1985) J Gen Physiol 85: 291-320
46. Kentish JC, Barsotti RJ, Lea TJ, Mulligan IP, Patel JR and Ferenzi MA (1990) Am J Physiol 258:H610-H615
47. Chamberlain BK, Volpe P and Fleischer S (1984) J Biol Chem 258:7540-7546
48. Rousseau E, Smith JS, Henderson JS and Meissner G (1986) Biophys J 50:1009-1014
49. Meissner G and Henderson JS (1987) J Biol Chem 262:3065-3073
50. Benevolensky DS, Menshikova EV, Watras J, Levitsky DO and Ritov VB (1987) Biomed Biochem Acta 46:S393-S398
51. Nabauer M, Ellis-Davies GCR, Kaplan JH and Morad M (1989) Am J Physiol 256:H916-H920
52. Leblanc N and Hume JR (1990) Science 248:372-376
53. Sham JSK, Cleeman L and Morad M (1992) Science 255:850-853

54. Radon DP, Cefali DC, Mitchell RD, Seiler SM and Jones LF (1989) Circ Res 64:779-789
55. Lindsay AR and Williams AJ (1991) Biochem Biophys Acta 1064:89-102
56. Ashley RH and Williams AJ (1990) J. Gen. Physiol. 95:981-1005
57. Nabauer M, Callewaert G, Cleemann L and Morad M (1989) Science 244:800-803
58. Anderson K, Anthony F, Liu Q-Y, Rousseau E, Erickson HP and Meissner G (1989) J Biol Chem 264:1329-1335
59. Smith JS, Coronado R and Meissner G (1986) J Gen Physiol 88:573-588
60. Morgan JP and Blinks JR (1982) Can J Physiol 60:524-528
61. Rousseau E, La Dine J, Liu Q-Y and Meissner G (1988) Arch Biochem Biophys 267:75-86
62. Imagawa T, Smith JS, Coronado R and Campbell KP (1987) J Biol Chem 262:16636-16643
63. Jones LR and Burrows SD (1983) Fed Proc Am Soc Exp Bio 42:1932-1936
64. Miyamoto H and Racker E (1982) J Memb Biol 66:193-201
65. Jones LR, Seiler SM and Winkle BV (1986) In: Entman ML, Winkel WBV (eds) Sarcoplasmic Reticulum in Muscle Physiology, CRC Press, Florida, Vol 1, pp 65-99
66. Jones LR and Cala SE (1981) J Biol Chem 256:11809-11818
67. Jones LR, Besch, Jr HR, Sutko JC and Willerson JT (1979) J Pharmacol Exp Ther 209:48-55
68. Berridge MJ (1987) Annu Rev Biochem 56:159-193
69. Bezprozvanny I, Watras J and Ehrlich BE (1991) Nature 351:751-754
70. Ehrlich BE and Watras J (1988) Nature 336:583-586
71. De Chaffoy D and Courecelles DE (1989) Mol Cell Biochem 88:65-72
72. Zhu Y and Nosek TM (1991) Pflugers Arch 418:1-6
73. Borgatta L, Watras J, Katz AM and Ehrlich BE (1991) Proc Natl Acad Sci USA 88:2486-2490
74. Nosek TM, Williams MF, Zeigler ST and Godt RE (1986) Am J Physiol 250:C807-C811
75. Feher JJ, Lebolt WR and Manson NH (1989) Circ Res 65:1400-1408
76. Thandroyen FT, McCarthy J, Burton KP and Opie LH (1988) Circ Res 62:306-314
77. Holmberg SR and Williams AJ (1992) Basic Res Cardiol 87:255-268
78. Otsu K, Willard HF, Khanna VK, Zorzato F, Green NM and Maclennan DH (1990) J Biol Chem 265:13472-13483
79. Brillantes AM, Allen P, Takahashi T, Izumo S and Marks AR (1992) Circ Res 71:18-26
80. De Meis L and Vianna AL (1979) Annu Rev Biochem 48:275-292
81. Jorgensen AO, Shen AC, Daly P and Maclennan DH (1982) J Cell Biol 92:409-416
82. Ikemoto N (1982) Annu Rev Physiol 44:297-317
83. Racker E (1972) J Biol Chem 247:8198-8200

84. Thorley-Lawson DA and Green NM (1973) Eur J Biochem 40:403-413
85. Inesi G and Scales D (1974) Biochem 13:3298-3306
86. Klip A, Reithmeier RAF and Maclennan DH (1980) J Biol Chem 255:6562-6568
87. Pick U and Racker E (1979) Biochemistry 18:108-113
88. Tada M and Katz AM (1982) Annu Rev Physiol 44:401-423
89. Tada M, Ohmori F, Yamada M and Abe H (1979) J Biol Chem 254:319-326
90. Kirchberger MA, Tada M and Katz AM (1974) J Biol Chem 249:6166-6173
91. Bidlack JM and Shamoo AE (1980) Biochem Biophys Acta 632:310-325
92. Sasaki T, Inui M, Kimura Y and Tada M (1992) J Biol Chem 267:1674-1679
93. Lamers JM and Stinis HT (1980) Biochem Biophys Acta 624:443-459
94. Tada M, Kirchberger MA and Katz AM (1975) J Biol Chem 250:2640-2647
95. Gertz EW, Hess ML, Lain RF and Briggs FN (1967) Circ Res 20:477-484
96. Lentz BR, Clubb KW, Barrow DA and Meissner G (1983) Proc Natl Acad Sci USA 80:2917-2921
97. Maire MI and Moller JV (1986) In: Entman ML, Winkle WBW (eds) Sarcoplasmic Reticulum in Muscle Physiology, CRC Press, Florida, Vol 1, pp 101-126
98. Murphy A (1976) Biochemistry 15:4492-4496
99. Kawakita M, Yasuoka K and Kaziro Y (1980) J Biochem 87:609-617
100. Hidalgo C, Petrucci DA and Vergara C (1982) J Biol Chem 257:208-216
101. Yu BP, Masoro EJ and Bertrand HA (1974) Biochemistry 13:5083-5087
102. Jones LR, Besch HR and Watanabe AM (1978) J Biol Chem 253:1643-1653
103. Herbette L, Messineo FC and Katz AM (1982) Annu Rev Pharmacol Toxicol 22:413-434
104. Katz AM, Repke AM, Corkedale DI and Schwartz S (1975) Cardiovasc Res 9:764-769
105. Watanabe AM and Besch HR (1974) J Pharmacol Exp Ther 191:241-251
106. Entman ML, Allen JC, Bornet EP, Gillette PC, Wallick E and Schwartz A (1972) J Mol Cell Cardiol 4:681-687
107. Messineo FC and Katz AM (1979) J Cardio Pharmacol 1:449-459
108. Littleton JM, Grieve SJ, Griffiths PJ and John GR (1980) Adv Exp Med Biol 126:7-19
109. Hara K and Kasai M (1977) J Biochem 82:1005-1017
110. Grossman W (1990) Circulation 81:III 1 - III 7
111. Harigaya S and Schwartz A (1971) Circ Res 25:781-794
112. Sordahl LA, Wood WG and Schwartz A (1970) J Mol Cell Cardiol 1:341-344
113. Sulakhe PV and Dhalla NS (1971) J Clin Invest 50:1019-1027
114. McCollum WB, Crow C, Harigaya S, Bajusz E and Schwartz A (1970) J Mol Cell Cardiol 1:445-457
115. Ito Y, Suko J and Chidsey CA (1974) J Mol Cell Cardiol 6:237-247

116. Tomlinson CW, Lee SL and Dhalla NS (1976) Circ Res 39:82-92
117. Movsesian MA, Bristow MR and Krall J (1989) Circ Res 65:1141-1144
118. Komuro I, Kurabayashi M, Shibazaki Y, Takaku F and Yazaki Y (1989) J Clin Invest 83:1102-1108
119. Nagai R, Zarain-Herzberg A, Brandl CJ, Fujii J, Tada M, Maclennan DH, Alpert N and Periasamy P (1989) Proc Natl Acad Sci USA 86:2966-2970
120. Mercadier JJ, Lompre AM, Duc P, Boheler KR, Fraysse JB, Wisnewsky C, Allen PD, Komajda M and Schwartz K (1990) J Clin Invest 85:305-309
121. Katoh Y, Komuro I, Yamaguchi H and Yazaki Y (1992) Jpn Circ J 56:694-700
122. Suzuki T and Wang JH (1986) J Biol Chem 261:7018-7023
123. Movsesian MA, Leveille C, Krall J, Colyer J, Wang JH and Campbell KP (1990) J Mol Cell Cardiol 22:1477-1485
124. Movsesian MA, Colyer J, Wang JH and Krall J (1990) J Clin Invest 85:1698-1702
125. Movsesian MA (1992) Basic Res Cardiol 87:277-284
126. Gwathmey JK and Morgan JP (1985) Circ Res 57:836-843
127. Afzal N, Ganguly PK, Dhalla KS, Pierce GN, Singal PK and Dhalla NS (1988) Diabetes 37:936-942
128. Dhalla NS, Pierce GN, Innes IR and Beamish RE (1985) Can J Cardiol 1:263-281
129. Tani M (1990) Annu Rev Physiol 52:543-559
130. Gwathmey JK, Copelas L, Mackinnon R, Schoen FJ, Feldman MD, Grossman W and Morgan JP (1987) Circ Res 61:70-76
131. Gwathmey JK, Slawsky MT, Hajjar RJ, Briggs GM and Morgan JP (1990) J Clin Invest 85:1599-1613
132. Kaufman RL, Homburger H and Wirth H (1971) Circ Res 28:346-357
133. Ginsberg R, Esserman LJ and Bristow MR (1983) Ann Intern Med 98:603-606
134. Fellenius E, Hansen CA, Mjos O and Neely JR (1985) Am J Physiol 249:H80-H87
135. Rupp H (1986) In: Rupp H (ed) The Regulation of Heart Function, Thieme Inc, New York, pp 234-348
136. Katz AM (1988) Am J Cardiol 62:3A-8A
137. Morgan JP, Erny R, Allen PD, Grossman W and Gwathmey JK (1990) Circulation 81 (Suppl 111):21-32
138. Afzal N and Dhalla NS (1992) Am J Physiol 262:H868-H874
139. Ganguly PK, Pierce GN, Dhalla KS and Dhalla NS (1983) Am J Physiol 244:E528-E535
140. Limas CJ (1978) Am J Physiol 234:H426-H431

CHAPTER 4

MEMBRANE DYSFUNCTION AND ABNORMAL SPONTANEOUS ACTIVITY: A STUDY IN EXPLANTED CARDIAC CELLS

Otto F. Schanne, Jihong Qu, George E. Haddad and Elena Ruiz-Petrich

Department of Physiology and Biophysics
University of Sherbrooke, Quebec, Canada

INTRODUCTION

Despite considerable recent advances in cellular cardiac electrophysiology relatively little is known about the changes in membrane function associated with abnormal spontaneous activity under pathological conditions (1,2). Even more the mechanism(s) of normal spontaneous activity are still somewhat controversial (3,4,5). Lately new results have shown the importance of the resting potential, the membrane resistance, and the accompanying role of developmental changes in the function of ionic channels for the development of spontaneous activity in cells that do not show spontaneous activity under physiological conditions (5). Most of these results were obtained from explanted ventricle cells that develop spontaneous activity after explantation.

It is the objective of this report to summarize recent developments from the work with cultured ventricle cells that could be relevant to further our understanding of the development of abnormal spontaneous activity in the whole heart. We will emphasize quantitative aspects and we will first summarize what is known about normal mechanisms of spontaneous activity and contrast these with what has been learned from the spontaneous activity that develops in cultured cells after explantation. Finally we will show how these results relate to features of arrhythmogenesis in the human heart.

NORMAL AND ABNORMAL CARDIAC PACEMAKER ACTIVITY

Since we are dealing with a subject that concerns cardiologists, pharmacologists as well as electrophysiologists the basic nomenclature describing phenomena related to abnormal spontaneous activity is somewhat confusing. Therefore we will begin with some definitions as they will be used in this report.

Definitions

(a) Spontaneous activity means contractile activity with or without accompanying electrical signal in the absence of an external stimulus. This term is quite general and applies equally to the activity of an isolated preparation of the SA node, a spontaneously contracting cluster of cultured ventricle cells and to the contraction of an isolated myocyte caused by Ca release from the sarcoplasmic reticulum in the absence of an action potential; (b) pacemaker activity or automaticity is the electrical manifestation of a pacemaker mechanism characterized by a diastolic depolarization (phase 4 depolarization) that develops independent of the preceding action potential. This is the type of electrical activity found during the diastole in the normal SA node and in Purkinje fibers; and (c) triggered activity is characterized by delayed afterdepolarizations (DADs) and early afterdepolarizations (EADs) that reach the threshold of the current determining the action potential upstroke, and that are triggered by the preceding action potential (6,7). DADs and EADs are often found coupled to the preceding action potential after digitalis and quinidine intoxication and when they reach threshold they will lead to extrabeats or sustained trains of action potentials.

Normal Pacemaker Activity

Normal pacemaker activity represents the activity of the sinoatrial (SA) node and that of Purkinje fibers. It is accepted today that pacemaker activity does not reflect the predominance of a single current but constitutes a complex interaction of several current systems that results in a net inward current driving the diastolic potential towards positive values. The isolation and measurement of these currents is difficult because many cells with pacemaker activity are very small (especially nodal cells) e.g. a cell with a surface of 1000 μm^2 and a specific membrane capacitance of 1 $\mu F \bullet cm^{-2}$ with a typical slope of the pacemaker potential of 50 $mV \bullet s^{-1}$ needs only a net inward current of 0.5 pA to drive the diastolic potential positive. This is beyond the resolution of present whole cell patch clamp amplifiers and therefore imposes a limit to a quantitative analysis of pacemaker activity. In spite of these limitations considerable progress has been made in our understanding of cardiac pacemakers and the present state of the art is summarized in recent reviews (3,4,8,9). For the purpose of this report, we will outline normal mammalian cardiac automaticity using as an example the isolated SA node cell and the Purkinje fiber strand.

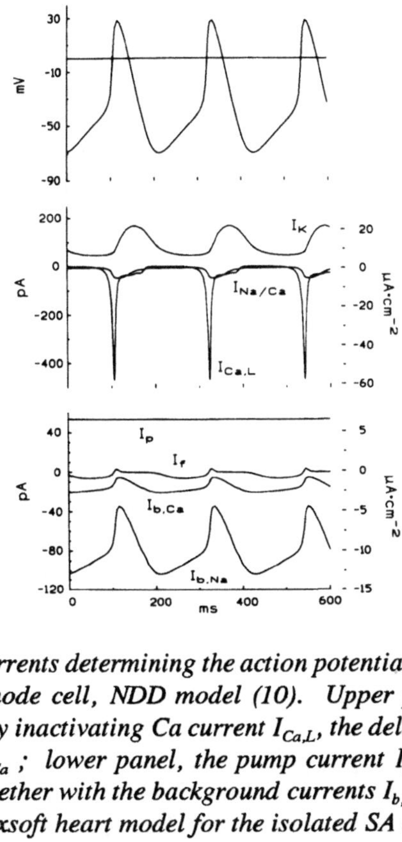

Figure 1. Simulated currents determining the action potential and pacemaker activity of a single sinoatrial node cell, NDD model (10). Upper panel, action potential; middle panel, the slowly inactivating Ca current $I_{Ca,L}$, the delayed rectifier I_K, and the exchanger current $I_{Na,Ca}$; lower panel, the pump current I_p, the hyperpolarization activated current I_f together with the background currents $I_{b,Ca}$ and $I_{b,Na}$. Simulations were made using the Oxsoft heart model for the isolated SA node cell (Oxsoft Heart, V. 3.2, Oxsoft Ltd., Oxford, England).

Sinoatrial Node. Figure 1 represents a computer generated synopsis of the single cell rabbit SA node action potential together with the time course of the currents involved in pacemaker activity. The calculations were made using the Noble, DiFrancesco, Denyer (NDD) model (10). To facilitate comparisons with other structures, we have added the current densities on the right side of the current tracings. The calculation of the current densities is based on a cell surface of 800 μm^2 that corresponds to a spherical cell with 16 μm diameter (10,11). The cells have a maximum diastolic potential of about -70 mV and the upstroke of the action potential is mainly determined by the slowly inactivating Ca current $I_{Ca,L}$, whereas the fast inward current I_{Na} is poorly developed (not shown). The main gated currents that determine the phase 4 depolarization are the inactivation of the delayed rectifier I_K and, to a lesser degree the activation of the hyperpolarization activated inward current I_f. A complete block of $I_{Ca,L}$ leads to the suppression of the action potential and a partial block entails a slight lengthening of the phase 4 depolarization around the threshold potential, Fig. 19-3 in ref. 3. The frequency of the SA node pacemaker is also influenced by the time independent background currents $I_{b,Na}$ and $I_{b,Ca}$ that are carried by Na and Ca respectively. Because of their linear current voltage relationship the time course of these currents reflects the changes in membrane potential. $I_{b,Na}$ and $I_{b,Ca}$ are different from the Na and Ca window currents that result from the overlap of the voltage range of the steady state activation and inactivation of I_{Na} and $I_{Ca,L}$. In addition, the pacemaker activity depends on non channel mediated currents (1) the pump current related to the electrogenic Na/K pump that extrudes 3 Na while uptaking 2 K ions, so contributing a voltage independent current component and (2) the electrogenic Na/Ca exchanger that normally pumps out 1 Ca^{2+} in exchange of the uptake of 3 Na^+. The exchanger is activated by an increase of Ca_i following the Ca influx during the upstroke of the action potential resulting in a transient inward current that does not contribute much to the diastolic depolarization.

A complete block of I_f results in a slowdown but not in a loss of automatic activity (5). Such an effect can be produced experimentally by exposing the cell to a low concentration of Cs known to block I_f rather selectively in the low mM range (9,12). A 60% block of the Na background conductance causes an inflection point during the phase 4 depolarization and doubling the same conductance leads to damped oscillatory potentials that stabilize at a depolarized level (5). These results are similar than those obtained by the group of Noble (12).

Purkinje Fibers. Figure 2 shows the automatic activity of Purkinje fibers together with the time course of the currents that determine their pacemaker mechanism (13). The upper panel of Fig. 2 shows the action potential with spontaneous activity and a maximum diastolic potential of about -95 mV. The action potential upstroke is determined by a well developed Na current (I_{Na}) and the fast repolarization phase following the action potential spike is determined by the transient outward current I_{to} known today to be composed of a Ca dependent and a 4-AP sensitive component (currents not shown in Fig. 2). The middle panel of Fig. 2 shows the time course of the gated currents $I_{Ca,L}$, I_K, and I_f together with the exchange current $I_{Na,Ca}$. In the lower panel the time course of the background currents $I_{b,Na}$ and $I_{b,Ca}$ together with the pump current I_p, and the inward rectifier I_{K1} are represented.

Compared to the SA node the pacemaker mechanism of the Purkinje fibers produces a much slower beating rate that is mainly determined by the slow activation of I_f and the voltage dependence of I_{K1}, not present in the SA node. Compared to the SA node, I_f is more important for the automatic activity in Purkinje fibers. A block of I_f abolishes the automaticity and the membrane stabilizes at the resting potential (5), whereas the same maneuver resulted only in a decrease of the beating rate in the SA node cell. The pacemaker mechanism in the Purkinje fiber is also

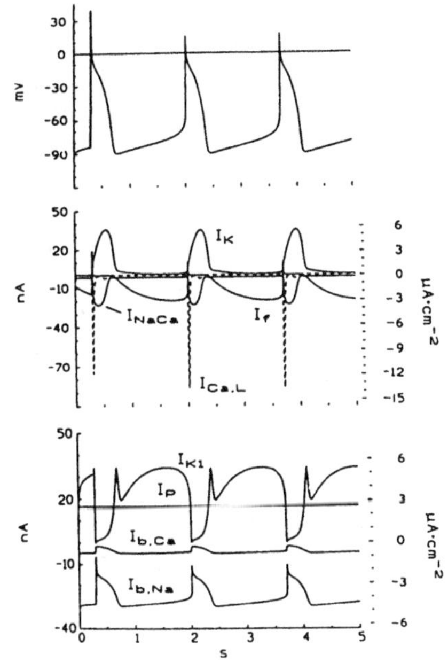

Figure 2. Simulated currents determining the action potential and the pacemaker activity of a multicellular Purkinje fiber preparation, DiFrancesco Noble model (13). Upper panel, action potential; middle and lower panel, ionic currents. Symbols are the same as for Fig. 1 except I_{K1}, inward rectifier. The current densities were calculated with a total membrane surface of 0.0063 cm^2 (10). Oxsoft heart model Purkinje fiber, Oxsoft Heart, V. 3.2 (13).

more sensitive to changes in $I_{b,Na}$ than the SA node cell. A decrease of 30% in the conductance of the background current leads to a drastic decrease in the beating rate whereas doubling of $g_{b,Na}$ results in EADs without loss of automaticity (5).

In summary the reactions of the pacemaker in the SA node and Purkinje fibers to experimental interventions can be traced to 2 basic differences. First the more negative membrane potential of Purkinje fibers relative to that of the SA node activates and inactivates to a different degree gated currents common to both structures and also modifies the driving force of the background currents. Second the presence of currents in Purkinje fibers that are absent in the SA node necessarily contributes to differences in the nature of their pacemaker mechanisms.

Abnormal Pacemaker Activity

Abnormal automaticity can be defined as pacemaker activity that differs quantitatively or qualitatively from normal activity. The problem sometimes arises as what to consider as the reference point for normal. In the case of a normally electrically silent preparation like a ventricle cell, there spontaneous activity will be considered abnormal. If one deals with a structure that normally is spontaneously active, the comparison is made between the activity observed under physiological conditions and that found under specific experimental conditions. If one wants to interpret results from pacemaker activity in different species, probably the best reference is the rabbit SA node or one of the Purkinje fiber preparations for which a maximum of information is available but not the human tissue where the mechanisms of automaticity are little known.

A quantitative difference in the pacemaker mechanism means that the currents underlying the pacemaker mechanism remain the same but their relative importance has changed with respect to that of the normal state. We showed above that such quantitative differences can lead to changes in the beating rate and even to the appearance of EADs.

Two pacemaker mechanisms are qualitatively different when different currents underlie their activity. The slope of the diastolic depolarization of cultured chick embryo ventricle cells decreases when exposed to Cs whereas the same parameter increases in explanted neonatal rat ventricle cells. This suggests that different currents determine the qualitatively different pacemaker mechanisms of the 2 preparations (14).

THE SPONTANEOUS ACTIVITY FOUND IN EXPLANTED VENTRICLE CELLS

Most of the available data about spontaneous activity in cultured cardiac cells was obtained from embryonic chick and neonatal rat hearts. Both preparations are quite well characterized (5,15,16) and we will discuss here the spontaneous activity in neonatal rat ventricle cells and complement the discussion with results from the embryonic chick heart where important differences exist.

The Development of Spontaneous Activity After Explantation

Ventricular cells from 1 day old rats were isolated with trypsin from the lower third of the heart to obtain cultures that were relatively free of conducting and pacemaker tissue (14). The cells were cultured in Eagle's Minimum Essential Medium with Hanks salts and supplemented with 20% fetal bovine serum (16,17). Small clusters of these cells were electrically excitable but did not show spontaneous

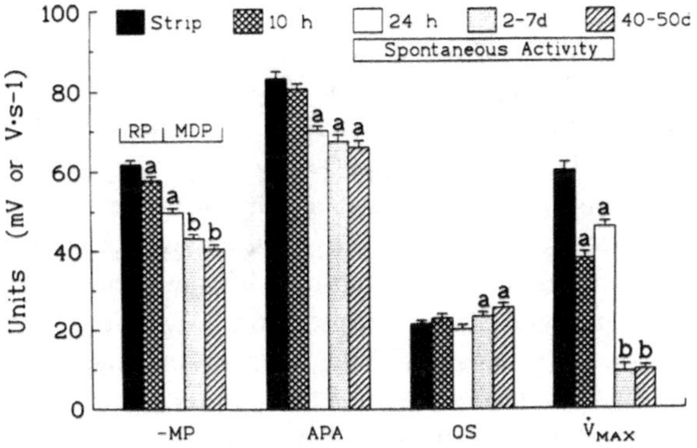

Figure 3. Changes with time after explantation in electrical parameters measured with microelectrodes at 36°C in clusters of cultured neonatal rat ventricle cells compared with the electrical parameters of isolated ventricle strips (column labelled strip). Time in culture is indicated as hours (h) or days (d) beside the symbols for the different columns. Note that the clusters cultured for 24 hrs and longer showed spontaneous activity. Clusters were derived from ventricles of 1-2 day old neonatal rats, dispersed with trypsin and cultured in Eagle's Minimum Essential Medium with Hanks salts supplemented with 20% fetal bovine serum. MP: membrane potential, i.e. resting potential (RP) in quiescent cells and maximum diastolic potential (MDP) in spontaneously active clusters; APA: action potential amplitude; OS: overshoot, and $\dot{V}_{max}$ *maximum rate of rise of action potential. Error bars are SEM, letters a or b over a column indicate a statistically significant difference of at least* $P < 0.05$ *with respect to columns labelled strip or 24 h respectively. Pooled results from our laboratory, 20 - 180 measurements per group (18,19,20,21).*

activity during the first 24 hrs in culture but later they started to contract and they could maintain their contractile activity for weeks. Figure 3 compares the electrical properties of isolated ventricle strips with those of cultured cells at different times after explantation. The properties of the cells 10 hrs after explantation compare roughly with those of the isolated strips except for a slight depolarization that is probably responsible for the decrease in $\dot{V}_{max}$. As suggested by the high rate of rise of the action potential, the action potential upstroke is TTX sensitive and determined by the fast inward current (5). This does not change when after 24 hrs spontaneous activity develops with a rather slow beating rate of about 20-30 bpm, that is accompanied by a further depolarization so that the maximum diastolic potential is more positive than the resting potential of the 10 hrs group. After day 2 there is a drastic decrease of the $\dot{V}_{max}$ and an increase of the overshoot of the action potential that has developed into a slow response action potential with an upstroke velocity sensitive to Mn and verapamil (14,18,21). These changes are accompanied by a further depolarization and an increase in the beating rate to about 60-80 bpm. Compared to the isolated ventricle, there is a significant increase in action potential duration at 0 mV and at 90% repolarization (22). From there on no significant changes in the electrical parameters and spontaneous activity occur for at least 40-50 days after explantation. In one case we observed spontaneous activity in a cluster for 6 months until the culture was lost by infection (17).

One can ask whether the appearance of spontaneous activity is limited to explanted neonatal rat ventricle cells in culture. The majority of published reports demonstrates that independent of the species from which the cultured cells were derived, spontaneous activity develops if the cells are cultured long enough.

The time course of the development of the spontaneous activity may vary with culture conditions or the type of cultured cells: myocytes, monolayers, cell clusters or reaggregates. It appears that spontaneous activity occurs earlier in isolated myocytes than in cell clusters but that the spontaneous activity is more stable in the latter. Reaggregates cultured for 3 days and originally derived from ventricles of 15 day old chick embryos do not show spontaneous activity but exhibit fast action potentials. After being exposed to normal culture conditions for 2 more days, they

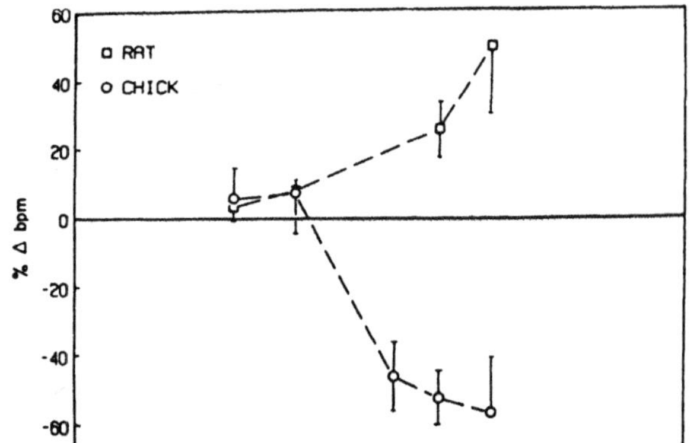

Figure 4. Percent changes in beating rate (% Δbpm) of explanted rat (open squares) and chick ventricle cells (open circles) exposed to Cs. Data points represent means ± SEM, 4-22 measurements per data point. Cell clusters and reaggregates were selected to have under control conditions beating rates between 55 and 65 bpm. Modified from 14.

beat spontaneously and show slow response action potentials (14). Figure 4 compares the response of the beating rate to an increase in extracellular Cs concentration of these cells with that of cultured neonatal rat ventricular cells: the beating rate of the rat cells increased whereas the beating rate of the embryonic chick heart cells was depressed at Cs concentrations > 2 mM. This result suggests a different pacemaker mechanism for the 2 cell types. It appears that in the rat block of an outward current by Cs increases the net inward current and the beating rate whereas in the chick embryo the decrease in the beating rate could be explained by a block of an inward current, thus decreasing the net inward current that drives the pacemaker. Cs above 2 mM blocks K currents that are outward in the potential range of the chick embryo cells so that the deactivation of the delayed rectifier could also be reduced resulting in a decrease of a net inward current. Recently the pacemaker in the chick embryo has been associated with the hyperpolarization activated current I_f, (16) but this current is blocked by Cs concentrations < 2 mM (12) where we did not find any influence of Cs on the beating rate in the same preparation. However, the result could be explained with a block of I_f provided that the current's sensitivity to Cs is reduced in the chick embryo.

Once the cells exhibit spontaneous activity and slow action potentials, their reactivity to metabolic inhibition is also changed. In the isolated ventricle of the neonatal rat exposure to dinitrophenol, glucose free medium, and 2-deoxyglucose causes a significant shortening of the action potential duration at 0 mV and 90% repolarization whereas in the cultured cells the action potential shortens at both levels

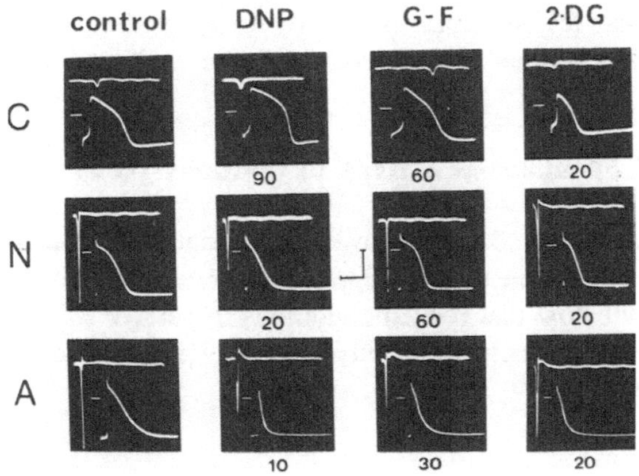

Figure 5. Effects of dinitrophenol (DNP), substrate deprivation (G-F), and deoxyglucose (2DG) on action potential configuration (lower traces) and maximum rate of rise (upper traces, downward deflection positive) of cultured (C), neonatal (N), and adult rat ventricle cells. Numbers below panels indicate time of exposure in min. Calibration: vertical 50 mV and 50 V• s^{-1}; horizontal 100 ms for panels (C) and (N), 20 ms for (A) and dV/dt throughout. From (22) with permission.

only in the presence of 2-deoxyglucose but this shortening is accompanied by a depolarization of 12 mV (Fig. 5) (22). This result suggests that the main source of energy in cultured cardiac cells is the glycolytic pathway and that its block leads to depolarization via the inhibition of an electrogenic Na/K pump. The importance of the glycolytic pathway as an energy source in cultured cardiac cells derived from chick embryos and neonatal rat heart cells has also been recognized by others (23,24,25), for a review see (26).

The results of Fig. 3 suggest that spontaneous activity in cultured neonatal rat ventricle cells appears some time after explantation and before the action potential becomes TTX insensitive. In fact spontaneous activity has been found associated with fast action potentials whether or not special measures were taken to prevent the disappearance of fast TTX sensitive action potentials (22,27,28,29,30,31). Moreover, the disappearance of the TTX sensitive action potentials does not necessarily mean a downregulation of the Na channels, because it can be reconciled with the persistence of functional I_{Na} channels when one assumes that I_{Na} is largely inactivated either by depolarization (Fig. 3) and/or by a shift in the voltage dependence of its gating parameters. This notion is supported by our measurements of I_{Na} between 2½ and 50 hrs after explantation that did not show a decrease in peak current density (5) and by earlier results based on $\dot{V}_{max}$ measurements from the same preparation (32).

It follows from the results discussed above that spontaneous activity in cultured cells (a) occurs after explantation and is not due to damage during the explantation process; (b) it is independent of whether the upstroke of the action potential depends on I_{Na} or $I_{Ca,L}$, and (c) spontaneous activity is a very sensitive index for subtle electrical changes that occur in the membrane after explantation.

Determinants of Spontaneous Activity in Cultured Ventricle Cells

The analysis of Figs. 3 and 5 showed that variations of the membrane potential and ionic currents determine the development of spontaneous activity in cultured cells and we will show below that related alterations in passive electrical properties and intracellular Ca homeostasis are equally important features in the development of spontaneous activity.

Passive Electrical Properties. Practically speaking the passive electrical properties represent the resistances and capacitances associated with biological structures and in the context of pacemaker mechanisms, the most interesting quantity is the specific membrane resistance R_m with the dimension $\Omega \bullet cm^2$. In a cell with a simple geometry R_m represents the product of the directly measurable input resistance R_{inp} (Ω) and the cell surface area measured in cm^2. However, apart from the most simple cell geometries, the relation between R_{inp} and R_m can be quite complicated as it is the case in multicellular cables or disk like structures (23,24).

R_m represents the inverse of the sum of the specific ionic conductances in a cell and it is in general voltage and time dependent. In the case where at the resting potential one ionic conductance dominates, R_m reflects this conductance. In ventricle cells the inward rectifier I_{K1} determines the resting potential and the membrane conductance at potentials negative to the threshold of the gated currents responsible for the upstroke of the action potential, i.e. R_{inp} can be used to determine the conductance of the inward rectifier or to follow R_m changes during the phase 4 depolarization. It has recently been shown with R_m measurements that in clusters of neonatal ventricle cells cultured for 2-6 days about 40% of the inwardly rectifying channels become electrically silent when spontaneous activity appears (35). This decrease with time in culture has been confirmed and it has also been shown that in neonatal rats I_{K1} increases sharply after day 9 post partum (36). Moreover, the modulation of the inward rectifier by developmental processes is also reflected in the increase of the sensitivity of I_{K1} to Ba with age: in the 1 day old neonatal rat 2.4 mM are needed to block the current but in the 7 day old animal only 200 μM are sufficient (37). As for cultured mammalian cardiac cells a close relationship between K conductance and spontaneous activity has long been postulated for embryonic chick heart cells (38). These results suggest a close association of I_{K1} with developmental and environmental modulators so that a decrease in I_{K1} leads to an unstable resting potential that facilitates pacemaker activity in cultured cells.

Resting Potential. In cardiac cells the resting potential V_m consists in general of 2 additive components, the diffusion potential V_{diff} (calculated with the Goldman Hodgkin Katz equation) and a potential component created by the activity of the electrogenic Na/K pump V_p (calculated with the Mullins and Noda equation, 4,39). The pump potential represents the voltage drop of the pump current density J_p across the specific membrane resistance R_m: $V_p = J_p \bullet R_m$. J_p depends on the turnover rate of the pump and is about 1 $\mu A \bullet cm^{-2}$. Because the J_p is quite independent of the membrane voltage (Figs. 1 and 2) any change in R_m will change the resting potential. If after explantation R_m increases this will (a) tend to depolarize the membrane at potentials positive to E_K because I_{K1} decreases and (b) the increasing R_m will tend to hyperpolarize the membrane. Therefore, although there is little change in the overall resting potential, the relative importance of V_p relative to V_{diff} increases. One of the consequences of this change is that the resting potential becomes more sensitive to metabolic inhibition, especially to interference with the ATP compartment related to the glycolytic pathway (40).

In cultured neonatal rat ventricle cells we found a ouabain sensitive pump potential of about 11 mV that resulted in a J_p of 0.85 $\mu A \bullet cm^{-2}$ considering an R_m of 13 $k\Omega \bullet cm^2$ (35,39). Assuming that the turnover rate of the pump does not change after explantation one obtains a V_p of 2.1 mV in freshly isolated myocytes having an R_m of 2.5 $k\Omega \bullet cm^2$ (35). Neglecting the pump potential the relative permeability P_{Na}/P_K has been estimated as 0.03 (41) in neonatal rats and as between 0.25-0.35 in spontaneously active cultured cells. This implies the existence of a Na background current whose amplitude can be estimated from the available permeability and

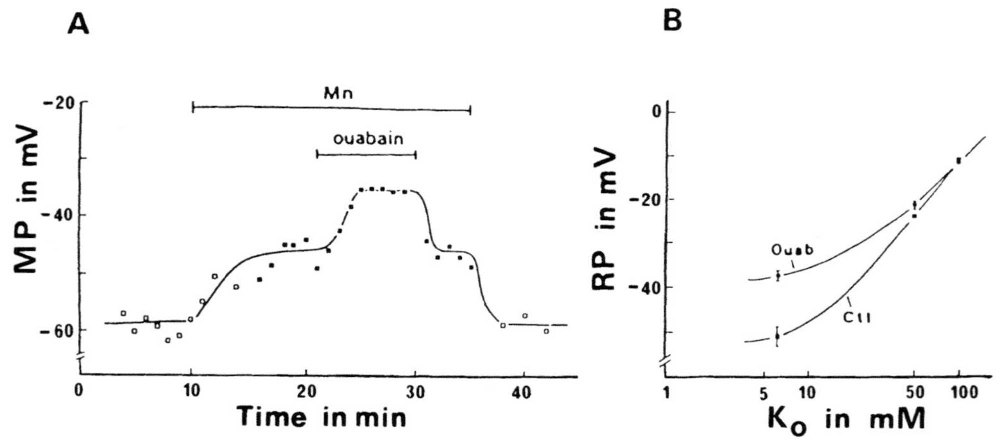

Figure 6. *Determination of diffusion and pump component of the resting potential in cultured neonatal ventricle cells exhibiting spontaneous activity.* ***(A)*** *Beating cells (open squares) were exposed to 4 mM Mn. Spontaneous activity was lost and a stable resting potential could be measured (full squares). After exposing the cells to 0.1 mM ouabain and Mn the resting potential stabilized again at an even more positive potential level (diffusion component of resting potential). The difference between the 2 levels of the resting potential represents the pump potential (about 12 mV). The final part of the experiment represents the washout demonstrating the reversibility of the procedure.* ***(B)*** *variation of the composite resting potential (full squares, marked Ctl) and its diffusion component (full circles, marked Ouab) with increasing concentrations of extracellular K. The curve labelled Ctl represents the Mullins and Noda equation and that labelled Ouab was fitted using the Hodgkin Goldman Katz equation resulting in a P_{Na}/P_K of 0.25, a pump coupling ratio of 2.0, and a K_i of 160 mM. In this experiment the cells were stopped by adding 10 mM Ca. The diffusion component of the resting potential was measured in the presence of 0.1 mM ouabain. Vertical bars SEM, 14-31 measurements per data point. MP: membrane potential, RP: resting potential. From (39) with permission.*

conductance data, Table 1. Examples of measurements that allow to determine the resting potential in spontaneously active cells and how to determine P_{Na}/P_K are shown in Fig. 6.

Considering the range of P_{Na}/P_K measured in explanted cells exhibiting spontaneous activity, Table 1 shows a 32 to 74% increase in a Na background conductance that translates into a 4 to 34% increase in the Na background current. It is interesting to note that with voltage clamp measurements, a Na window current (a time independent current resulting from the overlap of the voltage dependence of the steady state activation and inactivation) of about 0.4 $\mu A \bullet cm^{-2}$ at -43 mV has been estimated in isolated neonatal rat myocytes 44 hrs after explantation (42). Such a current density can drive a pacemaker potential of 400 $mV \bullet s^{-1}$ but in a cell with a surface of 1000 μm^2 a current of only 4 pA would flow. This current is too small to be directly measured with modern amplifiers in the whole cell patch configuration.

Table 1: Change in Membrane Conductances During Time in Culture

	P_{Na}/P_K	G_m	G_K	G_{Na}	J_{Na}
			$\mu S \bullet cm^{-2}$		$\mu A \bullet cm^{-2}$
Neonatal	0.03	400.0	388.3	11.7	1.23
Cultured	0.25	76.9	61.5	15.4	1.28
	0.35	76.9	57.0	19.9	1.65

G_m: specific membrane conductance, G_i: specific conductance for ion i, J_{Na}: current density of Na background current, P_{Na}/P_K: relative Na permeability. G_K and G_{Na} were calculated using P_{Na}/P_K and G_m values, J_{Na} was obtained from G_{Na} using +40 mV as the Na reversal potential, -65 mV as resting potential for neonatal, and -43 mV as that for cultured ventricle cells. Sources for data: G_m from (35), P_{Na}/P_K: (41) for neonatal and (39) for cultured cells.

Ionic Currents. In this section we briefly describe the currents found in neonatal rat cells, their modification during time in culture, and their importance for the appearance of spontaneous activity as far as it is known. Figure 7 summarizes changes in action potential configuration under the influence of channel blockers.

Figure 7A shows a control action potential measured in a ventricular strip at 36°C with a resting potential of -65 mV. The action potential has a fast upstroke of about 60 $V \bullet s^{-1}$ that is TTX sensitive. There is a pronounced phase 1 repolarization followed by a well developed plateau and a slower phase 3 repolarization. Such an action potential configuration suggests the presence of I_{Na} (fast rate of rise of action potential), I_{to} (phase 1 repolarization), $I_{Ca,L}$ (pronounced plateau phase), and that of one or more other K currents (phase 3 repolarization).

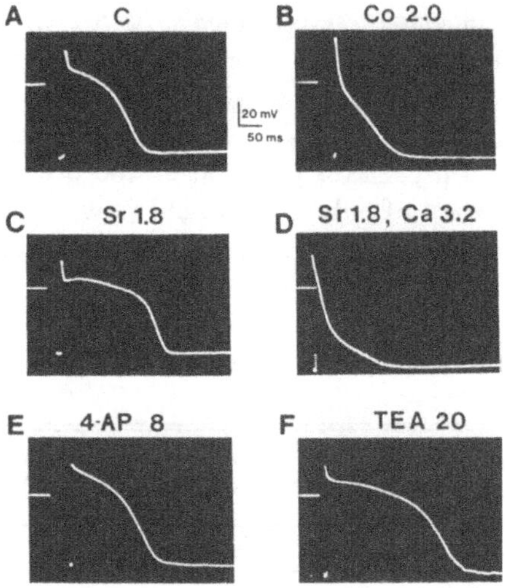

Figure 7: *Action potentials measured with microelectrodes in newborn rat ventricles after different experimental interventions. The numbers above the panels represent concentrations in mM. The horizontal lines at the left side of the panels represent the zero potential level. C: action potential in control medium (Hanks solution); Ca calcium; Co cobalt; Sr strontium; TEA tetraethylammonium. Note that in panels C and D Sr replaced Ca in the control medium and in panel D Ca was added once the Sr action potential with the long plateau had been established. Modified from (48).*

The presence of I_{Na} is suggested by the fast rate of rise of the action potential upstroke and its TTX sensitivity. The TTX sensitivity is one of the lowest known in cardiac muscle: 100 μM or sometimes more are needed to completely block I_{Na} (42) whereas in the heart muscle of 2 to 7 day old chick embryos a $K_{0.5}$ of 0.6 nM was found for TTX (43). I_{Na} has been measured in explanted neonatal rat ventricle cells by several groups (44,45,46). In our laboratory we followed changes of I_{Na} in 1 day old rats for up to 50 hrs in culture when the myocytes show spontaneous activity and slow action potentials. During this time the cell surface doubled and the normalized maximum current as well as the normalized maximum chord conductance increased slightly (42). We assume that the insensitivity of the action potential upstroke observed in spontaneously active cells with slow action potentials is because I_{Na} is largely inactivated by the depolarization at the potential level of these cells but that the phase 4 depolarization remains TTX sensitive because a Na window current persists in the voltage range of the maximum diastolic depolarization.

The transient outward K current I_{to} has been described in adult rat myocytes and the existence of this current in the newborn rat was doubted (47). The disappearance of the phase 1 repolarization under the influence of 4-AP suggests that I_{to} indeed exists in neonatal rat myocytes, Fig. 7E. Since then we have also reported voltage clamp measurements of this current in neonatal rat myocytes (48). This current is an important determinant of the action potential duration but changes during development are unknown and it seems not to be related to spontaneous activity.

$I_{Ca,L}$ exists in the neonatal rat and it is responsible for the pronounced action potential plateau. This notion is reenforced because Co and other $I_{Ca,L}$ blockers eliminate the plateau (Fig. 7B) and Sr enhances it (Fig. 7C) indicating that Sr can replace Ca as charge carrier but slows the $I_{Ca,L}$ kinetics (48,49). Figure 7D shows that when 3.8 mM Ca are added to a preparation when Sr acts as charge carrier there is a drastic reduction of the action potential duration at the plateau level that is more pronounced than that caused by Co. This result suggests that relatively high Ca_o not only competes with Sr as charge carrier but that the channel inactivation does not only depend on voltage but also on Ca (48,50). The hypotheses based on the interpretation of Figs. 7B-7D have been confirmed by voltage clamp measurements from our laboratory (48). These measurements showed that the $I_{Ca,L}$ channel in the neonatal rat has the following particularities compared to the benchmark for the cardiac $I_{Ca,L}$ channel, that of the adult guinea pig: (a) the Ca current is already saturated at 1.8 mM Ca_o; (b) an increase in Ca_o from 1.8 to 5.0 mM decreases the Ca permeability of the channel by 90%, and (c) the channel is more permeable for Sr than for Ba (48).

When cultured neonatal rat cells have developed spontaneous activity associated with slow action potentials, the beating rate and the phase 4 depolarization is sensitive to changes in Ca_o and Ca blockers (14,19,21,39). Voltage clamp measurements and threshold determinations of the action potential in spontaneously active cells indicate that $I_{Ca,L}$ in these cells is activated around -25 mV. This potential is too positive to allow a significant contribution of the L type current in the range of -43 to -25 mV where spontaneous activity occurs (Fig. 3). One must therefore envisage other Ca dependent mechanisms to explain this type of spontaneous activity and a possibility is triggered activity that depends on a periodic increase in cytoplasmic Ca, see discussion in next section.

The phase 3 repolarization is generally determined by K currents, mainly the delayed rectifier I_K and the inward rectifier I_{K1}. In the neonatal rat ventricle no slowly activating time dependent outward currents were found in the time and voltage range where repolarization and diastolic potentials occur (35,48). Nevertheless exposure of the ventricles to the rather unspecific K blocker TEA lengthened the action potential duration considerably indicating the presence of a K current in this potential range. Apart from I_{to} the only other K current available under physiological conditions in this preparation is I_{K1}. The importance of I_{K1} for the development of spontaneous activity was discussed in the previous section in context of the resting potential.

The absence of I_K renders the neonatal rat ventricle electrically a rather simple preparation and it is interesting that this current was also not found in human atrium and in adult rat ventricle (51). However, I_K is present in cultured chick embryo ventricles and it participates in the pacemaker activity that develops during culture in this preparation (16).

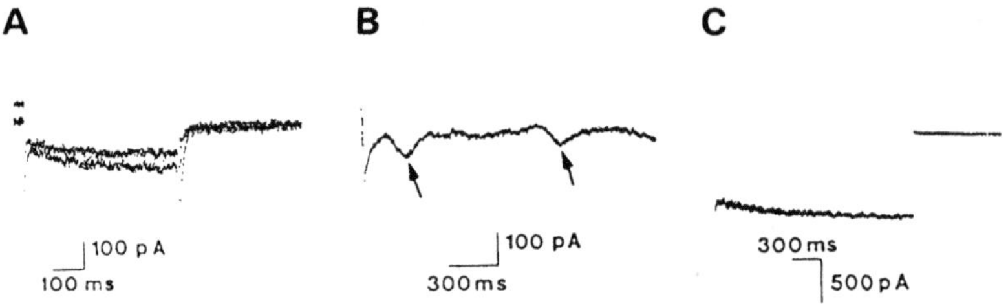

***Figure 8:** A-C voltage clamp records from currents involved in spontaneous activity. **A:** I_f type current elicited from a holding potential of -35 mV (trace with smaller amplitude at end of pulse) and after a 150 ms prepulse to +25 mV. In both cases the testpulse was to -85 mV. The myocyte contracted spontaneously before the voltage clamp was applied. **B:** I_{TI}s (arrows) elicited from a holding potential of -65 mV upon repolarization to the holding potential from a 400 ms testpulse to +10 mV. **C:** I_f type current recorded with a testpulse to -90 mV from a holding potential of -40 mV. The currents in B and C were recorded from the same myocyte that contracted spontaneously before the voltage clamp was applied. The experiments were performed at 36°C in a medium composed of Minimum Essential Medium with Hanks salts supplemented with 20% fetal bovine serum. From (52).*

Apart from the already mentioned modulating influence of the window currents on spontaneous activity, 2 other types of currents that are associated with spontaneous activity were found in spontaneously active neonatal rat ventricular myocytes, an I_f type current and the transient inward current usually associated with triggered activity (42,52). Figures 8A and 8C show recordings of the I_f type current in cells that were beating spontaneously in current clamp mode before the currents were recorded in voltage clamp configuration. Figure 8B shows recordings of classical I_{TI}s in normal medium, i.e. without any measures to raise internal Ca, normally a prerequisite to obtain DADs and their electrical substrate the I_{TI}s. When DADs reach threshold, they give rise to extrabeats or to sustained triggered activity that cannot be distinguished from pacemaker activity unless it stops spontaneously with a DAD (6,7). It follows from the previous discussion of pacemaker mechanisms that the finding of I_f and I_{TI}s in spontaneously active cells is not sufficient to explain the mechanism of spontaneous activity in explanted ventricle cells but it suggests the possibility that there is more than one mechanism that can give rise to spontaneous activity in explanted cells: a classical pacemaker mechanism with a contribution of I_f and triggered activity. However, if I_f participates in the spontaneous activity of explanted rat ventricle cells, an explanation must be found why the beating rate increases with increasing Cs_o instead of decreasing as would be expected from a block of a classical I_f. One explanation would be that the I_f type current found in neonatal rat heart cells is insensitive to Cs or, alternatively that the pacemaker with I_f involvement determines the initial phase of spontaneous activity as soon as it appears after explantation and where the beating rate is low. Later the initial pacemaker activity will be overdriven by sustained triggered activity with its higher beating rate.

Altered Ca Homeostasis. Figure 9 shows action potentials from a cell cluster where electrical activity stopped spontaneously while the microelectrode remained impaled. The action potential train stopped spontaneously with a DAD suggesting that the action potential train represented triggered activity. After a silent period of 15 s another action potential appeared following a slow diastolic depolarization and 10 s later the action potentials resembled those of the control record. Figure 9 shows that in the same cell cluster triggered activity can coexist with a conventional pacemaker mechanism. It further suggests that in slow beating cells a conventional pacemaker mechanism dominates whereas in cells with a faster beating rate triggered activity might be more important. Delayed afterdepolarizations in control medium have also been reported in cultured chick embryo heart cells (53). In the same preparation membrane potential oscillations and delayed afterdepolarizations could be provoked with measures that increase Ca_i (54) and exposure to ouabain elicited bigeminy and sustained triggered activity (55).

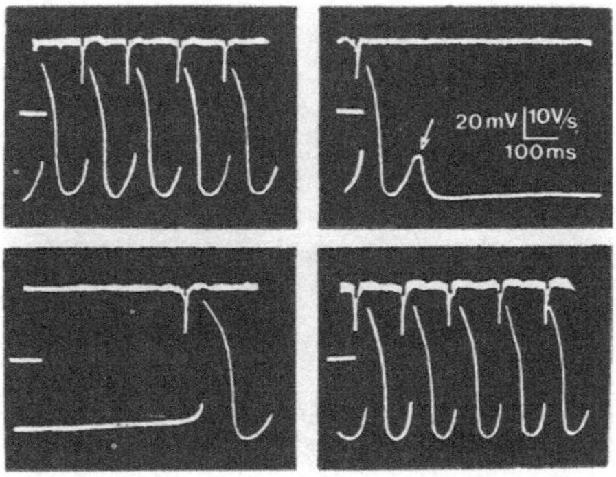

***Figure 9.** Appearance of a DAD when electrical activity ceased spontaneously in a cluster of neonatal rat ventricular cells 4 days after explantation. **(A)** Record in control medium (Minimum Essential Medium supplemented with 20% fetal bovine serum), **(B)** cessation of spontaneous activity followed by a DAD (arrow), **(C)** first action potential after a silent period of 15 s, and **(D)** 10 s after resumption of electrical activity the action potentials are identical to those of panel (A). Microelectrode recordings (lower trace) and inverted time derivative (upper trace).*

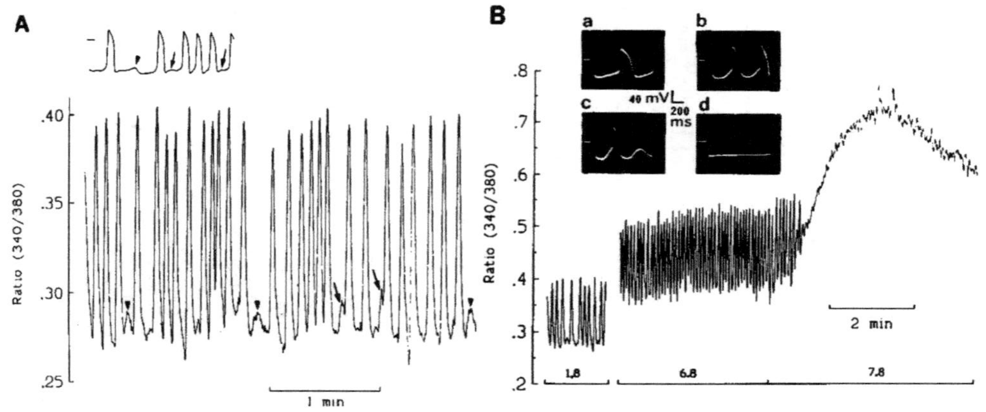

Figure 10. *Spectrofluorimetric measurements of Ca transients observed with Fura2-AM as indicator in a spontaneously active neonatal rat ventricle cell 30 hrs after explantation (cell R9250021).* ***A:*** *Ca transients in 1.8 mM Ca_o (Hanks solution supplemented with 20% fetal bovine serum, 26°C). The frequency of the transients is irregular with a mean rate of 7.4 min^{-1} corresponding to a beating rate of 38.5 bpm at 36°C (Q_{10} 5.2, unpublished results from our laboratory). The large amplitude transients are associated with action potentials and the low amplitude transients correspond to subthreshold DADs (arrowheads). The low amplitude transients have a common threshold at a ratio of 0.28 that corresponds to 0.5 μM Ca_i, the reported threshold for I_{TI} the current determining the DADs (57). Arrows indicate low amplitude DADs that just reach threshold and trigger action potentials. The inset represents nonstimulated action potentials and DADs measured in current clamp in a cluster of 4 cells 24 hrs after explantation. Same extracellular medium as for recording of Ca transients but temperature was 36°C. Arrow heads: low amplitude transients or DADs.* ***B:*** *Influence of increasing Ca_o on the Ca transients of the same cell as in A. Ca_o was gradually raised from 1.8 to 7.8 mM (only first and last 2 concentrations tested are shown, concentrations indicated above bottom line). After Ca_o was raised to 7.8 mM spontaneous activity ceased within 30 s accompanied by a drastic rise in Ca_i. Ten minutes after reaching its peak the ratio had returned to 0.54 but high frequency, high amplitude transients never recovered (not shown). Note the increase in basal Ca_i and frequency of the Ca transients with increasing Ca_o. The inset shows a microelectrode measurement in a cluster of neonatal rat ventricle cells cultured for 4 days and exposed to high Ca_o. (a) control in 1.8 mM Ca_o, (b) 10 min after exposure to 10 mM Ca_o, frames (b) to (c) were taken at 10 s intervals, note abrupt cessation of spontaneous activity in (c) followed by a resting potential of -42 mV in (d). Same experimental conditions for recording of transients and action potentials as in A. From ref. (42).*

As already mentioned the presence of triggered activity presupposes an increase in intracellular Ca that is a common denominator of the experimental manipulations designed to produce it (6,7). This increase in Ca_i has been rather loosely defined as Ca-overload and it has been recently redefined as "the occurrence of spontaneous Ca^{2+} release" (56). In explanted cells triggered activity appears in control medium after explantation and therefore we looked for evidence for spontaneous Ca release in explanted neonatal rat ventricle cells. Figure 10A shows spectrofluorimetrically recorded Ca oscillations in a spontaneously active cell using Fura-2 as indicator. The large amplitude oscillations can be associated with spontaneous action potentials whereas small amplitude transients (arrow heads) represent spontaneous Ca release insufficient to trigger an action potential. The transients marked by arrows were large enough to activate I_{TI}s to reach threshold. Whether or not the transients reached threshold they all started from a Ca level corresponding to a fluorescence ratio of 0.28. I_{TI} is activated by these transients and the Ca threshold for activation of I_{TI} is 0.5 μM Ca_i (57), suggesting that in the cell of Fig 10A the basal Ca concentration was about this value that is well above the 100-200 nM level generally found in resting heart cells (58,59). The inset shows the electrical correlate to these phenomena in an action potential recording from an isolated myocyte.

Figure 10B demonstrates the increase of the frequency of the Ca transients when Ca_o was increased until at a Ca_o of 7.8 mM the high amplitude transients ceased abruptly and basal Ca continued to rise. Moreover, the figure also demonstrates an increase in basal Ca_i with increasing Ca_o. The inset shows the loss of spontaneous activity in a cell cluster after an increase of Ca_o from 1.8 to 10 mM. The increase in beating rate is evident as well as the abrupt cessation of the action potentials followed by a DAD. A comparison of the Ca transients and the inset of Fig. 10B suggests that when spontaneous activity stops in high Ca_o, small amplitude Ca transients continue without causing a drastic depolarization or membrane potential oscillations.

The Nature of Spontaneous Activity in Explanted Cells

The spontaneous activity that develops in explanted neonatal rat ventricle cells is determined by classical pacemaker mechanism(s) and triggered activity (52). That the pacemaker appears in normally electrically silent ventricles makes it by definition an abnormal pacemaker and in addition, there are qualitative and quantitative differences between the abnormal pacemaker mechanisms developing in explanted chick and neonatal rat ventricles (14). The chick cells have a delayed rectifier not found in the rat and an important I_f type current (16) that may have an atypical Cs sensitivity (14). The relative importance of I_f in the chick as compared to the rat could be due to the more negative membrane potential found in the chick. Moreover, the involvement of Ca transients in the spontaneous activity of explanted cardiac cells reenforces the concept of the abnormal nature of this contractile activity.

Although many elements are missing to draw a complete picture of the events leading to spontaneous activity in explanted cells, some types of membrane dysfunction can be identified as being involved in the development of spontaneous activity: Factors involved in the destabilization of the membrane potential and in the occurrence of triggered activity.

Destabilization of the Membrane Potential. The decrease of g_{K1} after explantation indicates a downregulation of the I_{K1} channels in cultured cells, causes depolarization, and destabilizes the membrane potential in that it makes it more sensitive to non channel mediated currents like I_p and $I_{Na,Ca}$ (35,36). The mechanism of the downregulation is unknown but since there is also evidence that g_{K1} increases during maturation of the newborn animal (36), it appears that during development this channel is highly sensitive to regulatory mechanisms.

Development of Triggered Activity. Triggered activity is normally related to Ca loading of the cells (6,7) and for the development of spontaneous activity in explanted cells the conditions for the development of low amplitude Ca transients are of particular interest. In the presence of drastic blockade of $I_{Ca,L}$ by Co and I_{Na} by TTX we have consistently observed slow rhythmic contractions in part of the cells and cell clusters that spontaneously contracted before the blockers had been added (unpublished observations). In these cells slow and fast action potentials were blocked as well as the Ca influx through the $I_{Ca,L}$ channel. It is therefore highly probable that the contractile activity is based on spontaneous Ca release from the SR.

While there is substantial evidence for the involvement of Ca transients and triggered activity in the spontaneous activity observed in cultured ventricle cells, two questions remain unanswered: What is the condition that causes the Ca transients and which mechanism causes the first action potential and contraction after explantation. The occurrence of the spontaneous Ca transients are related to Ca overload and are even used as a definition of this condition (56) but in the case of cultured cells it is also possible that the threshold of the SR for spontaneous Ca release is lowered after explantation. Provided the Ca threshold for the activation of I_{TI} has not changed after explantation, the cell in Fig. 10A should have had a basal Ca concentration of about 0.5 μM that would indicate a considerably increased Ca_i (59) and suggest an increase of Ca_i after explantation by yet unexplained mechanisms. In the absence of measurements of Ca transients at different times after explantation and especially in cells that do not exhibit spontaneous activity it is difficult to decide whether synchronized multifocal Ca transients cause the first action potential after explantation or whether a classical pacemaker mechanism causes an action potential via a phase 4 depolarization. Such an action potential could then cause the synchronization of the Ca transients that lead to sustained triggered activity.

In summary, after explantation subtle changes in the surface membrane and eventually in the membrane of the SR will cause a depolarization, a destabilization of the membrane potential, and a change in the Ca handling of the cell. These changes result in the appearance of classical pacemaker mechanism(s) and low amplitude Ca transients that determine the spontaneous activity in explanted ventricle cells.

Spontaneous Activity in Cultured Cells: Are there Parallels to Clinically Observed Arrhythmias?

A cautious answer to this question is possible in spite of the scarcity of data available about the cellular electrophysiology of the human heart and the difficulties of interpretation because of the medication given to the patients from which the preparations were obtained.

There is a remarkable degree of agreement between results concerning the nature of abnormal automaticity in human atrium and ventricle cells and that found in explanted cardiac cells. This agreement pertains to the multiple mechanisms of abnormal automaticity coexisting in the same preparation and even in the same cell (60,61,62,63); to the decrease in K conductance especially g_{K1} as the cause of the reduced membrane potential (64,65,66,67); and to triggered activity as a component of abnormal spontaneous activity (60). In the light of these findings one can expect that the study of mechanisms of the abnormal spontaneous activity in cultured cells will produce results relevant to further our understanding of abnormal automaticity in the human heart.

Acknowledgements

We thank Mrs. Christiane Ducharme for expert help in processing the manuscript. The skill of Mr. Denis Chartier and Mr. Robert Dumaine in producing the art work is greatfully acknowledged. The results reported here were supported by grants from the Medical Research Council of Canada and the Quebec Heart Foundation to O. F. S. who is also an Edwards Foundation Career Investigator. G. E. H. held a studentship from the Canadian International Development Agency (Francophonie).

REFERENCES:

1. Benditt DG, Milstein S, Goldstein M, Reyes W and Gornick CC (1990) In: Zipes DP, Jalife J (eds) Cardiac Electrophysiology From Cell to Bedside. Saunders, Philadelphia, pp. 708-734
2. Marchlinski FE (1990) In: Zipes DP, Jalife J (eds) Cardiac Electrophysiology From Cell to Bedside. Saunders, Philadelphia, pp. 756-777
3. Noble D (1990) In: Zipes DP, Jalife J (eds) Cardiac Electrophysiology From Cell to Bedside. Saunders, Philadelphia, pp. 163-171

4. Baumgarten CM and Fozzard HA (1991) In: Fozzard HA, Haber E, Jennings RB, Katz AM, Morgan HE (eds). The Heart and Cardiovascular System. 2nd ed, vol 2, Raven Press, New York, pp 963-1002
5. Schanne OF, Dumaine R and Ruiz Petrich E (1993) In: Huizinga JD (ed) Pacemaker Activity and Intracellular Communication. CRC Press, Boca Raton, in press
6. Cranefield PF and Aronson RS (1988) Cardiac Arrhythmias: The Role of Triggered Activity and Other Mechanisms. Futura Publishing Company, Mount Kisco
7. Wit AL and Rosen MR (1991) In: Fozzard HA, Haber E, Jennings RB, Katz AM, Morgan HE (eds) The Heart and Cardiovascular System. 2nd ed, vol 2, Raven Press, New York, pp 2113-2164
8. Irisawa I and Giles WP (1990) In: Zipes DP, Jalife J (eds) Cardiac Electrophysiology From Cell to Bedside. Saunders, Philadelphia, pp 95-102
9. Campbell DL, Rasmussen RL and Strauss HC (1992) Annu Rev Physiol 54:279-302
10. Noble D, DiFrancesco D and Denyer JC (1989) In: Jacklet JW (ed) Cellular and Neuronal Oscillators. Marcel Dekker, New York, pp. 59-85
11. Irisawa H (1978) Physiol Rev 58:461-498
12. Noble D, Denyer JC, Brown HF and DiFrancesco D (1992) Proc Roy Soc Lond B 250:199-207
13. DiFrancesco D and Noble D (1985) Phil Trans Roy Soc B 307:353-398
14. Schanne OF, Boutin L and Derosiers J (1989) Can J Physiol Pharmacol 67:795-800
15. Clay JR and Shrier A (1981) J Physiol 312:471-490
16. Brochu RM, Clay JR and Shrier A (1992) J Physiol 454:503-515
17. Schanne OF (1972) Can J Physiol Pharmacol 50:523-532
18. Schanne OF, Ruiz-Ceretti E, Rivard C and Chartier D (1977) J Mol Cell Cardiol 9:269-283
19. Schanne OF, Ruiz-Ceretti E, Payet MD and Deslauriers Y (1979) J Mol Cell Cardiol 11:477-484
20. Payet MD, Bkaily G, Schanne OF and Ruiz-Ceretti E (1980) Can J Physiol Pharmacol 58:433-435
21. Schanne OF, St-Vincent M and Bkaily G (1983) Rev Can Biol Exptl 42:199-202
22. Yamamoto T, Ruiz-Ceretti E and Schanne OF (1987) Can J Physiol Pharmacol 65:252-256
23. Higgins TJC, Allsopp D, Bailey PJ and D'Souza EDA (1981) J Mol Cell Cardiol 13:599-615
24. Girardier L (1971) Cardiology 56:88-92
25. Doorey AJ and Barry WH (1983) Circ Res 53:192-201
26. Pinson A (1987) In: Pinson A (ed) The Heart Cell in Culture. Vol. 3, CRC Press, Boca Raton, pp 35-48

27. Hyde A, Blondel B, Matter A, Cheneval JP, Filloux B and Girardier L (1969) In: Akert K and Waser PG (eds) Progress in Brain Research. Vol 13, Elsevier, Amsterdam, pp 283-311
28. Athias P, Frelin C, Groz B, Dumas JP, Klepping J and Padieu P (1979) Pathol Biol 27:13-19
29. Lompre AM, Poggioli J and Vassort G (1979) J Mol Cell Cardiol 11:813-825
30. Robinson RG and Legato MJ (1980) J Mol Cell Cardiol 12:493-498
31. Robinson RB (1982) J Mol Cell Cardiol 14:367-370
32. Matsuki N and Hermsmeyer K (1983) Am J Physiol 245:C381-C387
33. Jack JJB, Noble D and Tsien RW (1975) Electric Current Flow in Excitable Cells, Clarendon Press, Oxford
34. Schanne OF and Ruiz-Ceretti E (1978) Impedance Measurements in Biological Cells. Wiley, New York
35. Schanne OF, Lefloch M, Fermini B and Ruiz-Petrich E (1990) Am J Physiol 258:H145-H152
36. Wahler GM (1992) Am J Physiol 262:C1266-C1272
37. Haddad G (1992) Metabolic Inhibition and the Background K^+ Current in 1-Day and 7-Day Rat Ventricular Myocytes Ph.D. Thesis, University of Sherbrooke
38. Pappano AJ and Sperelakis N (1969) Am J Physiol 217:1076-1082
39. LeFloch M, Schanne OF and Ruiz-Ceretti E (1989) Can J Physiol Pharmacol 65:2110-2116
40. Glitsch HG and Tappe A (1993) Pflügers Arch 422:380-385
41. Yamamoto T (1986) J de Physiol 81:191-194
42. Dumaine R (1993) Activité spontanée des cellules ventriculaires du rat nouveau-né en culture: rôle du sodium et du calcium. Ph.D. Thesis, University of Sherbrooke
43. Fujii S, Ayer PK Jr and DeHaan RL (1988) J Membrane Biol 101:209-223
44. Brown AM, Lee KS and Powell T (1981) J Physiol 318:479-500
45. Bodewei R, Hering S, Lenke B, Rosenstraukh LV, Undrovinas AI and Wollenberger A (1982) J Physiol 325:301-315
46. Cachelin AB, de Peyer JE, Kokubun S and Reuter H (1983) J Physiol 340:389-401
47. Josephson IR and Brown AM (1986) J Membrane Biol 94:19-35
48. Fermini B and Schanne OF (1991) Cardiovasc Res 25:235-243
49. Lux HD (1991) In: Hurwitz L, Partridge LD and Leach JK (eds) Calcium Channels: Their Properties, Functions, Regulation, and Clinical Relevance. CRC Press, Boca Raton, pp 9-19
50. Carbone E and Swandulla D (1991) In: Hurwitz L, Partridge LD and Leach JK (eds) Calcium Channels: Their Properties, Functions, Regulation and Clinical Relevance. CRC Press, Boca Raton, pp. 35-60.
51. Coraboeuf E and Escande D (1990) NIPS 5:28-31
52. Schanne O and Dumaine R (1992) J Mol Cell Cardiol 24: Suppl 1,S105

53. Sperelakis N (1972) In: De Mello WC (ed) Electrical Phenomena in the Heart. Academic Press, New York, pp 1-61
54. Nathan AD and Bhattacharyya ML (1984) Am J Physiol 247:H273-H282
55. Koidl B and Tritthart HA (1980) J Mol Cell Cardiol 12:663-673
56. Stern MD, Capogrossi MC and Lakatta EG (1988) Cell Calcium 9:247-256
57. Colquhoun D, Neher E, Reuter H and Stevens CF (1981) Nature 294:752-754
58. Gibbons WR and Zygmunt AC (1991) Fozzard HA, Haber E, Jennings RB Katz AM and Morgan HE (eds) The Heart and Cardiovascular System. 2nd ed, vol 2, Raven Press, New York, pp 1249-1279
59. Ahlemeyer B, Weintraut H, Antolovič R and Schoner W (1992) Eur J Biochem 205:269-275
60. Singer DH (1990) In: Rosen MR, Janse MJ and Wit AL (eds) Cardiac Electrophysiology: A Textbook. Futura, Mount Kisco, pp 247-263
61. Singer DH, Ten Eick RE and DeBoer AA (1973) In: Dreifus L and Likoff W (eds) Electrophysiologic Correlates of Human Tachyarrhythmias. Grune and Stratton, New York, pp 97-111
62. Cranefield PF (1975) The Conduction of the Cardiac Impulse: Slow Response and Cardiac Arrhythmias. Futura, Mount Kisco
63. Dangman KH, Danilo PJr, Hordoff AJ, Mary-Rabine L, Reder RF and Rosen MR (1982) Circulation 65:362-368
64. Ten Eick RE and Singer DH (1979) Circ Res 44:545-557
65. McCulloch JR, Baumgarten CM and Singer DH (1987) J Mol Cell Cardiol 54:65-73
66. Imanishi S and Arita M (1987) Jpn J Physiol 37:393-410
67. McCulloch JR, Chua WT, Rasmussen HH, Ten Eick RE and Singer DH (1990) Circ Res 66:191-201

CHAPTER 5

IONIC CHANNELS INVOLVED IN THE MYOCARDIAL RESPONSE TO METABOLIC STRESS

Elena Ruiz Petrich[1], Fabien deLorenzi, Shufen Cai, Otto F. Schanne

Département de Physiologie et Biophysique, Université de Sherbrooke, Canada

INTRODUCTION

In 1944, Erk and Schaefer (1) reported that anoxemia shortened and decreased the amplitude of monophasic action potentials recorded with suction electrodes in *in situ* hearts of guinea pigs and cats. These effects were accompanied by flattening of the plateau and were correlated with the shortening of the QT interval and the T wave alterations previously observed in human electrocardiograms. In 1954, Trautwein et al. (2) showed that intracellularly recorded action potentials underwent similar changes on transition from 95% to 0% oxygen in the fluid superfusing a ventricular muscle preparation. Within 15 min, the total action potential duration decreased by 20% whereas the plateau duration fell to 50% of the control. These early changes occurred without modification of the resting potential. In addition, it was later

[1]The author wishes to express her appreciation to Dr Jorge Suarez Estrade (University of Cuyo, Argentina) who introduced her to the exploration of the fascinating world of research in the field of cardiac hypoxia.

shown that action potential shortening also occurs within seconds of coronary ligation (3). This provides a suitable explanation for the deviation of the S-T segment in ischemia which is later followed by a T-Q segment depression arising from depolarization of the resting membrane (4,5). We have made a long way since these pioneer observations and our knowledge and understanding of the ionic basis of the heart electrical activity have considerably deepened in the last years through the development and application of the patch clamp technique (6) to isolated cardiac cells. Although this technique allows a precise determination of the ionic currents underlying the myocardial electrical activity under physiological or simulated pathological conditions, the state of the art is such that the responses observed in single cells cannot yet fully account for the effects observed in multicellular preparations or in the whole heart. Necessarily, a complete understanding of the mechanisms involved cannot be achieved unless the results of studies performed at different levels of organization are integrated.

In addition to the action potential shortening (classically recognized as the earliest and most profound effect of O_2 depletion, (7)) another equally important consequence of myocardial metabolic stress is a dramatic increase in the rate of K^+ efflux (8) leading to cytosolic K^+ depletion in well perfused tissues (high flow metabolic inhibition) and to a concomitant extracellular K^+ accumulation in ischemic preparations (9). Whether these two main events are casually related or not, it is undeniable that the enhanced K^+ movement will condition or modulate the electrical changes induced by inhibition.

Several methodological approaches have concurred to study the multifaceted effects of myocardial hypoxia and data on the metabolic changes, transmembrane ionic shifts, and bioelectrical phenomena complement each other in supplying an insight into the participating mechanisms although we are far from having achieved a complete understanding of all the factors involved and their regulation. A short review of the main metabolic changes occurring upon a sudden decrease in the energy supply to the myocardial cell is presented here because the electrical responses to metabolic inhibition are closely related to subsequent changes in cytosol composition. A turning point in our knowledge of these phenomena was the discovery by Noma in 1983 of K^+ channels that are electrically silent in the presence of normal ATP levels (10). Activation of these channels was soon afterwards related to the electrical effects of inhibition and this strongly stimulated further research aimed at the determination of their role in action potential changes and transmembrane K^+ fluxes during hypoxia and ischemia. The literature review in this chapter is by no means exhaustive and many valuable contributions had to be omitted because of space restrictions.

CARDIAC METABOLISM IN HYPOXIA AND ISCHEMIA

Most of the studies concerned with the cardiac effects of O_2 depletion were inspired by the fact that anoxia is a major component of ischemia. Consequently, more often than not, they combined hypoxia or anoxia with substrate depletion or inhibition of glycolysis (complete metabolic blockade). Although this situation mimics better the conditions found in ischemia, their maintenance leads to irreversible cell injury. On the other hand, a moderate degree of hypoxia with preservation of glycolytic processes mimics situations that are often encountered in vivo (cardiac or respiratory insufficiency and anemia) that are compatible with cell viability and can be chronically tolerated. It is therefore important to establish the difference between these two conditions and their metabolic consequences.

Cardiac ischemia is defined as an imbalance between oxygen and substrate supply to the heart and its energetic demands. Hypoxia, in its pure form (without reduction of coronary flow), implies an insufficient oxygen supply to meet the energy requirements of the tissue. The most important difference between ischemia and hypoxia is that in the latter substrate supply is maintained and removal of waste is adequate. The metabolic and physiological consequences of the persistence of coronary flow are striking. The myocardium depends on a constant supply of oxygen and substrate to support the production of high energy phosphate compounds (HEP) by oxidative phosphorylation in mitochondria. When the amount of oxygen reaching the myocardium decreases, anaerobic glycolysis is stimulated resulting in the synthesis of some, albeit markedly reduced, amounts of HEP and the production of lactate as the end product of glycolysis. Although insufficient to maintain an adequate level of contractile performance, the glycolytically derived HEP will be enough to sustain sarcolemmal integrity and vital transport functions thus assuring cell viability. In contrast, the potential for energy production via anaerobic glycolysis decreases rapidly with time in ischemia where the lack of waste removal results in accumulation of lactate and other metabolites leading to the development of acidosis (9,11). This in turn slows down the rate of anaerobic glycolysis (12) even when sizable amounts of glycogen and glycolytic intermediates are present in the tissue because of inhibition of glyceraldehide-3-phosphate dehydrogenase and probably other glycolytic enzymes by lactate, NADH, and H^+ (13). In contrast, in high flow hypoxia, lactate accumulation and acidosis do not occur and therefore anaerobic glycolysis continues at an accelerated rate as long as glycolytic substrate (exogenous glucose or endogenous glycogen) remains available. Consequently, depletion of HEP occurs much more slowly or not at all (depending on contractile activity and energy demands). Moreover, contractile activity is rapidly depressed in hypoxia which reduces the energy requirements of the tissue and, hence, the HEP depletion. This initial contractile failure is not due to a decrease in HEP or in the free energy change of ATP hydrolysis but rather to rising levels of P_i and $H^+{}_i$ and to the drastic shortening of action potential duration (14,15). Another dramatic difference between the effects of hypoxia and ischemia is the extracellular K^+ accumulation in the latter that leads to membrane depolarization and transformation of fast rising action potentials to depressed fast responses (16,17). These effects will entail electrical heterogenities during the diastolic interval with the subsequent flow of injury currents which, added to slow conducting action potentials, will contribute to the generation of the functional substrate for re-entry pathways and arrhythmias.

In contrast, K^+ leaking out of hypoxic cells is removed from the interstitial fluid and depolarization does not occur. Consequently, coordinated electrical activity is preserved and arrhythmias are rarely observed as a consequence of high flow hypoxia. The mechanism by which the resting potential is maintained despite the reduction of the transmembrane K^+ gradient under these conditions will be discussed later in this chapter.

Degradation of HEP Compounds

Much of our recent knowledge in this field results from the application of nuclear magnetic resonance spectroscopy to isolated hearts where the changes in nucleotides, P_i, and H^+_i can be studied in relation to the evolution of other parameters such as contractile function or electrical activity. Within a few seconds of inhibition of oxidative mitochondrial metabolism in the setting of anoxia or ischemia, mitochondrial electron transport ceases and anaerobic glycolysis is accelerated (18). This initial phase is common to hypoxia and ischemia and is facilitated by β adrenergic stimulation that contributes, together with the initial fall of ATP, to activate the conversion of phosphorylase b to phosphorylase a (11). As mentioned before, the glycolytic flux is soon slowed down in ischemia because of accumulation of metabolic byproducts. In hypoxia, a linear relationship between global ATP levels and lactate production exists (19). Although anaerobic glycolysis can produce no more than 7% of the HEP needed for the function of a working heart (20), the ATP thus synthesized can subserve important sarcolemmal functions such as phosphorylation of ionic channels (21) and energy dependent ionic transport systems (22).

The time course of ATP depletion, lactate accumulation, and degradation of adenine nucleotides to its subproducts is not the same in regional (following coronary occlusion) or global ischemia. It is slower and less drastic in the latter because the heart stops beating soon after cessation of flow and the energy demands markedly decrease (23). The ATP levels fall by 50% and a slight transient increase in ADP is observed whereas AMP progressively accumulates (23,24). Adenine metabolites also accumulate and are released to the circulation on reflow. In regional ischemia, the metabolic changes vary in the transmural direction: the ATP content falls by 65% and 30% after 30 min of ischemia in the subendocardial and subepicardial layers respectively. The opposite occurs with lactate and nucleotide byproducts accumulation (25). Creatine phosphate is almost completely depleted within the first 3 min and a 50% early increase in ADP occurs followed by a decrease that parallels the fall in ATP levels after 10 min (11). The accumulation of metabolic end products leads to a rapid development of acidosis (26) and the creation of a substantial osmotic load (27). The latter is considered to be of importance in the development of ischemic cell injury (28,29) whereas the degree of acidosis is a conditioning factor in the pathogenesis of myocardial reperfusion injury and arrhythmias (30).

In hypoxic hearts with glycolysis present phosphocreatine falls to 14% of the control within 4 min and P_i increases 5 fold in the same time (31). Developed pressure decreases to 35% of control by the sixth minute and resting tension increases. The intracellular pH transiently rises from 6.98 to 7.07 (2 min), falls to 6.85 by the 8 th min and then it remains stable at this level. Concomitant inhibition of glycolysis during anoxia or cyanide poisoning reduces contractile activity to 5% of control by the fifth minute. Intracellular pH increases initially from 7.03 to 7.1 and then slowly decreases to 7.0. Phosphocreatine decreases rapidly under these conditions and ATP falls by a factor of 2. These observations illustrate the importance of anaerobic glycolysis in the maintenance of ATP levels. Nevertheless, early contractile failure is not prevented and is in part attributed to the concomitant acidosis (31) because the Ca^{2+} transients are not altered at this early stage (32). Anaerobic glycolysis is not the only mechanism for ATP production in anoxia with persistence of glycolysis. Pyruvate can be metabolized in the mitochondria together with glutamate to alanine plus 2-oxoglutarate which can enter the Krebs cycle and form ATP by reduction of fumarate to succinate. Two moles of HEP can thus be formed from 1 mol pyruvate (33). The importance of anaerobic glycolysis is further illustrated by the established relationship between the onset of ischemic contracture and the level of glycolytic ATP synthesis that suggests subcellular compartmentation of ATP or metabolite channelling (34). Moreover, complete metabolic blockade leads to rapid intercellular uncoupling (35,36) most probably due to failure of Ca^{2+} homeostasis and a concomitant rise in cytosolic free Ca^{2+} concentration (37). This explains that complete metabolic blockade (hypoxia in the presence of 5 mM 2-deoxy-glucose) in perfused hearts leads to abolition of propagated action potentials without cell depolarization within 20 min (Lafond and Ruiz Petrich, unpublished).

Unlike the whole heart that responds in a gradual fashion to lowering of the oxygen tension, isolated cardiomyocytes respond with an almost all or none pattern: mitochondrial respiration does not stop and lactate production does not rise unless PO_2 has dropped below 1 torr (38). In addition, the time course of HEP depletion is much slower most probably because of the low energy requirements of quiescent myocytes (38,39). In anoxia, the ATP levels do not change before 30 min whereas at this time, PCr is reduced to 30%. After 60 min, ATP falls to 50% of control and moderate leak of cytosolic enzymes occurs. After 2 hrs of severe inhibition, contracture appears but it does not necessarily reflect irreversible cell damage because the concomitant rise in intracellular Ca^{2+} can be fully reversed at this stage (39). With exogenous substrate present, lactate production is maintained at a high level for 120 min, enzyme leakage is postponed and ATP levels are conserved (40). In agreement with these mild metabolic effects, the electrical and contractile activity of spontaneously beating neonatal cells is not largely modified up to 90 min of exposure to anoxia in the presence of glucose (41). Elimination of substrate leads to early deterioration (45 min) and finally abolition (60 min) of activity. Another effect of oxygen and substrate deprivation is abnormal lipid metabolism resulting in accumulation of compounds that are deleterious to sarcolemmal structure and functional integrity (42). ATP depletion is accompanied by decreases in phospholipids and triacylglicerol content whereas intracellular levels of palmitic and arachidonic acid are elevated. These changes are attenuated by the presence of glucose while ATP levels are conserved (42). Oxygen deprivation and depression of glycolysis are also associated with an enhanced intracellular formation of adenosine whose threshold is as low as 3 mm Hg but probably much higher in active multicellular preparations where diffusion of adenosine to the extracellular fluid may mediate important homeostatic responses (43).

Table 1. Cellular contents and concentrations of K and Na in ischemia and hypoxia

Condition	K_c	K_i	Na_c	Na_i	Remarks
ISCHEMIA					
Control	58.2 (1.0)	141.6	15.0 (1.2)	36.5	total issue
60 min	55.7 (1.1)	120.6	26.9 (1.3)	57.8	rabbit (46)
Control	ND	101.0 (5.6)	ND	13.0 (1.7)	ISE
HAN	ND	79.5 (5.8)	ND	14.5 (2.2)	dog (47)
HYPOXIA					
Control	59.3 (1.3)	127.0	18.3 (2.2)	38.0	total issue
N_2, 60 min	43.1 (1.8)	92.0	24.6 (1.6)	53.0	(48)
N_2, TTX	51.9 (0.9)	120.7	11.0 (1.0)	25.6	id
N_2, TEA	50.5 (2.1)	117.4	26.2 (1.3)	60.9	id
N_2, Ba^{2+}	49.7 (2.0)	115.6	24.9 (2.9)	57.9	id (49)
TOLB	31.7 (1.9)	73.7	24.9 (1.1)	57.9	id
GLIB	43.1 (1.9)	100.2	38.9 (5.8)	90.5	id
N_2	ND	90.1	ND	ND	ISE (47)

Cellular contents (Na_c, K_c) in mmol/kg wet weight; concentrations (Na_i, K_i) in mmol/L cell water. Means (SEM). Except for measurements with ion sensitive electrodes (ISE) all data for hypoxia were obtained from determinations of total water, volume of inulin space and total tissue content measured by spectrophotometry in samples of left ventricle free wall. For ischemia, the extracellular space was washed out with isotonic sucrose prior to removal of samples. Ischemia corresponds to cessation of coronary flow in perfused rabbit hearts; HAN indicates superfusion of dog muscle strips with hypoxic, acidotic, glucose free medium. Blockers added to perfusate during hypoxia; 9 μM TTX; 10 mM TEA; 40 μM $BaCl_2$; 1 mM tolbutamide; 30 μM glibenclamide. ND: no data.

CHANGES IN ELECTROLYTE DISTRIBUTION

Several methodological approaches have been used to study the effects of hypoxia and ischemia on the transmembrane distribution of ions with a double objective: the study of the ionic transport systems per se and/or the determination of the mechanisms underlying the electrical disturbances observed during metabolic stress. Before the direct measurement of the transmembrane fluxes became possible by the application of radioactive isotopes, changes in the myocardial ionic metabolism were inferred from the difference between the electrolyte concentrations in blood or saline buffers entering and leaving the heart (44) or measurements of myocardium electrolyte content under different experimental conditions (45). Although more sophisticated techniques were later developed that allowed the direct determination of ionic fluxes (labeled isotopes) and the cytosolic ionic activities (ion sensitive electrodes, fluorescent dyes), some key conclusions reached with that early work are still valid today, namely the early K^+ loss induced by hypoxia or ischemia that is unrelated to a depression of the Na:K pump (46-51).

Changes in K^+ Distribution. In perfused rabbit myocardium with glucose present K^+ loss starts within seconds of the onset of anoxia (51). This loss progresses with time in anoxia and is rapidly reversible by reoxygenation for anoxic periods up to 60 min (52). Moreover, it is totally attributable to increased K^+ efflux (45,51) and it is largely excitation dependent and non related to the energetic state of the cell evaluated by total tissue levels of ATP, ADP, PCr and creatine (52). Similar results have been obtained from measurements of cellular K contents and cytosolic K^+ activity (Table 1, refs 47,48). This table lists almost exclusively data from our laboratory because they will be later used as a frame work for the interpretation of some electrical parameters.

After 60 min of hypoxia, the decrease in K_c (16 mmol/kg ww) is not compensated by an equivalent increase in Na_c, supporting other observations on the dissociation between these effects that speaks against an involvement of the Na pump in their generation. The net K^+ loss estimated from these bulk K measurements (100 mmol/kg dry weight) compares well with a loss of 90 mmol/kg dry weight reported by Rau and Langer (52). As for the data for cytosolic K concentration, the control values correspond to those determined by ISE when corrected according to the activity coefficients reported by Lee and Fozzard for rabbit myocardium (53). Moreover, there is a very good agreement between the decrease in intracellular K concentration with ischemia or hypoxia whether estimated from tissue content or ISE. In embryonic cells, the K^+ loss is also related to increased efflux and is greater in 19 day old embryos than in 7 day old ones. In both, it is enhanced in the absence of glucose and counteracted by excess glucose (54). The magnitude of the K^+ loss in ischemia is comparable to that produced by hypoxia (Table 1). Moreover, the other two main components of ischemia, substrate depletion and acidosis do not induce per se sizable decreases in cytosolic K^+ activity (47) which indicates that lack of oxygen is the major stimulus for increased K efflux. Because of the restricted extracellular compartment of ischemic tissue, the enhanced K^+ efflux leads to K^+ accumulation that characteristically occurs in 3 phases: an initial rapid increase from basic levels of 4 mM K^+ to 8-12 mM within 10 min, a plateau phase where those levels are maintained for 10 min more followed by a second rise whose onset is associated to irreversible sarcolemmal damage (9,55-58). This extracellular K^+ accumulation is modulated by the frequency of stimulation and although attenuated it is still observed in quiescent preparations (59). The same is observed in hypoxia where depression of electrical activity by TTX reduces the effect of hypoxia on K_i (Table 1, ref. 48). The extracellular K^+ accumulation induces membrane depolarization in ischemic tissue (46,47,55-57) whereas no depolarization is observed in well perfused hypoxic preparations despite the decrease in transmembrane K^+ gradient (17,48).

Changes in Na^+ Distribution. Early determinations of total tissue Na content indicated that the cytoplasmic Na concentration increased with hypoxia and ischemia (46,48,54). This has not been confirmed by data obtained with ion sensitive electrodes (47 in Table 1, 57,58) which strongly supports the notion that the function of the Na:K pump is well preserved in ischemia and anoxia at times when there is a marked cellular K loss. This discrepancy can be explained by subcellular compartmentation of Na reflected by the low activity coefficient (γ^i_{Na} = 0.216) reported by Lee and Fozzard (53). It remains to be established whether this activity coefficient remains constant or not under conditions of metabolic inhibition. It has been shown that γ^i_{Na} decreases from 0.356 to 0.286 during hemorrhagic shock in rabbit skeletal muscle in vivo (60). We have used these values in an attempt to estimate whether the increases in Na_i listed in the table are an artefact resulting from a change in γ^i_{Na}. It does not look like because a^i_{Na} thus calculated is still 20 to 25% higher in hypoxia or in ischemia than in the controls. Moreover, the control levels correspond to those of Nakaya et al. (47) and the value found for the TTX experiments corresponds to what is expected for a strong reduction in Na^+ influx through the Na^+ channels (61). Despite the preservation of the Na:K pump function intracellular Na^+ may increase as a result of stimulation of the Na:H exchanger by acidosis resulting from the enhanced lactate production. Since the activity of the exchanger is strongly regulated by both, extracellular and intracellular pH, its contribution to Na^+ influx will vary with the duration of ischemia and the subsequent degree of acidosis. It is expected that the exchanger will be stimulated during high flow hypoxia because of the development of intracellular acidosis with normal extracellular pH. Since fast rising potentials are well preserved under these conditions (17) it is not surprising that the level of internal Na^+ will reflect the extra load due to the activity of the exchanger. In ischemia, the extracellular pH decreases progressively with time and the action potential is severely depressed because of resting membrane depolarization (56). Consequently, Na^+ influx is strongly diminished so that even if the Na:K pump is slightly depressed, the internal Na^+ level is not expected to rise. The slight increase we found may have been conditioned by a decrease in temperature that is almost unavoidable during 60 min of global ischemia by arrest of perfusion in an isolated heart (63).

Changes in Intracellular Ca^{2+}. Although a sharp increase in free cytosolic Ca^{2+} is associated with irreversible myocardial injury (64,65), little calcium loading occurs during hypoxia or ischemia that do not cause irreversible cell damage (66,67,68). However, it is currently accepted that a transient increase in cytosolic Ca^{2+} underlies reperfusion related arrhythmias as well as contractile dysfunction (69). It is proposed that sodium loading by reactivation of the Na:H exchanger leads to the Ca^{2+} increase (through the Na:Ca exchanger) which in turn induces transient inward currents and delayed after depolarizations that account for the production of extra beats and tachycardia. This hypothesis is validated by the fact that realkalization after acidotic perfusion reproduces the arrhythmias usually observed on restoration of flow after an ischemic episode and this effect is antagonized by inhibition of the Na:H exchanger with amiloride or its analogs (70).

Although it is generally accepted that internal Ca^{2+} levels are maintained in the absence of sarcolemmal damage, studies have been performed to investigate whether changes in the Ca^{2+} transient are responsible for the acute contractile failure occurring very early during metabolic inhibition. No correlation has been found in papillary muscles between the rapid fall in developed tension and the Ca^{2+} transient in severe hypoxia or CN^- poisoning when glycolysis can proceed (18). Contractile failure has then been attributed to the concomitant acidosis. On the other hand, neither the Ca^{2+} transient or the Ca^{2+} sensitivity of contractile activity is modified by anoxia in isolated rat myocytes where depression of contraction is attributed to the marked reduction in action potential duration (15). Similar observations were made in papillary muscles during simulated ischemia of short duration that depressed developed tension without altering the Ca^{2+} transients (71). On longer exposures, the Ca^{2+} transients undergo biphasic changes and most of the effects are mimicked by superfusion with lactate containing solutions. In summary, it appears that Ca^{2+} homeostasis is not affected during early hypoxia or ischemia but that anomalous Ca^{2+} movements in prolonged exposure or reperfusion can result from mechanisms related to the control of internal Na^+ or H^+. On the other hand, cell injury may occur by breakdown of Ca^{2+} homeostasis without relation to acidosis in severely energy depleted cells (72).

The Mechanism of Hypoxic K^+ Loss

The experimental evidence reviewed in the preceding section indicated that the cellular K^+ loss induced by hypoxia and ischemia was not due to Na pump depression but to an enhanced K^+ efflux. Two hypotheses were proposed to explain these findings: an increase in background K^+ conductance (17,48,52,54) and anion coupled K^+ efflux (58,73,74). The first one was consistent with the action potential shortening and the little effect of metabolic inhibition on the resting potential specially in the presence of glucose. The second one was supported by the observation that the time course of K^+ loss during ischemia parallels the loss of lactate and P_i in the coronary sinus blood (74).

We tested the first hypothesis by examining the effects of several blockers of K^+ channels as well as the influence of the external K concentration on the cellular K^+ and Na^+ contents in hypoxic hearts (Tables 1 and 2). The cellular K^+ loss is considerably attenuated by the unspecific K^+ channel blocker TEA, by abolition of electrical activity with 9 μM TTX, and by 40 μM $BaCl_2$ that selectively blocks the inward rectifier in rabbit myocytes (75). Conversely, the blockers of the K-sensitive

ATP channels do not alter (glibenclamide) or even increase (tolbutamide) the K^+ depletion, in agreement with data on K^+ accumulation in ischemic rabbit hearts (76,77). The cell Na content is insensitive to these treatments, except for the reduction found under TTX. In agreement with the role of a K conductance as a predominant pathway for enhanced K^+ efflux Table 2 shows that a high extracellular K concentration is as effective as K^+ channel blockers in counteracting K^+ depletion which is abolished by the combination of high K_o and cessation of electrical activity (K_o, 25 mM plus TTX). Moreover, lowering K_o increases the cellular K^+ loss. This effect is not due to Na pump inhibition because the addition of Rb counteracts the increase in Na content at low K but does not modify the K content (48). These findings do not support the hypothesis that the driving force for the increased K^+ efflux is the outward movement of anions, in agreement with data showing no relation between extracellular K^+ accumulation and formation of lactate and P_i (78).

Table 2. Influence of K^+_o and Ba^{2+} on cellular K and Na content and concentrations in rabbit myocardium.

Condition		K_c	K_i	Na_c	Na_i
	O_2	59.4 (2.1)	132 (4.6)	25.7 (2.3)	57 (5.7)
K_o 1.5	N_2	36.1 (2.3)	80 (3.9)	38.1 (2.3)	85 (6.8)
	N_2, Ba	27.2 (3.2)	43 (1.7)	38.0 (3.7)	70 (4.8)
	O_2	61.7 (2.3)	137 (1.4)	16.2 (1.4)	36 (5.0)
K_o 5.0	N_2	43.1 (1.8)	96 (3.0)	24.6 (1.6)	55 (4.5)
	N_2, Ba	49.7 (2.0)	106 (3.3)	24.9 (2.9)	58 (4.1)
	O_2	57.6 (2.0)	128 (3.8)	14.9 (1.3)	33 (1.0)
K_o 10	N_2	48.7 (2.2)	108 (2.4)	26.4 (2.9)	59 (3.8)
	N_2, Ba	48.8 (1.7)	92 (3.4)	39.2 (3.8)	73 (5.9)

Values are means (SEM) from 5 to 10 hearts per condition; O_2 and N_2 denote normoxia and hypoxia respectively. Subscript c indicates cell content in mmol/kg ww and subscript i corresponds to concentration in mmol/L cell water.

Moreover, we have also shown that cellular K^+ loss becomes insensitive to extracellular K^+ concentration in the presence of Ba^{2+} in hypoxia (49) where a dissociation between external K^+ and the diffusion-dependent component of the resting potential is observed (see below). These data also provide an alternative explanation for the plateau phase of extracellular K^+ accumulation during ischemia, attributed to a stimulation of the Na pump by high K^+ so that K^+ efflux is balanced by an equivalent K^+_o influx (55). At these K_o levels, the resting potential is equal or even slightly more negative than the K equilibrium potential (46) and the action potential is considerably depressed (17). Consequently, the driving force for K^+ is inward during most of the cycle and the time the myocardium spends in the depolarized state and hence the driving force for the outward movement of K^+ are strongly reduced. These factors added to the inwardly rectifying properties of g_{K1} most certainly oppose the outward movement of K^+. Moreover, given the properties of the Na pump, its binding site for K^+ saturates at 5 mM so no extra activation occurs above this level. Stimulation of the Na^+ pump could be involved, though, in a decrease of K^+_o sometimes observed during the plateau phase that occurs concomitantly with catecholamine release and an increased glycolytic flux in globally ischemic hearts (79).

THE ELECTRICAL EFFECTS OF METABOLIC INHIBITION

The effects of hypoxia on global cardiac electrical activity differ widely from those of ischemia. These differences arise mainly from the extracellular K^+ accumulation and the acidosis occurring in the latter. In high flow hypoxia, K^+ leaking out of the cells is washed out so that depolarization is not observed in arterially perfused preparations (17,49,80). In superfused preparations, a slight depolarization has been observed that can be explained by "trapping" of K^+ in the interstitial space between tightly packed cells (47,73). Because of the maintenance of the resting potential, fast rising action potentials are preserved and conduction is not delayed (17,81). In contrast, in severe metabolic depression of long duration cytosolic Ca^{2+} accumulation and subsequent acidosis lead to disruption of electrical coupling, conduction delays and eventually cessation of propagated electrical activity (36,37). The membrane depolarization resulting from the increased K_o during ischemia markedly depresses the Na^+ inward current, slows down conduction, prolongs the recovery of excitability beyond the action potential duration and shifts the activation threshold towards more positive potentials (56). In addition, the extra and intracellular acidosis exert profound effects on the steady state inactivation of the Na^+ current and depress the Ca^{2+} current thus enhancing the effects of high K_o and further deteriorating propagated electrical activity. These effects are extensively described in reference 56. We will therefore limit our discussion here to the mechanisms involved in the maintenance of the resting potential and the action potential shortening in hypoxic myocardium.

The Resting Potential

In a classic paper still widely quoted today McDonald and MacLeod (81) showed a dissociation between the resting potential (V_m) and the potassium equilibrium potential in anoxic guinea pig papillary muscles. After 8 hrs of exposure to anoxia in 5 mM glucose, V_m had decreased from -81 to -77 mV whereas E_K, estimated from measurements of total tissue K content had fallen to -47 mV. Moreover, cooling and rewarming of the preparations resulted in rapid shifts of V_m that could not be attributed to changes in E_K that lagged far behind the potential changes. The effects of cooling and rewarming were abolished by ouabain which lead to the proposition that the resting potential consisted of two components, one determined by E_K and the other one derived from the electrogenic activity of the Na pump. The already known contribution of electrogenic sodium extrusion to the membrane potential (82) was later on confirmed by several investigators in different cardiac preparations (83-87). The two components of the resting potential, V_d and V_p (the subscripts stand for diffusion generated and pump related) could be separated by inhibition of the pump with high ouabain concentrations. The sudden early depolarization occurring under ouabain was attributed to the abolition of V_p (88). With this approach, we estimated that V_p amounts to about -10 mV in normoxic rabbit hearts (87,89) and decreases to -6 mV during hypoxia in the presence of glucose (17). Despite the reduction of V_p, the resting potential remains at control levels during hypoxia. However, when plotted as a function of K_o/K_i, V_m was more negative than E_K for $K_o \geq 5$ mM (17). These observations and the notion that electrogenic Na^+ extrusion persists during metabolic inhibition were challenged on the basis of simultaneous measurements of V_m and a^i_K in superfused preparations that showed an inhomogenous K^+ distribution and a good correlation between V_m and E_K after 8 hrs of hypoxia (90). However, these results were not confirmed by others using the same methodology (91) and can be interpreted as artifacts resulting at least partly from inhomogenous washout of the interstitial space and the existence of a diffusional gradient for ions and small molecules between the core and the superficial layers of a superfused preparation which determines an unequal degree of metabolic depression across the preparation (92). In contrast, in perfused hearts, a step change from high to very low PO_2 levels

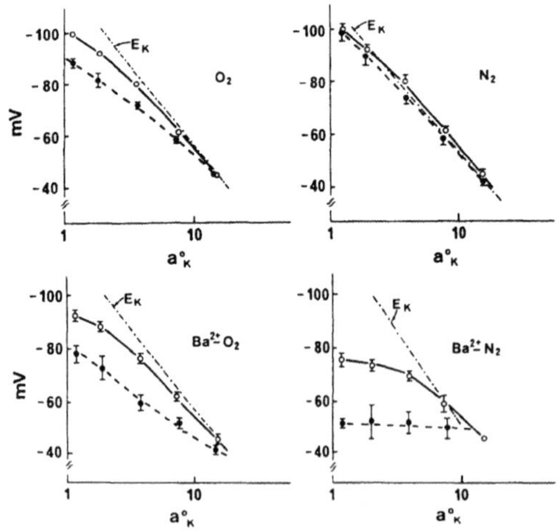

Figure 1. Variations of the resting potential (full lines) and its diffusion generated component (dashed curves) on the external K^+ activity (a^o_K) in normoxia (O_2) and hypoxia (N_2) with and without 40 μM $BaCl_2$. E_K potassium equilibrium potential. Means (SEM).

results in a homogenous anaerobiosis (93). Moreover, heterogeneities in ionic distribution between cells are unlikely in viable preparations where the cells remain coupled. Furthermore, even if local changes in transmembrane ionic distribution occur, the electrical consequences will affect the rest of the preparation because of electrotonic current spread (92) unless some of the cells in the preparation are irreversibly damaged.

An alternative analytical approach to meet the criticism that K_i calculated from measurements of bulk tissue K cannot be used to estimate E_K for single cells is to determine the zero potential intercept of the relationship V_m vs K_o for high K_o (e.g. between 15 and 100 mM) where $V_m \approx E_K$ in ventricular cells (80). If the curve is plotted against a^o_K the intercept will yield a valid estimate of a^i_K. The values thus obtained agree with direct determinations with ISE both for control and metabolic inhibition (47,48,52,80,94). This good correspondence suggests that the activity coefficient for internal K^+ ($\approx$ 0.7) does not vary with hypoxia.

The contribution of V_p to V_m varies with external K^+ in normoxia (Fig. 1, upper left): it decreases as K_o increases, an effect attributable to the influence of K_o on the background K conductance and hence on the membrane resistance R_m (75,83). In hypoxia, V_p decreases and both V_m and V_d approach and eventually reach E_K at high K_o (upper right panel in Fig. 1). These results are compatible with the postulated rise in a K^+ conductance that would induce the increase in K^+ efflux with hypoxia and ischemia (48,52,59). Although these effects are closely linked, their causal relationship in ischemia has not yet been conclusively established (95). However, the predominant role of g_{K1} as a major determinant of the maintenance of V_m in hypoxia is demonstrated in the lower panels of Fig. 1. Exposure to Ba^{2+} in normoxia produces a slight fall in V_m but a substantial decrease in V_d with a concomitant larger contribution of V_p to V_m (lower left panel). In the presence of Ba^{2+}, hypoxia produces a significant depolarization and V_d becomes insensitive to variations in K_o remaining at a stable level of $\approx$ -50 mV between 1.5 and 15 mM K_o (lower right panel in Fig. 1). The enhanced contribution of V_p to V_m in the presence of Ba^{2+} can be attributed to the increased R_m resulting from the blocking of g_{K1} (83) and illustrates the importance of electrogenic Na extrusion in the maintenance of the resting potential in conditions of depressed K^+ conductance such as observed in diseased human atria (96). Moreover, our data confirm once more the preservation of Na pump function in the presence of glycolysis despite the inhibition of ATP formation from oxidative phosphorylation as first reported by McDonald and MacLeod (81). The preferential fuelling of the pump by glycolytic ATP synthesis (22) supports the notion of subcellular ATP compartmentation (21,78) and explains the disappearance of the pump related component of V_m in hypoxia in the absence of glucose and its restitution by insulin, known to stimulate the Na pump independent of metabolic effects (97,98).

In order to verify the hypothesis that the maintenance of V_m in hypoxia depends on an increase in K^+ conductance at the resting level, we characterized I_{K1} in rabbit myocytes at 35°C. Figure 2 shows characteristic inward currents elicited by hyperpolarizing pulses and the effect of Ba^{2+} which corresponds to the classic

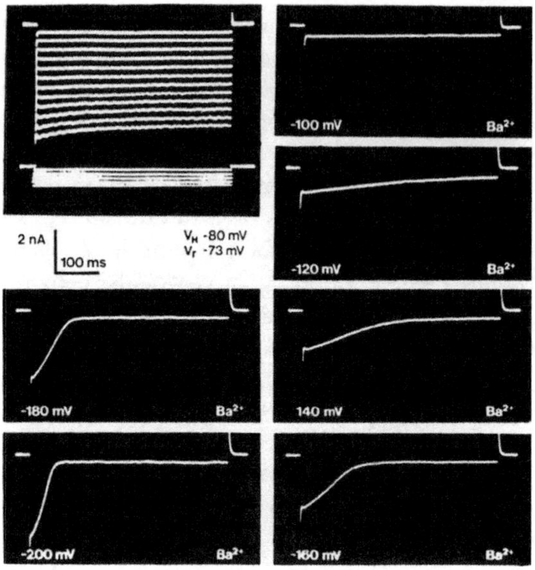

Figure 2. *Characterization of the inward rectifying current I_{K1} in rabbit ventricular myocytes in normal Tyrode at 35°C. Upper left: inward currents elicited by hyperpolarizing pulses (400 ms long) from a holding potential of -80 mV to test potentials between -90 and -200 mV. Note the relaxation at potentials more negative than -130 mV. The resting potential of the cell was -73 mV. Other panels: addition of 50 μM Ba^{2+} does not modify the instantaneous inward current but markedly enhances current relaxation and suppresses the inward rectification of the steady state current.*

description of a voltage and time-dependent blockade (99). In these experiments, the resting R_m was 4.3 ± 1.1 and 5.8 ± 1.2 kΩ • cm² on hyperpolarizations and depolarizations of 10 mV from a holding potential of -80 mV (n = 6, mean cell surface 12689 ± 2186 μm²). R_m increased to 23.4 ± 5.9 kΩ • cm² in the presence of Ba^{2+} and the asymmetry disappeared.

We isolated I_{K1} from other background currents by exposing the cells to Ba^{2+} and determined its K^+ dependence (Fig. 3A). As expected, the current rectifies in the inward direction and the slope conductance increases with K_o: 0.3, 0.9 and 1.54 mS/cm² at 1.5, 5.4 and 20 mM K_o respectively. Figure 3B illustrates the variation of the reversal potential E_{rev} with K_o and compares it to the K^+ dependence of V_d as determined in whole hearts by exposure to ouabain. The correspondence is striking because potential measurements in the heart are strongly influenced by spread of electrotonic current which in part derives from heterogeneities arising from regional differences in electrical properties (100,101). Our data validate the approach used to estimate the contribution of V_p to V_m because they demonstrate that the inwardly rectifying properties of I_{K1} do not interfere with the estimation of V_d by inhibition of the Na^+ pump with ouabain or low temperature. The right panel in Fig. 3 illustrates the current voltage relation for the Ba^{2+} sensitive current measured in cells previously exposed to 200 nM of the inhibitor carbonyl cyanide m-chlorophenyl hydrazone (CCCP) in the presence of 30 μM glibenclamide to block the ATP-sensitive current elicited by CCCP. In these experiments, E_{rev} for the global current in the presence of CCCP and glibenclamide shifted from -70 mV to -54 mV on addition of Ba^{2+}. This value corresponds to V_d observed in hypoxic hearts with Ba^{2+} (Fig. 1) and evidently corresponds to the reversal potential of the yet non identified membrane currents determining V_d in the absence of I_{K1}. The Ba^{2+} sensitive current in these cells shows a steeper slope conductance at the level of the resting potential than in control conditions. Besides, although the inward branch of the I/V curve is not modified, the outward branch levels off at a higher current level. This explains that Ba^{2+} can partly offset the shortening of the late repolarization in hypoxic hearts (49).

In summary, our findings confirm the hypothesis that an increase in g_{K1} constitutes an important mechanism responsible for the maintenance of the resting potential in hypoxic conditions despite the K^+ loss. They also explain the decrease in the size of the electrogenic component of V_m despite the preservation of the Na pump function. In ischemia, the depolarization results from the reduced transmembrane K^+ gradient which cannot be offset by the pump potential because of the increase in resting membrane conductance induced by anoxia and the high K_o. In addition, it is

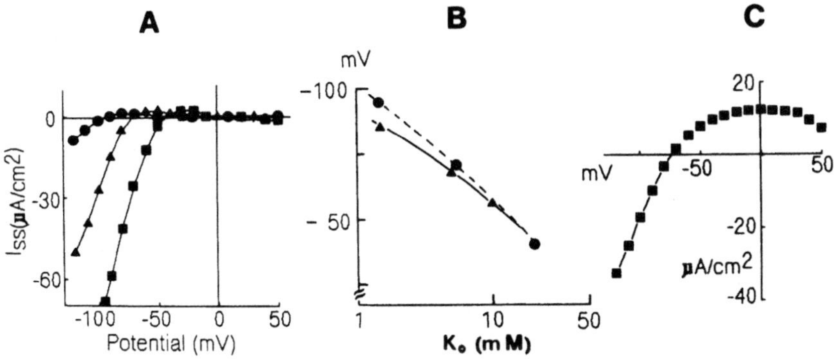

Figure 3. The Ba^{2+} sensitive steady state current (I_{ss}) determines the diffusion component of the resting potential in rabbit myocardium. A: current/voltage curves for the difference current (I_{ss} control - I_{ss} with Ba^{2+} at 1.5 (●), 5.4 (▲), and 20 (■) mM K_o. The currents were measured at the end of 400 ms long hyperpolarizing and depolarizing pulses (10 mV steps) between -120 and +40 mV. B: the reversal potential for the Ba^{2+} sensitive current (●) as a function of K_o compared to the variation of V_d (▲) with K_o as determined in whole hearts after inhibiting the Na pump with 10^{-4} M ouabain (80). C: I/V curve for the difference current measured in the presence of CCCP and glibenclamide before and after addition of 50 μM Ba^{2+}.

expected that the pump would be inhibited rather early during ischemia because of the cessation of glycolysis (11).

The Action Potential Duration

Although it has long been recognized that a marked shortening of the action potential is the most striking effect of metabolic inhibition (7), the mechanism of this response remains controversial, specially when data obtained in single cells are compared to observations performed in multicellular preparations. McDonald and MacLeod did not find a correlation between total cell ATP and action potential duration (APD) but concluded that the latter was related to the metabolic state of the cell because a high glucose concentration antagonizes the anoxia induced changes in action potential morphology and that the reduction in amplitude and duration reflected the influence of metabolism on the slow inward Ca^{2+} current (19,81). This view was supported by later studies showing that Ca^{2+} dependent action potentials were inhibited by metabolic depression whether they were induced by stimulation of c-AMP mediated phosphorylation or not (102-104). A protective effect for the myocardial cell was attributed to the metabolic control of Ca^{2+} influx through voltage dependent channels (105). This view was supported by the finding that anoxia decreased the Ca^{2+} current in rat trabecula (106). In contrast, Vleugels et al. found in cat papillary muscles that the main effect of hypoxia was a large increase in a time independent current and that the slow Ca^{2+} current, estimated as the difference between the early inward peak and the current at 200 ms did not vary (107). These opposite results could reflect a greater severity of inhibition in the rat preparation where an increase of the internal resistance during anoxia indicated intercellular uncoupling and a certain degree of Ca^{2+} overload. In cat ventricle, the hypoxia induced current shifted the global membrane current outwardly and net inward currents were suppressed. These findings explained the apparent contradiction between the sensitivity of Ca^{2+} dependent action potentials to metabolic inhibition and the preservation of I_{Ca} because the net membrane current becomes outward over most of the voltage range where these potentials occur. Therefore, unexcitability results from the absence of net inward current instead of abolition of I_{Ca}. The predominant role of an outward current as determinant of APD and overshoot reduction during hypoxia was further confirmed by the moderate sensitivity of depressed fast responses to hypoxia (17, 108) and the attenuation of APD decline by TEA in guinea pig ventricles (109). The stimulation of an outward current by metabolic inhibition without early changes in the Ca^{2+} current was later confirmed in isolated guinea pig myocytes exposed to 0.1 mM DNP (110) where artifacts arising from ion accumulation phenomena (107) or intercellular uncoupling (106) can be avoided. This current was not inhibited by 1 mM Ba^{2+} that blocked I_{K1}. Similar effects were reported for inhibition with 1 mM cyanide and FCCP (carbonyl cyanide 4-trifluoromethoxyphenyl hydrazone). Under cyanide, the slope conductance at the level of the resting potential was 7 fold higher than in the control (111). The effects were partly counteracted by increasing the glucose concentration to 20 mM.

The discovery of adenosine 5'triphosphate sensitive K^+ channels (K^+-ATP) in cardiac cells by Noma (10) and the finding that cyanide poisoning could induce their opening (112) led to the proposition that activation of these channels is responsible for the increased outward current and K^+ efflux during inhibition. These channels have also been found in skeletal muscle (113), neurons (114), and insulin secreting cells (115) where they play a physiological role in the control of insulin secretion by glucose blood levels. They are activated by a fall in cytosolic ATP whereas in the presence of normal ATP concentrations channel opening is prevented by ATP binding to the inner mouth of the channel. Inhibition of opening by ATP is independent of phosphorylation. Half maximum activation is obtained with 0.1 mM ATP in inside out patches (10) and with 0.5 mM ATP in open cell attached patches (116). These low ATP concentrations are not attained early during ischemia and hypoxia (117) and it was shown that the time course of APD shortening in ferret hearts exposed to cyanide is much faster than bulk ATP depletion (118). These observations gave rise to several questions concerning the validity of direct extrapolation of data obtained in cell free patches to intact cells and eventually to multicellular preparations specially because of modulation of channel activity by other cytosolic nucleotides and the possibility of functional compartmentation of ATP (119).

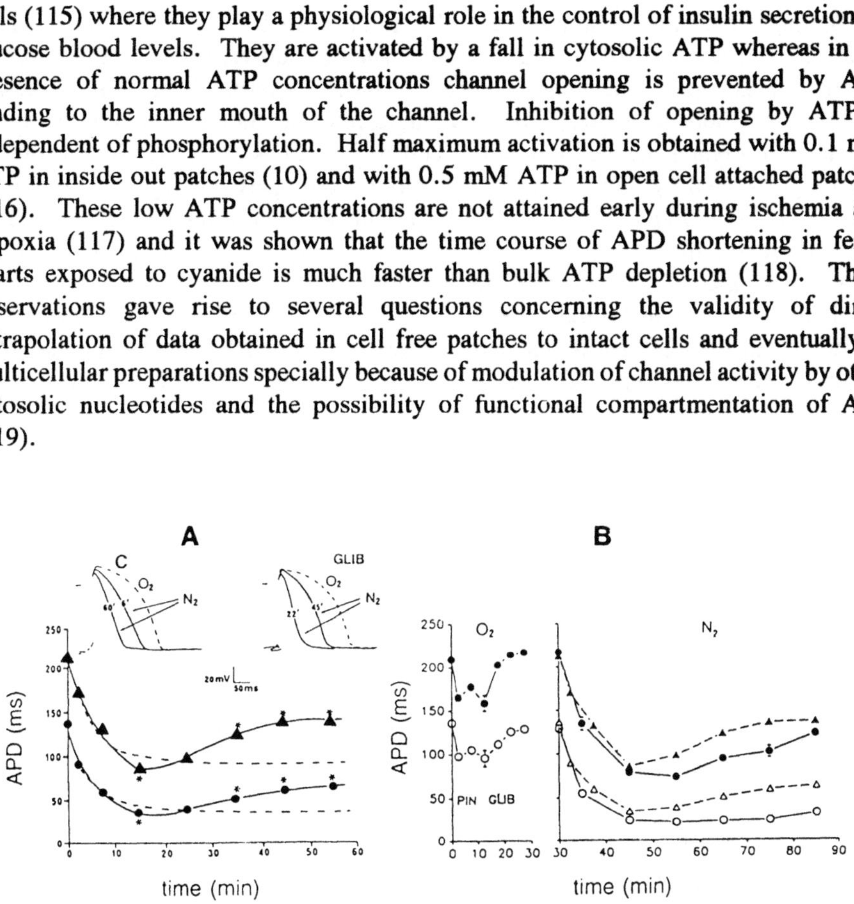

Figure 4. A, Top: Action potentials recorded in normoxia and hypoxia without (left) and in the presence of 30 μM glibenclamide (right). Bottom: Change in APD at 25% (circles) and 95% repolarization (triangles) with time in hypoxia in the presence of glibenclamide. Dashed lines: control without glibenclamide. B: Effects of 30 μM glibenclamide on the APD changes induced by 100 μM pinacidil in normoxia (O_2) and hypoxia (N_2). Full symbols: APD 95%, empty symbols, APD 25%. The dashed curves represent the effect of glibenclamide without pinacidil. Five hearts per protocol.

Inhomogeneities of ATP distribution (120) and variability of channel sensitivity to ATP (121) have also been invoked to support the activation of these channels in early metabolic stress despite the preservation of bulk ATP levels.

Since the direct and simultaneous determinations of the ATP level in close proximity to the channels and of the current resulting from their activation is not possible in intact cells or tissues, their role in determining the action potential shortening in metabolic inhibition as well as in interstitial K^+ accumulation has been assessed with the use of oral antidiabetic agents (sulfonylureas), considered as specific blockers of the channels (76,122). Inconsistent results have been found with this approach both in multicellular and single cell preparations. While the action potential shortening induced in a single cell by lack of ATP in the pipette solution was fully reversed by glibenclamide, this blocker only partly antagonized the effect of DNP (122). In hypoxic guinea pig papillary muscles, glibenclamide prevented but did not reverse the decline in APD whereas it fully reversed the DNP induced current in single cells of the same species (76).

Figure 4A shows that glibenclamide did not modify the time course of early APD shortening in hypoxic hearts but partly reversed it after 20 min of hypoxia (123). A similar effect was reported in guinea pig papillary muscles under complete

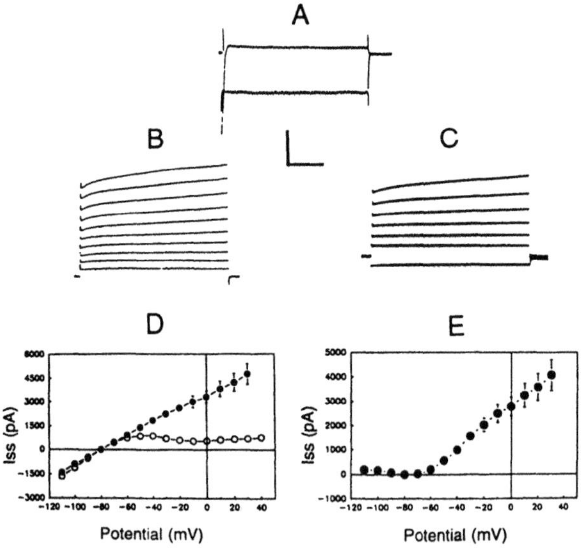

Figure 5. Membrane currents in the presence of 3 X 10^{-5} M BRL 38227 and 2 mM cyanide. Holding potential -80 mV. A: control records at -120 and 40 mV, 450 ms pulses; B: 10 mV steps from -70 to 20 mV with cyanide; C: 20 mV steps from -100 to 40 mV with BRL 38227. D: I/V curves for the current at the end of the pulses in control (empty circles) and under BRL 38227 at the steady state of the effect (full circles); E: difference current obtained from the I/V curves in D. Calibration: 1.5 nA in A and B, 2 nA in C.

metabolic blockade (124). The delayed onset of action of glibenclamide on APD decline was explained by diffusion delays and by the drug binding only to open channels. We tested this hypothesis by investigating the effects of glibenclamide on the action potential changes induced by the K-ATP channel activator pinacidil (125). At 100 μM, pinacidil in normoxia reduced APD to levels comparable to those in hypoxia and glibenclamide reversed this effect within few minutes. In addition, even if pinacidil enhanced the APD shortening during hypoxia, the presence of glibenclamide did not slow it down and only produced a late, partial recovery of APD (Fig. 4B). Similar observations were made by Nakaya et al. (126) in guinea pig myocytes. These experiments suggest that the late onset of glibenclamide action is not due to diffusional delays but probably to a late participation of $I_{K\text{-}ATP}$ in the response to hypoxia.

Since species differences exist in the effect of glibenclamide on K^+_o accumulation in globally ischemic hearts (76, 126,127) and most of the work on the role of K-ATP channels in the response to metabolic inhibition has been carried out in guinea pig cells, we further characterized the effects of glibenclamide and low concentrations of Ba^{2+} on the currents induced by K-ATP channel openers and by metabolic inhibition with cyanide (CN^-) or carbonyl cyanide m-chlorophenylhydrazone (CCCP). Figure 5 shows that cyanide and BRL 38227, another activator of K-ATP channels (128) produce similar increases in the outwardly directed steady state current without changing significantly the steady state inward current (Fig. 5D). The BRL induced current strongly rectifies in the outward direction at potentials more positive than -60 mV, does not show an inward directed component (Fig. 5E), and is fully reversed by addition of glibenclamide (Fig. 6).

The action potential changes accompanying these effects were studied in the current clamp configuration (Fig. 7). The large outward shift in current produced by BRL 38227 (lemakalim) rapidly results in inexcitability (Fig. 7A) but glibenclamide restores the action potential, although the duration is decreased probably because of a certain degree of rundown of I_{Ca} since the pipette solution did

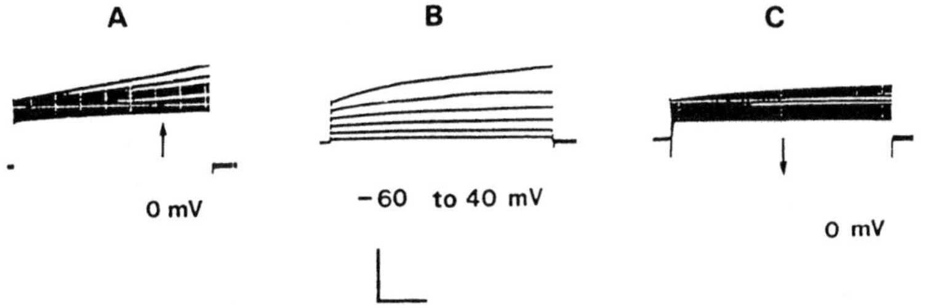

Figure 6. Effect of glibenclamide on the BRL 38227 induced current. A and C show the time course of changes in outward current produced by exposure to BRL 38227 (A) and its reversal by glibenclamide (C), the arrows indicate the direction of the change. Step depolarizations from -80 to 0 mV at a rate of 0.2 Hz. ***B:*** *effect of BRL 38227 at the steady state for indicated test potentials. Calibration 1 nA in A & C, 2 nA in B, 40 ms.*

not contain ATP or cAMP. However, in the presence of lemakalim and glibenclamide, addition of cyanide produced a new increase in outward current and action potential shortening (Fig. 7B). In the absence of a K^+-ATP channel activator, glibenclamide did not prevent an increase in outward current on addition of CCCP and only partly reversed the effect of CCCP even at a very low concentration (Fig. 8A and B). This current increase was not a leak because it was reversible on washout. These data show that in rabbit myocytes, glibenclamide is an effective and specific blocker of the K^+-ATP channels but its presence does not fully prevent the activation of an outward current by inhibition of oxidative phosphorylation. In the absence of single channel data, it is difficult to establish the nature of the glibenclamide insensitive current, which is partly blocked by the Ba^{2+} concentration that inhibits I_{K1} regardless whether the current is triggered by CCCP or cyanide (Fig. 9). Also, the inward rectification for the remaining global current is strongly diminished which decreases the slope conductance at the level of the reversal potential. In addition, the reversal potential shifts from -72 to -55 mV. This value compares well to the diffusion dependent component of V_r under conditions of hypoxia in the presence of Ba^{2+} (Fig. 1) and corresponds to E_{Cl} for in vitro rabbit myocardium (129). It is therefore possible that the transmembrane Cl^- gradient plays a key role in the determination of the resting potential in conditions of depressed resting K^+ conductance. In addition, with glibenclamide and CCCP present, the Ba^{2+} sensitive current shows an increased slope conductance at the resting level (Fig. 3C) suggesting that under these conditions, Ba^{2+} is blocking a current component other than $I_{K\text{-}ATP}$.

When compared to available data in the literature, our results agree with some of them in that glibenclamide did not prevent and only partly reversed the effects of metabolic inhibition on action potential duration and the time-independent outward

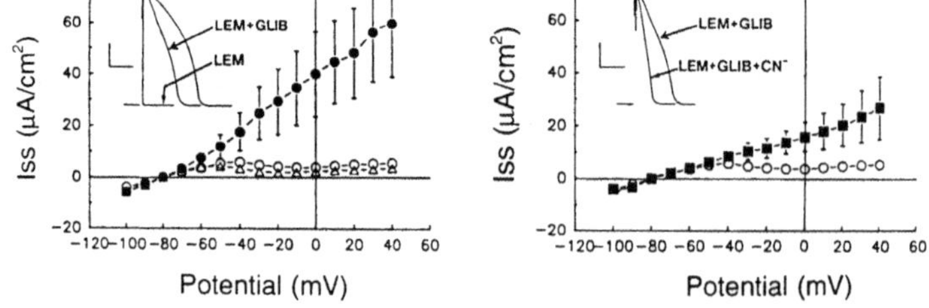

Figure 7. A: The increase in outward current by lemakalim (full circles) leads to inexcitability. Glibenclamide restores electrical activity and reduces the outward current (empty circles) to control levels (triangles). B: Addition of cyanide (squares) to the solution containing lemakalim and glibenclamide.

current both in whole hearts and single cells (122,123,124,126). They disagree however with other reports (76,120,121,130,131,132) specially concerning the ability of glibenclamide to prevent APD shortening and increased K^+ efflux although it is recognized that the effects are more modest in intact myocardium than in single cells (133). This decreased sensitivity to glibenclamide has been attributed to rising levels of free cytosolic ADP concentration. In cell free patches, the block of K^+-ATP channels by glibenclamide was incomplete in the presence of an ADP concentration considered equivalent to what can be found during ischemia or substrate free hypoxia (100 μM) as estimated from the creatine kinase reaction assuming that this reaction is at equilibrium. However, the ADP_i accumulation is comparable in hypoxia with or without substrate but glibenclamide is much less effective in blocking the early increased K^+ efflux and APD reduction in the presence of glucose (123,134). Moreover, in the absence of substrate and in ischemia, it is unlikely that the creatine kinase reaction is at equilibrium because when ATP is not resynthesized ADP does not accumulate and degrades to AMP and P_i (24). In addition, it has been shown that during the first hour of substrate free anoxia where most of the degradation of CrP and ATP occurs in rat ventricular myocytes, 94% of ATP is completely hydrolyzed splitting off the three phosphate groups and only 6% is turned into free ADP (135). Consequently, it is unlikely that ADP_i accumulation could explain the insensitivity of early APD reduction to glibenclamide shown in Fig. 5A which we rather attribute to a delayed opening of K^+-ATP channels because of preservation of subsarcolemmal ATP pools derived from glycolytic sources (21,119). In addition to ourselves (123), several other investigators have proposed that a mechanism other than opening of K^+-ATP channels has to be invoked to explain the early effects of metabolic stress on K^+ movements (95,126,136). On the other hand, receptor desensitization to glibenclamide by ADP_i or a channel activator may be at least in part responsible for the partial reversal in late hypoxia and the decreased effect in the presence of pinacidil (Fig. 5).

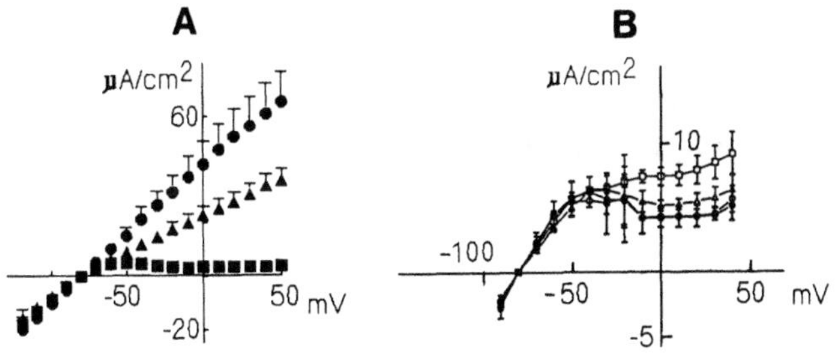

Figure 8. Effect of glibenclamide on CCCP induced current. A: I/V curves in control condition (squares), with 200 nM CCCP (circles), and after addition of 30 μM glibenclamide (triangles). B: circles: control (full) and with 30 μM glibenclamide (empty) added before exposure to 50 μM CCCP (squares); triangles: washout.

In perfused hearts, a combination of Ba^{2+} and Cs^+ at concentrations that blocked I_{K1} as well as $I_{K\text{-}ATP}$ only partly counteracted the APD shortening at plateau levels (123). Since the plateau duration results from a delicate balance of inward and outward currents and the response of these systems to metabolic stress may depend on the degree of inhibition, their contribution to plateau duration may vary with the progression of the hypoxic state. The transient outward current I_{to} sensitive to 4 aminopyridine, responsible for decreases in APD with stimulation rates below 1 to 2 Hz (137) does not appear to participate in the hypoxia-induced APD reduction because the rate dependence of APD is lost in hypoxia (123). A depression of $I_{Ca,L}$ in early hypoxia with preservation of glycolysis seems unlikely (109,110,131) although it has been shown that $I_{Ca,L}$ is depressed when feline myocytes are exposed to 1 mM cyanide in substrate-free solution. The decrease in $I_{Ca,L}$ and APD under these conditions are greater in epicardial than in endocardial cells (138). On the other hand, endogenous catecholamine release in intact hearts may contribute to the preservation of the Ca^{2+} current specially when hypoxia is combined with high levels of K^+_o (17).

Another element that will markedly influence the current system(s) participating in the response to hypoxia or ischemia is the regional heterogeneities observed in the expression of ionic channels (100,101) as well as in their response to experimental challenges (139,140). The electrical activity of epicardium is more severely disturbed than that of endocardium by metabolic inhibition. This can be in part attributed to the greater sensitivity of $I_{Ca,L}$ mentioned above, but most probably results from the different properties of K^+-ATP channels whose K_d for ATP is 24 μM in cells derived from endocardial layers and 98 μM for epicardial cells (141). It is difficult to assess the impact of these regional heterogeneities on the global electrical activity of the heart where the extensive electrical coupling of ventricular

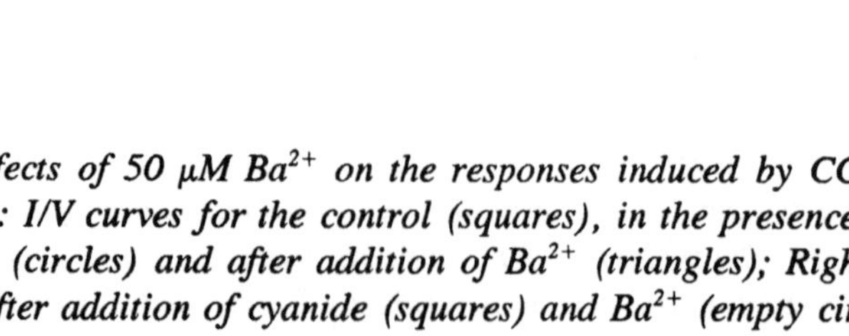

Figure 9. Effects of 50 μM Ba^{2+} on the responses induced by CCCP and 2 mM cyanide. Left: I/V curves for the control (squares), in the presence of CCCP and glibenclamide (circles) and after addition of Ba^{2+} (triangles); Right: control (full circles) and after addition of cyanide (squares) and Ba^{2+} (empty circles).

muscle will attenuate the influence of electrical disturbances arising from a greater sensitivity of different current systems to changes in cytosolic composition. This property as well as the absence of extra myocardial influences in single cell experiments markedly limits the extrapolation of data from the latter to the whole heart in vitro and even more in situ.

There are still other currents that could contribute to APD shortening whose activation depends on the disturbances of electrolyte distribution or sympathetic stimulation arising from metabolic stress. Isenberg et al. (110) suggested that an increased level of Ca^{2+}_i could participate in the activation of an outward current. A Ca^{2+} dependent K^+ conductance remains to be conclusively proven in ventricular cells but it has been shown that Ca^{2+}_i modulates a component of I_{to} (142) as well as a chloride current (143). These currents are enhanced by β adrenergic stimulation through the transient increase in Ca^{2+}_i resulting from an enhanced $I_{Ca,L}$. It is probable that these currents participate in the response to hypoxia with preservation of glycolysis because the stimulation of Na:H exchange by excess production of lactate may induce increases in Ca^{2+}_i thru changes in the function of the Na:Ca exchanger. They would play a lesser role in ischemia because of depression of the Na:H exchanger by acidosis.

A sodium activated K^+ current was first described in cardiac cells by Kameyama et al. (144) but its potential importance was for some time discounted because in cell free patches it activated far beyond the physiological range for Na^+_i. However, this current can be partly activated near the physiological range of Na^+_i in guinea pig myocytes (145). The single channel conductance shows inward rectification for $K_i < K_o$ and outward rectification for $K_i > K_o$. The open probability depends on Na^+_i and Na pump activity and it has been proposed that this channel is responsible for the action potential shortening produced by digitalis (146). Its possible role in hypoxia or ischemia induced APD shortening has not been investigated but it is possible that it becomes active in hypoxia with preservation of glycolysis and on reflow after ischemia because of Na^+ loading resulting from activation of Na^+:H^+ exchange.

A time independent Cl^- current activated by β adrenergic stimulation has been found in rabbit and guinea pig myocytes (147,148). Its activation induces action potential shortening and slight depolarization, as expected from the E_{Cl} value of $\approx$ -50 mV in intact cells (129). This current could contribute to the early APD decline during hypoxia or ischemia because of the sympathetic stimulation and catecholamine release occurring under these conditions (149). This "non myocardial" mechanism could explain the exquisite sensitivity of the plateau duration to hypoxia: its shortening is discernible within seconds of switching from oxygenated to hypoxic solution in perfused hearts. It would also provide the basis for the dissociation observed between the energetic state of the cell and APD reduction but it would not explain the early increase in K^+ efflux. However, it has been recently noted that APD reduction and K^+ efflux in ischemia although linked to a common origin, may not be causally related (95).

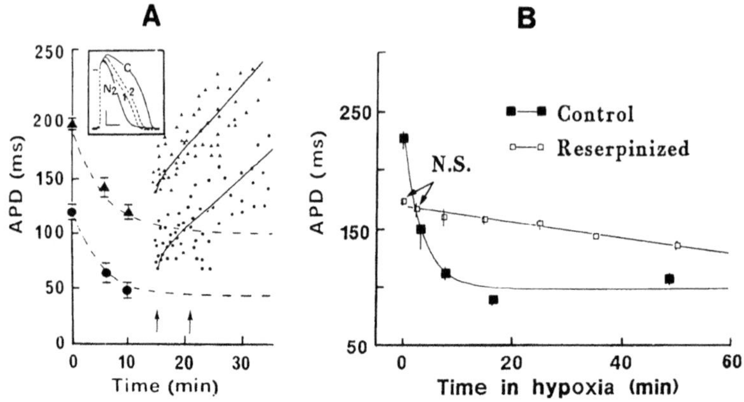

Figure 10. Time course of changes in action potential duration (APD) with time of hypoxia and the effects of DIDS and pretreatment with reserpine. A: dashed lines, control. Symbols: APD-25 (circles) and APD-95 (triangles), means ± SEM at 5 and 10 min of hypoxia before addition of 100 μM DIDS (1st arrow) and individual impalements afterwards. Second arrow, interruption of regular, paced activity. Inset: action potential traces in control (c), after 15 min of hypoxia (N_2), and after 1 and 2 min of addition of DIDS (1,2). B: APD decline during hypoxia in non treated (control) and reserpine pretreated hearts (Leblanc N. and Ruiz Petrich E., unpublished).

We have recently investigated the possible contribution of a chloride current to APD shortening by determining the effects of two blockers of chloride transport (DIDS and SITS derivatives of stilbene disulfonic acid) on hypoxic hearts. When added after 15 min of hypoxia, once the action potential had shortened to a steady level, DIDS produced a fast APD lengthening followed, within minutes, by a marked depression in $\dot{V}_{max}$ with the subsequent conduction delays and loss of regular, paced propagated action potentials (Fig. 10A). Similar effects were observed with SITS. Given the non specificity of these blockers for chloride channels and their high toxicity, these results are not conclusive but their fast onset of action allows the attribution of the very early effect to a Cl^- channel block (150). Moreover, catecholamine depletion by pretreatment with reserpine abolishes APD shortening by hypoxia at high K_o (Fig. 10B). This indirect evidence is still inconclusive and the role of this catecholamine mediated activation of a Cl^- current remains to be proven. Nevertheless, it opens an interesting question and a promising research field. It could imply that the long latency of the response of single cells to metabolic stress is not only due to their lower energetic requirements but also to the absence of a neural regulating response. It would also show that although the unifying hypothesis of a single mechanism (the opening of the K-ATP channels) underlying the main electrical effect of metabolic inhibition is very attractive, it is not sufficient to explain the global response. Moreover, it is too simplistic and even naive to think that the preservation of coordinated electrical activity, a main determinant of cardiac function, depends on an unique line of defense when facing a stress that not only compromises contractile performance but cell survival as well.

Acknowledgements

Original results from studies supported by grants from the Medical Research Council and the Quebec Heart Foundation to Elena Ruiz Petrich. The expert technical assistance of Mr Denis Chartier and the competent secretarial support of Ms Denise Girardin are also gratefully acknowledged.

REFERENCES

1. Erk C and Schaefer H (1944) Pflügers Arch 248:515-520
2. Trautwein W, Gottstein V and Dudel J (1954) Pflügers Arch 260:40-60
3. Samson WE and Scher AM (1960) Circ Res 8: 780-787
4. Sano T, Masaru O and Shimamoto T (1956) Circ Res 4:444-450
5. Scher AM (1962) In: Hamilton WF and Dow P (eds) Handbook of Physiology, Section 2, Circulation vol 1 American Physiological Society, Washington, pp 287-322
6. Hamill OP, Marty A, Neher E, Sakmann B and Sigworth FJ (1981) Pflügers Arch 391:85-110
7. Carmeliet E (1978) Circ Res 42:577-587
8. Hercus VM, McDowall JS and Mendel D (1955) J Physiol 129:177-183
9. Hill JL and Gettes LS (1980) Circ Res 61:768-778
10. Noma A (1983) Nature 305:147-148
11. Reimer KA and Jennings RB (1986) In: Fozzard HA, Haber E, Jennings RB, Katz AM, Morgan HE (eds) The Heart and Cardiovascular System Scientific Foundations, vol 2, Raven Press, New York, pp 1133-1201
12. Rovetto MJ, Whitmer JT and Neely JR (1973) Circ Res 32:699-711
13. Rovetto MJ, Lamberton WF and Neely JR (1975) Circ Res 37: 742-751
14. Kentish JC and Allen DG (1986) J Mol Cell Cardiol 18:879-882
15. Stern MD, Silverman HS, Houser SR, Josephson RA, Capogrossi MC, Nichols CG, Lederer WJ and Lakatta ED (1988) Proc Natl Acad Sci USA 85:6954-6958
16. Gettes LS (1986) In: Fozzard HA, Haber E, Jennings RB, Katz AM, Morgan HE (eds) The Heart and Cardiovascular System, Scientific Foundations, vol 2, chap 57, Raven Press, New York, pp 1317-1341
17. Ruiz-Ceretti E, Ragault P, Leblanc N and Ponce Zumino AZ (1983) J Mol Cell Cardiol 15:845-854

18. Whitman G, Kieval R, Wetstein L, Seeholzer S, McDonald G and Harken A (1983) J. Surg Res 35:332-339
19. McDonald TF and MacLeod DP (1971) Pflügers Arch 325:305-322
20. Kobayashi K and Neely JR (1979) Circ Res 44:166-175
21. Weiss J and Hiltbrand B (1985) J Clin Invest 75:436-447
22. Glitsch HG and Tappe A (1993) Pflügers Arch 422:380-385
23. Van Blisen M, Van Der Vusse GJ, Coumans WA, deGroot MJM, Willemsen PHM and Reneman RS (1989) Am J Physiol 257 (Heart Circ Physiol 26): H47-H54
24. Poston JM and Parenteau GL (1992) Arch Biochem Biophys 295:35-41
25. Murry CE, Reimer KA, Hill ML, Yamasawa I and Jennings RB (1985) Fed Proc 44: 823 (Abstract)
26. Steenbergen C, Delleuw G, Rich T and Williamson JR (1977) Circ Res 41:849-858
27. Jennings RB, Reimer KA and Steenbergen C Jr (1986) J Mol Cell Cardiol 18:769-780
28. Vander Heide RS, Rim D, Hohl CM and Ganote CE (1990) J Mol Cell Cardiol. 22:165-181
29. Armstrong SC and Ganote CE (1991) J Mol Cell Cardiol 23:1001-1014
30. Karmazyn M and Moffat MP (1993) Cardiovasc Res 27:915-924
31. Allen DG, Morris PG, Orchard CH and Pirolo JS (1985) J Physiol 361:185-204
32. Allen DG and Orchard CH (1983) J Physiol 339:107-122
33. Taegtmeyer H, Peterson MB Ragavan VV, Ferguson AG and Lesch M (1977) J Biol Chem 252:5010-5018
34. Kingsley PB, Sako EY, Yang MQ, Zimmer SD, Ugurbil K, Foker JE and From AHL (1991) Am J Physiol 261:H469-H478
35. Dahl G and Isenberg G (1980) J Membrane Biol 53:63-75
36. Wojtczak J (1979) Circ Res 44:88-95
37. Jongsma HJ and Gros D (1991) NIPS 6:34-39
38. Piper HM, Schwartz P and Siegmund B (1989) In: Piper HM and Isenberg G (eds) Isolated Adult Cardiomyocytes, vol 1, Structure and Metabolism, Chap 11, CRC Press, Florida, pp 259-285
39. Piper HM, Koop A, Nool T, Schlüter KD and Siegmund B (1990) In: Vereecke J, van Bogaert PP and Verdonck F (eds) Ionic Currents and Ischemia, Leuven University Press, pp 17-49
40. Piper HM, Spahr R, Hütter JF and Spieckermann PG (1984) In: Piper HM and Spieckermann PG (eds) Adult Heart Muscle Cells. Isolation Properties and Applications, Springer Verlag, New York, pp 143-147
41. Fantini E, Athias P, Courtois M, Khatami S, Grynberg A and Chevalier A (1990) Can J Physiol Pharmacol 68:1148-1156
42. Hagve T-A, Sprecher H and Hohl CM (1990) J Mol Cell Cardiol 22:1467-1475
43. Smolenski RT, Schrader J, deGroot H and Deussen A (1991) Am J Physiol 260:C708-C714

44. Ruiz-Petrich ECC, Ponce Zumino AZ, Caballero OH and Suarez JRE (1962) Rev Soc Argent Biol 38:311-319
45. Goerke J and Page E (1965) J Gen Physiol 48:933-948
46. Nguyen-Thi A, Ruiz-Ceretti E and Schanne OF (1981) Can J Physiol Pharmacol 59:876-883
47. Nakaya H, Kimura S and Kanno M (1985) Am J Physiol 249:H1078-H1085
48. Leblanc N, Ruiz-Ceretti E and Chartier D (1987) Can J Physiol Pharmacol 65:861-866
49. Ruiz-Petrich E, deLorenzi F and Chartier D (1991) Cardiovasc Res 25:17-26
50. Gerlings ED, Miller DT and Gilmore JP (1969) Am J Physiol 216:559-562
51. Rau EE, Shine KI and Langer GA (1977) Am J Physiol 232: H85-H94
52. Rau EE and Langer GA (1978) Am J Physiol 235: H537-H543
53. Lee CO and Fozzard HA (1975) J Gen Physiol 65:695-708
54. Vleugels A, Carmeliet E, Bosteels S and Zaman M (1976) Pflügers Arch 365:159-166
55. Weiss J and Shine KI (1982) Am J Physiol 242:H619-H628
56. Gettes LS (1986) In: Fozzard HA, Haber E, Jennings RS, Kartz AM, Morgan HE (eds) The Heart and Cardiovascular System, Scientific Foundations vol 2, chap 57, Raven Press, New York, pp 1317-1341
57. Kleber AG (1983) Circ Res 52:442-450
58. Kleber AG (1984) J Mol Cell Cardiol 16:389-394
59. Weiss J and Shine KI (1986) Am J Physiol 250:H982-H991
60. Chiao JJC, Minei JP, Shires III GT and Shires GT (1990) Am J Physiol 258:R684-R689
61. January CT and Fozzard HA (1984) Circ Res 54:652-665
62. Wallert MA and Frohlich O (1989) Am J Physiol 257:C207-C213
63. Samson JP, Reisin I, Ruiz-Ceretti E and Schanne OF (1977) J Mol Cell Cardiol 9:39-50
64. Reimer KA, Jennings RB and Hill ML (1981) Circ Res 49:901-911
65. Shen AC and Jennings RB (1972) Am J Physiol 67:417-433
66. Shine KI, Douglas AM and Ricchuiti N (1976) Am J Physiol 231:1225-1232
67. Shine KI, Douglas AM and Ricchuiti NV (1978) Circ Res 43:712-720
68. Shine KI, Weiss J and Bersohn M (1985) In: Stone HL and Weglicki WB (eds) Pathobiology of Cardiovascular Injury, Martinus Nijhoff Publishing, Boston, pp 180-189
69. Tani M (1990) Ann Rev Physiol 52:543-559
70. Duan J and Moffat MP (1992) Adv Exp Med Biol 311:435-436
71. Allen DG, Lee JA and Smith GL (1989) J Physiol (London) 410:297-323
72. Hayashi H, Miyata H, Noda N, Kobayashi A, Hirano M, Kawai T and Yamazaki N (1992) Am J Physiol 262:C628-C634
73. Gaspardone A, Shine KI, Seabrooke SR and Poole Wilson PA (1986) J Mol Cell Cardiol 18:389-399
74. Mathur PP and Case RB (1973) J Mol Cell Cardiol 5:375-393
75. Kameyama K, Kiyosue T and Soejima M (1983) Jpn J Physiol 33:1039-1056

76. Wilde AAM, Escande D, Schumacher CA, Thuringer D, Mestre M, Fiolet JWT and Janse MJ (1990) Circ Res 67:835-843
77. Wilde A, Schumacher C, Fiolet J, Opthof T and Janse M (1992) J Mol Cell Cardiol 24 (supp 1), (Abstract) no P-09-47
78. Weiss JN, Lamp ST and Shine KI (1989) Am J Physiol 256:H1165-H1175
79. Wilde AAM, Peters RJG and Janse MJ (1988) J Mol Cell Cardiol 20:887-896
80. Leblanc N and Ruiz-Ceretti E (1987) Can J Physiol Pharmacol 65:246-251
81. McDonald TF and MacLeod DP (1973) J Physiol (London) 229:559-582
82. Page E and Storm SR (1965) J Gen Physiol 48:957-972
83. Sperelakis N (1979) In: Handbook of Physiology. The Cardiovascular System sec 2, vol 1, part 1, chap 6, American Physiological Society, Bethesda MD, pp 187-267
84. Isenberg G and Trautwein W (1974) Pflügers Arch 350:41-54
85. Glitsch HG, Pusch H, Schumacher TH and Verdonck F (1982) Pflügers Arch 394:256-263
86. Daut J and Rudel R (1980) J Physiol (London) 305:22
87. Ruiz-Ceretti E, Nguyen-Thi A, Schanne OF and Caillé J-P (1981) Am J Physiol 240:C28-C34
88. Marmor MF and Gorman ALF (1970) Science 167:65-67
89. Ruiz-Ceretti E, Caillé J-P and Dorticos F (1982) J Mol Cell Cardiol 14:301-305
90. Baumgarten CM, Cohen CJ and McDonald TF (1981) Circ Res 49:1181-1189
91. Guarnieri T and Strauss HC (1982) J Clin Invest 69:435-442
92. Lee JA and Allen DG (1988) Pflügers Arch 413:83-89
93. Steenbergen C, Deleeuw G, Barlow C, Chance B and Williamson JR (1977) Circ Res 41:606-615
94. Dresdner K and Boyden P (1986) Biophys J 49:346a (Abstract) no T-Pos 284
95. Vanheel B and de Hemptinne A (1992) Cardiovasc Res 26:1030-1039
96. Imanishi S and Arita M (1987) Jpn J Physiol 37:393-410
97. Ruiz-Ceretti E, deLorenzi F, Lafond JS and Chartier D (1988) Can J Physiol Pharmacol 66:202-206
98. Moore RD (1983) Biochim Biophys Acta 737:1-49
99. Imoto Y, Ehara T and Matsuura H (1987) Am J Physiol 252:H325-H333
100. Litovsky S and Antzelevitch C (1988) Circ Res 62:116-126
101. Litovsky SH and Antzelevitch C (1989) J Am Coll Cardiol 14:1053-1066
102. Schneider JA and Sperelakis N (1974) J Surg Res 16:389-403
103. Shigenobu K, Asano T and Kasuya Y (1981) J Pharmacol Dynam 4:317-328
104. Wahler GM and Sperelakis N (1984) Can J Physiol Pharmacol 62:569-574
105. Sperelakis N And Schneider JA (1976) Am J Cardiol 37:1079-1085
106. Payet MD, Schanne OF, Ruiz-Ceretti E and Demers J-M (1978) J Physiol, Paris 74:31-35
107. Vleugels A, Vereecke J and Carmeliet E (1980) Circ Res 47:501-508
108. Alvarez J, Dorticos F and Morlans J (1981) J Physiol, Paris 77:807-812
109. Asano T, Shigenobu K and Kasuya Y (1985) Jap J Pharmacol 38:65-72

110. Isenberg G, Vereecke J, Van der Heyden G and Carmeliet E (1983) Pflügers Arch 397:251-259
111. Van der Heyden G, Vereecke J and Carmeliet E (1984) In: Piper HM and Spieckermann PG (eds) Adult Heart Muscle Cells. Isolation, Properties and Applications, Springer Verlag, New York, pp 93-96
112. Noma A and Shibasaki T (1985) J Physiol (London) 363:463-480
113. Spruce AE, Standen NB and Stanfied PR (1985) Nature 316:736-738
114. Ashford MLJ, Sturgess NC, Trout NJ, Gardner N.J. and Hales CN (1988) Pflügers Arch 412:297-304
115. Cook DL and Hales CN (1984) Nature 311:271-273
116. Kakei M, Noma A and Shibasaki T (1985) J Physiol (London) 363:441-462
117. Allen DG and Orchard CH (1987) Circ Res 60:153-168
118. Elliott AC, Smith GL and Allen DG (1989) Circ Res 64:583-591
119. Weiss JN and Lamp ST (1989) J Gen Physiol. 94:911-935
120. Nichols CG and Lederer WJ (1990) J Physiol (London) 423:91-110
121. Findlay I and Faivre J-F (1991) FEBS 279:95-97
122. Fosset M, De Weille JR, Green RD, Schmid-Antomarchi H and Lazdunski M (1988) J Biol Chem 263:7933-7936
123. Ruiz-Petrich E, Leblanc N, deLorenzi, F, Allard Y and Schanne OF (1992) Br J Pharmacol 106:924-930
124. Carmeliet E, Storms L and Vereecke J (1990) In: Zipers DP and Jalife J (eds) Cardiac Electrophysiology from Cell to Bedside, chap 12, Saunders Co, pp 103-108
125. Arena JP and Kass RS (1989) Am J Physiol 257:H2092-H2096
126. Nakaya H, Takeda Y, Tohse N and Kanno M (1991) Br J Pharmacol 103:1019-1026
127. Gasser RNA and Vaughan-Jones RD (1990) J Physiol (London) 431:713-741
128. Sanguinetti MC, Scott AL, Zingaro GJ and Siegel PKS (1988) Proc Natl Acad Sci USA 85:8360-8364
129. Caillé J-P, Ruiz-Ceretti E and Schanne OF (1981) Am J Physiol 240:C183-C188
130. Deutsch N, Klitzner TS, Lamp ST and Weiss JN (1991) Am J Physiol 261:H671-H676
131. Friedrich M, Benndorf K, Schwalb M and Hirche Hj (1990) Pflügers Arch 416:207-209
132. Benndorf K, Friedrich M and Hirche Hj (1991) Pflügers Arch 419:108-110
133. Venkatesh N, Lamp ST and Weiss JN (1991) Circ Res 69:623-637
134. Weiss JN, Venkatesh N and Lamp ST (1992) J Physiol (London) 447:649-673
135. Siegmund B, Koop A and Piper HM (1989) Pflügers Arch 413:435-437
136. Belles B, Hescheler J and Trube G (1987) Pflügers Arch 409:582-588
137. Ruiz-Petrich E and Leblanc N (1989) Can J Physiol Pharmacol 67:780-787
138. Kimura S, Bassett AL, Furukawa T, Furukawa N and Myerburg J (1991) Circ Res 84:768-777
139. Kimura S, Nakaya H and Kanno M (1982) J Cardiovasc Pharmacol 4:658-667

140. Gilmour RF and Zipes DP (1980) Circ Res 46:814-825
141. Furukawa T, Kimura S, Furukawa N, Bassett AL and Myerburg RJ (1991) Circ Res 68: 1693-1702
142. Hiraoka M and Kawano S (1989) J Physiol 410:187-212
143. Zygmunt AC and Gibbons WR (1991) Circ Res 68:424-437
144. Kameyama M, Kakei M, Sato R, Shibasaki T, Matsuda H and Irisawa H (1984) Nature 309:354-356
145. Rodrigo GC and Chapman RA (1990) Exper Physiol 75:839-842
146. Luk H-N and Carmeliet E (1990) Pflügers Arch 416:766-768
147. Harvey RD, Clark CD and Hume JR (1990) J Gen Physiol 95:1077-1102
148. Hume JR and Harvey RD (1991) Am J Physiol 261:C399-C412
149. Karwatowska-Krynska E and Beresewicz A (1983) J Mol Cell Cardiol 15:523-536
150. Ruiz Petrich E, Ponce Zumino A, Yergeau P and Chartier D (1992) J Mol Cell Cardiol 24 (supp 1) (Abstract) no 0-23-5

CHAPTER 6

ROLE OF Na/H EXCHANGE AND $[Ca^{2+}]_i$ IN ELECTROPHYSIOLOGICAL RESPONSES TO ACIDOSIS AND REALKALIZATION IN ISOLATED GUINEA PIG VENTRICULAR MYOCYTES

Margaret P. Moffat,
Jianmin Duan and Christopher A. Ward

University of Western Ontario, London, Canada

INTRODUCTION

It is well known that reperfusion of the ischemic heart results in a rapid increase in the intracellular calcium concentration ($[Ca^{2+}]_i$) (1). However, the mechanisms responsible for the increase in $[Ca^{2+}]_i$ remain unclear (1). In this regard, since the intracellular pH is much lower than that of the perfusate during early reperfusion (2), a pH gradient develops instantaneously across the cardiac cell membrane. This pH gradient activates the sodium/hydrogen (Na^+/H^+) exchange, leading to a rapid accumulation of intracellular sodium ($[Na^+]_i$). The increase in $[Na^+]_i$ may, in turn, result in an increase in $[Ca^{2+}]_i$ via sodium/calcium (Na^+/Ca^{2+}) exchange (reviewed in 1, 3, 4). Therefore, the realkalization-induced activation of Na^+/H^+ exchange, and subsequent accumulation of $[Na^+]_i$ may explain, at least in part, the reperfusion-induced increase in $[Ca^{2+}]_i$ (reviewed in 1, 3, 4). This hypothesis has been supported recently by several lines of evidence. Firstly, it has been reported that the increase in $[Na^+]_i$ occurs prior to and is closely correlated to the subsequent increase in $[Ca^{2+}]_i$ (5). Secondly, the increases in $[Na^+]_i$ and $[Ca^{2+}]_i$ associated with reperfusion can be attenuated either by reperfusion with an acidotic buffer (6), or by inhibitors of the Na^+/H^+ exchanger (5, 7). Finally, improved ventricular recovery and a reduction in reperfusion associated contracture has been demonstrated using amiloride or amiloride analogues in rat and guinea pig hearts (8, 9, 10). Conversely, administration of lactate to isolated rat hearts prior to the induction of ischemia significantly reduces recovery of function upon reperfusion; an effect which can be reversed by inhibitors of Na^+/H^+ exchange (11). In addition to salutary effects on function, the antiarrhythmic effects reported to occur in reperfused cardiac tissues treated with amiloride and its analogues may be mediated by the inhibition of Na^+/H^+ exchange (12, 13, 14). Nevertheless, since the maintenance of calcium homeostasis in the heart is extremely complex, many other alterations, which occur in response to ischemia and reperfusion, may contribute to the observed increase in $[Ca^{2+}]_i$ (1). These include depressed sodium/potassium ATPase activity, and increased calcium influx through the voltage dependent slow calcium channel (1). Thus, the exact contribution of the Na^+/H^+ exchange to the elevation in $[Ca^{2+}]_i$ during early reperfusion remains to be elucidated. In this regard, it has been shown previously that realkalization causes a transient positive inotropic effect in cardiac ventricular muscle (15, 16). This response could result from realkalization-induced electrophysiological changes, altered calcium homeostasis or changes in the sensitivity of the myofibrils as a result of changes in pH_i. Collectively, the effects of changes in pH_i during reperfusion of the ischemic myocardium have been termed the pH paradox (4). The present chapter describes the electrophysiological effects of lactate acidosis and realkalization in single cardiac myocytes and their potential relationship to Na^+/H^+ exchange and alterations in $[Ca^{2+}]_i$.

EFFECTS OF LACTATE ACIDOSIS AND REALKALIZATION ON ELECTROPHYSIOLOGICAL ACTIVITY

We chose to study the effects of lactate acidosis in isolation from the other components of ischemia, such as hypoxia, in order to determine more precisely the role of pH changes in the response to ischemia and reperfusion. Although the role of lactate in the pH changes associated with ischemia remains controversial (17), some studies suggest that it may play an important role in the response to these conditions (5, 11, 18). Lactate is produced by the ischemic heart as a result of anaerobic metabolism and previous studies have shown that glycogen depletion prior to ischemia and, as a result, a reduction in lactate production can protect the heart upon reperfusion (18). In addition, lactate has been shown to cause electrophysiological changes (19) and to reduce recovery of the post-ischemic heart (11).

We studied these effects in single isolated guinea pig ventricular myocytes subjected to 5 min of acidosis followed by 30 min of realkalization. The acidotic solution consisted of oxygenated, HEPES buffered Tyrode's solution containing no substrate (glucose, 0) with the addition of 20 mM d,l lactate and pH adjusted to 6.8 with NaOH. Measurements of transmembrane activity were made using standard microelectrode techniques.

Transmembrane activity with respect to resting membrane potential and action potential duration (APD_0, APD_{30} and APD_{60}) remained stable throughout a 50 min perfusion with normal HEPES solution and values for these parameters are similar to those reported by other laboratories for this preparation (Table 1).

Table 1: Action Potential Characteristics of Isolated Guinea Pig Ventricular Myocytes Perfused with HEPES Solution.

Time (min)	Resting Potential (-mV)	APD_0 (msec)	APD_{30} (msec)	APD_{60} (msec)
0	89 ± 2	181 ± 30	219 ± 28	236 ± 37
% of Time Zero				
10	100 ± 1	105 ± 4	103 ± 2	103 ± 2
20	100 ± 1	104 ± 3	102 ± 2	102 ± 2
30	100 ± 1	105 ± 4	107 ± 3	102 ± 2
40	100 ± 1	103 ± 2	103 ± 3	102 ± 2
50	100 ± 1	105 ± 4	103 ± 4	104 ± 3

APD = Action Potential Duration; APD_0, APD_{30}, APD_{60} = APD at 0, -30 and -60 mV, respectively. n = 7.

Representative recordings from a single cell under control conditions following 0, 25 and 50 min of perfusion are shown in Figure 1 (Panels A-C).

The electrophysiological responses to 5 min of lactate acidosis followed by 30 min of realkalization are summarized in Table 2.

Lactate acidosis caused a small, but significant hyperpolarization of the membrane potential following 5 min ($p<0.05$). Despite the small magnitude of the response, it occurred consistently in all cells exposed to acidic conditions. Further, a significant shortening of the APD_0 to 65% of the control value (time=0 min) was seen at this time. No significant changes in APD were seen at either -30 or -60 mV. The shortening of APD in the plateau range of potentials could be the result of attenuation of an inward current, augmentation of an outward current or both (20-22). Nonetheless, the coincident increase in resting membrane potential at this time suggests that an increase in outward current, most likely carried by potassium, could underlie the observed changes in membrane potential and APD. In this regard, internal dialysis of guinea pig ventricular cells with an acidotic solution has been shown to be associated with an increase in outward current at positive potentials as well as an increase in potassium conductance at potentials near to the resting potential

(20). These investigators proposed that the increase in outward current at positive potentials was responsible for the shortening of the APD in response to acidosis (20). In addition, they tentatively identified the increase in potassium conductance at more polarized potentials as the inward rectifier current (20). Both of these findings are consistent with our results for the effects of lactate acidosis on the electrophysiological activity of isolated guinea pig ventricular myocytes. However, it must be noted that our observations were made using standard microelectrode techniques and, therefore, do not provide direct evidence for the role of alterations in specific ionic conductances in the response of these cells to acidosis.

Table 2: Effects of Lactate Acidosis (5 min) and Realkalization (30 min) on Action Potential Characteristics of Isolated Guinea Pig Ventricular Myocytes.

Time (min)	Resting Potential (-mV)	APD_0 (msec)	APD_{30} (msec)	APD_{60} (msec)
0	91 ± 1	165 ± 29	203 ± 32	221 ± 31
% of Time Zero				
Acidosis				
2.5	102 ± 1	109 ± 3	108 ± 2	107 ± 2
5.0	106 ± 1*	65 ± 9*	90 ± 6	95 ± 4
Realkalization				
2.5	99 ± 8	51 ± 8*	64 ± 6*	76 ± 2*
10	96 ± 4	72 ± 15	65 ± 12*	75 ± 12
20	100 ± 2	76 ± 16	81 ± 14	84 ± 11
30	100 ± 3	105 ± 18	88 ± 13	95 ± 6

for abbreviations see Table 1. n = 8 *$p < 0.05$

Membrane potential recovered rapidly upon realkalization, however APD_0 was still significantly shorter than control (time=0 min) after 2.5 min ($p<0.05$, Table 2). APD_0 remained abbreviated at 10 and 20 min, however, these values were not statistically different from control. Full recovery of APD_0 occurred following 30 min of realkalization. In contrast to the lack of effect of lactate acidosis, realkalization resulted in significant shortening of both APD_{30} and APD_{60} following 2.5 min ($p<0.05$, Table 2). In the case of APD_{30}, this value remained significantly abbreviated following 10 min ($p<0.05$, Table 2). The underlying cause(s) of the persistent abbreviation of APD_0 and the realkalization-induced reduction in APD at more polarized potentials is uncertain but is likely related to activation of outward currents. Since early realkalization is also associated with a rapid increase in $[Na^+]_i$ and $[Ca^{2+}]_i$ (1, 5, see below), it is possible that the effects on APD represent activation of one or more cation-dependent outward currents and/or calcium-dependent inactivation of inward current. Our results cannot discriminate between these possibilities, but prevention of the increase in $[Ca^{2+}]_i$ does substantially attenuate the effects of realkalization on changes in APD (see below).

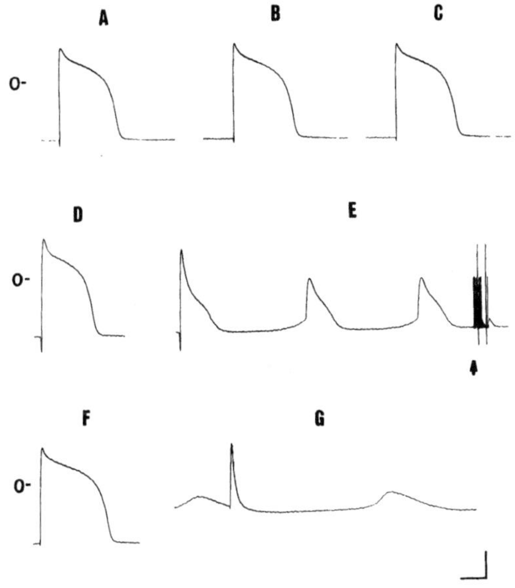

Figure 1: Representative recordings of action potentials obtained from isolated guinea pig ventricular myocytes perfused with well oxygenated, HEPES buffered Tyrode's solution following equilibration for 30 min (Panel A, control) and following 25 and 50 min of continued perfusion (Panels B and C, respectively). Electrophysiological responses of two cells prior to lactate acidosis (Panel D and F) and following 1 (Panel E) and 3 min (Panel G) of realkalization.

Realkalization also resulted in abnormal transmembrane activity. Delayed after depolarizations (DADs) were observed in four of eight preparations studied and occurred early during realkalization. An example of abnormal activity which is likely related to the development of DADs is illustrated in Figure 1E. In the experiment illustrated, repetitive, unstimulated activity coupled to driven activity terminated spontaneously (arrow) but was reinitiated by a subsequent stimulus (Figure 1E). The driven beat was followed by another spontaneous action potential. When the paper speed was increased termination of this spontaneous activity was accompanied by the appearance of a prominent, subthreshold DAD (Figure 1E). In addition to DADs, severe depolarization and spontaneous slow responses were observed in two of eight preparations (Figure 1G). Previously, we have reported the appearance of both DAD and depolarization-induced automaticity in response to simulated ischemia in a variety of preparations (23,24). Our present results suggest that acidosis *per se* may be sufficient to produce these responses upon return to perfusion with normal solution.

ROLE OF Na^+/H^+ EXCHANGE IN THE ELECTROPHYSIOLOGICAL EFFECTS OF ACIDOSIS AND REALKALIZATION

To study the role of Na^+/H^+ exchange in the electrophysiological responses to lactate acidosis and realkalization, we examined the effects of the potent and specific inhibitor hexamethylene amiloride (HMA). Under control conditions neither HMA (8μM) nor methanol (HMA vehicle) had any significant effects on transmembrane electrical activity (data not shown). Acidic conditions resulted in hyperpolarization of the resting membrane potential irrespective of the presence or absence of HMA (Figure 2). However, HMA treated cells remained hyperpolarized throughout realkalization and resting membrane potential was significantly greater than in control cells from 10 to 30 min ($p<0.05$, Figure 2). If an increase in potassium conductance underlies the hyperpolarizing effects of acidosis, as suggested by Sato et al (20), then delay in the recovery from the acid load, which would be expected to be a consequence of inhibition of Na^+/H^+ exchange with HMA, could prolong the effects of lactate acidosis on membrane potential.

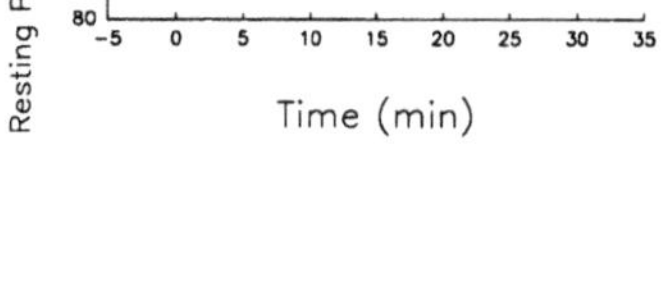

Figure 2: *Time-dependent effects of HMA (8 uM) on the response of resting membrane potential to 5 min of lactate acidosis followed by 30 min of realkalization. n=8. * p<0.05.*

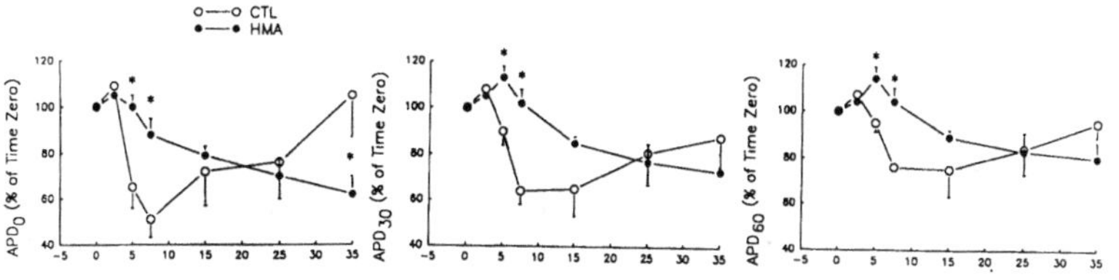

Figure 3. *Time dependent effects of HMA (8 uM) on the APD_0, APD_{30} and APD_{60} of isolated guinea pig ventricular cells exposed to 5 min of lactate acidosis followed by 30 min of realkalization. n=8. *p<0.05.*

In contrast to its lack of effect on acidosis-induced changes in the membrane potential, HMA prevented the shortening of APD_0 following 5 min ($p<0.05$, Figure 3) and caused a small increase in both APD_{30} and APD_{60} at this time (Figure 3). Although the increase in the latter two parameters was small (15%), both were significantly longer than in control cells in which a small reduction in APD_{30} and APD_{60} occurred in response to acidic conditions ($p<0.05$, Figure 3).

The mechanism underlying this effect of HMA is uncertain as it has been reported that Na^+/H^+ exchange is not active under some ischemic conditions (25). However, findings from other laboratories suggest that inhibition of the exchanger during ischemia may protect the reperfused heart by limiting the elevation in sodium and calcium contents during the ischemic period (26). If the enhancement of outward current at positive potentials, reported to occur in response to acidosis (20), is related to increased levels of one or the other of these cations, the ability of exchange inhibitors to prevent their accumulation could account for the actions of HMA on APD in our experiments. Alternatively, a further decrease in pH_i in the presence of HMA during lactate acidosis may cause depression of outward current, as has been suggested in other studies (20). Regardless, the divergent response of membrane potential and APD to HMA in cells exposed to lactate acidosis suggests that the phenonema underlying these two effects may differ with respect to their dependence on the activity of Na^+/H^+ exchange.

Following 2.5 min of realkalization, HMA significantly attenuated the further decrease in APD_0 in comparison to control cells and completely abolished the early shortening of APD_{30} and APD_{60} ($p<0.05$, Figure 3). Thereafter, APD at all voltages decreased and values for these parameters did not differ from untreated cells from 5 to 25 min of realkalization. Surprisingly, HMA prevented the recovery of APD_0 seen following 30 min and APD_0 was significantly shorter than in the untreated cells at this time. A similar effect was noted for APD_{30} and APD_{60} in HMA treated cells, however values for these parameters were not statistically different from control cells at the 30 min time point. The reasons for the abolition by HMA of APD shortening associated with realkalization are uncertain, however, like the effects of this drug during acidosis, they may be related to attenuation of sodium and calcium accumulation reported to occur at this time (1, 5). Indeed, our own results (see below) suggest that HMA does prevent the accumulation of calcium during the early minutes of realkalization. Identification of the current(s) responsible for HMA-sensitive APD shortening during realkalization is beyond the scope of the present study, but it is likely that $[Na^+]_i$-stimulated Ca^{2+} accumulation via Na^+/Ca^{2+} exchange may contribute to an increase in the outward current under these conditions. Such an interpretation is compatible with the suggested role of coupled Na^+/H^+ and Na^+/Ca^{2+} exchanges in the functional and electrophysiological effects of reperfusion and the protective effects of amiloride and its analogues against reperfusion-associated dysfunction (reviewed in 1, 3, 4). Importantly, HMA completely abolished the occurrence of DADs and depolarization-induced slow potentials during realkalization. Since both of these types of abnormal electrophysiological activity are believed to be related to increased $[Ca^{2+}]_i$ (27), it is likely that HMA is acting through a mechanism similar to that described above. Definitive identification of the currents underlying the complex response of ventricular myocytes to acidosis and realkalization will require the use of voltage and current clamp techniques.

EFFECTS OF LACTATE ACIDOSIS AND REALKALIZATION ON $[Ca^{2+}]_i$:ROLE OF NA^+/H^+ EXCHANGE

To examine the role of $[Ca^{2+}]_i$ in the response to lactate acidosis and realkalization, we used suspensions of isolated myocytes which had been loaded with the calcium-sensitive fluorescent compound INDO1-AM. Preliminary experiments were conducted to determine $[Ca^{2+}]_i$ under resting and stimulated conditions. Under our experimental conditions, measurement of $[Ca^{2+}]_i$ was highly reproducible under resting conditions ($[Ca^{2+}]_i=92\pm15$ nM, n=4). In addition, the cells displayed reproducible, concentration-dependent responses to several pharmacologic agents ($BaCl_2$, isoproterenol, KCl, and ouabain, data not shown). In the presence of 10 nM of isoproterenol and 0.4 mM of $BaCl_2$, used to induce spontaneous cell beating, $[Ca^{2+}]_i$ increased to 131 ± 11 nM, n=4). This combination of $BaCl_2$ and isoproterenol was used in all subsequent experiments. Lactate acidosis *per se* caused a small and transient increase ($[Ca^{2+}]_i=108\pm17$ nM, $P>0.05$), followed by a rapid recovery of the calcium signal under resting (unstimulated) conditions. When cell beating was induced under acidotic conditions, $[Ca^{2+}]_i$ (Figure 4A, 135 ± 20 nM) was very similar to that observed under unstimulated conditions. Similarly, preincubation of cells in the normal HEPES (control) solution at 37°C for 5 min did not change the maximum $[Ca^{2+}]_i$ when the cells were resuspended in normal HEPES buffer containing isoproterenol and $BaCl_2$ (Figure 4B). However, preincubation of cells in acidic HEPES solution induced a significant increase in the calcium signal when the cells were resuspended in normal HEPES buffer containing isoproterenol and $BaCl_2$ (Figure 4C). This effect was attenuated if isoproterenol and $BaCl_2$ were added to the media either 1 min (Figure 4D), or 4 min (Figure 4E) following resuspension. Compared to the time control (treatment B), an increase of 63% and 49% in $[Ca^{2+}]_i$ was observed in cell suspensions to which isoproterenol and $BaCl_2$ were added immediately upon realkalization or following 1 min of realkalization, respectively. However, this effect was not significant if cell beating was induced 4 min after resuspension.

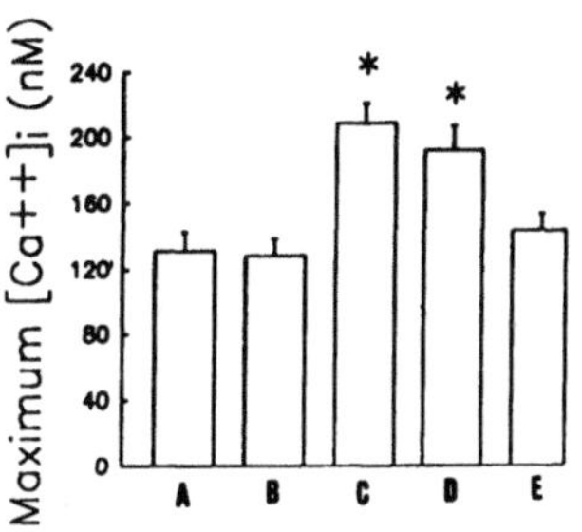

*Figure 4. Effects of realkalization on maximum $[Ca^{2+}]_i$ of myocytes measured using INDO1-AM. Treatment A represents cells exposed to acidic HEPES buffer for 5 min in the presence of $BaCl_2$ and isoproterenol. Treatment B represents cells preincubated with normal HEPES buffer for 5 min in the presence of $BaCl_2$ and isoproterenol and then resuspended in normal HEPES buffer in the presence of $BaCl_2$ and isoproterenol (time control). Treatments C through E represent cells preincubated in acidic buffer for 5 min in the presence of $BaCl_2$ and isoproterenol, then resuspended in normal HEPES buffer with $BaCl_2$ and isoproterenol added immediately (C) or at 1 and 4 min (D and E) following resuspension, respectively. *$p<0.05$ compared to treatments A and B.*

Methanol alone (vehicle control) had no significant effect on $[Ca^{2+}]_i$ in either normal cells or in those which were made acidic followed by realkalization (data not shown). In addition, HMA (up to 8 μM) only slightly decreased maximum $[Ca^{2+}]_i$ ($P>0.05$) in control cells. However, when HMA was present during both acidic preincubation and realkalization, it inhibited the realkalization-induced increase in maximum $[Ca^{2+}]_i$ in a concentration-dependent manner (Figure 5, left panel). If HMA (4 μM) was present during acidic preincubation (Figure 5, right panel, treatment B), or during realkalization (Figure 5, right panel, treatment C) only, the inhibitory effect on the maximum increase in $[Ca^{2+}]_i$ was less but remained statistically significant.

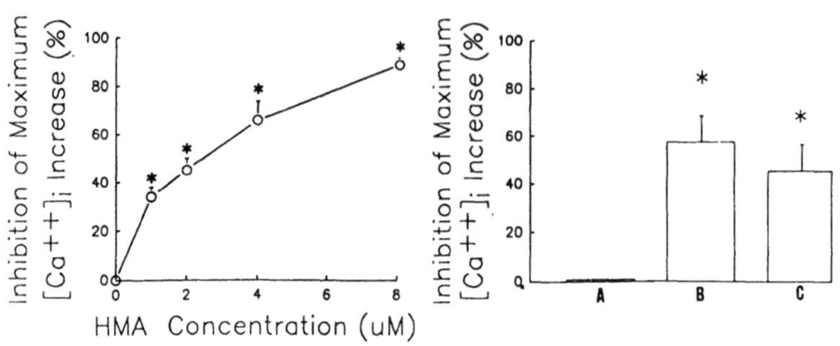

*Figure 5: (left panel) The concentration dependent inhibition of the increase in $[Ca^{2+}]_i$ by HMA in cells preincubated in acidic buffer for 5 min in the presence of $BaCl_2$ and isoproterenol and resuspended in normal HEPES buffer with $BaCl_2$ and isoproterenol added immediately upon resuspension. (right panel) Time dependence of the inhibitory effects of HMA (4 uM) on the increase in $[Ca^{2+}]_i$ in cells preincubated in acidic buffer for 5 min and resuspended in normal HEPES buffer. Treatments A through C represent untreated cells (A, without HMA) and cells treated with HMA during acidosis only (B), or during realkalization only (C). *$p<0.05$*

The present study provides direct evidence that the realkalization-mediated increase in $[Ca^{2+}]_i$ may be the mechanism underlying many of the observed electrophysiological changes. Although several previous studies have attempted to address the mechanisms underlying the acidosis-mediated increase in $[Ca^{2+}]_i$, a definitive cause has not been established with certainty because of the complexity of the problem (28). Under the present experimental conditions, lactate acidosis itself caused a transient increase in $[Ca^{2+}]_i$ in cells under resting conditions, but failed to induce a significant change in $[Ca^{2+}]_i$ when spontaneous beating was initiated during the acidic period. These observations may be explained by the complex interactions between the different effects of acidosis on the level of $[Ca^{2+}]_i$. While changes in the activity of Na^+/H^+ and Na^+/Ca^{2+} exchangers might increase $[Ca^{2+}]_i$, inhibition of slow inward current and of SR calcium release under these conditions could offset these alterations (28). The increase in $[Ca^{2+}]_i$ observed upon realkalization in the present study, however, was critically dependent on the time when spontaneous beating was initiated. This observation provides two important pieces of information. Firstly, the mechanism(s) responsible for the increase in $[Ca^{2+}]_i$ may require cell activation and secondly, the mechanism(s) underlying the increase in $[Ca^{2+}]_i$ may be slowly reversible under unstimulated conditions. Both of these suggestions are consistent with the hypothesis that Na^+/H^+ exchange, and subsequent Na^+/Ca^{2+} exchange activation, represent the mechanism underlying the realkalization-induced increase in $[Ca^{2+}]_i$. Although Na^+/Ca^{2+} exchange can carry Ca^{2+} across the cell membrane in both directions, Ca^{2+} influx via the exchange (reverse mode) may occur only under depolarized conditions where $E_{Na/Ca}$ is negative to the membrane potential

(29,30). This can occur in unstimulated cells which are loaded with Na^+ and are severely depolarized, a condition similar to that of cells which developed depolarization-induced slow responses in the present study. However, it is also thought to occur during the plateau phase of action potential (29). This may explain the dependence of the realkalization-induced effects on $[Ca^{2+}]_i$ on spontaneous beating in well polarized myocytes. It has been demonstrated in several studies that intracellular acidosis activates the Na^+/H^+ exchange (3, 31, 32) and leads to the accumulation of intracellular Na^+ (3, 26, *33*). The accumulation of Na^+, in turn, activates the Na^+/K^+ ATPase (34-36). The increased activity of the Na^+/K^+ pump would relieve the sodium load of the cell and explain the slow reversibility of the realkalization-induced increase in $[Ca^{2+}]_i$ under resting conditions. Therefore, the fast recovery from the acid load observed in the isolated myocytes may be related to the relative frequency of cell beating.

Another possible explanation for the time-dependence of the realkalization effects may be related to the intracellular Ca^{2+} buffering capacity. Since acidosis may lower the intracellular Ca^{2+} buffering capacity (28), $[Ca^{2+}]_i$ may be increased more if cells are activated immediately following acidosis.

The present study addressed the potential relationship between electrophysiologic responses and $[Ca^{2+}]_i$ caused by realkalization. Our results suggest that the HMA-sensitive increase in $[Ca^{2+}]_i$ upon realkalization may represent one of the mechanisms underlying some of the electrophysiological responses, particularly those related to the development of DADs and of depolarization-induced spontaneous activity. In particular, it is important to note that these responses occurred in cells which were well oxygenated, suggesting acidosis *per se* may be both sufficient and necessary to precipitate many of the electrophysiologic changes and alterations in $[Ca^{2+}]_i$ associated with reperfusion.

The potential involvement of the Na^+/H^+ exchange in the realkalization-induced increase in $[Ca^{2+}]_i$ is further supported by the inhibitory effects of HMA, a Na^+/H^+ exchanger inhibitor. In addition, since HMA was effective when it was present either during acidosis or realkalization only, it is likely that Na^+ accumulation via the Na^+/H^+ exchange mechanism occurs both during acidosis and realkalization. These observations provide strong evidence that Na^+/H^+ exchange is activated under these conditions and could, in turn, stimulate Na^+/Ca^{2+} exchange, leading to an increase in $[Ca^{2+}]_i$ upon realkalization. Since realkalization occurs early during reperfusion, this process may play an important role in reperfusion-induced calcium overload and cardiac dysfunction. Therefore, acidic reperfusion and amiloride analogues may be protective as has been reported in a number of studies (5-10). In addition, the paradoxical positive inotropic response reported to occur early during reperfusion is also consistent with this interpretation (37).

We conclude that the HMA sensitive Na^+/H^+ exchange and subsequent Na^+/Ca^{2+} exchange may play an important role in the realkalization-induced increase in $[Ca^{2+}]_i$ and electrophysiological responses.

Acknowledgement

This study was supported by a grant-in-aid to M.P.M. from the Medical Research Council of Canada (MT 9300). M.P.M. is a career investigator of the Heart and Stroke Foundation of Ontario.

REFERENCES

1. Tani M (1990) Ann Rev Physiol 52:543-559
2. Jacobus WE, Taylor GJ, Hollis, DP and Nunally RL (1977) Nature 265:756-758
3. Lazdunski M, Frelin C and Vigne P (1985) Eur J Biochem 149:1-4
4. Karmazyn M and Moffat MP (1993) Cardiovasc Res 27:915-924
5. Tani M and Neely JR (1989) Circ Res 65:1045-1056
6. Panagidopoulos S, Daly MT and Nayler WG (1990) Am J Physiol 258:H821-H828
7. Meng H and Pierce GN (1991) J Pharmacol Exp Ther 256:1094-1100
8. Karmazyn M (1988) Am J Physiol 255:H608-H615
9. Karmazyn M, Ray M and Haist J (1993) J Cardiovasc Pharmacol 21:172-178
10. Moffat M and Karmazyn M (1993) J Mol Cell Cardiol, in press
11. Karmazyn M (1993) Br J Pharmacol 108:50-56
12. Dennis SC, Coetzee WA, Cragoe EJ and Opie LH (1990) Circ Res 66:1156-1159
13. Duff HJ, Brown CE, Cragoe EJ and Rahmberg M (1991) J Cardiovasc Pharmacol 17:879-888
14. Duan J and Moffat MP (1992) Adv Exp Med Biol 311:435-436
15. Serur JR, Skelton R, Bodem R and Sonnenblick EH (1976) J Mol Cell Cardiol 8:823-836
16. Fry CH and Poole-Wilson PA (1981) J Physiol 313:141-160
17. Dennis SC, Gevers W and Opie LH (1991) J Mol Cell Cardiol 23:1077-1086
18. Neely JR and Grotyohann LW (1984) Circ Res 55:816-824
19. Ellis D and Nioreaud J (1987) J Physiol 383:125-141
20. Sato R, Noma A, Kurachi Y and Irisawa H (1985) Circ Res 57:553-561
21. Irasawa H and Sato R (1986) Circ Res 59:348-355
22. Yatani A and Goto M (1983) Jpn J Physiol 33:403-415
23. Duan J and Moffat MP (1990) Naunyn-Schmiedeberg Arch Pharmacol 342:342-348
24. Moffat MP and Tsushima RG (1989) Can J Physiol Pharmacol 67:929-935
25. Imai S, Shi AY, Ishibashi T and Nakazawa M (1991) J Mol Cell Cardiol 23:505-517
26. Anderson SE, Murphy E, Steenbergen C, London RE and Cala PM (1990) Am J Physiol 259:C940-C948
27. January CT and Fozzard HA (1988) Pharmacol Rev 40:219-227

28. Orchard CH and Kentish JC (1990) Am J Physiol 258:C967-C981
29. Ravens U and Wettwer E (1989) J Cardiovasc Pharmacol 14 (suppl 3):S30-S35
30. Reiter M (1988) Pharmacol Rev 40:189-217
31. Kim D and Smith TW (1987) Am J Physiol 253:C137-C146
32. Kim D and Smith TW (1988) J Physiol 398:391-410
33. Piwnica-Worms D, Jacob R, Horres CR and Lieberman M (1985) J Gen Physiol 85:43-64
34. Lazdunski M, Frelin C and Vigne P (1985) J Mol Cell Cardiol 17:1029-1042
35. Piwnica-Worms D, Jacob R, Shigeto N, Horres CR and Lieberman M (1985) J Mol Cell Cardiol 85:43-64
36. Rasmussen HH, Cragoe EJ and Ten Eick RE (1989) Am J Physiol 256:H256-H264
37. Shouten VJA, Los GJ, Kuypers CJ, Brinkman J and Huysmans HA (1991) Am J Physiol 260:H89-H99

CHAPTER 7

CARDIAC CELL LINES: ELECTROPHYSIOLOGICAL PROPERTIES AND IMPLICATIONS IN CARDIAC DISEASE

Adrian Sculptoreanu[*]

Dept of Pharmacology, University of Washington, Seattle, Washington, USA

INTRODUCTION

Permanent cell lines with or without inducible differentiation have been a valuable asset in the study of skeletal and smooth muscle physiology, biochemistry, and myogenesis (1-12). Immortalized cell lines from dysgenic mouse skeletal muscle (13) may become excellent transient systems for expression of skeletal muscle calcium channels (14). Cardiac cell lines have been more difficult to obtain and to date only a small number has been thoroughly characterized (15-23).

The development of the heart follows two stages: *hyperplasia* is the proliferative stage of heart development and takes place during embryogenesis. During the postnatal stage of development, cardiomyocytes stop dividing but continue to enlarge, the result is an increase in cardiac mass (*hypertrophy*, see Ref. 24).

There is ample evidence which suggests that terminally differentiated cardiomyocytes have only a limited ability to reenter cell cycle. Consequently, one

[*] Present address:
Lady Davis Institute for Medical Research, McGill University, Montreal, Quebec H3T 1E2

of the strategies used to develop cardiac cell lines was based on the ability of certain proto-oncogenes to induce cardiac tumors (15, 23, 25) and maintain mitogenesis in isolated cells. However, some mitogens and some oncogenes are known to profoundly alter expression of muscle specific genes, particularly ion channels (26, 27).

A possible inference is that cardiomyopathies of viral and genetic origin may result from changes in ion channel expression in the cardiac tissue (see Bkaily, this book). Indeed, proto-oncogene (myc) expression has been observed in hamsters with idiopathic cardiac myopathy (28). Similarly, Swain et al. (29) reported that expression of c-myc nuclear oncogene in transgenic mice was accompanied by cardiac hyperplasia, though in this instance it appeared to be limited to embryonic development. This chapter will summarize the strategies used in obtaining cardiac cell lines and emphasize published electrophysiological properties of these clonal lines.

STRATEGIES USED TO PRODUCE CONTINUOUS CELL LINES WHICH MAINTAIN CARDIAC PROPERTIES

At the turn of the century Burrows (30) and Carrel (31) reported serial culturing of cardiac cells derived from an adult chicken. Subsequently, serious attempts to produce stable continuous lines of cardiac cells have only been sporadic (17, 32, 33). More recently, the demonstrated ability of certain oncogenes to induce tumors in mammalian (15, 25, 34) and avian (19, 35) heart has rekindled the interest in immortalizing cardiac cell lines.

Gigardi cell line (17) is probably fibroblastic in origin and will not be discussed further. Three methods have been successfully employed to produce cardiac cell lines: serial passaging, transfection of cells with oncogenes and controlled differentiation of multipotent stem cell lines.

Selective Serial Passaging

Kimes and Brandt took advantage of the successful use of the selective serial passage method to establish clonal smooth muscle (36) and clonal skeletal muscle (37) cell lines and produced several clonal cell lines derived from embryonic BDIX rat heart. One of these cell lines, designated as H9c2, was initially judged to be more skeletal than cardiac in behavior. The spindle shaped cells which grew in organized arrays in confluent cultures eventually formed multinuclated myotubes reminiscent of primary cultures of skeletal muscle. At this stage, the cells expressed skeletal muscle specific enzymes: myokinase and creatine phosphokinase. Electrophysiologically, the H9c2 cells triggered regenerative action potentials in response to iontophoretic application of acetyl choline (ACh). Characteristic of skeletal muscle nicotinic receptor, the response to ACh was blocked by curare and -

neurotoxin (33). More recently, electrophysiological (18, 22) and biochemical (18) studies questioned the skeletal muscle characteristics of this cell line. This point will be further expanded in a later section.

Oncogenes and Oncogene Induced Cardiac Tumors

Induction of Cardiac Tumors. Two groups (38, 39) have independently produced transgenic mice using fusion genes of an SV40-T antigen linked to either atrial natriuretic factor (ANF) promoter (23, 34, 39) or protamine promoter (15, 25, 38). In both cases the animals developed right atrial tumors even though both atria expressed the T-antigen. The reason for this may be a yet unidentified trophic factor. Although in the first instance (39) atria were specifically targeted with the ANF promoter, the development of right atrial tumors was an unexpected result in the protamine-SV40 transgene experiments (38). Cardiac rhabdomyosarcomas of avian heart were induced by *in ovo* injection of quail embryos with a myelocytomytosis virus strain (MC29).

Restoration of Proliferative Ability by Oncogenes. Another group (40) has demonstrated that expression of SV40 T-antigen in primary cultures of neonatal rat heart subsequent to infection with a human adenovirus, restores the proliferative ability of the cells in cultures. The cell cultures expressing the T-antigen continued spontaneous beating activity suggesting maintenance of a differentiated state.

Establishment of Cardiac Cell Lines. The ability to produce transgenic animals which develop cardiac tumors led to establishment of mammalian (38) and avian (19) clonal cell lines which maintain a differentiated state and some degree of proliferation potential in cultures.

Though the quail heart cell lines (19) lose their differentiated state when actively dividing, a precontractile differentiated state can be reestablished in co-cultures with mammalian 3T3 cells when the heart cells have slower division rates. This observation stresses the inverse relationship between the dividing rate of cells and the extent of cellular differentiation.

Gardside and Hauschka (38) derived permanent mouse cardiac muscle cell lines (MCM1, MCM2) from right atrial tumors of transgenic mice. The proliferation rate and differentiation state of these cell lines is highly sensitive to cell culture density. The cells survive passaging in transplants of high density monolayer cultures. Highly differentiated, spontaneously contracting cell clusters can be generated from these monolayer cultures. The monolayer cultures retain several cardiac characteristics: expression of cardiac specific sodium channels (21), ANF production, cardiac muscle specific muscarinic receptors (38) and response to -adrenergic agonist, isoproterenol (21), and the action of cAMP-phosphodiesterase inhibitor-the IBMX (21). The muscarinic receptors were shown to be the cardiac subtype by pharmacological and immunological analyses, and were functionally active as demonstrated by their ability to activate phospholipase C (M. Morton and N. Nathanson, personal communication).

Controlled Differentiation of Pluripotent Embryonic Stem Cells. Rudnicki et al. (20) reported that P19 embryonic carcinoma cells, a multipotent stem cell line (41) can be induced to differentiate into striated muscle cells by treatment with dimethyl sulfoxide (DMSO). Differentiated cells were mononucleated and had many cardiac characteristics, including: expression of atrial and ventricular specific myosin, spontaneous contractile activity, and junctional complexes connecting adjacent cells (20).

This cell line appears to have some value in studies of cardiac muscle differentiation and regulation of muscle specific gene expression. Since the induced cardiac phenotype seems to continue differentiating with time in culture (20), it would be interesting to know if expression of cardiac specific ion channels follows a specific temporal pattern.

ELECTROPHYSIOLOGICAL PROPERTIES OF ESTABLISHED CARDIAC CELL LINES

Expression of ion channels has been studied in only two clonal cardiac cell lines. The electrophysiological characteristics of these cell lines will be briefly reviewed here.

The MCM1 Cell Line

This cell line was derived from atrial tumor of a transgenic mouse as described in a previous section. Whole cell patch clamp technique was used to study the electrophysiological properties of single cells in low density long term cell cultures (21). In physiological solutions the dominant current was a sodium current which was completely tetrodotoxin insensitive (21). Relatively small delayed outward potassium currents were also observed in these cells and could be largely inhibited by Cs^{+} substitution of K^{+} ions or 10 to 20 mM of tetraethyl ammonium (TEA). Functional expression of L-type calcium channel currents was not detected in these cells after complete substitution of sodium with TEA^{+} and using 10 mM Ca^{++} as a charge carrier (personal observations, 21). With respect to ion channel expression, the MCM1 cells resemble early embryonic heart cells (42, 43). Similar to MCM1 cells, cells in cultures from 3-day-old chick embryo heart have small potassium currents, large TTX-insensitive sodium currents, and lack any measurable calcium currents despite the spontaneous contractile activity (personal observations). In mammalian heart, TTX-insensitive sodium currents are progressively replaced by sodium currents which are blocked by micro-molar concentrations of TTX during

embryogenesis and postnatal development (personal observations).
Abnormal expression of embryonic type TTX-insensitive sodium channels has been associated with Duchenne skeletal muscle dystrophy (44) and genetic cardiomyopathy of Syrian hamster (45, see Bkaily, this book). The MCM1 cell line may provide a useful system in which to study factors controlling expression of ion channels in muscle development and certain muscle pathologies.

The H9c2 Cell Line

Reassessments of H9c2 cell line as cardiac. Hescheler et al (18) questioned the early characterization of H9c2 cell line as having a skeletal muscle phenotype (33). Although the H9c2 cells lost some cardiac phenotype properties such as expression of gap junctions, caveolae, t-tubules or myofibrils, they maintained other cardiac-like characteristics. These included the presence of a rich rough endoplasmic reticulum, and the failure to fuse in multinuclated myotubes as earlier reported (33). H9c2 cells also had lectin binding pattern which was cardiac specific (18) and expressed a complex of G-proteins typical of striated muscle cells (18). Hescheler et al. (18) also reported expression of L-type calcium channel currents which kinetically and pharmacologically seemed to resemble cardiac L-type Ca-channel currents. Expression of these currents required high cell densities and long times in culture. The cardiac origin of L-type Ca-channel currents was confirmed in single channel studies, based on single channel conductances (22), though this later study also detected skeletal muscle specific Ca-channel currents.
Therefore, it appears that, in many respects, the H9c2 cell line has lost some of the cardiotypic properties that are still present in MCM1 cell line (see also Ref. 22). Whether the H9c2 cells also express embryonic heart type, TTX-insensitive Na-channel currents as reported for MCM1 cell line (21), remains to be established.

CONCLUDING REMARKS

The discovery that myocytes from atrial tumors of transgenic mice can maintain proliferative ability while at the same time retaining cardiac phenotype characteristics has led to the development of a new cardiac cell line (MCM1). Electrophysiologically, the MCM1 cardiac cell line appeared to have reverted to an early embryonic stage of development. However, it is not yet clear whether the lack of measurable L-type Ca channel expression in these cells is due to an altered balance of modulatory mechanisms of channel function (protein kinases and phosphatases) or takes place at the level of gene regulation. The expression of one proto-oncogene, c-myc, has been previously reported to delay appearance of L-type calcium channel currents in the BC3H1 muscle cell line (26). Over-expression of c-myc proto-oncogene, which normally stimulates DNA-polymerase and is turned off in adult heart, has been reported to precede the development of cardiac lesions in certain strains of cardiomyopathic Syrian hamster (28) and accompany cardiac hypertrophy (46). At the same time the pathogenesis of Syrian hamster cardiomyopathy may be the result of abnormal expression of cardiac ion channels (see Bkaily, this book). These recent observations may encourage future development and use of cardiac cell lines in several areas of investigation: studies of cardiac myogenesis, pathogenesis of genetic and viral cardiomyopathies, factors controlling expression of cardiac specific ion channels, and techniques to control regeneration of damaged cardiac tissue.

Acknowledgments

The author thanks Drs. W. A. Catterall, T. Scheuer and S. Hauschka for helpful and critical comments on the manuscript and Dr. W. A. Catterall for his support and advice during my post doctoral fellowship in his laboratory. During preparation of this chapter, the author was supported by a Research Fellowship from the Heart and Stroke Foundation of Canada and by NIH Research Grant PO-1 HL 44948, William A. Catterall , Principal Investigator.

REFERENCES

1. Antin PB and Orahl CP (1991) Dev Biol 143:111-121
2. Blau HM, Pavlath GK, Hardeman EC, Chiu CP, Silberstein L, Webster SG, Miller SC, and Webster C (1985) Science 230:758-766
3. Brennan TJ, Edmondson DG, and Olson EN (1990) J Cell Biol 110:929-937
4. Caffrey JM, and Farach MC (1988) Mol Pharmacol 34:518-526
5. Lim RW, and Hauschka SD (1984) Dev Biol 105:48-58
6. Miller JB (1990) J Cell Biol 111:1149-1159
7. Muscat GEO, Gobius K, and Emery J (1991) Mol Endocrin 5:802-814
8. Nacatsuji Y, Hidaka K, Tsujino S, Yamamoto Y, Mukai T, Yanagihara T, Kishimoto T, and Sakoda S (1992) Mol Cell Biol 12:4384-4390
9. Poyner DR, Andrew DP, Brown D, Bose C, and Hanley MR (1992) Br J Pharmacol 105:441-447
10. Schreier T, Kedes L, and Gahlmann (1990) J Biol Chem 265:21247-21253
11. Giannatasio B, Jones SW, and Scarpa A (1991) J Gen Physiol 98:987-1003
12. Zheng W, Rampe D, and Triggle DJ (1991) Mol Pharmac 40:734-741
13. Pinçon-Raymond M, Vicart P, Bois P, Chassande O, Romey G, Varadi G, Li ZL, Lazdunski M, Rieger F, and Paulin D (1991) Dev Biol 148:517-528
14. Tanabe T, Beam KG, Powell JA, and S Numa (1988) Nature 336:135-139
15. Behringer RR, Pechon JJ, Messing A, Gartside CL, Hauschka SD, Palmiter RD, and Brinster RL (1988) PNAS USA 85:2648-2652
16. Borchgrevniok PC and Ryan MP (1988) Br J Pharmacol 95:614-618

17. Girardi, AJ, Warren J, Goldman C, and Jefferies (1958) Proc Soc Exp Biol Med 98:18-23
18. Hescheler J, Meyer R, Plant S, Krautwurst D, Rosenthal W, and Schultz G (1991) Circ Res 69:1476-1486
19. Jaffredo T, Bachnou N, and Dieterlen-Liévre F (1991) Exp Cell Res 92:481-491
20. Rudnicki MA, Jackowski G, Saggin L, and McBurney MW (1990) Dev Biol 138:348-358
21. Sculptoreanu A, Morton M, Gartside CL, Hauschka SD, Catterall WA, and Scheuer T (1992) Am J Physiol 262:C724-C730
22. Sipido KR and Marban E (1991) Circ Res 69:1487-1499
23. Steinhelper ME, Lanson NA, Dresdner KP, Delcarpio JB, Wit AL, Claycomb WC, and Field LJ (1990) Am J Physiol 259:H1826-H1834
24. Zak R (1984) Growth of the Heart in Health and Disease, New York Raven Press, pp. 1-24
25. Gardner DG, Camargo MJF, Behringer RR, Brinster RL, Baxter JD, Atlas SA, Laragh JH, and Descheper CF (1992) Am J Physiol 262:E524-E531
26. Caffrey JM, Brown AM, and Schneider MD (1987) Science 236:570-573
27. Estacion M (1990) J Membr Biol 113:169-175
28. Deguchi Y, Zuma J, Mamaguchi T, Kurimoto T, Sawamura A, Awata N, Kishimoto S, and Omishi S (1988) J Mol Cell Cardiol 20:801-809
29. Swain JL, Stewart TA, and Leder P (1987) Cell 50:719-727
30. Burrows MT (1910) J Am Med Ass 55:2057-2072
31. Carrel A (1914) J Exp Med 20:1-14
32. Schneider H, Shaw MW, Muirhead EE, and Smith A (1965) Exp Cell Res 39:631-636
33. Kimes BW, and Brandt BL (1976) Exp Cell Res 98:367-381
34. Field LJ (1988) Science 239:1029-1033
35. Saule S, Mérigaud JP, Al-Moustafa AM, Ferré F, Rong PM, Amouyel P, Quatannens B, Stehélin D, and Dieterlen-Liévre F (1987) PNAS USA 84:7982-7986
36. Kimes BW and Brandt BL (1976) Exp Cell Res 98:349
37. Yaffe D (1968) PNAS USA 61:477
38. Gardside CL and Hauschka SD (1991) in The Development and Regenarative Potential of Cardiac Muscle, Edts Oberpriller, Harwood Acad Press NY pp 365-384
39. Steinhelper ME and Field LJ (1991) in The Development and Regenarative Potential of Cardiac Muscle, Edts Oberpriller, Harwood Acad Press NY pp 385-397
40. Sen A, Dunnmon P, Henderson SA, Gerard RD, and Chien KR (1988) J Biol Chem 263:19132-19136
41. McBurney MW and Rogers BJ (1982) Dev Biol 89:503-508
42. Bkaily G, Jacques D, Yamamoto T, Sculptoreanu A, and Payet MD (1988) Can J Pharmacol 66:1017-1022

43. Bkaily G, Jacques D, Sculptoreanu A, Yamamoto T, Carrier D, Vignault D, and Sperelakis N (1991) J Mol Cell Cardiol 23:25-39
44. Bkaily G, Jasmin G, Tautu C, Prochek L, Yamamoto T, Sculptoreanu A, Peyrow M, Jacques D (1990) Muscle & Nerve 13:939-948
45. Jasmin G, Pasternac A, Bkaily G (1990) In: Calcium Channels: their Properties, Functions, Regulation and Clinical Relevance (Edt) Hurwitz L, Parridge D, and Lech J Telford Press
46. Starsken NF, Simpson PC, Bishopric N, Coughlin SR, Lee WMF, Escondo JA, Williams LT (1986) PNAS USA 83:8348-8350

CHAPTER 8

FETAL-LIKE SLOW Na^+ CHANNELS IN DUCHENNE'S MUSCULAR DYSTROPHY

Ghassan Bkaily, Adrian Sculptoreanu and Gaétan Jasmin

*Department of Physiology and Biophysics,
Faculty of Medicine, University of Sherbrooke,
Sherbrooke, Quebec, Canada
and Department of Pathology,
Faculty of Medicine, University of Montreal
Montreal, Quebec, Canada*

Pathophysiology of Duchenne's muscle dystrophy (DMD)

DMD is one of the most common and severest form of SM dystrophy that affects humans (for review see 16, 17). The basic problem in DMD is the lack of knowledge of the causes of the disease and the exact nature of events during initiation and progression of the segmental necrosis as a result, patients with DMD still face a break future (16). Increased $[Ca]_i$ in DMD has been well established. If it were known what caused this increase of $[Ca]_i$, whether it be accelerated transport through voltage-dependent channels, alteration in Na^+, or the activity of sarcolemmal Ca-ATPs, it would then be possible to develop specific drugs to prevent the accumulation of Ca^{2+} ions. Increased $[Ca]_i$ could stimulate several pathways including certain proteases, phospholipase A2 and calmodulin (which increases in dystrophic muscle) (18). The latter was reported to activate Ca^{2+} channel in heart (19).

Necrosis of SM in the dystrophic hamster can be delayed by treatment with Ca^{2+} -antagonist drugs (20, 21). Unfortunately, Ca^{2+} channel blockers of different chemical classes and, possibly, mechanisms of action have been ineffective in alleviating the clinical condition (16). It is possible that the lack of effect of Ca^{2+}-antagonist drugs in DMD could be due to a change in the pharmacology of the Ca^{2+} channels in DMD and/or to activation of a TTX- and Mn^{2+} insensitive slow I_{Na}. The results in the human myotube of DMD (11) and young embryonic heart (9, 10) support this idea. Results using the patch clamp technique on the myotube of a human DMD culture (11), the kinetics (but not the pharmacology) of both the Ca^{2+} and Na^{+} channels seem to be different when compared to normal SM myotube. The presence of a TTX- and Mn^{2+} - insensitive slow I_{Na} in DMD seems to resemble the slow I_{Na} found in early embryonic heart (9, 10). Animal models of muscular dystrophy have been useful in judging the process of muscle weakness, wasting and necrosis, as well as in testing treatment modalities which might interfere with or alleviate the dystrophic process (21-27).

Animal models and DMD

The debate continues regarding the relevance of different animal models to human forms of muscular dystrophy (MD) and the controversy will probably continue until specific biochemical defects that can be used as a basis of comparison between the human and animal forms of MD are identified (28). Most of the animal models of muscular dystrophy show early abnormalities in $[Ca]_i$ accumulation and leakage of HRP as in DMD (for review, see ref. 28). The hamster and mouse show changes of potential relevance toward an understanding of the pathogenesis of individual muscle fiber breakdown in human neuromuscular diseases. Bulfield et al. (29) discovered a x-linked mutant in an mdx mouse which had a mild form of muscular dystrophy. It seems that the gene locus affected was homologous to the one responsible for DMD (29). Also using this animal model, Marshall et al. (29) reported a similarity between the dystrophic mdx mouse, the dystrophic hamster and DMD.

Cultured myotube of SM

The use of the cultured myotube in the study of SM electrophysiology would be desirable due to the absence of connective tissue in the cell culture. This allows direct access to the sarcolemma without enzymatic treatment. Also, the fast I_{Na} can be better clamped and the seal is more easily achieved. After fusion to form myotubes, the cells are treated for 1 day with 10^{-6}M colchicine, which causes them to assume a spherical shape and are then returned to normal medium before they are used (30). This spherical shape of the cells allows the use of the whole-cell voltage clamp and patch clamp techniques. Electrical recordings have first been reported in cultured SM cells of the chick (31), L6 cells (32), rat (6, 7, 14, 15, 33), mouse embryo (34), hamster (18) and adult human (35-37). These preparations maintained their electrophysiological properties for several weeks while in culture. The use of human SM cells in culture eliminates the investigator's dependency on the availability of fresh biopsy specimens. Several samples can be made from one biopsy specimen. This increases the reproducibility of data from the same patient and allows the use of several experimental conditions.

Electrophysiology of normal skeletal muscle

The transmembrane resting potential (E_m) and the action potential (AP) are a direct result of the special properties of the cell membrane which include its selective ion permeability and ion-pumping and ion-exchanging properties (for review 1-3). Electrical excitation in skeletal muscle (SM) involves voltage- and time-dependent changes of permeability to Na^+, K^+ and Ca^{2+}. It induces a fast transient inflow of Na^+ into the cells followed by a slow transient inflow of Ca^{2+} and later by an outflow of K^+. As a result of these ionic movements, the AP is generated. The E_m of normal SM is about -80 mV in mammals and about -90 mV in amphibians. The E_m and excitability are not immediately dependent on the Na-K pump, however they are ultimately dependent on it (2). Any agent that affects the E_m has important repercussions on the AP.

Voltage-dependent Ca^{2+} channel

The voltage-dependent Ca^{2+} channel (I_{Ca}) of SM fibers is localized in the T-tubule system of the cell membrane. The SM Ca^{2+} channel is the object of intense efforts at the level of biochemical purification and was the first Ca^{2+} channel to be cloned and sequenced by the group of Numa (12). The physiological role of the plasma membrane L-type Ca^{2+} channel is not known. The slow Ca^{2+} channel behaves in much the same way as the fast Na^+ channel with respect to inactivation, the main difference being the voltage range over which the slow channel is inactivated: -45 to 0 mV for the Ca^{2+} channel. Another major difference is that the slow Ca^{2+} channel inactivates much more slowly than the fast Na^+ channel. The permeability sequence of the Ca^{2+} channel is $Ba^{2+} > Sr^{2+} > Ca^{2+} > Mn^{2+} > Mg^{2+}$. Other I_{Ca} types in SM that are more likely to be opened during action potentials were reported (14). In addition to the classical slowly-activating current, a I_{Ca} that activates at more negative potentials, is activated and inactivated quickly and is not blocked by dihydropyridine drugs (14, 15). The significance of this current in adult fibers is questionable since it seems to disappear during development. An understanding of the role of the Ca^{2+} channel in SM is beyond the scope of muscle physiology alone. The functional role of SM plasma membrane Ca^{2+} channels (for the time being) has been brought into sharper focus by biochemical evidence which has shown that the Ca^{2+} channel density is unusually high in SM. Therefore, the plasma membrane Ca^{2+} channel could play an important role in SM.

The slow inward current carried by barium ions in normal human myotube is activated near -20 to -15 mV, the peak I/V curve was near +5 mV and the current reversed at +50 mV (11). This current was blocked by verapamil, Mn^{2+}, (-)D888 and cobalt. Also, a TTX- and Mn^{2+} insensitive slow I_{Na} was reported in these cells (11).

The fast Na^{2+} channel (I_{Na})

The fast Na^{2+} channel (I_{Na}) is inactivated over a voltage range of -80 to -50 mV. The I_{Na} can be described by a m^3 h relationship. The half-inactivation potential of this channel ranges from -70 to -80 mV. The permeability sequence of the Na^+ channel is: Na^+ > Li^+ > hydroxylammonium > hydrazinium > ammonium > guanidinium > K^+. The fast I_{Na} has two gating mechanisms: inactivation (I-gate) and activation (A-gate). These two processes can be dissociated chemically and pharmacologically. Inactivation is blocked by tetrodotoxin (TTX) and reduced by external application of 0.7 mM N. bromoacetamide or by reduction of the $[pH]_i$ to pH 4. The activation process is little affected, perhaps because the gate molecules are deeply buried inside the membrane (4). Using the patch clamp technique, it was demonstrated that the gating kinetics of individual Na^+ channels are heterogenous in SM, and that they may occasionally be changed in a single channel (5). The single channel current conductance was found to be 18 pS (6). Weiss and Horn (7) reported the presence of two classes of I_{Na} in the SM of the developing rat. One class had a lower single-channel conductance, activated at more hyperpolarized voltages and was more resistant to TTX than the other. The TTX-resistant I_{Na} resembled the fast transient, slow I_{Na} found, in young embryonic heart (8-10). Also, this channel was found in normal and Duchenne's muscular dystrophy (DMD) human myotube (11). The channel was blocked by the Ca^{2+} blocker (-)D888. However, the same channel in DMD myotube, was insensitive to (-)D888 (11).

Fetal slow channel in DMD

A sodium channel which is resistant to the specific fast Na^+ current (I_{Na}) blocker, TTX, has been described as an immature (8-10) and primitive (38) element of excitable cells. Once the cells are mature, I_{Na} becomes TTX-sensitive (2, 7, 8, 16). This change in the sensitivity to TTX could be explained by a gradual shift in the affinity of the channels (30) or by the existence of two classes of channels (7, 8, 38) during the ontogeny of the cell. A TTX-resistant I_{Na} was found in many cell types. In neurons, but not in early embryonic heart (8, 10), this TTX-resistant I_{Na} is inhibited by the inorganic calcium blocker Mn^{2+} (39, 40). Electrophysiological studies have indicated that the APs in DMD did not differ from those in normal skeletal muscle (36) and that all current types had similar characteristics in normal and DMD myotubes (41). However, calcium channels were reported to be absent in human skeletal muscle cultures (41).

Using the whole-cell voltage-clamp technique, it was indicated that similarities in the characteristics of the whole Na^+ current may exist in normal and DMD myotube cultures. However, the steady-state inactivation curves of the whole I_{Na} seem to be different in DMD myotube cultures when compared to those of normal myotube cultures and this difference is statistically significant (11).

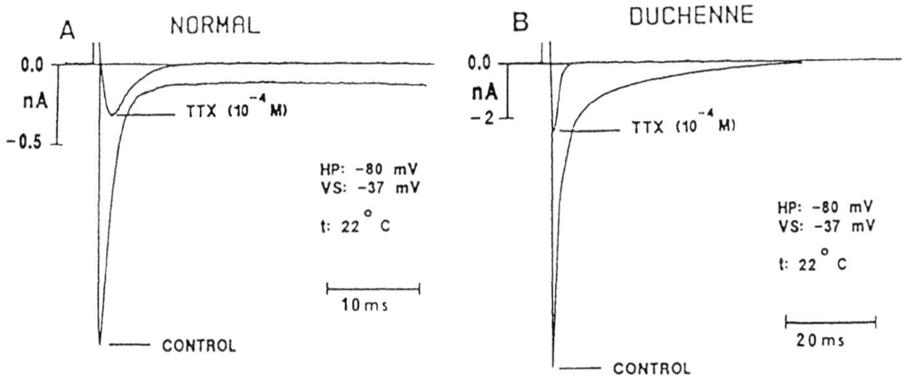

Figure 1. *Effect of TTX on the whole Na^+ current in normal and DMD myotube cultures (A) the whole Na^+ current of a normal myotube culture (lower trace) was recorded normal from a holding potential (HP) of -80 mV to a voltage step (VS) of -37 mV; the inward current showed early and late components. Adding 10^{-4}M TTX decreased the early peak I_{Na} amplitude and completely blocked the late long lasting component. No further decrease was seen after 5 minutes in the presence of TTX. (B) in DMD myotube cultures, the early whole Na^+ current peak amplitude (lower trace) was decreased by 10^{-4}M TTX. Another late component was completely blocked by the toxin. No further decrease of the early component was observed after 3 minutes of superfusion with TTX (upper trace). (Modified from Bkaily et al. (11)).*

In both normal and DMD myotube cultures (Fig. 1) there is an early and a late distinct component of the whole Na^+ current. The late component of the whole Na^+ current may represent the tubular membrane current (42-44). Since TTX largely decreased the amplitude of the whole early Na^+ current and completely blocked the late I_{Na}, it is possible that the TTX-resistant Na^+ channel may be a surface sodium current. The presence of a late Na^+ current and of a slow inward current in normal and DMD myotube cultures may suggest a functional tubular system in these preparations.

The TTX-resistant Na^+ current in normal myotube cultures is sensitive to calcium blockers (Fig. 2A). This is quite reminiscent of that reported in neurons

(45). The presence of this channel in the adult human myotube is not unusual. However, the presence of the TTX- and Mn^{2+} -insensitive I_{Na} only in DMD myotube cultures is unusual because this current has never been detected in normal myotube cultures and was reported to be present (or active) in early embryonic heart cells (8-10) and cardiomyopathic hamster (Chapter 2). The TTX- and Mn^{2+} -resistant I_{Na} in DMD myotube (Fig. 2B) seems to share some characteristics with the immature I_{Na} in young embryonic hearts (8-10).

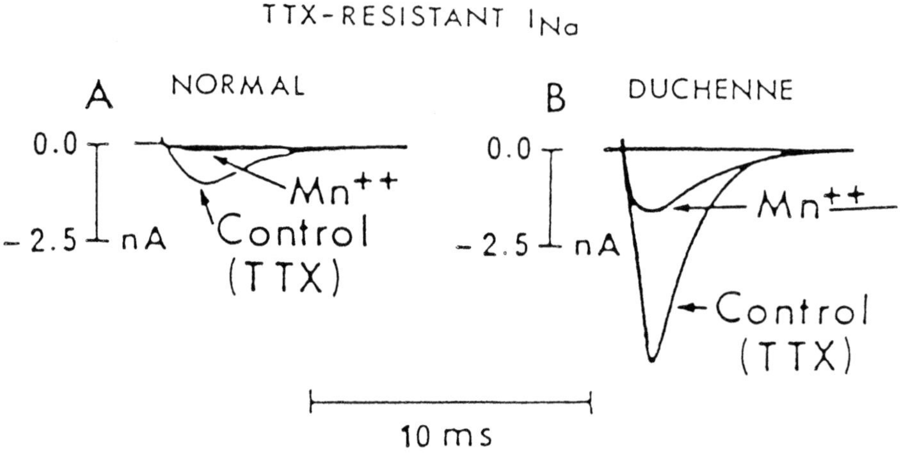

Figure 2. (A-B) Whole-cell Na^+ currents in normal (A) and Duchenne's (B) skeletal muscle myotubes treated with 10^{-4}M TTX for 30 minutes. (A) In normal myotube cultures, the addition of 2 mM Mn^{2+} completely blocked the TTX-resistant I_{Na} within 2 minutes. (B) TTX-resistant I_{Na} current trace in a DMD myotube (lower trace); addition of 2 mM Mn^{2+} decreased this current and no additional change was detected for a higher concentration of Mn^{2+} (4 mM). HP: holding potential; temperature: 22°C. (Modified from Bkaily et al. (11)).

In DMD myotube cultures, the threshold potential peak I_{Ba}/V curve amplitude, and reversal potential of the slow inward current were shifted toward more positive potentials (Fig. 3). This could be partially due to the shift of the inactivation curve to a more positive potential. The bell-like shape of the slow inward current inactivation curve in DMD (Fig. 4) is unusual and may suggest that complex kinetics of the slow channel may take place in DMD. The fact that the reversal potential of I_{Ba} in DMD myotube cultures is more positive than that of normal myotube cultures suggested that the selectivity of the slow channel may undergo a change during the development of DMD. Also, the fact that Mn^{2+} and (-)D888 blocked the slow inward barium current in both normal and DMD myotube cultures suggested that the pharmacology of this channel seemed to undergo no change during the development of DMD. Since the slow inward Ba^{2+} current in DMD is blocked by Mn^{2+}, it is more likely that the TTX- and Mn^{2+}- insensitive Na^+ channel protein is different from that of the slow Ca^{2+} channel protein.

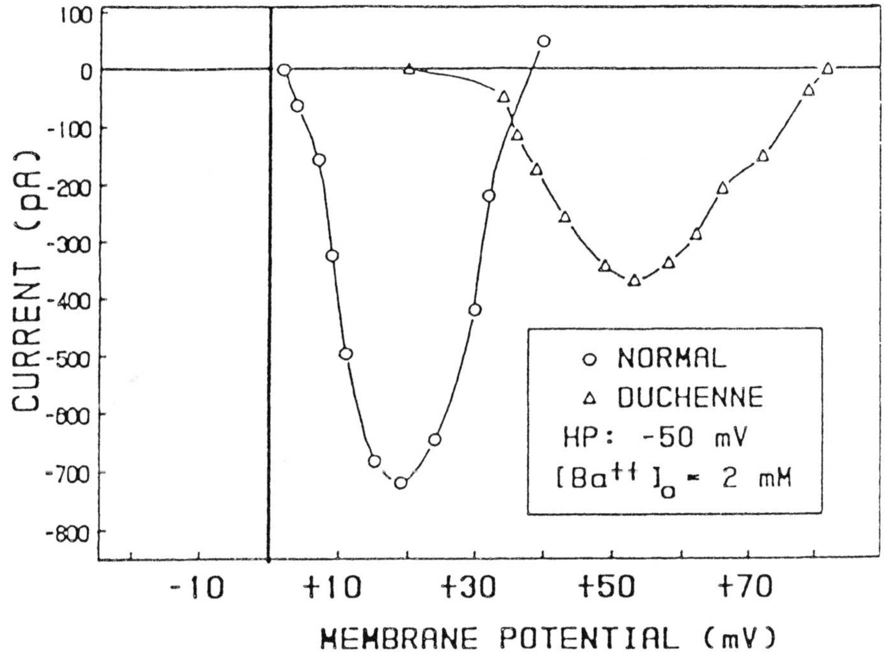

Figure 3. *Current/voltage (I/V) relationship of the slow inward Ba^{2+} current in normal (open circles) and DMD (open triangles) myotube cultures; the threshold potential, the peak amplitude of the I/V curve and the reversal potential of I_{Ba} in a normal myotube (open circles) are different from those found in a DMD myotube culture (open triangles). Currents were measured at the peak amplitude. (Modified from Bkaily et al. (11)).*

At least three alternative explanations concerning the origin of the TTX- and Mn^{2+} -resistant I_{Na} in DMD myotubes can be put forward: 1) the TTX- and Mn^{2+} -resistant Na^+ channel that is activated during the development of DMD may derive from the same gene as that coding the TTX-resistant Na^+ channel; 2) it is possible that the TTX- and Mn^{2+} -resistant Na^+ channel in DMD myotubes may result from a minor structural modification (38) of the otherwise normal TTX-resistant Na^+ channel (or the TTX-sensitive Na^+ channel) which becomes less sensitive to certain calcium blockers; and 3) the TTX- and Mn^{2+} -insensitive Na^+ channel may derive from a gene that is mainly expressed in DMD or in the early fetal development of the SM. Because of recent results showing the presence of a TTX- and Mn^{2+} -insensitive I_{Na} (8-10) in hereditary cardiomyopathy (46), we are inclined to favor the third alternative. The pharmacology of the slow inward current is the same in both normal and DMD myotube cultures. However, the slow channel kinetics is quite different in DMD and normal myotube cultures; also, in DMD myotubes this channel seems to exhibit complex kinetics.

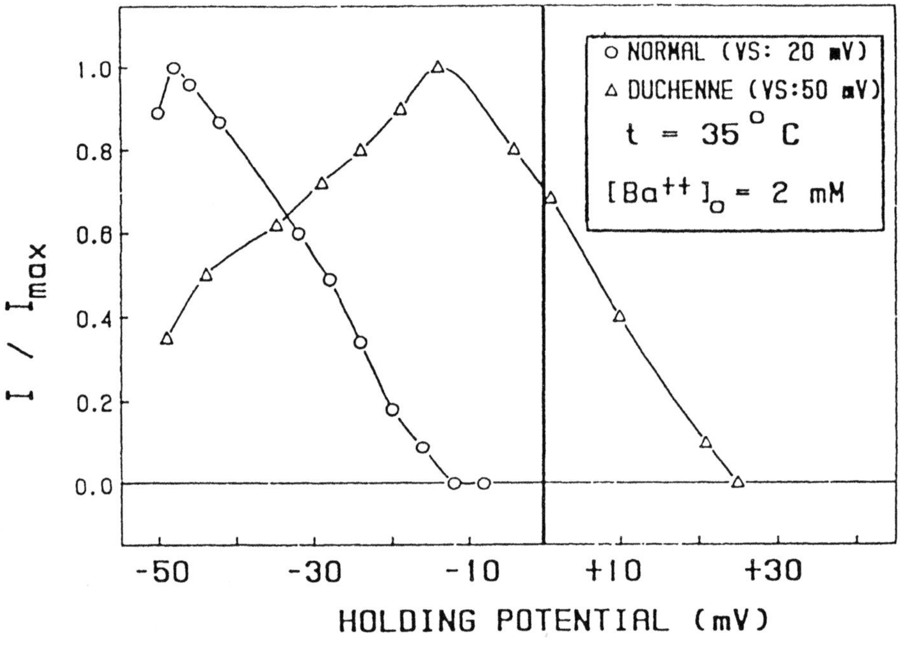

Figure 4. Steady-state inactivation of the slow Barium current in normal (curve in open circles) and DMD (curve in open triangles) myotube cultures. The shape of the slow Ba^{2+} current inactivation curve of a normal myotube culture (open circles) is completely different from that found in a DMD myotube culture (open triangles). Currents were measured at the peak amplitude. (Modified from Bkaily et al. (11)).

Dystrophin, the protein product of the DMD gene was reported to be associated with the triad structure (22) and tightly bound to the surface membrane of skeletal muscle (47). A possible relationship between the expression of the TTX- and Mn^{2+} -insensitive I_{Na} in DMD and the abnormalities reported in the human dystrophies should be not excluded. Also, because dystrophin (29) was reported to be associated with the triad structure and interfered with the movement of Ca^{2+}, in this case, it is likely that dystrophin abnormalities in human dystrophies may account in part for the change in Ca^{2+} channel properties in DMD.

Acknowledgements

This chapter is based on work supported by the Medical Research Council of Canada (MT 16496). Dr Bkaily is a FRSQ-Merck Frosst professor. These studies were performed in my laboratory by A. Sculptoreanu, M. Peyrow and Miss D. Jacques.

REFERENCES

1. Renaud J-F and Schmid A (1986) J Physiol (Paris) 81:177-184
2. Sperelakis N and Fabiato A (1985) In: The Thorax (Roussos C, Macklem PT eds) p.45-113
3. Stefani E and Chiarandini DJ (1982) Ann Rev Physiol 55: 357-372
4. Nonner W, Spalding BC and Hille B (1980) Nature 284: 360-363
5. Patlak JB, Ortiz M and Horn R (1986) Biophys J 49: 773-777
6. Sigworth FJ and Neher E (1980) Nature 287: 447-449
7. Weiss RE and Horn R (1986) Science 233: 361-364
8. Bkaily G, Jacques D and Yamamoto T et al (1988) Can J Physiol Pharmacol 66: 1017-1022
9. Bkaily G, Peyrow M and Yamamoto T et al (1988) Mol Cell Biochem 80: 59-72
10. Bkaily G, Jacques D, Sculptoreanu A et al (1991) J Mol Cell Cardiol 23: 25-39
11. Bkaily G, Jasmin G, Constantin T et al (1990) Muscle and Nerve 13: 939-948
12. Tanabe T, Takeshima H, Mikami A et al (1987) Nature 328: 313-318
14. Beam KG, Knudson CM and Powell JA (1986) Nature 320: 168-170
15. Cognard et al (1986) Proc Natl Acad Sci USA 83: 517-521
16. Elbrink J, Malhotrask and Elleker MG (1987) TIPS 8: 109-113
17. Sperelakis N (1985) In: The Thorax (Roussos C, Macklem PT eds) p 115
18. Klamut HJ, Kotraba JA and Strickland KP (1987) Muscle and nerve 10: 69-76
19. Bkaily G and Sperelakis N (1986) J Cyclic Nucleotide and Prot Phosphoryl Res 11: 25-34
20. Proschek L and Jasmin G (1983) Muscle and Nerve 5: 26-32
21. Jasmin G, Solymoss B and Prochek L (1979) In: Muscular Dystrophy and Other Inherited Desease of Skeletal Muscle in Animals (Harris JB ed) Ann NY Acad Sci, New York 317: 338-398
22. Moxley RT, Brook MH, Fenickel GM et al (1987) Muscle and Nerve 10: 22-23
23. Kometani K, Isugeno H and Yamada K (1990) Jap J Physiol 40: 541-549
24. Bakker AJ, Head SI, Williams DA and Stephenson DG (1993) J Physiol (Lond) 460: 1-13

25. Tautu C and Jasmin G (1981) Muscle and Nerve 4: 149-154
26. Lucas-Héron B, Mussini J-M and Olivier B (1989) J Neurol Sc 90: 299-306
27. Shimahara T, Bournaud R, Inoue I and Strube C (1992) J Physiol (Paris) 86: 117-121
28. Mendell JR, Higgings R, Sakenk Z and Cosmos E (1979) In: Molecular Dystrophy and Other Inherited Diseases of Skeletal Muscle in Animals (Harris JB, ed) NY Acad Sci, New York, pp 409-430
29. Hoffman EP, Hundson CM, Campbell KP and Kunkel LM (1987) Cell 51: 919-928
30. Frelin C, Vijverberg HPM and Romey G et al (1984) Pflügers Arch 402: 121-126
31. Saito K, Ozawa E (1986) J Cell Physiol 129: 289-294
32. Kidokoroy (1975) J Physiol (Lond) 244: 129-143
33. Romey G, Hugues M, Schmid-Antomarchi H and Lazdunski M (1984) J Physiol (Paris) 79: 259-264
34. Nelson P, Christian C and Nirenberg M (1976) Proc Natl Acad Sci 73: 123-127
35. Merickel M, Gray R, Chauvin P and Appel S (1981) Proc Natl Acad Sci USA 78: 648-652
36. Rothman SM and Bischoff R (1983) Ann Neurol 13: 176-179
37. Tahmonsh AJ (1983) Neurology 33: 311-316
38. Anderson PAV (1987) J Exp Biol 133: 231-249
39. Fukuda J and Kameyama M (1980) Brain Res 182: 191-197
40. Ikeda SR and Schofield GG (1987) J Physiol (Lond) 339: 255-270
41. Trautmann A, Delaporte C and Marty A (1986) 406: 163-172
42. Adrian RH and Peachy LD (1973) J Physiol 235: 103-131
43. Caillé J, Ildefonse M and Rougier O (1978) 374: 167-177
44. Heiny JA and Vergara J (1982) 80: 203-230
45. Bossu JL and Fltez A (1984) Neuroscience Lett 51: 241-246
46. Jasmin G, Pasternack A and Bkaily G (1991) In: Calcium Channels: Their Properties, Function, Regulation and Clinical Relevance (Hurwitz L, Partridge D, Leach J eds) Telford Press, New Jersey pp 295-307
47. Salviati G, Betto R, Ceoldo S et al (1989) Biochem J 258: 837-841

CHAPTER 9

SOME SMOOTH MUSCLES POSSESS FAST SODIUM CHANNELS

Nicholas Sperelakis, Zhiling Xiong, Yoshihito Inoue,
Yusuke Ohya, Keiichi Shimmamura
and David Bielefeld

Department of Physiology and Biophysics,
University of Cincinnati, College of Medicine,
Cincinnati, Ohio 45267-0576 USA

INTRODUCTION

Most smooth muscle (SM) cells normally do not possess fast Na^+ channels, and inward current for the action potential (AP) is carried primarily through slow (L-type) Ca^{2+} channels. However, it has been demonstrated that fast Na^+ channels appear in myometrial SM cells during the latter half of pregnancy in the rat, presumably under steroid hormone control (4, 7). In addition, we have recently found functioning fast Na^+ channels in rat and human intestinal (colonic) SM cells (23) and in rat portal vein cells (Xiong & Sperelakis, unpublished) under normal conditions. This review-type article provides a brief summary of (a) the changes in fast Na^+ channels (and slow Ca^{2+} channels) in myometrial SM cells (from longitudinal layer) during pregnancy in the rat (4, 7); (b) the existence of functional fast Na^+ channels in colonic SM cells (from circular layer) of the proximal (ascending) colon in both adult rat and human (13) and (c) the presence of functional fast Na^+ channels in vascular SM cells from the rat protal vein (Xiong & Sperelakis unpublished).

UTERINE SMOOTH MUSCLE CELLS

Fast Na^+ Channels in Uterine Smooth Muscle Cells from Pregnant Rat

The existence of fast Na^+ channels has been reported in pregnant rat myometrium (4, 5, 7) and in cultured pregnant human myometrial SM cells (15).

Single SM cells were obtained from the longitudinal layer of pregnant and postpartum rats (5, 9, 14, 18 and 21 days of gestation, and 1-day and 5 day

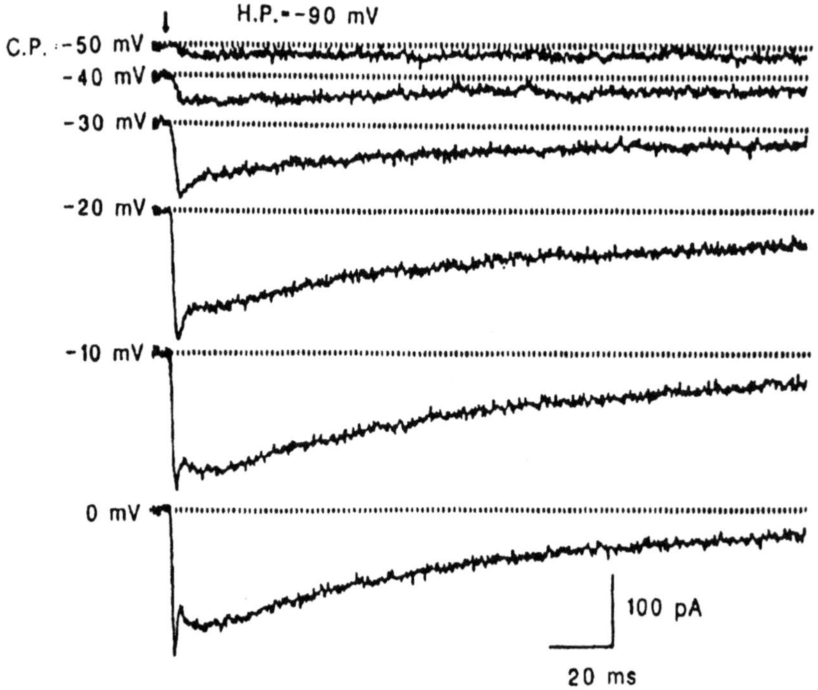

Figure 1. Two types of inward current recorded from isolated single myometrial cells from pregnant rat using the patch pipette technique. Depolarizing potential steps (-50 to 0 mV) were applied from a holding potential (HP) of -90 mV. The arrow indicates the beginning of voltage step that continues to the end of the trace. The bath contained K^+ -free solution with 150 mM Na^+ and 2 mM Ca^{2+}. The pipette solution contained high Cs^+. The cell capacitance was 80 pF. CP, command potential. (Taken with permission from ref. 7).

postpartum) using enzymatic digestion. Whole-cell voltage clamp was performed at room temperature (20-22°C). Membrane capacitance was determined from hyperpolarizing voltage ramp pulses of 0.2 V/s, applied from a holding potential (HP) of 0 mV. To isolate the inward currents, the pipette was filled with high Cs^+ solution of the following composition (in mM): 130 Cs^+, 10 Na^+, 5 ATP, 5.2 Mg^{2+}, 10 EGTA, 30 Cl^-, 112 glutamate, 10 HEPES, pH 7.2. The bath solution contained 3 mM 4-aminopyridine. For recording the Na^+ current, the bath solution also contained: 0 Ca^{2+}, 0 K^+, 0 EGTA, 150 Na^+, 2 Mn^{2+}, 10 HEPES, 154 Cl^-, 10 glucose, pH 7.3. For the recording of both Na^+ and Ca^{2+} currents, 2 mM $CaCl_2$ was used instead of $MnCl_2$. For isolating the Ca^{2+} channel current, the bath solution contained: 0 Na^+, 0 K^+, 100 Tris, 50 TEA^+, 2 Ca^{2+} (or Ba^{2+}), 10 HEPES, 154 Cl^-, 10 glucose, pH 7.3.

Two types of inward currents (fast and slow) were recorded from freshly-isolated single SM cells of pregnant (18-day) rat uterus. Figure 1 shows typical current traces of the fast and slow inward currents, evoked by command potentials to -50 mV, and above from a HP of -90 mV. Note the initial fast current followed by a slower current (e.g. at command pulses to -10 and 0 mV). When external Na^+ was replaced with $Tris^+$ and TEA^+, only slow current ($I_{Ca(L)}$) remained (Fig. 2A). In contrast, when Ca^{2+} was replaced with Mn^{2+}, only the fast inward current ($I_{Na(f)}$) was observed (Fig. 2B). This fast Na^+ current was very sensitive to tetrodotoxin (TTX): 0.1 μM nearly completely suppressed the Na^+ current (Fig. 2C). The dose-response relationship for the inhibitory action of TTX on the Na^+ current gave a K_i value of 27 nM. I/V relationships for both inward $I_{Na(f)}$ and $I_{Ca(L)}$ currents are shown in Fig. 2D. The current density of $I_{Na(f)}$ is about one half of that of $I_{Ca(L)}$.

Steady-state inactivation curves were obtained using a double-pulse protocol (Fig. 3). The inactivation curve for fast Na^+ current was more negative, compared with the slow Ca^{2+} current. The voltage for half-inhibition of the Na^+ current was -64 mV, and that for the Ca^{2+} current was -39 mV.

The steady-state window current for I_{Na} ($I_{Na(win)}$) was determined by plotting the activation and inactivation curves (Fig. 4). The region of overlap represents $I_{Na(win)}$, and gives the voltage range in which a small sustained inward current occurs: -55 to -25 mV. Since this voltage range over which $I_{Na(win)}$ exists is close to the resting potential of -50 to -60 mV, this window current may contribute to automaticity as a background inward (depolarizing) current.

Gain of Fast Na^+ Channels During Pregnancy

The change in fast Na^+ current during pregnancy was examined in rat yometrial SM cells during much of gestation and 1-day and 5-day postpartum. A fraction of the cells did not possess $I_{Na(f)}$. For example, at day 14 of gestation, two types of cells were found: (a) cells with fast [I_{Na}(-) cell] (Fig. 5). The I_{Na}(+) cells displayed both fast and slow inward currents (Fig. 5A). In contrast, the I_{Na}(-) cells showed only a slow inward Ca^{2+} current (Fig. 5B).

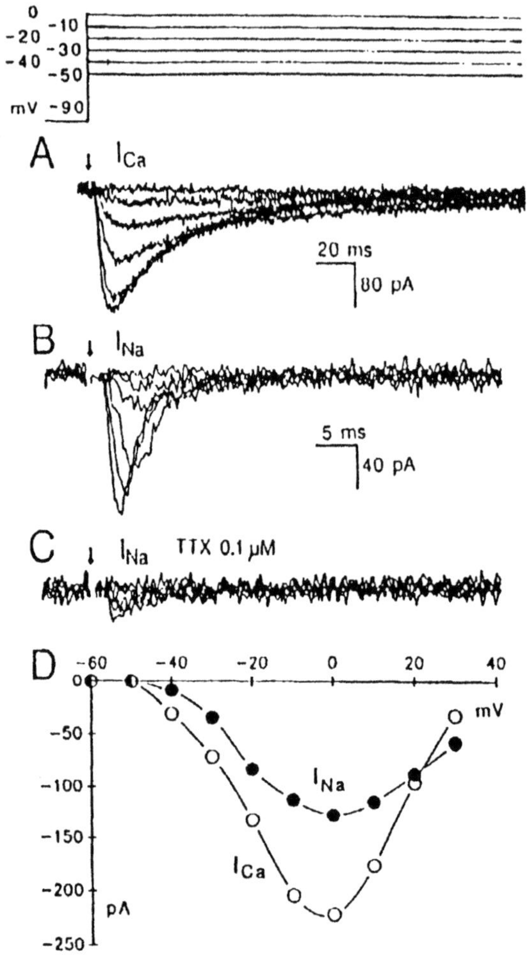

Figure 2. *Isolated fast Na^+ and slow Ca^{2+} inward currents.* ***A:*** *the slow Ca^{2+} inward current (I_{Ca}) recorded in a Na^+-free, bath solution containing 2 mM Ca^{2+}.* ***B:*** *The fast Na^+ current (I_{Na}) recorded in a Ca^{2+}-free, K^+-free bath solution containing 150 mM Na^+.* ***C:*** *the supporession of the Na^+ current by 0.1 μM TTX. Depolarizing voltage steps were applied in 10 mV increments from -50 to 0 mV from a HP of -90 mV.* ***D:*** *The current/voltage relationships for Ca^{2+} and Na^+ currents. The peak current amplitudes are plotted against command potentials. The pipette solution was the same as that used in Fig. 2. All data were obtained from one cell (capacitance of 76 pF). (Taken with permission from ref. 7)*

The probability of the existence of I_{Na} increased almost linearly during gestation (Fig. 6A). At day 5, none of the cells examined possessed I_{Na} , and at day 9 only 17% of the cells possessed I_{Na}. At day 14, I_{Na} could be detected in almost 50% of cells, and at days 18-21, the probability increased to ~85%. At 1 day post-partum, I_{Na} disappeared completely.

The averaged current densities, normalized by cell capacitance, were calculated from all cells examined (including those cells in which no I_{Na} was recorded) (Fig. 6B). The averaged density of I_{Na} increased almost linearly as term approached, and quickly decreased after parturition. The current densities of I_{Na} for only those cells that possessed visible I_{Na} did not change during gestation (Fig. 6C).

It was proposed that a Na^+ channel protein is synthesized and incorporated into the myometrial cell membrane during gestation. Once the gene for the Na^+ channels is turned on, the synthesis must proceed very quickly, because the averaged current densities obtained from $I_{Na}(+)$ cells did not change during gestation (Fig. 6C). That is, a larger and larger fraction of the cells gain fast Na^+ channels during pregnancy, in an apparent all-or-none manner, reaching nearly 100% of the cells near term.

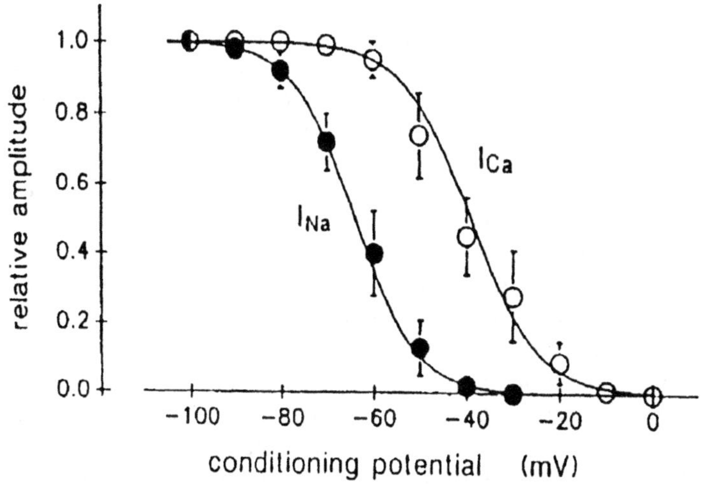

Figure 3. Steady-state inactivation curves for Ca^{2+} (unfilled circles) and Na^+ (filled circles) currents. Conditioning pulses of various ampltiudes were applied for 5s, followed by a test pulse to -10 mV (for Ca^{2+} current) or to -20 mV (for Na^+ current). The two curves plotted were obtained by fitting data to the Boltzmann distribution. Each point represents the mean ± SE of 2-4 values from 4 cells. (Taken with permission from ref. 7).

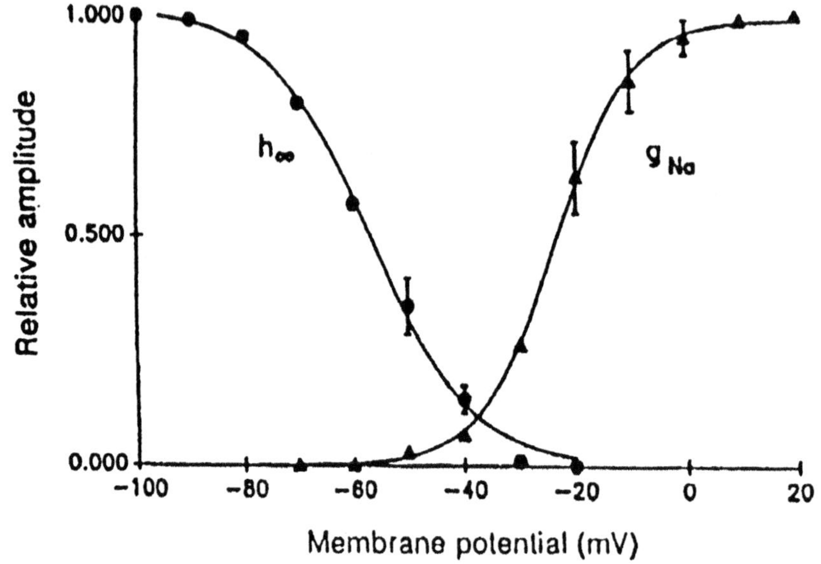

Figure 4. Steady-state activation and inactivation curves for ($I_{Na(f)}$. For the inactivation curve, a conditioning pulse to various amplitudes was applied for 2 sec before a test pulse to 0 mV. For the activation curve, data were obtained by measuring peak amplitude of I_{Na} elicited by various command potentials from a HP of -90 mV. Two curves were obtained by fitting data to the Blotzmann distribution. Each point represents mean ± SE (n = 7 (for inactivation) for 5 (for activation). (Taken with permission from ref. 12.)

A preliminary saxitoxin (STX) binding study (Tomsig et al., unpublished observations) supports the gain of fast Na^+ channels during pregnancy. The specific binding of STX to uterine smooth muscle increases during gestation, from a very low amount on day 5, to maximal values on days 18-21, and decreases again in postpartum. The general trend is consistent with the changes in $I_{Na(f)}$ observed in the voltage-clamp experiments.

A functional role for these newly-gained fast Na^+ channels may be proposed. **(a)** The insertion of fast Na^+ channels into the cell membrane during gestation should allow for faster and more complete propagation over the entire uterus by increasing max dV/dt of the AP. This factor, combined with the increase in the number of gap junctions (6), may result in faster propagation of excitation, and hence, more forceful contractions because of the series elastic element. **(b)** The increased $[Na]_i$ produced by Na^+ influx through fast Na^+ channels during excitation may cause elevation of [Ca]i by the Na^+/Ca^{2+} exchange system (reverse mode) to potentiate myometrial contraction (Savineau et al., 1987). (c) It is possible that Na^+ current may contribute to the pacemaker potential because a substantial amount of $I_{Na(win)}$ (or slowly-inactivating inward Na^+ current ($I_{Na(si)}$)) exists in myometrial cells (Fig. 4).

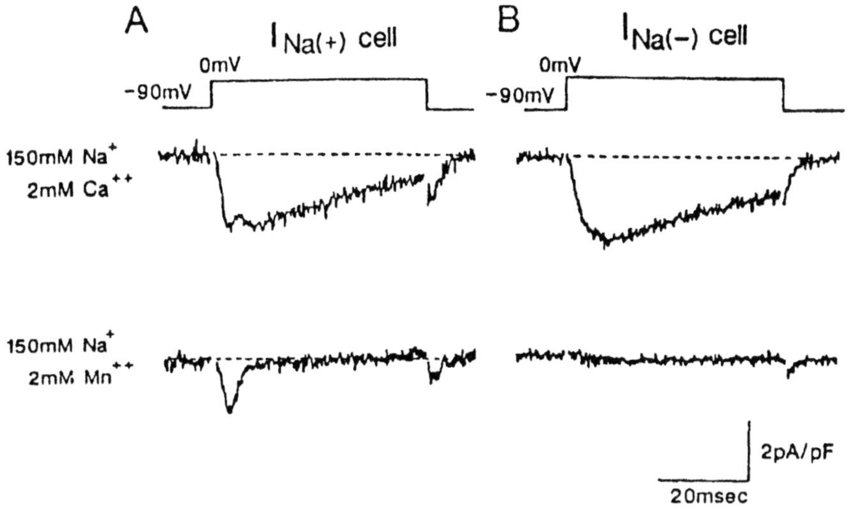

Figure 5 *Typical current traces recorded from two types of myometrial smooth muscle cells at day 14 of gestation: A: A cell with fast* Na^+ *current* $[I_{Na}(+)]$. ***B:*** *A cell without fast* Na^+ *current* $[I_{Na}(-)]$. *Depolarizing pulses to 0 mV were applied from a HP of -90 mV. The cell capacitances were 59.6 (A) and 68.0 pF (B). The bath solution was* K^+ *free and contained 150 mM* Na^+ *and 2 mM* Ca^{2+}, *isolated fast* Na^+ *current was obtained using* Ca^{2+}*-free bath solution containing 2 mM* Mn^{2+} *(to block any* Ca^{2+} *channels).* ***A:*** *Two types of inward currents (fast* Na^+ *and slow* Ca^{2+} *current) are present.* ***B:*** *Only slow inward* Ca^{2+} *current was present in the control solution (top trace), and no inward current was present in the* Ca^{2+}*-free solution (bottom trace). (Taken with permission from ref. 4.)*

Fast Na^+ Channels in Uterine Myosarcoma Cells

A fast inward current was found to be present in a cultured sarcoma cell line (SK-UT-1B) derived from a human uterine myometrial tumor (1). Since it was dependent on Na^+ ion in the bath and was completely blocked by TTX (1 μM), this inward current appeared to be a fast Na^+ current (I_{Na}). The mean value of I_{Na} was 11.3 $\pm$ 1.8 (n = 20). The potential for peak current (from the I/V curve) was 0 mV, the threshold potential was about -45 mV, and the extrapolated reversal potential (E_{rev}) was about +60 mV. I_{Na} disappeared after about 1 day when the sarcoma cells were replated into a culture medium containing low (1 %) or no (0%) serum (fetal calf). Therefore, it appeared that some factor present in fetal calf serum was required for the sarcoma cells to possess fast Na^+ channels.

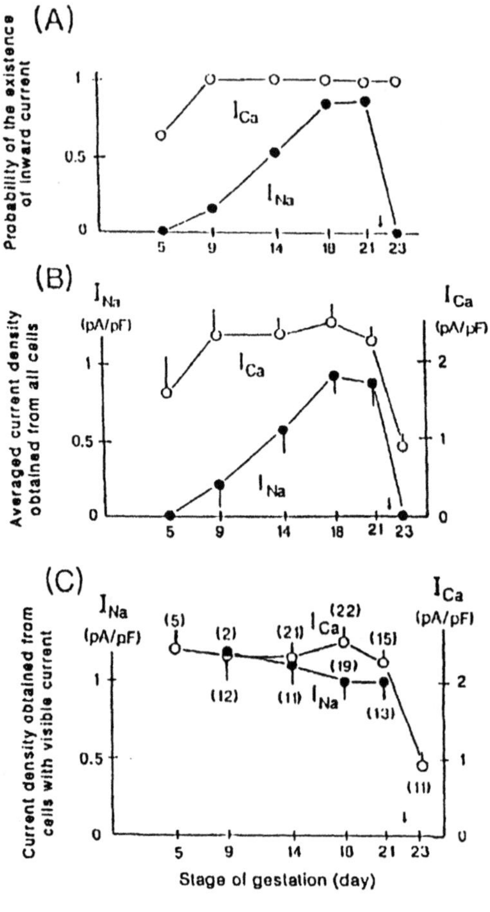

Figure 6. *Changes in the averaged densities for Na^+ and Ca^{2+} channels in myometrial smooth muscle cells at different stages of gestation.* ***A:*** *The change in the probability of the existence of the inward Na^+ and Ca^{2+} currents during gestation. Probability (the fraction of cells with visible inward current) was plotted against the days of gestation. Note that the fraction of cells exhibiting I_{Ca} was very high (1.0) beginning on day 9 of gestation; in contrast, the fraction of cells possessing an I_{Na} was 0 at day 5 and increased dramatically to > 0.8 at day 18-21; the fraction became zero at 1-day postpartum.* ***B:*** *The averaged current densities (normalized by cell capacitance) for Na^+ and Ca^{2+} currents obtained from all cells examined were plotted against the days of gestation. These values include those cells in which there was zero I_{Na} or I_{Ca}. Each point represents mean $\pm$ SE.* ***C:*** *The current densities for Na^+ and Ca^{2+} currents obtained from only those cells that displayed visible inward current were plotted against the days of gestation. Each point represents the means $\pm$ SE. Note that the densities of I_{Na} and of I_{Ca} were similar in the cells from all stages of gestation examined; the Ca^{2+} current density was decreased 1 day postpartum. Modified from ref. 4, by addition of unpublished data for 1-day postpartum.*

Increase in Ca^{2+} Channels During Pregnancy

Two types of Ca^{2+} channels (L and T) exist in pregnant human myometrium (3). On the other hand, only one type of Ca^{2+} channel (the L-type) was found in pregnant rat myometrium (4, 7).

All myometrial cells on day 9 of pregnancy and later possessed I_{Ca}, but some cells at day 5 did not display visible I_{Ca} (Fig. 6A). Thus, the probability of existence of Ca^{2+} channels changed less markedly during gestation than that of the fast Na^{+} channels. The averaged density of I_{Ca} of the cells at day 5 of gestation was smaller than those of cells at later stages of gestation (Fig. 6B). At 1 day post-partum, the average current density decreased dramatically, although all cells examined possessed I_{Ca} (Fig. 6B). The current density obtained from cells with visible I_{Ca} also did not change during gestation; however, it decreased dramatically after parturition (Fig. 6C).

K+ Currents

Voltage-dependent K^{+} currents were recorded in Ca^{2+}-free, Na^{+}-free bath solution containing 6 mM K^{+}, and the pipette was filled with high-K^{+} solution (130 mM). The presence of $I_{K(ATP)}$ and $I_{K(Ca)}$ channels could not be expressed because the pipette contained EGTA (10 mM) and ATP (5 mM). About two-thirds of all cells examined exhibited only a single component of IK, namely a slow sustained component, and the remaining one-third exhibited a second component, namely an initial fast transient component. A high dose of forskolin (30 μM) inhibited the slow component, but had little or no effect on the fast component. 4-aminopyridine inhibited primarily the fast transient component, whereas TEA inhibited primarily the slow sustained component.

Effects of Tocolytic and Uterotonic Agents on Ion Channels

Beta-agonists are clinically used for prevention and treatment of preterm labor. A high dose of ISO (10 μM) did not produce significant change in I_{Ca}, I_{Na}, or I_K. Thus, beta-agonists must relax the uterine muscle by mechanisms other than inhibition of the Ca^{2+} channels or Na^{+} channels. Sakai et al. (9) have suggested that phosphorylation by cA-PK closes gap junction channels.

Mg^{2+}, applied in the bath dose-dependently inhibited the Ca^{2+} current, with half-inhibition at 12 mM Mg^{2+}. This concentration of Mg^{2+} is not much different than the plasma levels attained during Mg^{2+} tocolysis, and so the effect of Mg^{2+} can be largely explained by a direct action on the Ca^{2+} slow channels.

Nifedipine inhibited the Ca^{2+} channel currents dose-dependently, with half-inhibition at 3.3 nM. Inhibition by nifedipine was greater at higher (more positive) HPs and higher command potentials. This direct action of nifedipine on the

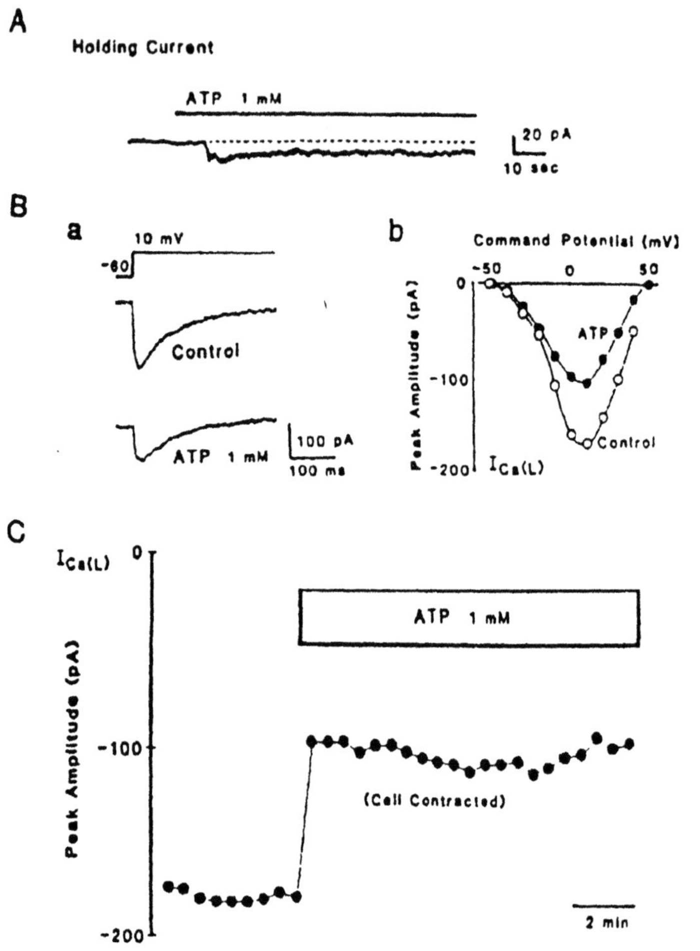

*Figure 7. Effect of ATP on I_{Ca} in longitudinal, myometrial smooth muscle cell from 18-day pregnant rat. The patch pipette contained 130 mM Cs^+ and nystatin (200 μg/ml) and the bath solution, 2 mM Ca. **A:** The change in holding current by the application of ATP (1 mM). **Ba:** $I_{Ca(L)}$ induced by a voltage step from -60 to +10 mV was decreased by ATP. **Bb:** The I-V relationship for $I_{Ca(L)}$ in the absence (unfilled circles) and presence (filled circles) of ATP. **C:** The typical time course of peak I_{Ca} change after ATP application. (Shimamura and Sperelakis, unpublished data.)*

Ca^{2+} slow channels can explain its tocolytic action.

Adenosine, a purinergic (P_1) receptor agonist which inhibits vascular SM contraction by elevating intracellular cAMP levels, had no effect either on $I_{Ca(L)}$ or on $I_{Na(f)}$ of uterine SM cells.

Oxytocin, often used in order to induce uterine contractions in cases of prolonged labor, unexpectedly decreased I_{Ca} slightly (12). Inhibition was not observed when Ba^{2+} was used as the charge carrier, suggesting the slight decrease in I_{Ca} was due to inhibition of the Ca^{2+} channel by Ca^{2+} release and influx. It has been reported that oxytocin increases IP_3 and DAG production, and IP_3-induced Ca^{2+} release from the SR could account for its stimulation of contraction. Therefore, stimulation of uterine contraction by oxytocin cannot be explained by a direct effect on the Ca^{2+} channel. However, it was reported (5) that oxytocin enhanced I_{Ca} in pregnant (18-day) rat myometrial SM cells, cultured with 1 μM estradiol. Substantial oxytocin receptors are present at 18 days pregnancy in the rat. I_{Na} and I_K were not affected by oxytocin in 18-day pregnant rat myometrium; therefore, these currents may not play an important role in oxytocin-induced uterine contraction.

ATP, an agent known to produce contraction of uterine SM cells, was added to the bath (1 - 1,000 μM) to activate the purinergic P_2 receptors. At 1,000 μM, ATP rapidly produced marked inhibition (about 40%) of $I_{Ca(L)}$ (Fig. 7B, C). ATP concentrations as low as 10 μM and 1 μM produced a similar, but smaller, inhibition of $I_{Ca(L)}$. Since the cell was simultaneously contracted, the inhibition of $I_{Ca(L)}$ may be mediated in part by IP_3-induced Ca^{2+} release from the SR and consequent Ca^{2+} inhibition of the Ca^{2+} channels. Inhibition of $I_{Ca(L)}$ is probably also caused by a decrease in the net driving force for Ca^{2+} influx (due to elevation of $[Ca]_i$). In addition, the holding current was shifted by ATP (Fig. 7A), representing a net inward sustained current, reflecting ATP activation of a receptor-operated non-selective cation channel, which allows Na^+ and Ca^{2+} entry (2).

A phorbol ester (phorbol 12,13-dibutyrate; PDB), a good activator of PK-C, when added to the bath (0.03 - 0.3 μM), substantially increased $I_{Ca(L)}$ within 5 min. (Fig. 8). This stimulation was reversed by addition of H-7, a non-specific protein kinase inhibitor (Fig. 8B), and by staurosporine (0.1 μM), a more specific inhibitor of PK-C.

Figure 9 summarizes the possible mechanisms by which uterine contractility can be modulated. In contrast to vascular SM, neither ISO nor adenosine, which produce elevation of cyclic AMP, affected I_{Ca} and I_{Na}. Therefore, no arrow can be drawn between cA-PK/cG-PK and the Ca^{2+} slow channel. Although oxytocin, which produces DAG and should activate PK-C, did not stimulate $I_{Ca(L)}$ (actually produced a small inhibition), an arrow can be drawn between PK-C and the Ca^{2+} slow channel based on the stimulation produced by phorbol ester and its prevention/reversal produced by PK-C inhibition. Oxytocin may fail to exhibit a stimulation of $I_{Ca(L)}$ because of the predominance of the inhibition produced by the $[Ca]_i$ increase. Fast Ca^{2+} channel (T-type) current could not be found in rat uterine muscle, as it was in human myometrium (3). Isoproterenol and oxytocin had no effect on $I_{Na(f)}$. Very little is known about the effects of phosphorylation by the various PKs on the

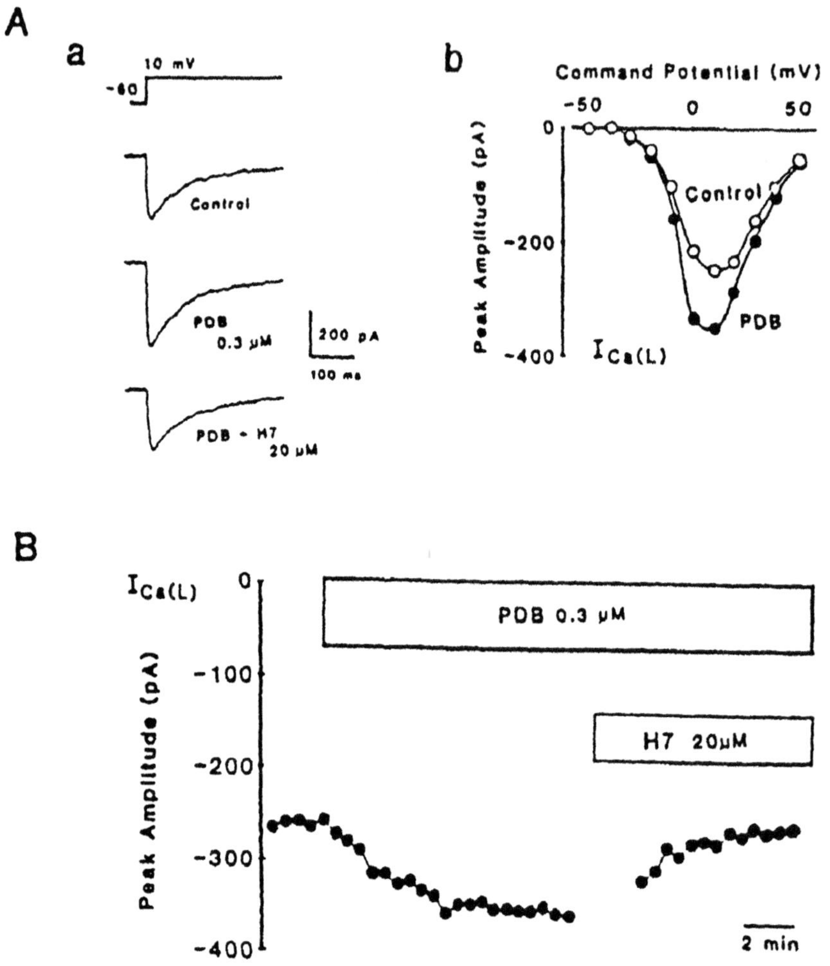

Figure 8. *Effect of phorbol 12,13-dibutyrate (PDB) on I_{Ca} in longitudinal, myometrial smooth muscle cell from 18-day pregnant rat. The patch pipette contained 130 mM Cs^+ and nystatin (200 μg/ml), and the bath solution, 2 mM Ca.* ***Aa:*** *$I_{Ca(L)}$ induced by a voltage step from -60 to +10 mV was increased by PDB (0.3 μM). This effect was reversed by the presence of H7 (20 μM).* ***Ab:*** *The I-V relationship for $I_{Ca(L)}$ in the absence (unfilled circles) and presence (filled circles) of PDB.* ***B:*** *The typical time course of peak $I_{Ca(L)}$ change after application of PDB alone or in combination with H7. (Shimamura and Sperelakis, unpublished data.)*

regulation of (a) the K^+ channels, (b) the Ca/Na exchange system, and (c) the sarcolemmal Ca pump. Therefore, it is unclear through what mechanism(s) beta-agonists relax uterine smooth muscle. In contrast, oxytocin may stimulate uterine contraction through increasing IP_3 production and thereby, Ca^{2+} release from the SR. Some agents may also alter the Ca^{2+} sensitivity of the contractile proteins, myosin light chain kinase, and phosphatases.

COLONIC SMOOTH MUSCLE CELLS

Whole-cell voltage clamp of freshly-dispersed single smooth muscle cells from adult rat and human colons gave evidence for the existence of a fast Na^+

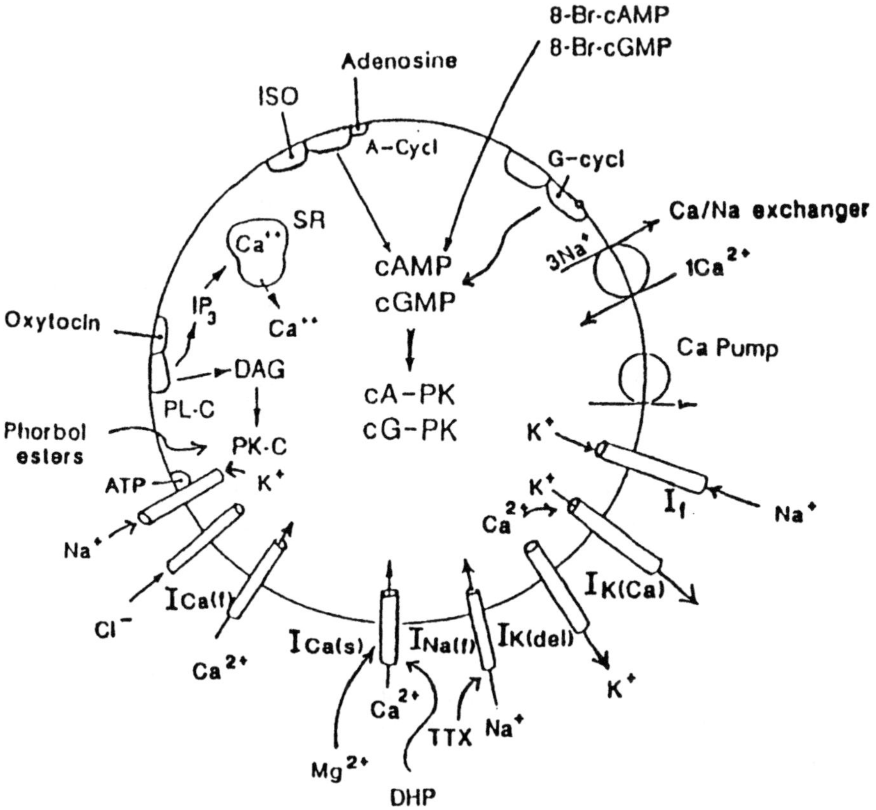

Figure 9. Diagrammatic representation of a uterine smooth muscle cell from a pregnant rat myometrium. As shown, cAMP and cGMP do not produce relaxation via inhibition of the Ca^{2+} slow channels; PK-C also has no effect on $I_{Ca(L)}$. As depicted, Mg^{2+} ion and dihydropyridine (DHP), Ca antagonist drug, do inhibit $I_{Ca(L)}$. The other ion channels, exchangers, and pumps known to be present are also depicted. (Modified from ref 12.)

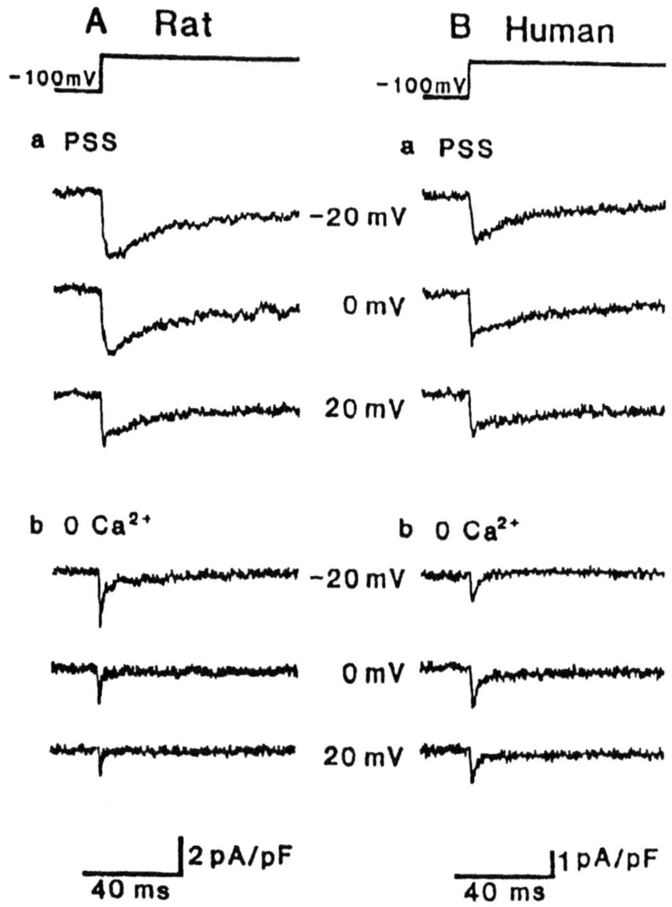

Figure 10. *Two components of inward current recorded from one colonic smooth muscle cell from the rat (A) and one from the human (B). Three different command potential steps (-20, 0, and +20 mV) were applied from a HP of -100 mV. The pipettes were filled with a high Cs^+ solution. The bath contained PSS (Aa & Ba) and 0 Ca^{2+} solutions (Ab & Bb). The cell capacitance was 64.0 pF for the rat cell (240 day male), and 90 pF for the human cell (50 year man). Note that in PSS there were two components of the inward current: an initial fast component and a delayed slow component, and that in 0 Ca^{2+} solution, the second component disappeared, leaving only the initial transient component. (14.)*

channel current (Fig. 10). With normal physiological salt solution (PSS) plus 4-aminopyridine (3 mM) in the bath, and high Cs^+ solution in the pipette to inhibit outward K^+ currents, an inward current possessing fast and slow components was observed when the cell membrane was depolarized, more positive, to -20 mV from a HP of -100 mV. When Ca^{2+} ion was removed from the PSS or when nifedipine (10 μM) and Ni^{2+} (30 μM) were simultaneously applied, the slow component disappeared and the fast component remained. The fast current component became almost completely inactivated within 10 ms. This fast component was dependent on $[Na]_o$ and was inhibited by TTX (IC_{50} of 130 nM in the rat and 14 nM in the human) (Fig. 11).

These results suggest that the slow component of inward current was a Ca^{2+} channel current, whereas the fast component was a TTX-sensitive fast Na^+ channel current. The threshold voltage, the voltage for peak current, and the reversal potential for the fast Na^+ current were, respectively, about -50, -20, and +50 mV in the rat, and -40, 0, and +60 mV in the human. The incidence of cells possessing fast Na^+ currents depended on the region of the colon: in rat proximal colon, the incidence was 64%; in distal colon, it was 10%. In humans, the incidence in the ascending colon was 73%, and descending colon was 22%. The densities of fast Na^+ and Ca^{2+} currents were 3.2 and 4.5 pA/pF in the rat; 1.0 and 1.4 pA/pF in the human, respectively. The ratio of both current densities (Na^+ vs Ca^{2+}) was 0.71, in both the rat and human. Therefore, the major ion channels associated with the generation of inward currents in the circular SM cells of the rat and human colon are voltage-dependent Ca^{2+} channels and fast Na^+ channels. The fast Na^+ current may facilitate propagation of excitation. Evidence for the presence of the fast Na^+ current in intestinal (rat ileum) SM cells was also given by Smirnov et al. (11).

Developmental changes of Ca^{2+} channel currents were observed in single SM cells freshly isolated from the circular layer of distal colon from the rat. Under physiological conditions, the averaged total Ca^{2+} current density increased markedly from 1.25 pA/pF in the newborn rat to 6.46 pA/pF in the 60-day old rat, then gradually declined thereafter with age. There were two types of Ca^{2+} channel currents: one type, the T-type, possessed more negative threshold potentials (-65 mV) and inactivated quickly. The voltage for peak current was about -15 mV, and the reversal potential was about +65 mV. This current was highly sensitive to a low concentration of Ni^{2+} (30 μM), but was resistant to nifedipine, diltiazem, Cd^{2+}, and TTX. In contrast, the other type, the L-type, possessed more positive threshold potentials (-40 mV) and inactivated more slowly. The voltage for peak current was 0 mV, and the reversal potential was about +65 mV. This current was insensitive to a low concentration of Ni^{2+}, but highly sensitive to nifedipine, diltiazem, and Cd^{2+}. The current densities of both $I_{Ca(L)}$ and $I_{Ca(T)}$ increased during development (Fig. 12). $I_{Ca(T)}$ remained high during aging, whereas $I_{Ca(L)}$ declined.

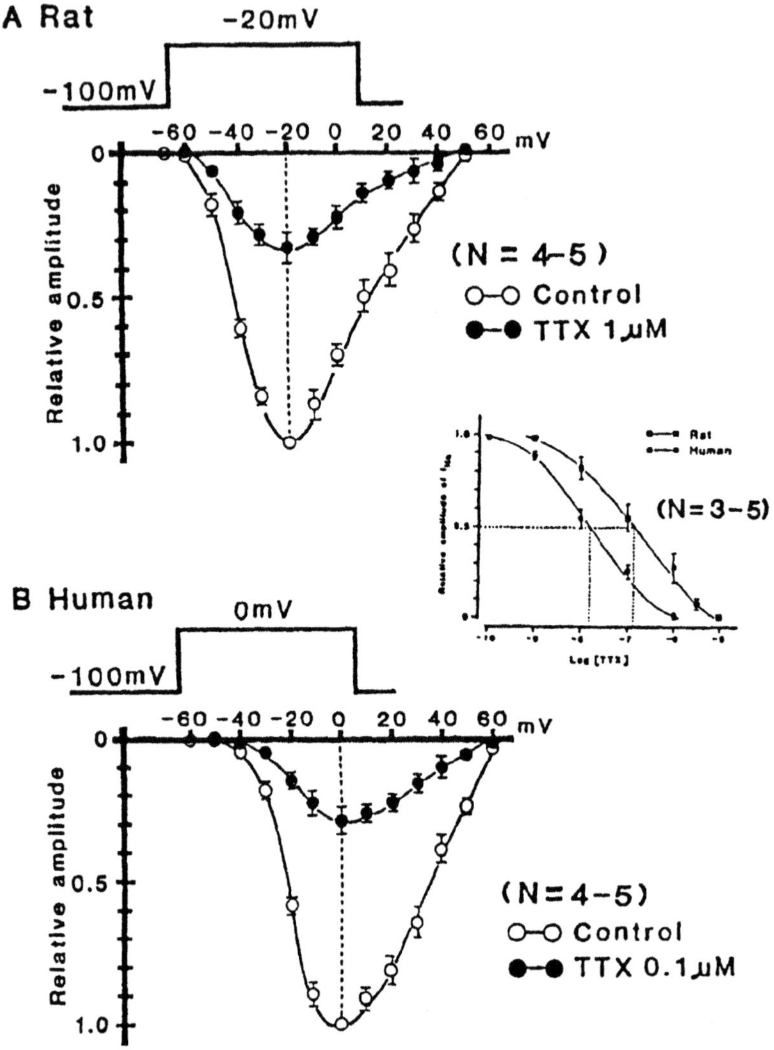

Figure 11. Effects of tetrodotoxin (TTX) on fast INa. (A) and (B) are the averaged I/V relationships in the absence (open circles) and presence (filled circles) of TTX (1 μM in rat and 0.1 μM in human). The peak amplitude of control current (without TTX) is normalized as 1.0. Each point represents 4 to 5 values from 5 different cells. Note that the voltage for peak $I_{Na\,(f)}$ was -20 mV in the rat, and 0 mV in the human. The dose/response relationship for TTX action on peak $I_{Na\,(f)}$ was also presented in the middle of the figure. As shown, the IC_{50} was 130 nM for the rat (filled squares), and 14 nM for the human (filled circles). (13.)

VASCULAR SMOOTH MUSCLE CELLS

In rat portal vein cells, evidence for the existence of fast Na^+ channels was found in about 35% of cells examined. In the rabbit portal vein, a much smaller fraction of cells possessed $I_{Na(f)}$. In cells containing $I_{Na(f)}$, there were two inward currents: an initial fast transient current and a second, slow more-sustained current (Fig. 13A). The first component was dependent on $[Na]_o$ and was blocked by TTX (Fig. 13A). The corresponding I/V curves for the two currents in rat cells are given in Figure 13B. The current density for I_{Na} in rat cells was about 2 pA/pF (compared to about 3 pA/pF for $I_{Ca(L)}$). Similar findings which demonstrated the presence of $I_{Na(f)}$ in rat portal vein cells were made by Okabe et al. (8).

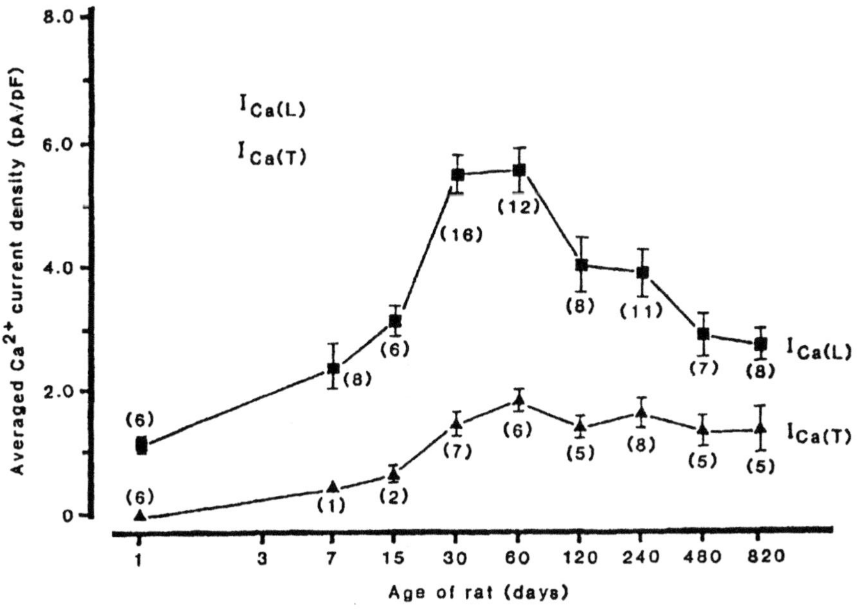

Figure 12. Changes in averaged T-type and L-type Ca^{2+} current densities in colonic smooth muscle cells at various ages of rat. The HP was -100 mV. Current densities (normalized by cell capacitance) were plotted aganist age (days). Filled squares: $I_{Ca(L)}$ density; filled triangles: $I_{Ca(T)}$ density. Each point represents mean ± SE for the number of cells indicated in parentheses. Note that both $I_{Ca(L)}$ and $I_{Ca(T)}$ densities increased during development; with aging, $I_{Ca(L)}$ density declined; whereas $I_{Ca(T)}$ density was constant. (14)

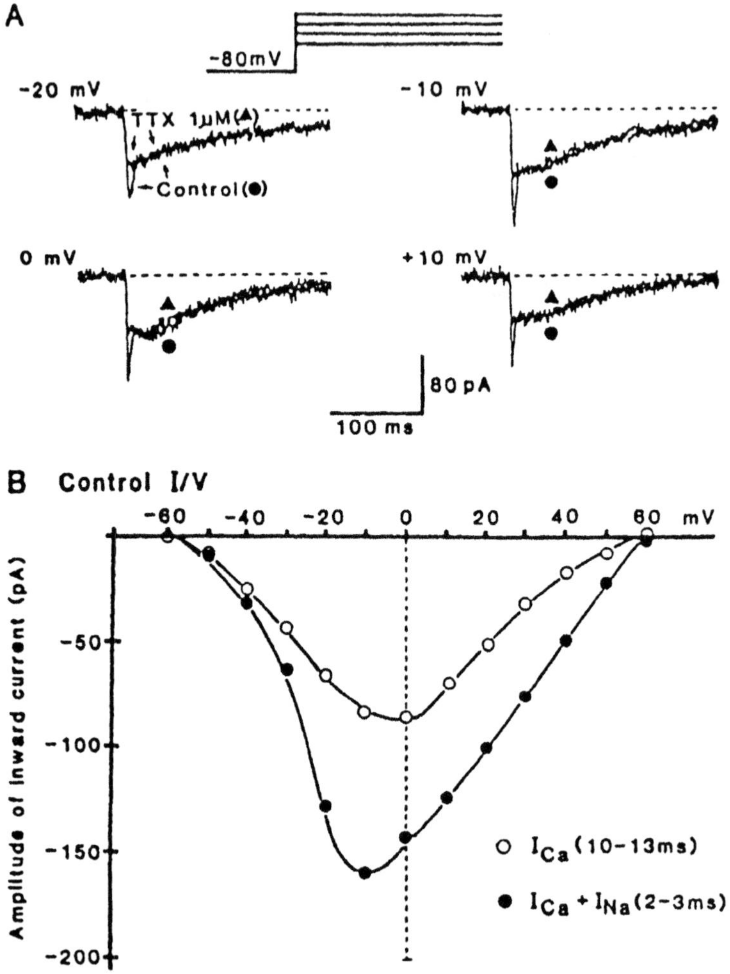

Figure 13. Two types of Ca^{2+} channel currents recorded from freshly-dispersed single smooth muscle cells from the rat portal vein. The pipette contained high Cs^+ and the bath contained PSS; the HP was -80 mV. A: The membrane currents evoked by 4 different CPs in the absence (filled circles) and presence (filled triangles) of TTX (1 μM). B: The amplitudes of inward currents (without TTX) at 2-3 ms (filled circles) and 10-13 ms (open circles) plotted at various membrane potentials. Values of (B) were from the control part of (A). (Xiong et al., unpublished data).

SUMMARY AND CONCLUSIONS

Most SM cells normally do not possess fast Na^+ channels, but inward current is carried through two types of Ca^{2+} channels: slow (L-type) and fast (T-type). Whole-cell voltage clamp was done on single SM cells isolated from three types of tissue: **(a)** the longitudinal layer of pregnant rat uterus, **(b)** the circular layer of rat and human colon, and **(c)** vascular SM cells from the rat and the rabbit portal vein and cultured cell lines (A10 and A7r5). Depolarizing pulses, applied from a HP of -60 to -100 mV, evoked two types of inward current, fast and slow. The fast inward current decayed within 5 - 10 msec, depended on $[Na]_o$, and was inhibited by TTX, with $K_{0.5}$ values of 27 nM (uterine), 14 nM (human colonic), and 130 nM (rat colonic). The slow inward current decayed slowly, was dependent on $[Ca]_o$ (or Ba^{2+}), and was inhibited by nifedipine. These results suggest that the fast inward current is a fast Na^+ channel current, and that the slow inward current is a Ca^{2+} slow channel current. A fast-inactivating Ca^{2+} channel current was not evident in pregnant uterine SM, but was in colonic SM and vascular SM (guinea pig mesenteric artery). Thus, the ion channels which generate inward currents in these SM cells are primarily TTX-sensitive fast Na^+ channels and dihydropyridine-sensitive slow Ca^{2+} channels.

In pregnant rat myometrial SM cells, the number of fast Na^+ channels increased during gestation. The averaged current density increased from 0 on day 5, to 0.19 on day 9, 0.56 on day 14, 0.90 on day 18, and 0.86 pA/pF on day 21. This almost linear increase occurs because of an increase in the fraction of cells which possess fast Na^+ channels. The Ca^{2+} channel current density also was higher during the latter half of gestation. These results indicate that the fast Na^+ channels and Ca^{2+} slow channels in myometrium become more numerous as term approaches, and that the fast Na^+ current may be involved in the spread of excitation.

The fast Na^+ current was present in sarcoma cells derived from uterine myometrium and cultured in media containing 5% or 10% fetal calf serum, but disappeared after 1 day in low (1%) or no (0%) serum.

In uterine SM cells, isoproterenol (ISO, beta-agonist) did not affect either $I_{Ca(L)}$ or $I_{Na(f)}$, whereas Mg^{2+} ($K_{0.5}$ of 12 mM) and nifedipine ($K_{0.5}$ of 3.3 nM) depressed $I_{Ca(L)}$. Oxytocin had no effect on $I_{Na(f)}$ and actually depressed $I_{Ca(L)}$ (but not I_{Ba}) to a small extent. Therefore, the tocolytic action of beta-agonists cannot be explained by an inhibition of $I_{Ca(s)}$, whereas that of Mg^{2+} and Ca antagonist drugs can be so explained. The stimulating action of oxytocin on uterine contractions cannot be explained by a stimulation of $I_{Ca(L)}$. Phorbol ester (PDB, 0.1 μM) increased $I_{Ca(L)}$ by about 30%. ATP (1 μM - 1 mM) applied externally decreased $I_{Ca(L)}$ by up to 50%, concomitant with induction of contraction, suggesting Ca^{2+} inhibition of the Ca^{2+} channels.

In colonic SM cells, the threshold voltage, the voltage for peak current, and the reversal potential for the fast Na^+ current were, respectively, about -50, -20, and +50 mV in the rat, and -40, 0, and +60 mV in the human. The incidence of cells possessing fast Na^+ currents depended on the region of the colon: in the rat, the incidence was 64% in the proximal colon, and 10% in distal the colon. In humans, the incidence was 73% in the ascending colon and 22% in the descending colon. The densities of fast Na^+ and Ca^{2+} currents were, respectively, 3.2 and 4.5 pA/pF in the rat, and 1.0 and 1.4 pA/pF in the human. The ratio of current densities (I_{Na}/I_{Ca}) was 0.71 in both rat and human cells. The fast Na+ current may facilitate propagation of excitation.

In vascular SM cells, about 35% of the cells tested from rat possessed a $I_{Na(f)}$, whereas in the rabbit, very few cells had a $I_{Na(f)}$. The current densities were approximately 2 pA/pF for $I_{Na(f)}$ and 3 pA/pF for $I_{Ca(L)}$ in the rat. In the rabbit, $I_{Ca(L)}$ was about 4 pA/pF.

In summary, some SM cells (e.g., colonic and some vascular) normally possess functional fast Na^+ channels, whereas in others (e.g., uterine), such current appears during pregnancy (presumably under hormonal control). Even though the relatively low resting potential in most SM cells would cause most of the fast Na^+ channels to remain in a continual state of inactivation, even a small amount of Na^+ current can contribute to propagation of excitation. In addition, the Na^+ window current would allow a sustained influx of Na^+ over the window voltage range through continuously active fast Na^+ channels, which would contribute to automaticity. With respect to Ca^{2+} currents, some SM cells (e.g., pregnant rat uterine) do not possess T-type Ca^{2+} channels, whereas others do (intestinal, some vascular). In the rat colon, there is an increase in both $I_{Ca(T)}$ and $I_{Ca(L)}$ during development. In uterine SM cells, the L-type Ca^{2+} current was greater during the latter half of pregnancy.

REFERENCES

1. Bielefeld DR and Sperelakis N (1993) Nature (submitted).
2. Honoré E, Martin C, Mironneau C and Mironneau J (1989) Am J Physiol 257:C297-C305
3. Inoue Y, Nakao K, Okabe K, Isumi H, Kanda S, Kitamura K and Kuriyama M (1990) Am J Obstet Gynecol 162:1090-1098
4. Inoue Y and Sperelakis N (1991) Am J Physiol 260:C658-C663
5. Martin C, Arnaudeau S, Jmari K, Rakotoarisoa L, Sayet I, Dacquet C, Mironneau C and Mironneau J (1990) Mol Pharmacol 38:667-673
6. Miller SM, Garfield RE and Daniel EE (1989) Am J Physiol 256:C130-C141
7. Ohya Y and Sperelakis N (1989) Am J Physiol 257:C408-C412
8. Okabe K, Kajioka S, Nakao K, Kitamura K and Kuriyama H (1990) J Pharmacol Exp Therap 252:832-839
9. Sakai N, Tabb T and Garfield RE (1992) Am J Obstet Gynecol 167:472-480
10. Savineau J, Mironneau J and Mironneau C (1987) Gen Physiol Biophysics 6:535-560

11. Smirnov SV, Zholos AV and Shuba MF (1992) J Physiol (London) 454:549-571
12. Sperelakis N, Inoue Y and Ohya Y (1992) Can J Physiol Pharmacol 70:491-500
13. Xiong ZL, Sperelakis N, Noffsinger A and Fenoglio-Preiser, C (1993) Eur J Physiol Pflugers Arch 423:485-491
14. Xiong ZL, Sperelakis N, Noffsinger A and Fenoglio-Preiser, C (1993) Am J Physiol 265:C000-C000
15. Young RC and Herndon-Smith L (1991) Am J Obstet Gynecol 164:175-181

CHAPTER 10

REGULATION OF Ca^{2+} CHANNELS IN VASCULAR SMOOTH MUSCLE CELLS BY CYCLIC NUCLEOTIDES AND BY G-PROTEIN GATING

Nicholas Sperelakis, Zhiling Xiong
and John Lorenz

Department of Physiology and Biophysics
University of Cincinnati, College of Medicine
Cincinnati, OH

INTRODUCTION

In myocardial cells, cyclic AMP (cAMP) and cyclic GMP-dependent protein kinase (PK-A) stimulate slow Ca^{2+} channel activity, whereas cyclic GMP (cGMP) and cGMP-dependent protein kinase (cG-PK) inhibit channel activity (reviewed in 18). Phosphorylation by protein kinase C (PK-C) also stimulates activity of the myocardial slow Ca^{2+} channels (5). Intracellular ATP also modulates slow Ca^{2+} channel activity in myocardial cells, ATP being obligatory for channel activity (7, 14). Gating of myocardial Ca^{2+} slow channels by G_s-protein (GTP-activated alpha subunit) has also been demonstrated (19).

In vascular smooth muscle (VSM) cells, intracellular ATP also modulates activity of the slow Ca^{2+} channels (11, 12), and phosphorylation by either cAMP or cGMP-mediated pathways inhibits slow Ca^{2+} channel activity (1) and the Ca^{2+}-dependent slow action potentials (APs) (15). The present article summarizes our more recent studies on the regulation of the slow (L-type) Ca^{2+} channels of several types of VSM cells: (1) cultured A-10 and A7r5 cell lines derived from rat aortic VSM, and (2) portal vein VSM cells from rat and rabbit. These recent studies

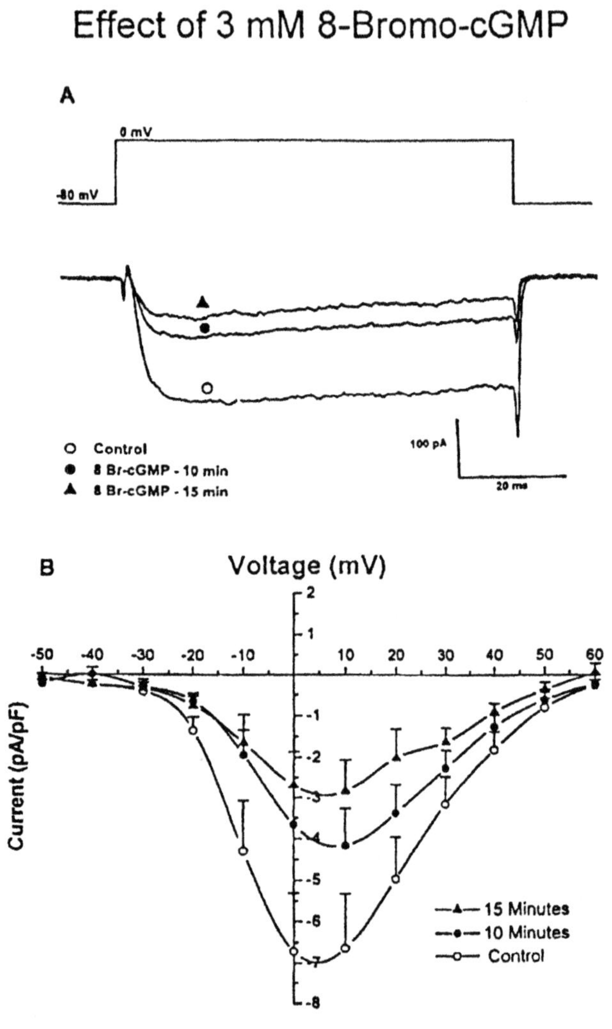

Figure 1. *Effect of 3 mM 8-Br-cGMP on the slow* Ca^{2+} *current of A7r5 cells.* ***A:*** *Individual tracings from a single experiment showing cGMP inhibition at 10 min (filled circles) and 15 min (filled triangles) compared to the control current (unfilled circles).* ***B:*** *The I/V curves (mean* $\pm$ *SE, n=7) for control, and at 10 and 15 min after cGMP.*

include the introduction of protein kinase A (PK-A, catalytic subunit) and protein kinase G (PK-G) inside the cell via diffusion from the patch pipette used to perform the whole-cell voltage clamp measurements.

There are two types of Ca^{2+} channels in VSM cells; slow (L-type) and fast (T-type). The properties of these two types of Ca^{2+} channels are rather similar to those in myocardial cells, with the exception of the regulatory mechanisms which are different. The two types of Ca^{2+} currents can be distinguished by differences in their kinetics, voltage ranges of activation and inactivation, and sensitivities to pharmacological agents. The slow (sustained) Ca^{2+} channel is sensitive to dihydropyridine Ca^{2+} agonists and antagonists, prefers Ba^{2+} over Ca^{2+} as a charge carrier, and activates at more positive membrane potentials (high threshold) than the fast (transient) Ca^{2+} channel. The fast Ca^{2+} channel activates at more negative membrane potentials (low threshold), has an equal preference for Ca^{2+} and Ba^{2+}, and is insensitive to the dihydropyridines. Both cAMP and cGMP elevation have been implicated in the relaxation of VSM in response to some vasodilators, such as PGI_2, EDRF, ANF and nitroprusside.

REGULATION OF Ca^{2+} CHANNELS BY CYCLIC NUCLEOTIDES VIA PHOSPHORYLATION

Cyclic Nucleotides

Agents that activate adenylate cyclase, such as isoproterenol and forskolin, depressed or abolished the Ca^{2+}-dependent APs and hyperpolarized the membrane in cultured rat aortic reaggregates (15). These effects are mimicked by membrane permeable analogs of c-AMP (dibutyryl or 8-bromo cAMP). 8-Br-cGMP also depressed or abolished the APs. Atrial natriuretic factor (ANF), which activates the membrane-bound guanylate cyclase, had inhibitory effects. These results suggest that cAMP and cGMP decrease the inward Ca^{2+} current (and/or increase an outward K^+ current) in the VSM cells.

In voltage-clamp experiments on cultured single cells from rabbit aorta, the dibutyryl analogs of cAMP and cGMP suppressed the inward Ba^{2+} current carried through the Ca^{2+} slow channels (1). The 8-Br analogs of cAMP and cGMP and nitroprusside also enhanced the delayed-rectifier outward K^+ current. Thus, these effects of cyclic nucleotides on the Ca^{2+} and K^+ channels can explain their inhibitory effects on the APs.

8-Br-cGMP and 8-Br-cAMP depressed I_{Ca} in cultured A10 cells and A7r5 cells, which are cell lines derived from rat aortic VSM cells (Figs. 1, 2). 3 mM 8-Br-cGMP reduced I_{Ca} to less than half of the control, and 3 mM 8-Br-cAMP depressed I_{Ca} by about 50% (Lorenz et al., submitted). Forskolin, an agent that directly stimulates adenylate cyclase to evaluate cAMP levels, also depressed Ca^{2+} channel current (carried by Ba^{2+}) (Fig. 3).

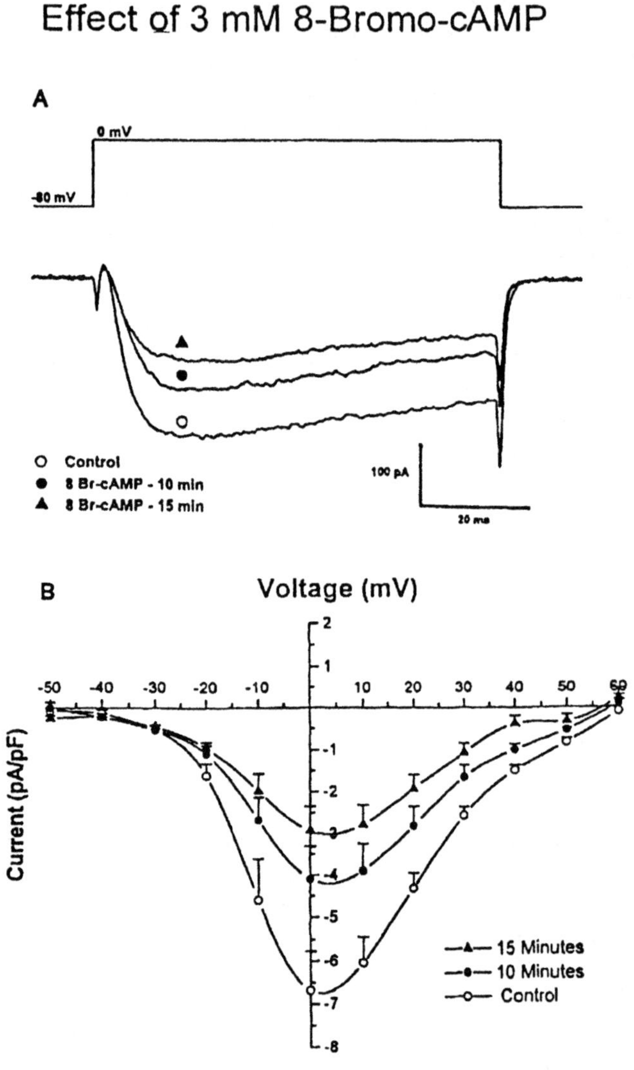

Figure 2. Effect of 3 mM 8-Br-cAMP on the slow Ca^{2+} current of A7r5 VSM cells. A: Individual tracings from a single experiment showing cAMP inhibition at 10 min (filled circles) and 15 min (filled triangles) compared to the control current (unfilled circles). B: The I/V curves (means $\pm$ SE, $n=5$) for control, and at 10 and 15 min. after 3 mM 8-Br-cAMP.

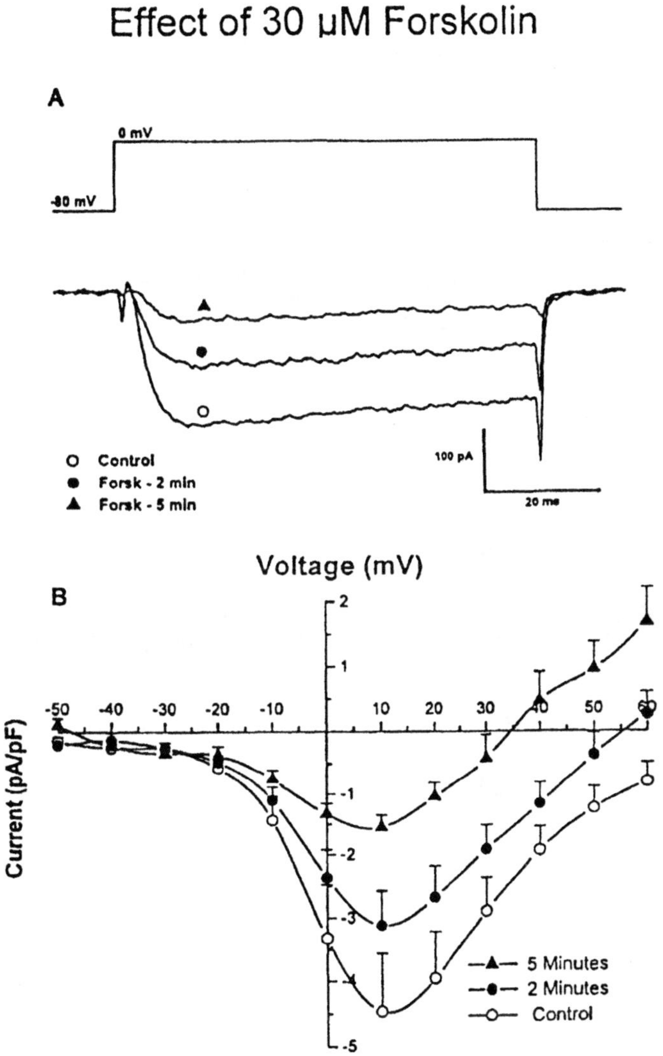

Figure 3. Effect of 30 μM forskolin (0.1% DMSO as vehicle) on the slow Ca^{2+} *current of A7r5 VSM cells.* ***A:*** *The individual tracings from a single experiment showing forskolin inhibition at 2 min (filled circles) and 5 min. (filled triangles) compared to the control current (unfilled circles).* ***B:*** *The I/V curves (mean ± SE, n=6) for control and at 2 and 5 min after forskolin.*

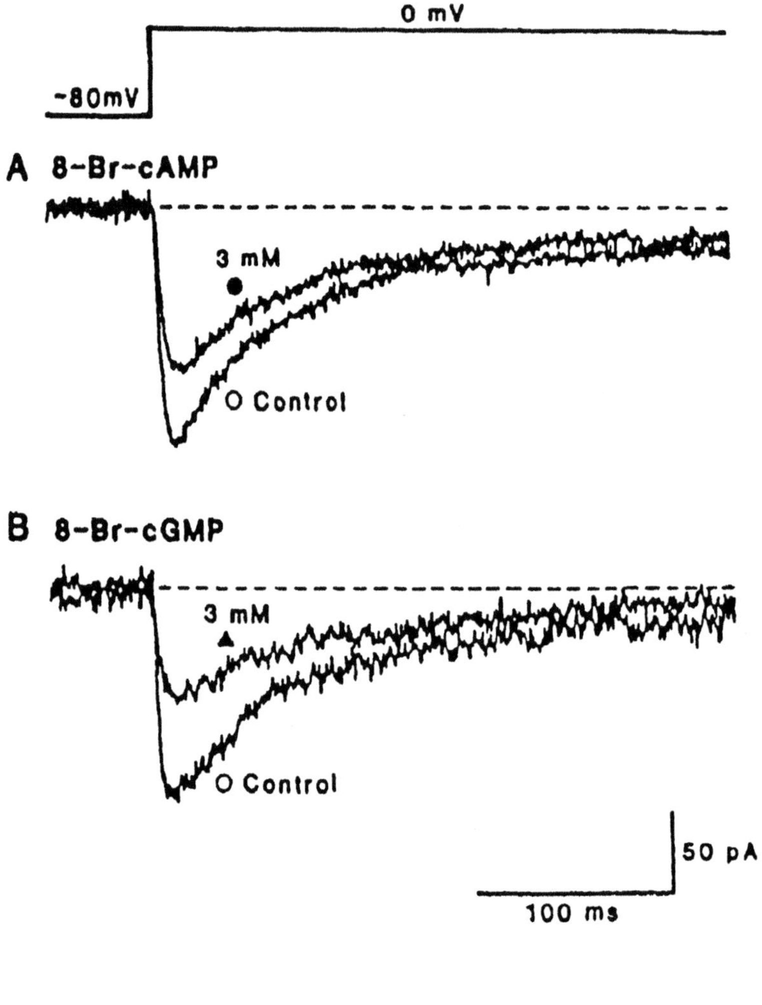

Figure 4. Effect of 8-Br-cAMP and 8-Br-cGMP on I_{Ca} in single smooth muscle cell from rabbit portal vein. The pipette contained high Cs^+ and the bath contained physiological salt solution (PSS). I_{Ca} was evoked by voltage step from a HP of -80 mV to 0 MV. A: Current traces recorded in the absence (control) and presence of 8-Br-cAMP (3 mM): B: Current traces recorded in the absence (control) and presence of 8-Br-cGMP (3 mM). (A) and (B) were from different cells. (Xiong et al., unpublished data).

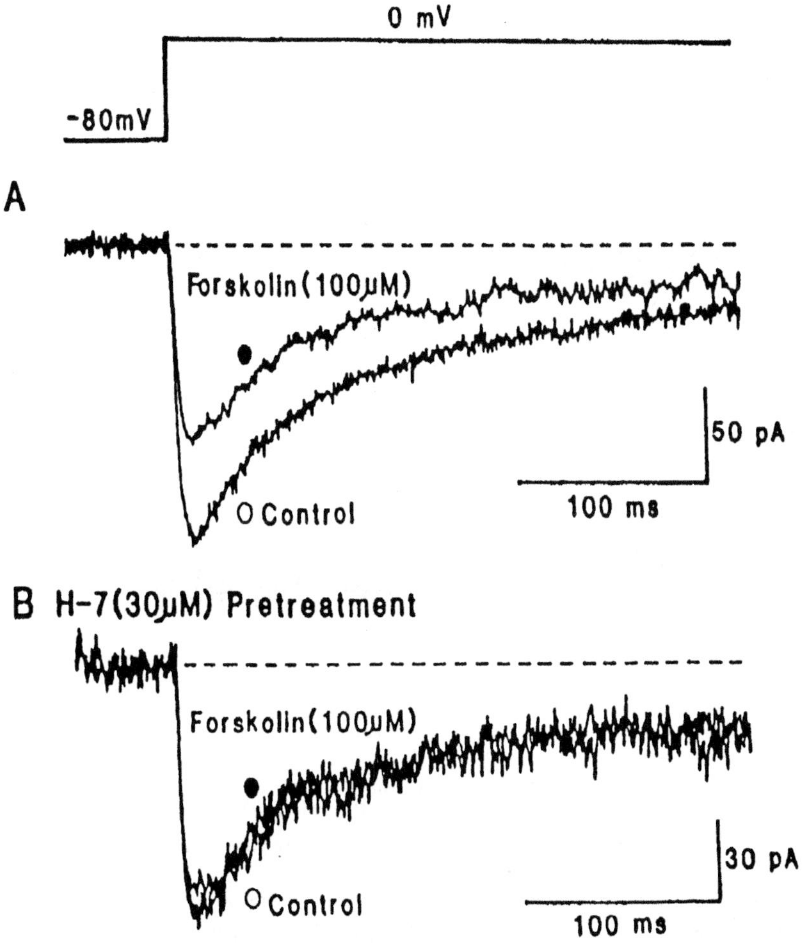

*Figure 5. Effect of forskolin on I_{Ca} in single smooth muscle cell from rabbit portal vein. The experimental conditions were the same as Fig. 4. **A:** Current traces recorded in absence (open circle) and presence (filled circle) of forskolin (100 μM). **B:** I_{Ca} recorded from another cell after pre-treatment with H-7 (30 μM). Note that the inhibitory effect of forskolin disappeared after pre-treatment with H-7. (Xiong et al., unpublished data.)*

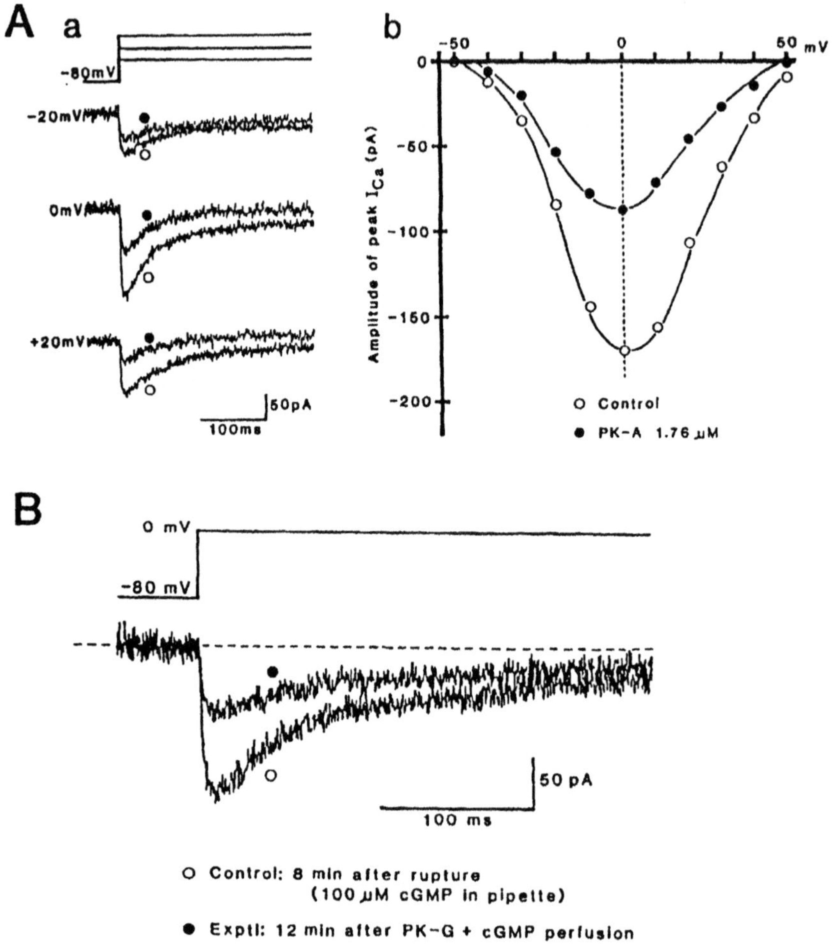

Figure 6. *Effect of cAMP-dependent protein kinase (PK-A, catalytic subunit)* **(A)** *and cGMP-dependent protein kinase (PK-G)* **(B)** *on I_{Ca} in single smooth muscle cell from rabbit portal vein. The pipette contained high Cs^+ solution and the bath contained PSS.* ***Aa:*** *Typical cell showing Ca^{2+} current traces in the absence (open circles) and presence (filled circles) of 1.76 μM PK-A at 3 CPs (-20, 0 and +20 mV) from a HP of -80 mV. PK-A was applied intracellularly via the perfusion technique.* ***Ab:*** *I-V curve of I_{Ca} before (open circles) and after (filled circles) the application of 1.76 μM PK-A. Note that PK-A inhibited I_{Ca} at all potentials. (Aa) and (Ab) were obtained from the same cell.* ***B:*** *I_{Ca} traces evoked by depolarizing the membrane to 0 mV from a HP of -80 mV in the absence (open circles) and presence (filled circles) of PK-G (50 nM) plus cGMP (100 μM). PK-G was applied by intracellular perfusion. (Xiong et al., unpublished data.)*

In freshly-isolated cells from the portal veins of rabbits and rats, 8-Br derivatives of both cAMP and cGMP substantially depressed $I_{Ca\ (L)}$ (Fig. 4). Forskolin also was able to inhibit $(_{Ca(L)})$ (Fig. 5A), and this action was prevented by pretreatment with H-7, an inhibitor of protein kinases (Fig. 5B).

Protein Kinases

Introduction of PK-A (cat. subunit) and PK-G into the cell via the patch pipette (perfused) also produced marked inhibition of $I_{Ca(L)}$ in freshly isolated cells from the portal veins of rabbits and rats. 1.76 μM PK-A inhibited $I_{Ca(L)}$ by about 50% within 10 min. (Fig. 6A). Similar inhibition was produced by 50 nM PK-G (Fig. 6B). These results provide evidence that the inhibitory effects of both cyclic nucleotides on $I_{Ca(L)}$ are exerted via the respective protein kinases and phosphorylation.

Phosphorylation Model

A phosphorylation model for regulation of Ca^{2+} slow channels in VSM cells is proposed. cAMP and cGMP have the same effect, namely both inhibit the Ca^{2+} slow channels (Fig. 7). A second regulatory protein, when phosphorylated by protein kinase C (PK-C), stimulates the Ca^{2+} slow channels, as depicted (Fig. 7).

MECHANISMS OF ACTION OF SOME VASOACTIVE SUBSTANCES

Cyclic Nucleotides Mediated Actions

Figure 8 summarizes the roles played by cyclic nucleotides and PK-C in regulating the intracellular Ca^{2+} concentration ($[Ca]_i$) in VSM cells. Activation of A-cyclase or G-cyclase by appropriate membrane receptors (e.g., beta-adrenergic, prostacyclin, ANP) and G coupling proteins or directly (by agents like nitroprusside, nitric oxide free radical, or forskolin) produces elevation of cAMP and cGMP and activation of PK-A and PK-G. These kinases can phosphorylase the Ca^{2+} slow channel (to inhibit channel opening) and the K^+ channels (to stimulate channel opening). Both mechanisms would inhibit the Ca^{2+}-dependent APs, and thereby diminish Ca^{2+} influx and lower $[Ca]_i$ cG-PK and PK-A also phosphorylase 12
e sarcolemmal Ca^{2+}-ATPase and stimulate the Ca^{2+} pump in the surface membrane, thus lowering $[Ca]_i$. Lowering $[Ca]_i$ inhibits contraction and produces vasodilation.

The cyclic nucleotides also phosphorylase the myosin light-chain kinase (MLCK), which diminishes its Ca^{2+} sensitivy, and they have been reported to inhibit phospholipase C (PL-C) and, therefore, production of IP_3 and DAG. These factors would contribute to the vasodilation.

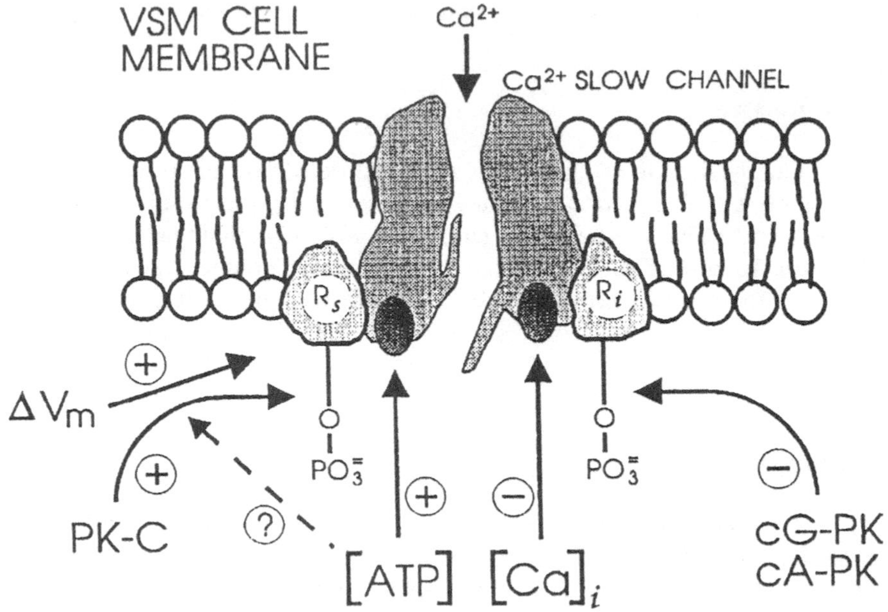

Figure 7. Diagrammatic model of a Ca^{2+} slow channel protein embedded in the lipid bilayer of the cell membrane of a vascular smooth muscle cell. Depicted are two regulatory proteins, inhibitory and stimulatory, associated with the channel protein. As shown, both regulatory proteins have phosphorylation sites which affect the function of the regulatory proteins. Phosphorylation of the inhibitory protein by cAMP-protein kinase (PK-A) or by cGMP-PK (PK-G) inhibits the function of the ion channel; e.g., decreases probability of channel opening at depolarized potentials. Phosphorylation of the stimulatory protein by PK-C stimulates the Ca^{2+} slow channel; eg. increases probability of opening. Thus, channel function is regulated by cyclic nucleotides and by diacylglycerol.

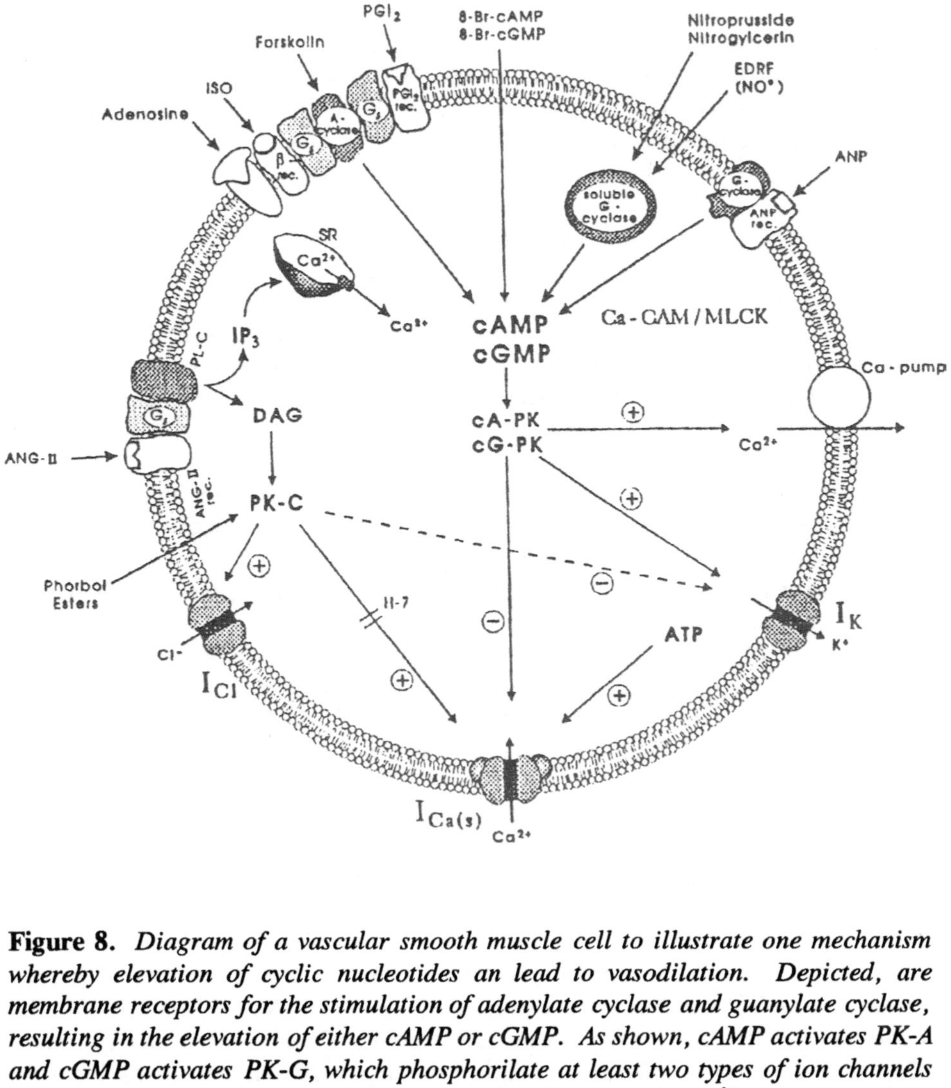

Figure 8. *Diagram of a vascular smooth muscle cell to illustrate one mechanism whereby elevation of cyclic nucleotides an lead to vasodilation. Depicted, are membrane receptors for the stimulation of adenylate cyclase and guanylate cyclase, resulting in the elevation of either cAMP or cGMP. As shown, cAMP activates PK-A and cGMP activates PK-G, which phosphorilate at least two types of ion channels and a Ca pump in the sarcolemman. Phosphorylation of the* Ca^{2+} *slow channel (or an associated inhibitory-type regulatory protein) produces inhibition of the channel, e.g., decreases probability of opening at depolarized potentials. Also depicted is the fact that phosphorylation by PK-C stimulates the* Ca^{2+} *slow channel. The delayed rectifier* K^+ *channel is stimulated by phosphorylation with PK-A and PK-G, as illustrated. Also depicted is stimulation of the sarcolemmal* Ca^{2+} *pump by PK-A and PK-G.*

Activation of angiotensin-II receptors stimulates IP_3 and DAG production (Fig. 8). DAG activates PK-C to phosphorylase the Ca^{2+} slow channels and stimulate Ca^{2+} influx and raise $[Ca]_i$. Phorbol esters act, like DAG, to directly activate PK-C. IP_3 acts on the SR to release Ca^{2+} (via the Ca-release channels) and elevate $[Ca]_i$. Raising $[Ca]_i$ potentiates contraction and produces vasoconstriction. The following section elaborates on the actions of angiotensin.

Modulation of Ca^{2+} Channels by Angiotensin

Agonists, such as norephinephrine (NE) or angiotensin-II (Ang-II), may induce contraction of VSM via several mechanisms, including (a) stimulation of Ca^{2+} influx through receptor-operated channels (ROCs); (b) stimulation of Ca^{2+} entry through voltage-dependent Ca^{2+} channels, opened indirectly by the depolarization resulting from an increase in membrane conductance for other ions; and (c) release of intracellular Ca^{2+} from store sites (for review: 3, 7, 9, 16).

In cultured VSM cells from rat aorta, Ang-II, applied by bolus perfusion, elicited a transient depolarization of up to 30 mV, which sometimes triggered an AP (22). The Ang-II-induced depolarization disappeared in Na^+-free solution, suggesting that it was primarily due to an increase in Na^+ conductance, with resulting Na^+ influx. Continuous superfusion of cultured aortic reaggregates with Ang-II (10^{-6}M) also produced depolarization, which triggered long-lasting APs (8). The membrane resistance was decreased by Ang-II, consistent with an increase in conductance for an inward depolarizing current. Low concentrations of Ang-II (10^{-9}M) caused a marked prolongation of ongoing Ca^{2+}-dependent APs without depolarizing the membrane.

Ang-II also greatly stimulated the Ca^{2+} slow channel current in isolated single cells from guinea pig portal vein (13) (Fig. 9). This effect of Ang-II was mediated via a G-protein, based on inhibition of the Ang-II effect by GDP-βS and stimulation of the basal I_{Ca} by GTP-γS. However, this G-Protein was not sensitive to pertussis toxin or to cholera toxin, hence is not G_i-like or G_s-like.

In cultured VSM cells from rabbit aorta, Ang-II caused a substantial increase in the inward Ca^{2+} channel current (carried by Ba^{2+}) (2). Ang-II (1-5 x 10^{-8}M) also depressed the delayed-rectifier outward K^+ current. These effects of Ang-II on the Ca^{2+} and K^+ channel currents can explain the depolarization and AP prolongation found in cultured rat aortic cells. These observations suggest that agonists control vascular tone, at least in part, by modifying the Ca^{2+} channel and K^+ channel activities, and thereby changing $[Ca]_i$.

REGULATION OF Ca^{2+} CHANNELS BY ATP

In freshly isolated cells from guinea-pig small mesenteric arteries (resistance vessels), using the whole-cell voltage clamp technique, injection of ATP (0.3-5 mM), using an intracellular perfusion method, modified only the slow Ca^{2+} current, but did not affect the fast Ca^{2+} current (11) (Fig. 10). When the production of ATP was inhibited by adding cyanide and 2-deoxy-D-glucose to the bath (in absence of glucose), the Ca^{2+} slow current was abolished within 10 min, whereas the fast Ca^{2+} current was only slightly affected. These results indicate that the slow Ca^{2+} channel is metabolically dependent. The ATP dependence of the activity of the Ca^{2+} slow channels was also shown at the single-channel level (12). Therefore, there may be an obligatory binding site for ATP on the inner surface of the channel which must be occupied for channel activity. The same appears to be true for slow Ca^{2+} channels of myocardial cells (17, 14).

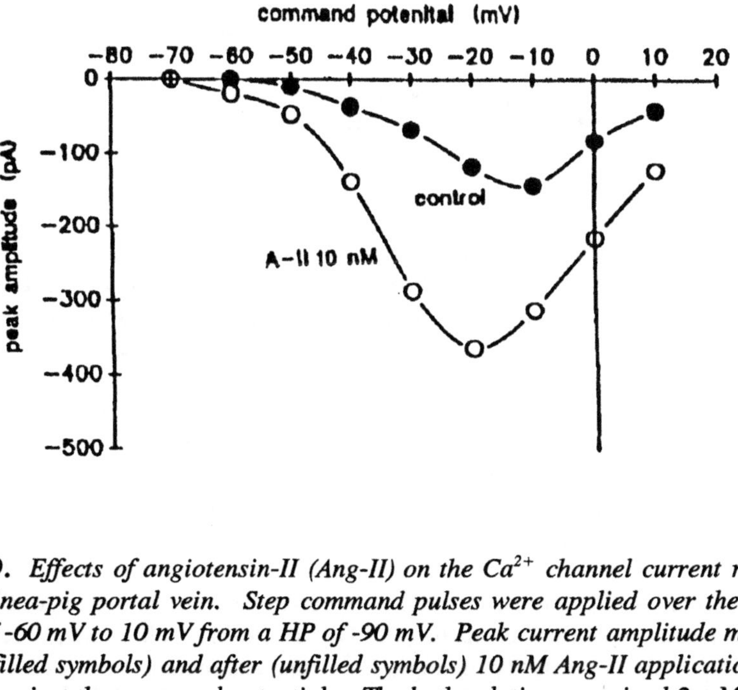

Figure 9. *Effects of angiotensin-II (Ang-II) on the Ca^{2+} channel current recorded from guinea-pig portal vein. Step command pulses were applied over the voltage range of -60 mV to 10 mV from a HP of -90 mV. Peak current amplitude measured before (filled symbols) and after (unfilled symbols) 10 nM Ang-II application, were plotted against the command potentials. The bath solution contained 2 mM Ba^{2+} as the charge carrier. The pipette solution contained high Cs with 5 mM ATP and 0.1 mM GTP. (13).*

DIRECT MODULATION OF Ca^{2+} CHANNELS BY G-PROTEIN

The direct regulation of ionic channels by GTP-binding protein (G-protein) has been identified in several tissues (for review: 4). In some neurons, G-protein (G_o) is thought to mediate the modulation (inhibition) of Ca^{2+} channels produced by

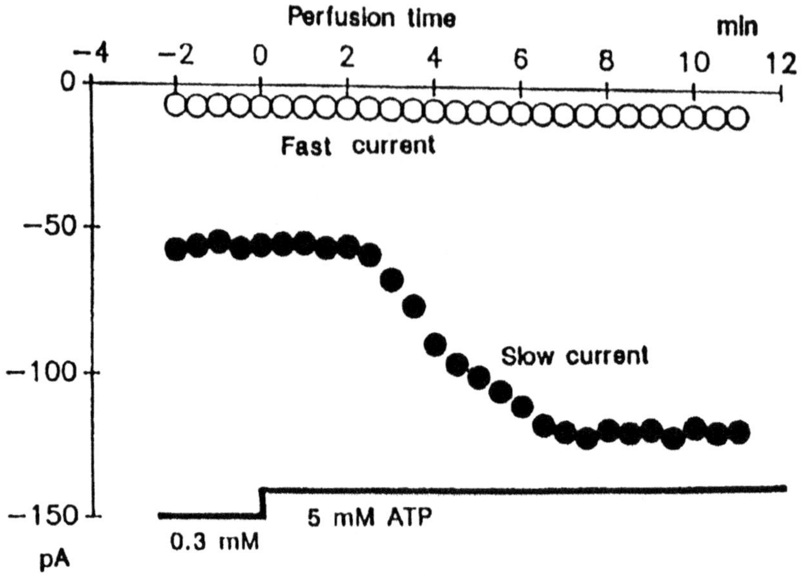

Figure 10. Effects of intracellularly-perfused ATP on the separated fast and slow Ca^{2+} channel currents. The concentration of ATP was increased from 0.3 mM to 5 mM using the intracellular perfusion technique. The time after beginning of the perfusion is indicated on the abscissa. The slow current was isolated by using a HP of -40 mV to voltage-inactivate the fast channels; command potentials to +30 mV were applied every 30 sec. The fast current was recorded using a HP of -80 mV and a command potential to -20 mV. The peak magnitudes of the fast (open circles) and slow (filled circles) Ca^{2+} channel currents were plotted as a function of time following perfusion with 5 mM ATP. The fast and slow currents were obtained from two different cells. (Data taken from 11).

receptor activation (6). In cardiac cells, G-protein (G_s) was reported to directly enhance the activity of Ca^{2+} channels (20, 21). In freshly-isolated single cells from guinea-pig portal vein, activity of single Ca^{2+} channels (conductance of 18 - 22 pS) were recorded in the cell-attached patch configuration (with 100 mM Ba^{2+} and Bay-K-8644 in the pipette and a high K^+ solution in the bath) (10). After opening one end of the cell, GTP-γS (0.1 mM) was applied to the bath (access into the cell interior) to enhance the channel activity. This observation suggests that a G-protein may be one of the factors regulating Ca^{2+} channels in VSM cells also, but it cannot exclude the possibility that the G-protein acts indirectly.

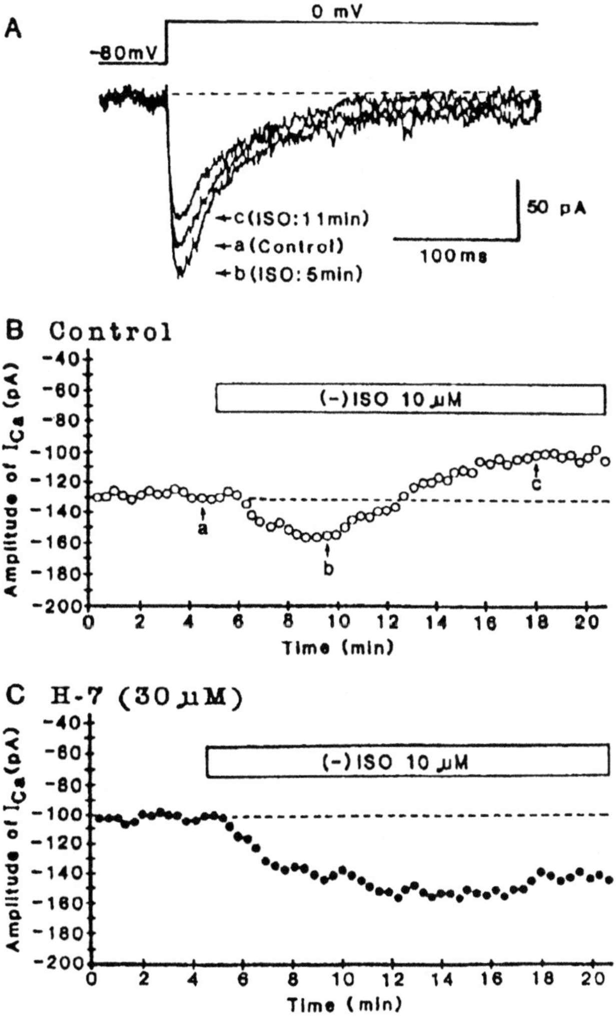

Figure 11. *Effect of isoproterenol (ISO, 10 μM) on I_{Ca} in single smooth muscle cell from rabbit portal vein. A: Current traces recorded in the absence (a, control) and presence of ISO (b, 5 min: c, 11 min). B: The time-course of I_{Ca} during the application of ISO. (A) and (B) were from the same cell. Note that ISO initially increased I_{Ca}, followed by a decrease. C: The time-course of I_{Ca} recorded from another cell during the application of ISO, after pretreatment with H-7 (30 μM). Note that isoproterenol only produced a sustained increase in I_{Ca}. (Xiong et al., unpublished data.)*

When isoproterenol (ISO) (10 μM) was added to whole-cell voltage clamped rabbit portal vein cells, there was a dual effect produced (Fig. 11A, B). First, there was a transient increase in $I_{Ca(L)}$, followed by a sustained decrease in $I_{Ca(L)}$. When H-7 was applied as a pretreatment (to inhibit PK-A), the second phase was prevented; namely, only a sustained increase was produced (Fig. 11C). These results indicated that the second inhibitory effect was mediated by cAMP and PK-A, and phosphorylation, whereas the first stimulatory effect was not. This was consistent with the first stimulatory effect being mediated by direct G-protein gating.

SUMMARY AND CONCLUSIONS

In several types of VSM cells and cultured cell lines, both 8-Br-cAMP and 8-Br-cGMP added into the bath substantially depress $I_{Ca(L)}$ (by about 50%). Agents that elevate intracellular cAMP, such as forskolin and isoproterenol, have the same effect. Several types of vasodilators, such as ANP, nitric oxyde, EDRF, nitroprusside, prostacyclin, and adenosine, act by elevating the cAMP or cGMP level. These cyclic nucleotides act by activating their respective protein kinases, namely PK-A and PK-G, which then phosphorylase several membrane proteins. This was demonstrated by injecting PK-A (catalytic subunit) and PK-G into the cell via the patch pipette, and finding inhibition of $I_{Ca(L)}$, and showing that the PK inhibitor H-7 blocked these actions.

The membrane proteins that are probably phosphorylated include ion channels, namely the slow Ca^{2+} channels and the delayed rectifier K^+ channels, and the sarcolemmal Ca-ATPase/pump. These phosphorylations all lead to a lowered $[Ca]_i$: (1) inhibition of the activity of the L-type Ca^{2+} channels, thereby inhibiting Ca^{2+} influx during depolarizing excitation; (2) stimulation of the activity of the K^+ channels ($I_{K(del)}$), thereby terminating the AP excitation prematurely and deactivating any Ca^{2+} channels that were activated; (3) stimulation of Ca^{2+} efflux via the Ca pump.

Thus, in VSM cells, both cAMP and cGMP act in the same direction on slow Ca^{2+} channels, namely to inhibit their activity. This is in sharp contrast to cardiac muscle, where cAMP and cGMP act in opposing directions: cAMP stimulates $I_{Ca(L)}$ and cGMP inhibits $I_{Ca(L)}$. In skeletal muscle (frog), both cyclic nucleotides act in the same direction, but to stimulate $I_{Ca(L)}$. In uterine smooth muscle cells (rat), neither cyclic nucleotide has any demonstrable effect on $I_{Ca(L)}$.

The vasoconstrictor, angiotensin-II, depolarized, prolonged the Ca^{2+}-dependent AP_s, stimulated $I_{Ca(L)}$, and inhibited $I_{K(del)}$ in VSM cells. The latter two effects on Ca^{2+} and K^+ channels, which can explain the AP prolongation and increased Ca^{2+} influx, presumably occurs via increased DAG production and activation of PK-C. It was shown that H-7 blocked the stimulation of $I_{Ca(L)}$ produced by phorbol esters. Angiotensin also increases IP_3 production, which acts as a second messenger to activate the Ca^{2+}-release channels of the sarcoplasmic reticulum (SR), and thereby elevate $[Ca]_i$ and promote contraction.

It was demonstrated that the slow Ca^{2+} channels require ATP for activity, with half-inhibition occurring at about 0.3 mM ATP. The fast Ca^{2+} channels did not require ATP. The relatively high $K_{1/2}$ value suggests that this ATP requirement is independent of any phosphorylation, and therefore, could reflect an obligatory binding site on the slow Ca^{2+} channel protein. A similar requirement has been suggested for the slow Ca^{2+} channel of cardiac muscle cells.

ISO (10 μM) produced a dual effect on $I_{Ca(L)}$: initially there was an increase, followed by a sustained decrease. Pretreatment with the protein kinase inhibitor, H-7, prevented the second effect of ISO (inhibition of $I_{Ca(L)}$), suggesting that the initial effect of ISO (stimulation of $I_{Ca(L)}$) was not due to phosphorylation (but perhaps reflected G-protein gating). In contrast, forskolin (10-100 μM) only had an inhibitory action on $I_{Ca(L)}$ (ca. 30% inhibition), and this action was prevented by H-7. Therefore, the first stimulatory effect of ISO is independent of any phosphorylation, and many reflect a more direct modulation of the slow Ca^{2+} channels by G_s-protein.

REFERENCES

1. Bkaily G, Peyrow M, Yamamoto T, Sculptoreanu A, Jacques D and Sperelakis N (1988a) Mol Cell Biochem 80:59-72
2. Bkaily G, Peyrow M, Sculptoreanu A, Jacques D, Chahine M, Regoli D and Sperelakis N (1988b) Pflügers Arch 412:448-450
3. Bolton TB (1979) Physiol Rev 59:606-718
4. Brown AM and Birnbaumer L (1988) Am J Physiol 254:H401-H410
5. Dosemeci A, Dhallan RS, Cohen NM, Lederer WJ and Rogers TB (1988) Circ Res 62:347-357
6. Hescheler J, Hameyama M, Trautwein W, Mieskes G and Soling HD (1987) Eur J Biochem 365:216-266
7. Johansson B and Somlyo AP (1980) In: Handbook of Physiology, Sect 2, The Cardiovascular System, Am Physiol Soc pp 301-323
8. Johns DW and Sperelakis N (1990) Eur J Pharmacol 187:183-191
9. Kuriyama H, Ito Y, Suzuki H, Kitamura T and Itoh T (1982) Am J Physiol 243:H641-H662
10. Ohya Y and Sperelakis N (1988) The Physiol 31:A88
11. Ohya Y and Sperelakis N (1989a) Circ Res 64:145-154
12. Ohya Y and Sperelakis N (1989b) Pflügers Arch 414:257-264
13. Ohya Y and Sperelakis N (1991) Circ Res 68:763-771
14. O'Rourke B, Blackx PH and Marban E (1992) Science 257:245-248
15. Ousterhout JM and Sperelakis N (1987) Eur J Pharmacol 144:7-14
16. Sperelakis N and Ohya Y (1989) In Physiology and Pathophysiology of the Heart, 2nd edition, Kluwer Academic Press, Boston pp 773-811
17. Sperelakis N and Schneider JA (1976) Am J Cardiol 37:1079-1085
18. Sperelakis N, Tohse N and Ohya Y (1992) In Excitation-contraction coupling in skeletal, cardiac and smooth muscle, Frank GP ed. Plenum Press, New York pp 163-187

19. Yatani A, Imoto Y, Codina J, Hamilton SL, Brown AM and Birnbaumer L (1988) J Biol Chem 263:9887-9895
20. Yatani A, Codina J, Imoto Y, Reeves JP, Birnbaumer L and Brown AM (1987a) Science 238:1288-1292
21. Yatani A, Seidel CL, Allen J and Brown AM (1987b) Circ Res 60:523-533
22. Zelcer E and Sperelakis N (1981) Blood Vessels 18:263-279

CHAPTER 11

ROLE OF THE R-TYPE Ca^{2+} CHANNEL IN CARDIOVASCULAR PHYSIOPATHOLOGY

[1]Ghassan Bkaily, [2]Pedro d'Orléans-Juste,
[1]Radha Naik, [1]Demetri Economos, [1]Shimin Wang,
[3]Elias Abdulnour and [4]Jean-Luc Ardilouze

Departments of [1]Physiology and Biophysics,
[2]Pharmacology, [3]Surgery and [4]Medicine (Endocrinology Service)
Faculty of Medicine, University of Sherbrooke,
Sherbrooke, Quebec, Canada

Three Types of Voltage-Dependent Ca^{2+} Channels in VSM Cells

Early in the 1980's, it became apparent that different types of Ca^{2+} currents coexisted in several kinds of excitable cells (1-21). In vascular smooth muscle (VSM), at least two types of Ca^{2+} currents are present (9, 13, 20, 22-24); L-type (for long lasting), and T-type (for transient) (3, 10, 21). Compared to the L-type Ca^{2+} current, the T-type current activates and inactivates relatively rapidly at more negative test potentials. The L-type Ca^{2+} channel inactivation is both voltage and Ca^{2+} dependent (i.e. biexponential), however, the T-type Ca^{2+} current inactivation is monoexponential and has little voltage dependency. The L-type Ca^{2+} channels are about three times more permeable to Ba^{2+} than to Ca^{2+}, while the T-type Ca^{2+} channels are equally permeable to the two ions (2, 10, 14, 21). The L-type Ca^{2+} current is sensitive to classical Ca^{2+} channels blockers and the T-type I_{Ca} is sensitive to low concentrations of Ni^{2+}. The N-type Ca^{2+} current which has kinetic properties intermediate to the L- and T-type currents has been described only in nerve cells and this type of channel seems to be non-functional in VSM cells in general (for review see 1-3, 5, 7, 11, 15, 17).

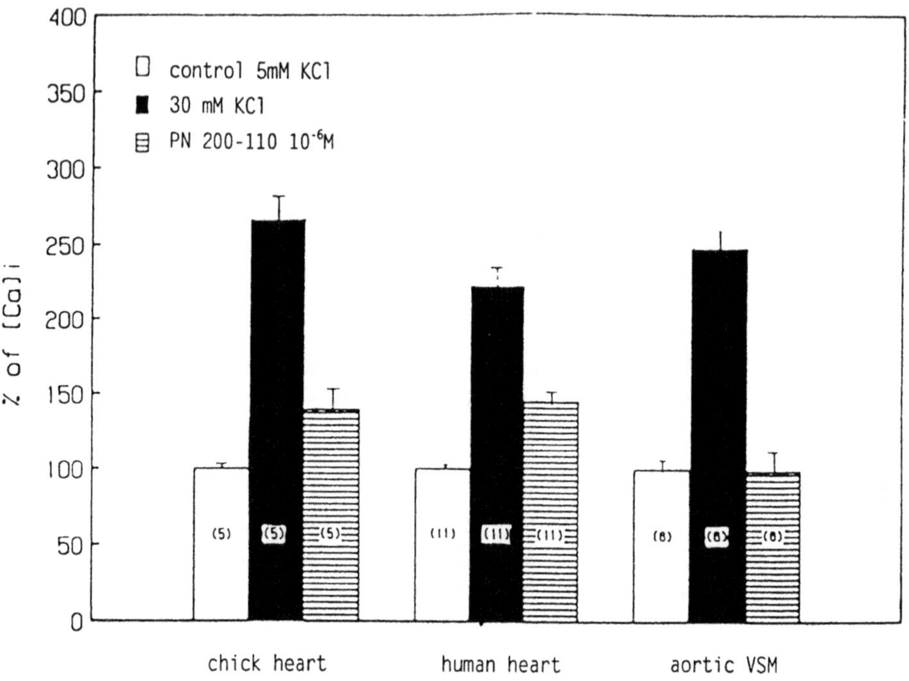

Figure 1. *Blockade of the sustained increase of $[Ca]_i$, induced with 30 mM extracellular K^+, by PN 200-110 in 10-day-old single chick heart cells (n = 5), 20 week-old single fetal human heart single cells (n = 11) and rabbit aortic VSM cells (n = 6). Vertical bar is SEM. (Modified from Bkaily et al. (25)).*

Very recently, a new type of Ca^{2+} channel called the R-type (for resting, 1, 25) was also found in the aortic VSM of the rabbit as well as in human renal artery single cells (1, 7, 25), vascular endothelial cells (26) and ventricular heart cells (25). This type of Ca^{2+} channel is activated during sustained depolarization of the cell membrane or in the presence of a specific vasoactive agent (1, 6-7, 25-26).

Using the microfluorometry and video imaging Fura-2 Ca^{2+} measurement techniques (18, 45), a sustained depolarization of the cell membrane of the VSM cell induced a transient increase of $[Ca]_i$ that was followed by a sustained component (25). The transient increase of$[Ca]_i$ is due to activation of the L-type Ca^{2+} channel that is blocked by calcium blockers such as nifedipine (25). However, the sustained increase of $[Ca]_i$ is due to opening of the R-type Ca^{2+} channel which is insensitive to inorganic and organic L-type Ca^{2+} channels blockers, with the exception of isradipine (PN 200-110) (Fig. 1). The R-type Ca^{2+} channel does not decay during a sustained depolarization as do the T and L-type Ca^{2+} channels. As can be seen in

Figure 1, in both chick and human heart cells, stimulation of the R-type Ca^{2+} channel with sustained depolarization of the cell membrane by 30 mM K^+ caused 1.6- and 1.2-fold sustained increases of $[Ca]_i$, respectively, over the control levels. In both chick and human heart cells, only isradipine significantly diminished the sustained increase of $[Ca]_i$ induced by a sustained depolarization of the cell membrane. As for heart cells, stimulation of the R-type Ca^{2+} channel in aortic VSM cells with a sustained depolarization by a 30 mM K^+ solution caused a 1.5-fold increase of $[Ca]_i$ and isradipine (but not nifedipine) completely returned the $[Ca]_i$ to the basal control level.

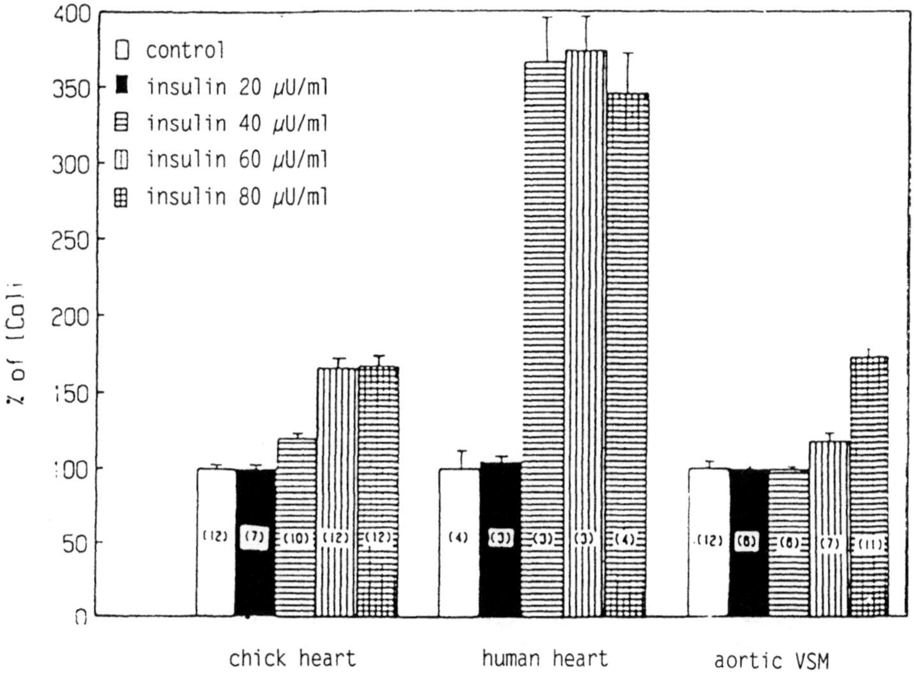

Figure 2. Effect of different concentrations of insulin on the basal resting level of $[Ca]_i$ in 10-day-old chick heart cells (n =12), 20-week-old fetal human heart cells (n = 4) and rabbit aortic VSM cells (n = 12). Vertical bar in SEM. (From Bkaily et al. (25)).

Activation of the R-Type Ca^{2+} Channel by Insulin

The R-type Ca^{2+} channel in VSM cells is activated from a resting membrane level by several hormones such as insulin. Insulin, at a concentration which is near the physiological concentration (20 to 30 μU/ml), has no significant effect on the R-type Ca^{2+} channel (25). However at pathological concentrations (60 to 80 μU/ml),

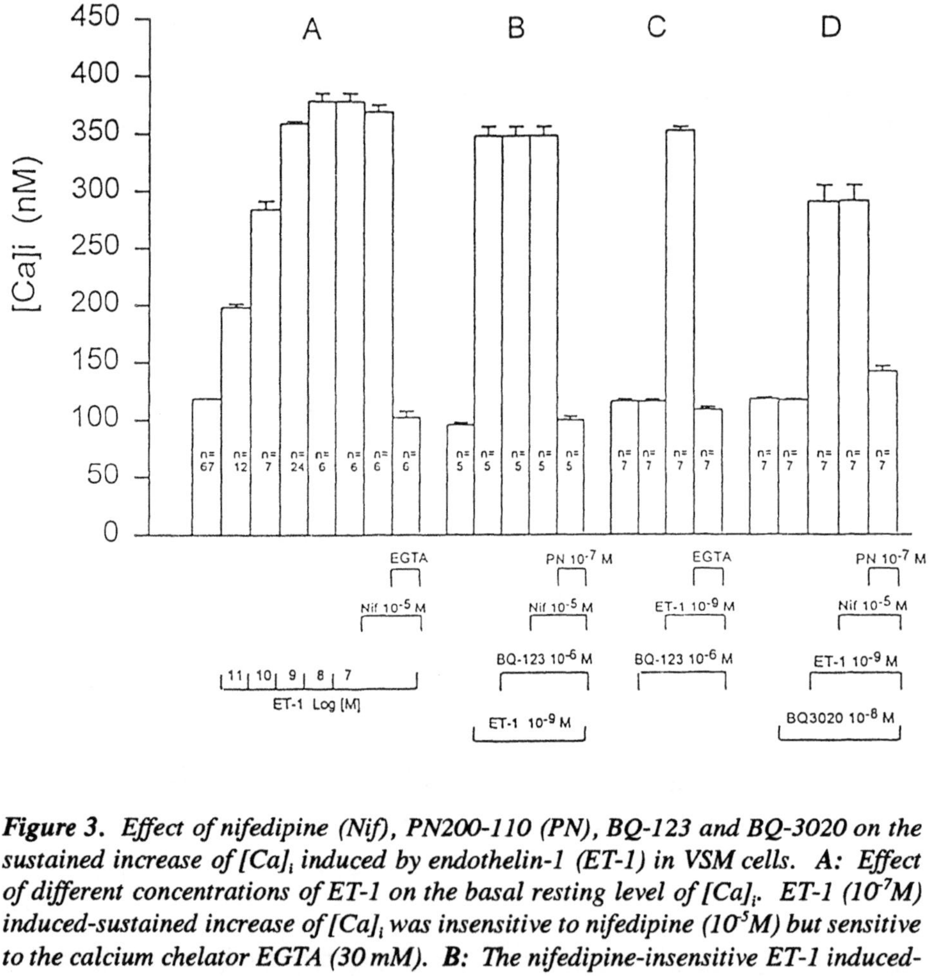

Figure 3. *Effect of nifedipine (Nif), PN200-110 (PN), BQ-123 and BQ-3020 on the sustained increase of [Ca]$_i$ induced by endothelin-1 (ET-1) in VSM cells.* ***A:*** *Effect of different concentrations of ET-1 on the basal resting level of [Ca]$_i$. ET-1 (10^{-7}M) induced-sustained increase of [Ca]$_i$ was insensitive to nifedipine (10^{-5}M) but sensitive to the calcium chelator EGTA (30 mM).* ***B:*** *The nifedipine-insensitive ET-1 induced-increase of [Ca]$_i$ was insensitive to the ET_A receptor antagonist BQ-123 but sensitive to PN200-110 (10^{-7}M).* ***C:*** *BQ-123 did not prevent the ET-1 induced-increase of [Ca]$_i$. Addition of EGTA blocked the effect of ET-1 on [Ca]$_i$.* ***D:*** *The selective ET_B agonist BQ-3020 had no effect on [Ca]$_i$ and did not prevent ET-1 from inducing a sustained increase of [Ca]$_i$ that was insensitive to nifedipine but sensitive to PN 200-110. Vertical bars represent s.e.m. and n is the number of experiments. In B-D, the substances were added sequentially after a long steady-state effect was reached. (From Bkaily et al., (27).*

this hormone activates only the R-type Ca^{2+} channel, leading to a sustained increase of $[Ca]_i$ in a dose dependent manner (Fig. 2). Once this channel is activated by a high insulin concentration (80 μU/ml), depolarization of the cell membrane does not further increase the sustained increase of $[Ca]_i$ induced by the hormone. The absence of effect of sustained depolarization on the R-type Ca^{2+} channel in the presence of a high concentration of insulin is due to the fact that the R-type channel is already fully activated by the hormone. Insulin at 80 μU/ml does not affect the T and L-type Ca^{2+} channels and does not depolarize the cell membrane (25). The effect of insulin on $[Ca]_i$ seems not to be mediated via activation of calcium release from intracellular pools and not mediated via activation of second messengers. However, the effect of insulin on the R-type Ca^{2+} channel seems to be mediated via a GTP-binding protein, probably a G_{IR} (28).

Activation of R-Type Ca^{2+} Channel by ET-1

Other hormones such as endothelin-1 (ET-1) were found to activate the R-type Ca^{2+} channel, which lead to a sustained increase of $[Ca]_i$ (27, 29). ET-1 is a potent vasoconstrictor peptide which is released by endothelial cells (E-C) (21). It has been shown to produce a contraction that has a different pattern and pharmacology from that induced by activation of the L-type Ca^{2+} channel activator Bay K 8644 (30). Thus, ET-1 produces prolonged contraction that is due to continual stimulation of the R-type Ca^{2+} channels which are responsible for the sustained increase of $[Ca]_i$, without affecting the T and L types Ca^{2+} channels and the resting membrane potential of VSM cells (30). ET-1 was found to induce a concentration dependent sustained increase of $[Ca]_i$ that was insensitive to the L-type Ca^{2+} channel blocker nifedipine, but sensitive to isradipine (Fig. 3A). The effect of ET-1 on the R-type Ca^{2+} channel did not seem to be mediated via activation of a known receptor (Fig. 3B-D): however, it seemed that ET-1 activation of the R-type Ca^{2+} channel was mediated via a specific G-protein that also seemed to mediate the effect of insulin on this channel (27).

As shown in figure 3, ET-1 at a concentration as low as 10^{-11}M significantly ($P < 0.005$) induced a sustained increase of $[Ca]_i$. The L-type Ca^{2+} channel blocker nifedipine had no effect on the Ca^{2+} chelator EGTA which decreased the level of $[Ca]_i$ to below the control basal resting level (Fig. 3A). The EC_{50} of ET-1 on $[Ca]_i$ was estimated at 10^{-10}M. Figures 3B and C show that a sustained increase of $[Ca]_i$ induced by ET-1 is not sensitive to the ET_A receptor antagonist BQ123, but sensitive to the R-type Ca^{2+} blocker isradipine (PN) and the Ca^{2+} chelator EGTA. Figure 3D shows that the ET_B receptor agonist BQ-3020 had no effect on the basal control level of $[Ca]_i$ and application of ET-1 induced a sustained increase of $[Ca]_i$ that was insensitive to the L-type Ca^{2+} blocker, nifedipine, but sensitive to the R-type Ca^{2+} blocker isradipine (PN). Also, endothelin-3, a selective agonist of ET_C receptors and natural agonist of ET_B receptors, had no effect on the R-type Ca^{2+} channel in VSM cells.

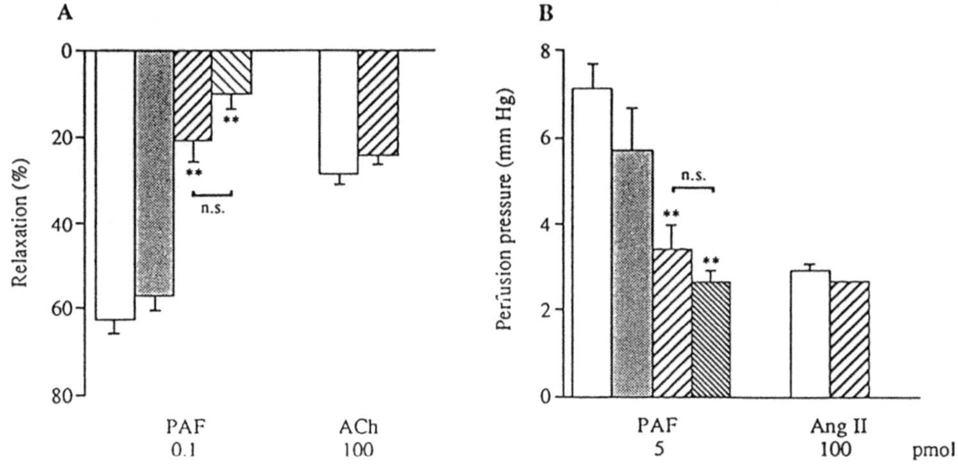

Figure 4. *Effect of PAF (0.1 and 5 pmol) or the arterial and venous sides, respectively in the absence (□), or presence of nifedipine (0.1 μM) (□), or nifedipine (0.1 μM) and (+) isradipine (1 μM) (▨), or nifedipine (0.1 μM), (+) isradipine (1 μM) and WEB-2170 (arterial: A: 0.1 μM: ▧: venous: B: 0.1 nM). The same mixture of antagonists was tested against ACh (100 pmol) on the arterial vasculature and Ang II (100 pmol) on the venous mesenteric vasculature. Each column represents the mean ± s.e. mean of 5 to 8 determinations. n.s.: not-significant, **P < 0.01 when compared to control values (□). (Modified from Claing et al. (40)).*

Activation of R-type Ca^{2+} channel by PAF

Platelet activating factor (PAF) is a lipid mediator with numerous biological activities related to inflammatory and immune (31-33) as well as respiratory, cardiovascular, reproductive and nervous system physiology (34). Inflammation for example, leads to the production of a variety of substances by EC and inflammatory cells. One of these substances is the potent mediator of inflammation, PAF (32). PAF can be produced by an activation of several types of cells such as neutrophils, monocytes, platelets, basophils, macrophages and EC (35-37). Stimulation of the inflammatory cells induces a prompt release of cytokines, PAF and eicosanoids. These products seem to be found in the immediate inflammatory microenvironment. Inflammatory mediators may cause EC to secrete not only ET-1, but also PAF. In the vascular system, the release of PAF has been associated, in several conditions, with increased vascular permeability. The increase of vascular permeability by PAF is due to heterogeneity of the vasoactive properties and the endothelium dependent action of PAF. Thus, PAF seems to induce an endothelium-dependent relaxation in the mesenteric arterial vasculature and vasoconstrict the venous side. Figure 4 illustrates that pretreatment with the L-type Ca^{2+} channel blocker, nifedipine (0.1 μM), did not affect the vasodilator properties of PAF on the arterial side nor the vasoconstrictive properties of the same agent. In contrast, in the presence of nifedipine, treatment with isradipine (+ or - enantiomer) significantly reduced the vasoactive effect of PAF on the arterial and venous mesenteric vasculature, without affecting the vasodilator properties of acetylcholine on the arterial side and the veno-constrictive properties of angiotensin II. The contribution of venoconstriction in plasma leakage induced by PAF has been suggested to be due to the complex phenomenom involved in capillary extravasation of both post-capillary disjunction of endothelial cells and constriction (38-39). The effect of PAF on EC seems to be

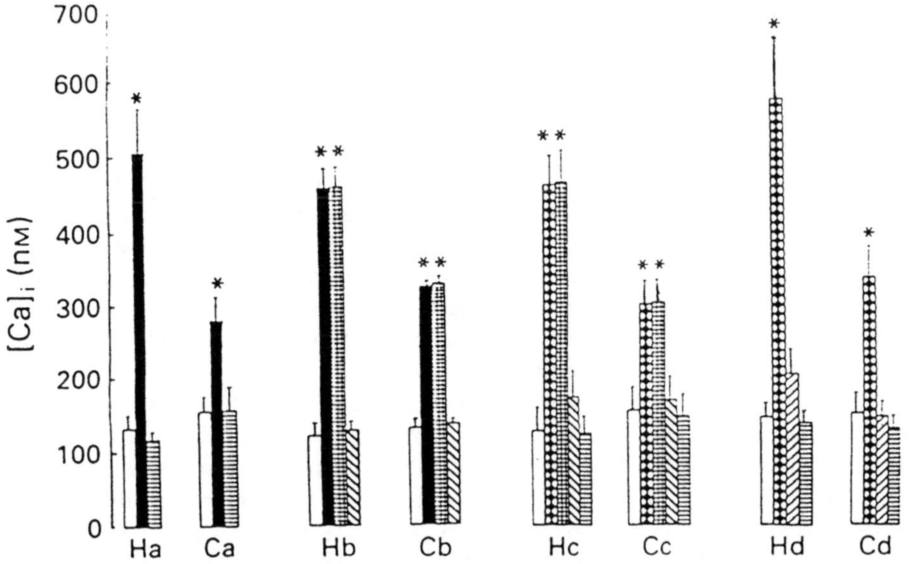

*Figure 5. High $[K]_o$ (30 mM) depolarization (■) and PAF (10^{-9}M, ▣) -induced a sustained increase of $[Ca]_i$ in both human (H) and canine (C) vascular endothelial cells that was completely blocked by 30 mM EGTA (▤) (a, c and d). The high $[K]_o$- or PAF-induced sustained increase of $[Ca]_i$ in both preparations was insensitive to 10^{-5}M nifedipine (▦), but sensitive to 10^{-7}M PN200-110 (▧) (Hb, c and Cb, c) and the PAF receptor antagonist, WEB2170 10^{-5}M, ▨) (Hd and Cd for PAF only; effects against high $[K]_o$ are not shown). The control steady-state $[Ca]_i$ basal level was □. The number of experiments (n) in Ha, Ca, Hb, Cb, Hc, Cc, Hd and Cd are respectively 5, 5, 4, 4, 3, 4, 4 and 3. *$P < 0.01$ compared to control values only. Substances were added sequentially at intervals of approximately 7 min in the order indicated. (From Bkaily et al. (26)).*

mediated in part by the activation of the R-type Ca^{2+} channel, which leads to a sustained increase of $[Ca]_i$ (Fig. 5). The sustained increase of $[Ca]_i$ by PAF may induce retraction of endothelial cells and thus increase the junctional gap between the endothelial cells, causing an increase in vascular permeability and infiltration of leucocytes which may add to intimal VSM cell proliferation. The latter event is considered to be important in the early events in heterogenesis.

Conclusion

It is apparent from the known biophysical properties of the L-type Ca^{2+} channel that it cannot necessarily play a role in maintaining the resting level of $[Ca]_i$ or in the sustained increase of $[Ca]_i$ that takes place in several physiological and pathological conditions. However, this type of channel could contribute to the resting or sustained level of $[Ca]_i$ if its window current increased due to changes in its kinetics. However, from the data presented in this chapter and the work published by our group, the R-type Ca^{2+} channel seems to be highly implicated in the determination of the basal resting level of $[Ca]_i$ as well as in the sustained increase of $[Ca]_i$ that is responsible for sustained contractions of VSM and the secretion of several hormones. The data presented in this chapter highly suggest that hormones such as insulin, ET-1 as well as PAF induce a sustained increase of $[Ca]_i$ via activation of the steady-state R-type Ca^{2+} channel in VSM cells. The high circulation or local release of each of these hormones seems to modulate the circulating level of the local secretion of another hormone via a common mechanism that seems to be related in part to the sustained increase of $[Ca]_i$. For example, stimulation of the PAF receptor may lead endothelial and VSM derived ET-1 release. It is also worthy of notice that specific activation of the R-type Ca^{2+} channels by PAF is indeed linked to physiological events in the rat mesenteric vasculature, such as the release of EDRF and venoconstriction. Also, insulin has been shown to stimulate synthesis of ET-1 by vascular EC (41-42) which leads to its increase in the circulation (43). The induction of ET-1 synthesis by insulin may represent a link between hyperinsulinemia and hypertension. From the data presented in this chapter, both insulin and ET-1 induce a sustained increase of $[Ca]_i$ in VSM via activation of the R-type Ca^{2+} channel. This type of channel seems to be a major mechanism by which intracellular calcium may be altered in a sustained manner. Such an alteration of intracellular calcium via the opening of the R-type channel may play an essential role in the pathophysiology of diabetes and its complications (for review see refs 44-46). Accumulated evidence supports the fact that a sustained increase of $[Ca]_i$, probably via activation of the R-type Ca^{2+} channel, is the pathological second messenger for several vascular pathologies including atherosclerosis. Also, accumulated evidence supports the concept that calcium is a universal regulatory signal which once having reached a certain sustained level inside the cell is capable of accounting for the myriad of aberrant cellular processes characterizing the atherosclerotic artery. The data presented in this chapter as well as those recently reported by our group highly suggest that the accumulation of $[Ca]_i$ via activation of the R-type Ca^{2+} channel in VSM cells by some leucocyte released factors, as well as by the vascular endothelial vasoconstrictive factor ET-1, may underly the long term etiology of VSM diseases such as the hypertrophy and proliferation of VSM cells seen in atherosclerosis. If the sustained increase of $[Ca]_i$ is atherogenic, it should be possible to ameliorate or prevent atherosclerosis by using compounds such as isradipine, that prevent or block the transport of Ca^{2+} via the R-type Ca^{2+} channel.

Acknowledgements

This research was supported by a Program Grant, PG-12350, of the Medical Research Council of Canada on Immuno-Cardiovascular Interactions. Dr Bkaily is a Merck-Frosst-FRSQ Professor and Dr D'Orléans-Juste is a scholar of the HSFC.

REFERENCES

1. Bkaily G (1991) In: Ion Channels of Vascular Smooth Muscle Cells and Endothelial Cells, Sperelakis N and Kuriyama H (eds) Elsevier, NY pp 185-198
2. Bean BP (1989) Annu Rev Physiol 51: 367-387
3. Bean BP (1985) J Physiol (London) 86: 1-30
4. Bean BP, Sturek M, Duga A and Hermsmeyer K (1986) Cir Res 59: 229-235
5. Benham CD, Hess P and Tsien RW (1987) Cir Res 61: 111-116
6. Bkaily G (1990) in: Frontiers in Smooth Muscle Research, Sperelakis N and Woods JD (eds) Allan R Liss Inc, NY pp 507-515
7. Bkaily G (1991) in: In Vitro Methods in Toxicology, Jolles G and Cordier A (eds) Acad Press (London), in press pp 289-334.
8. Bkaily G, Caillé J-P, Payet MD, Peyrow M, Sauvé R, Renaud J-F and Sperelakis N (1988) Can J Physiol Pharmacol 66: 731-736
9. Bkaily G, Peyrow M, Sculptoreanu A, Jacques D, Regoli D and Sperelakis N (1988) Pflugers Arch 412: 448-450
10. Fox AP, Nowycky MC and Tsien RW (1987) J Physiol (London) 394: 149-172
11. Hess P (1988) Can J Physiol Pharmacol 66: 1218-1223
12. Tsien RW, Lipscombe D, Madison DV et al (1988) Trends Neurosci 11: 431-437
13. Loirand G, Pacoud P, Mironneau C and Moronneau J (1986) 407: 566-568
14. McCleskey EW, Fox AP, Feldman D and Tsien RW (1986) 124: 177-190
15. Miller RJ (1987) Science 235: 46-52

16. Nelson MT, Patlak JB, Worley JF and Standen NB (1990) Am J Physiol 259: C3-C18
17. Ohya Y and Sperelakis N (1991) Cir Res 68: 763-771
18. Reuter H and Porzig H (1988) Nature 336: 113-114
19. Reuter H (1987) Experientia 43: 1173-1175
20. Sperelakis N and Ohya Y (1988) In: Physiology and Pathophysiology of the Heart, 2nd edition, Sperelakis N (ed) Kluwer Acad Press (Boston) pp 773-811
21. Tsien RW, Hess P, McCleskey EW and Rosenberg RL (1987) Annu Rev Biophys Chem 16: 265-290
22. Friedman ME, Suarez-Kuratz G, Kackorowski GJ, Katz GM and Reuben J-P (1986) Am J Physiol 205: H699-H703
23. Sturek M and Hermsmeyer K (1986) Science 233: 475-478
24. Wang R, Karpinski E and Pang PKT (1989) Am J Physiol 256: H361-H368
25. Bkaily G, Economos D, Potvin L, Ardilouze J-L, Marriott C, Corcos J, Bonneau D and Fong CN (1992) Mol Cell Biochem 117: 93-106
26. Bkaily G, D'Orléans-Juste P, Naik R et al (1993) Br J Pharmacol 110: 518-520
27. Bkaily G, Naik R, D'Orléans-Juste P et al (1994) Am J Hypertension (in press)
28. Nakazawa K, Matsuka N, Shigenobu K and Kasuya Y (1987) Pflugers Arch 408: 112-119
29. Bkaily G, Wang S, Economos D et al (1991) In: Ion Channels of Vascular Smooth Muscle, Cells and endothelial cells. Sperelakis N, Kuriyama H (eds) Elsevier NY pp 199-207
30. D'Orléans-Juste P, Finet M, DeNucci G, Vane jr (1989) J Cardiovasc Pharmacol 13: S19-S22
31. Braquet P, Rola-Pleszczynski M (1987) Immunol Today 8: 345-352
32. Lacasse C, Rola-Pleszczynski M (1991) J Leucocyte Biol 49: 245-252
33. Müller E, Dupuis G, Turcotte S, Rola-Pleszczynski M (1991) Biochem Biophys Res Commun 181: 1580-1586
34. Braquet P, Touqui L, Shen TY, Vargaftig BB (1987) Pharmacol Rev 39: 97-145
35. Kroegel C, Yukawa T, Westwick J, Barnes PJ (1989) Biochem Biophys Res Commun 162: 511-521
36. Hwang S-B (1988) J Biol Chem 263: 3225-3233
37. Honda Z-I, Nakamura M, Miki M et al (1991) Nature 349: 342-346
38. Heffener JE, Cook JA, Haluska PV (1989) J Clin Invest 84: 757-764
39. Sirois MG, Plante GE, Braquet P, Sirois P (1990) Br J Pharmacol 101: 896-900
40. Claing A, Bkaily G, Berthiaume N et al (1994) Br J Pharmacol (in press)
41. Hattori Y, Kasai K, Nakamura T et al (1991) Metabolism 40: 165-169

42. Oliver FJ, De La Rubia G, Feener EP et al (1991) J Biol Chem 266: 23251-23256
43. Wolpert HA, Steen SN, Istfan NW, Simonson DC (1993) Metabolism 42: 1027-1030
44. Levy J, Gavin RJ, Sowers JR (1994) Am J Med 96: 260-273
45. Epstein M, Sowers JR (1992) Hypertension 19: 403-418
46. Sowers JR (1992) J Clin Pharmacol 32: 529-535

CHAPTER 12

PASSIVE ENDOTHELIAL TRANSPORT: STUDIES IN EXPERIMENTAL ARTERIAL HYPERTENSION, DIABETES MELLITUS AND CHRONIC RENAL FAILURE

Gérard E. Plante and Mouna Chakir

Departments of Medicine,
Physiology and Pharmacology,
University of Sherbrooke, Sherbrooke, Québec,
Canada

1. PASSIVE ENDOTHELIAL TRANSPORT

a) Definitions and Description

In normal adult organisms, water occupies approximately 60% of total body weight, and is distributed among three major compartments, vascular, interstitial and cellular, representing 5%, 15% and 40%, respectively (33). Absolute and/or relative abnormalities in water and solute distribution, not surprisingly, are found in a variety of diseases, until now considered as specific and distinct morbid entities (22, 31, 41). The primary etiology of disease may vary, from physical and chemical injury, neoplasia, ischemia, nutriment deficiency and metabolic disorders, to infection and immune aggression. However, the basis of healing and/or progression of disease has a common denominator which we would propose to localize in the interstitial matrix (25).

The endothelial cell represents the site where exchanges of fluid and solutes between the vascular volume and the interstitium of various organs are taking place, establishing the critical equilibrium between the external *milieu* and intracellular space responsible for the maintenance of cell life (3). The transport of fluid and solutes across the endothelium is either transcellular and/or paracellular. The transcellular transport requires ionic pumps, channels, complex membrane fusion mechanisms (pino-endocytosis) or other specific or non-specific carriers. In most instances, transcellular transport is energy dependent. In contrast, paracellular transport is mostly passive (46). Although it may benefit from active transcellular transport processes, or indirectly from hemodynamic factors, paracellular transport can be considered as an energy-saver process. For the purpose of the following discussion, we define passive endothelial paracellular transport as **endothelial permeability**.

b) Ultrastructural Basis of Passive Endothelial Transport

Endothelial cell layers are heterogeneous in nature, either in larger blood vessels, such as the arteries and veins, or in capillary networks. The ultrastructural characteristics of endothelial cells lining the large blood vessels have not been as well defined as those of capillary beds. In the latter, the structural nature varies from fenestrated endothelial cells, which characterize the glomerular and intestinal mucosa microcirculation, to the continuous cerebral capillaries which establish the so-called very tight blood-brain barrier. The ultrastructure of endothelial cells sitting between these two extremes (fenestrated and continuous tight brain capillary) varies by a number of characteristics, including their thickness, the shape of their tight junctions (21), and probably other elements. These characteristics determine altogether the hydraulic conductivity of capillaries, commonly utilized in studies looking at permeability, which represents, in fact, the physiological consequence of their specific ultrastructure. In continuous type of endothelia, such as the cerebral, skin, skeletal muscle, pulmonary and cardiac, hydraulic conductivity (expressed in $cm^3sec^{-1}dyne^{-1} \times 10^{-13}$) ranges between 3 and 860, compared to 13,000 and 15,000 for fenestrated endothelia. It is of interest that epithelia from different organs possess similar ultrastructural characteristics which define comparable differences in their hydraulic parameters (46).

c) Physiological Basis of Passive Endothelial Transport

The indirect energy required for passive endothelial paracellular transport to occur is derived from two sources: **first**, the hemodynamic parameters in a given capillary network (pre- and post-capillary resistances) which determine the capillary hydrostatic and, in part, the oncotic pressures (3, 4); **second**, the contraction of endothelial cells, *per se*, and perhaps some yet undefined changes in the structure of the tight junction complex, which determine endothelial cell dysjunction (54). The arteriolar and venular vessels of the microcirculation contain smooth muscle cells responding to a variety of vaso-active autacoïds and hormones. Terminal nerve endings releasing neuro-hormones have also been described on these structures. In addition, the presence of pericytes, which appear to behave in a fashion similar to the mesangial cells of the glomerulus, suggest additional control mechanisms of the microcirculation. Among the recently identified vaso-active substances capable of modulating pre- and/or post-capillary resistances, suffice it to mention angiotensin II, vasopressin, bradykinin, atrial natriuretic peptides, endothelins, nitric oxide, and metabolites of arachidonic acid (15, 51, 55).

Changes in endothelial cell permeability result essentially from alteration of the paracellular transfer of fluid and solutes. Although small molecules such as ions and some organic substances cross the endothelial barrier through the paracellular route, the latter is mostly utilized by macromolecules such as albumin, since these are not normally passively transported across endothelial cells unless conveyed by pino-endocytosis (46). Cell-cell attachment elements, in particular the tight junction complexes, play a critical role in permitting the passage of fluid and solutes (including macromolecules) between endothelial or epithelial cells. In addition to these ultrastructural requirements, for endothelial permeability to macromolecules to develop, another and perhaps more critical factor which is of functional nature, is the contraction of endothelial cell (54). This physiological phenomenon is not a unique property of skeletal or smooth muscle cells. In fact, a variety of other cell types (epithelia, white blood cells, mesangial cells, and spermatozoids) have the ultrastructural basis for contraction, in particular the cytoskeletal system, which is mainly composed of tubulin. A number of morphological studies have been reported, in which endothelial cell contraction was highly correlated with enhancement of permeability to macromolecules. Among autacoids and hormones reputed to modulate endothelial cell permeability, let us mention platelet activating factor (47, 48), bradykinin (29, 41), endothelins 1 and 3 (23), the vaso-active intestinal and atrial natriuretic peptides, nitric oxide (34), and finally a variety of other cytokines.

2. SIGNIFICANCE OF PASSIVE ENDOTHELIAL TRANSPORT

a) Passive Endothelial Transport in Capillaries

Heterogeneity of capillary networks. Not only do the ultrastructural characteristics of endothelial cells differ from one capillary network to the other, but also the density of capillaries vary enormously in different tissues (53). Comparative studies express, in different manners, the degree of organ perfusion: absolute and relative blood flow per gram of tissue or as a percentage of cardiac output (16, 53). Two splanchnic organs, the gastro-intestinal tract (including the liver) and the kidney share the top positions in terms of absolute flow (1.5 and 1.3 liter/min, respectively), yet the kidney is by far, the most highly perfused organ in terms of relative flow (420 ml/100g/min). The two largest organs, the skeletal muscle and the skin, which represent 49% and 6% of total body weight, respectively, receive a considerable amount of blood (1.3 liter/min), yet the relative perfusion of these organs, containing approximately half of the total body water, reaches less than 13 ml/100g/min. These data illustrate that the organs in contact with the external *milieu* (gastro-intestinal tract and kidney for absorption-secretion functions) are much more perfused, than

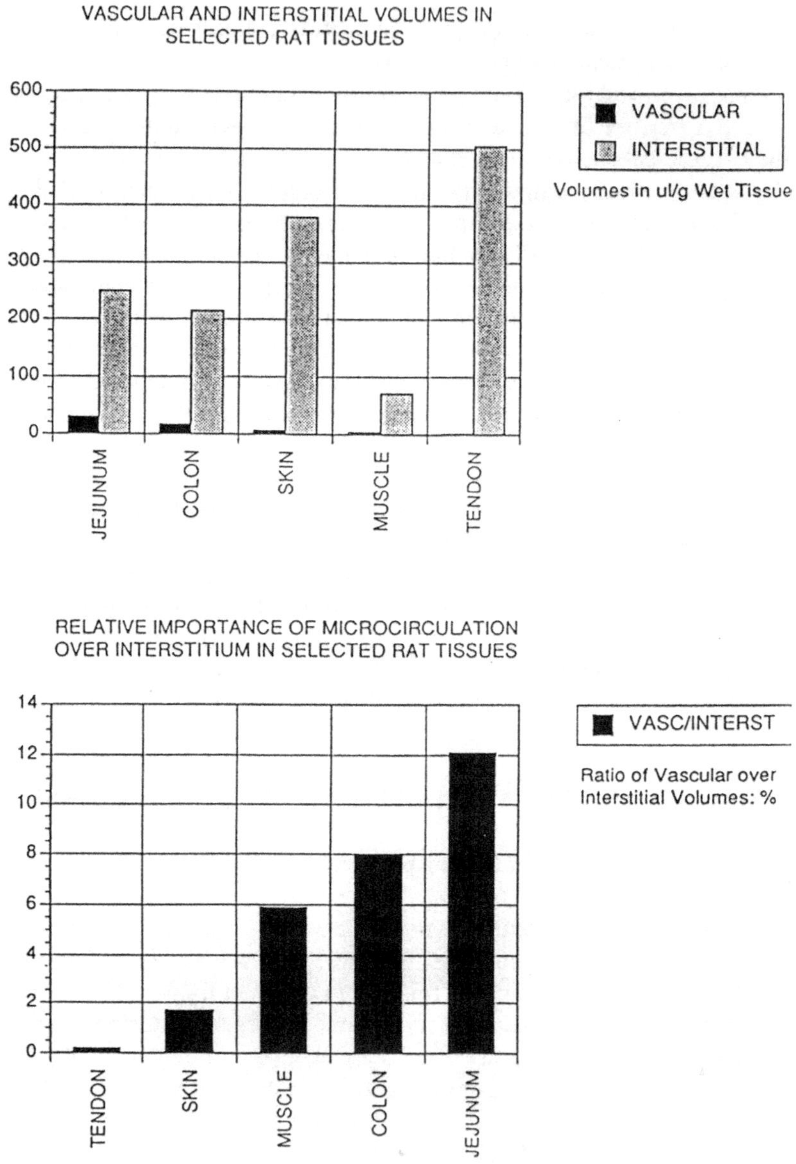

Figure 1. *Vascular and interstitial volumes in selected rat tissues, showing* ***first****, the heterogeneity of each volume distribution in different organs,* ***second****, the relative smallness of the vascular volume. The vascular/intertsitial volume ratio illustrates clearly the relative interstitial perfusion of individual organs. Modified from Wigg et al. (ref. 53).*

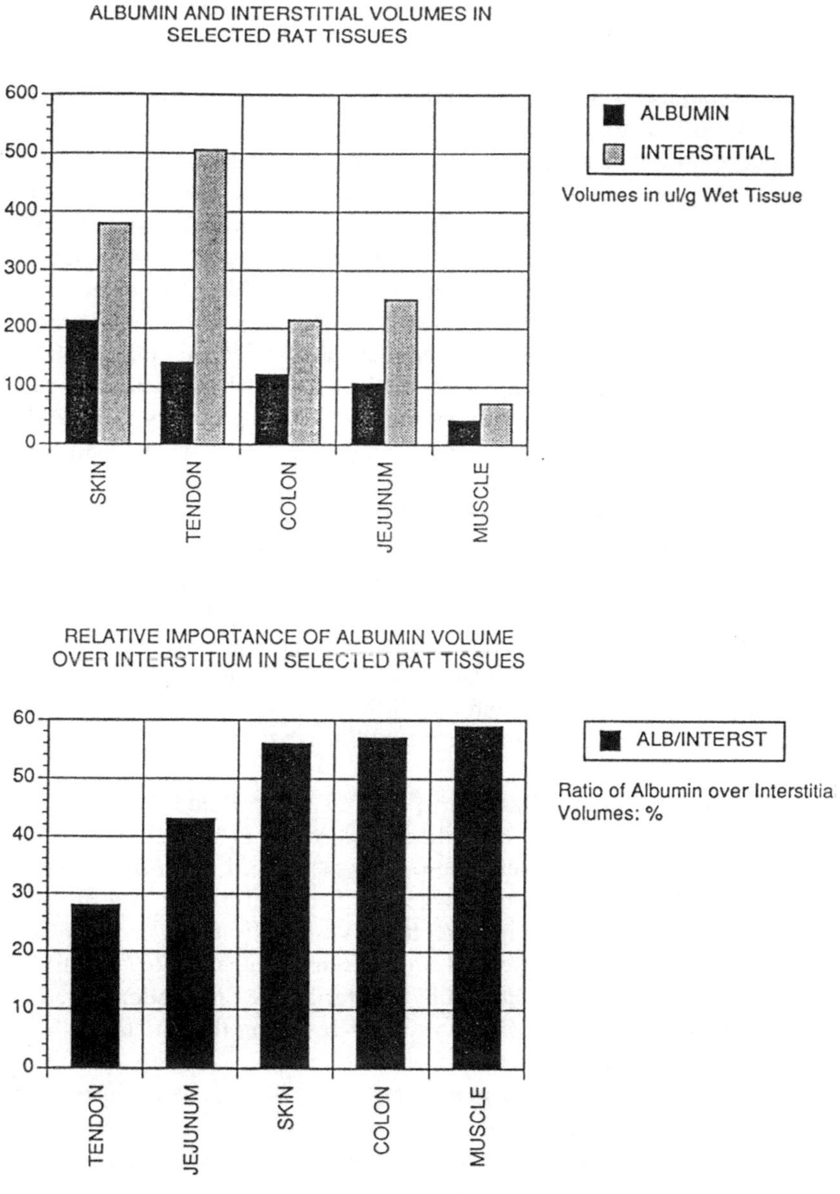

***Figure 2.** Albumin and interstitial volumes in selected rat tissues, showing again, **first**, the heterogeneity of each volume distribution in different organs, **second**, the relative homogeneity of the albumin/interstitial volume ratio in all organs examined but the tendon, indicating the large space (50%) occupied by the "carrier" protein. Modified from Wigg et al. (ref. 53.).*

those almost entirely exposed to the *milieu intérieur* (skeletal muscle and skin for storage-reservoir type of functions).

Heterogeneity of the interstitial matrix. Next to the capillaries of the different microcirculation networks, is the interstitial fluid compartment, the composition of which is as heterogeneous as are the permeability characteristics of capillaries, as illustrated in Figure 1. The interstitium contains fibrillar tissue, amorphous substance, and a variety of cells (mostly fibroblasts, and hematopoietic cells). The fibrillar component consists mainly of two types of structural proteins: dispersed or well-oriented collagen, and fibronectin (32). The latter protein, in conjunction with a variety of anchor molecules (integrins), distributed on the cellular surface and/or basement membranes of epithelia and endothelia, plays a critical role in maintaining the different components of the interstitial space together (45) .The amorphous substance consists of two types of molecules: proteoglycans and albumin. These molecules contain a high electrically negative charge which confers peculiar properties to the interstitial *milieu.*, including the blood vessel wall (7, 8). In fact, the hydraulic properties of the interstitial fluid compartment (12, 20), as well as the albumin exclusion phenomenon (6), are determined by the composition of proteoglycans (18), of this mostly important volume. The significance of interstitial albumin is not entirely understood at this point in time, yet its eventual role in the traffic of albumin-bound substances, such as hormones and xenobiotics (19), from the microcirculation to the cellular mass, must be seriously considered. The characteristic profile of albumin and interstitial volumes (5, 51), as well as the ratio of these volumes in representative organs is illustrated in Figure 2.

b) Passive Endothelial Transport in Larger Blood Vessels

This subject, although of critical importance in several areas of vascular pathophysiology, has not been thoroughly examined, primarily because of methodology difficulties. Two approaches are being utilized to look at endothelial permeability to macromolecules in larger blood vessels: **first**, isolated endothelial cell culture at confluence; **second**, isolated entire (from endothelium to adventicia) blood vessels, where permeability is studied by measuring the distribution of radio-active albumin in histologic sections of the vessel wall. Using the former method, some investigators confirmed similar or dissimilar effects of a number of autacoids on albumin permeability, already demonstrated in capillaries. More interestingly, in recent studies, it was demonstrated that the albumin traffic across the wall of rabbit aorta is heterogeneous and modulated by a variety of pharmacological agents (17). This model will certainly facilitate the exploration of the large blood vessel wall in a variety of pathophysiological conditions. If the abnormalities of capillary permeability to albumin documented in the rat, with several antihypertensive drugs (24), are confirmed in larger blood vessels, important break-through will probably arise, particularly in the field of atherogenesis research.

3. EXPERIMENTAL ARTERIAL HYPERTENSION

a) Rationale and Description of the Experimental Model

The distribution pattern of vascular and interstitial fluid in hypertensive patients remains controversial: a reduction or no change in plasma and/or total extracellular volumes were reported in such patients (27, 30, 49). Recently, selective augmentation of interstitial space was reported in 12-day-old spontaneously hypertensive rats (SHR). The mechanisms responsible for such abnormalities in body fluid compartments have not been elucidated, however. Transcapillary escape of albumin has been reported in hypertensive patients, but in one such report, this abnormality was attributed to the nonspecific elevation of arterial blood pressure (35). Because of the unique heterogeneous response of specific vascular beds to injection of vaso-active agonists (47, 48), and because of potential disturbances in the permeability characteristics of microcirculation in the pathophysiology of hypertension, we decided to examine the distribution of Evans blue dye (EB: a marker of albumin) in selected organs of SHR and Wistar-Kyoto rats (WKY) of different ages (41).

Six groups of SHR were compared to six paired WKY groups. The first four paired groups included rats of different age and blood pressure values: 4, 8, 12, and 16 weeks. The two last pairs only included rats of 4 and 8 weeks of age. All studies were conducted in awake animals, fasted the day before experiments, but allowed free water intake. Systolic blood pressure, injection of EB, sacrifice of animals and dye extraction in selected tissues were described in papers from our group, and repeatedly validated in our own and other laboratories (39, 44, 47, 48, 49).

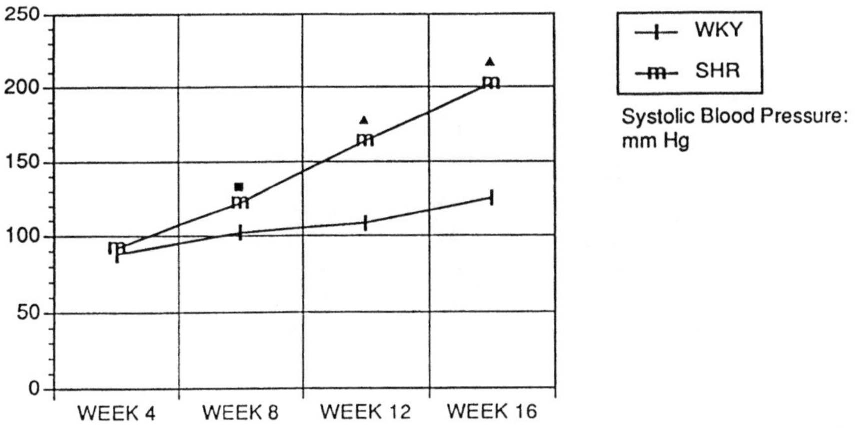

Figure 3. Systolic blood pressure in the Wistar-Kyoto (WKY) and the spontaneously hypertensive rat (SHR) at 4, 8, 12 and 16 weeks of age. Statistical significance between groups is indicated by squares ($P<0.05$) and triangles ($P<0.01$). Reproduced with permission from The Journal of Clinical Investigation.

In the last two groups, specific bradykinin (BK) receptor antagonists ([Leu^8]des-Arg^{9}BK or D-Arg^0[Hyp^3,D-Phe^7]BK, respectively) were injected individually (25ug/kg) and simultaneously with EB in order to assess the enventual contribution of BK to the abnormalities in capillary permeability.

b) Passive Endothelial Transport in Hypertension

Figure 3 illustrates the profile of systolic blood pressure in the two populations of rats examined at 4, 8, 12, and 16 weeks. It is important to note that elevation of systolic pressure in the SHR was only apparent from week 8. Therefore, at week 4, systolic blood pressure in the SHR and WKY groups were comparable. The distribution of EB in the lung, heart, duodenum, liver, pancreas, spleen, and skeletal muscle was comparable in both SHR and WKY, and for each age groups.

On the contrary, renal EB accumulation was significantly higher ($P<0.01$) in the SHR than in the WKY, at 4, 8, 12, and 16 weeks, by 31, 63, 51, and 71 ug/dry tissue, respectively. Of interest, the values measured at 4 weeks, at the time no diffference in blood pressure was detectable, renal EB was already higher in the SHR than in the WKY, suggesting an abnormality in capillary permeability to albumin preceded elevation of blood pressure in the SHR. These results are depicted in Figure 4. As shown, there was a tendency for EB values to decrease progressively with age in the WKY, from 151 to 116 ug/g dry tissue, whereas in the SHR, renal EB extravasation remained higher, in the range of 182 to 204 ug/g dry tissue.

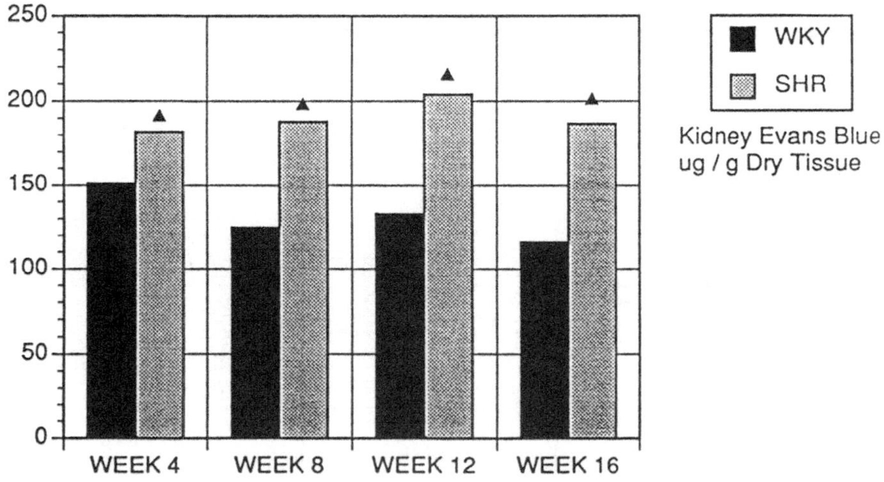

Figure 4. *Selective alteration in renal capillary permeability to albumin in the pre-, and the hypertensive stages of the SHR compared to the WKY of comparable age. Statistical significance as in Figure 3. Reproduced with permission from The Journal of Clinical Investigation.*

The effect of B1 and B2 bradykinin receptor antagonists on the kidney EB distribution is illustrated in Figure 5. In 4 and 8-week-old animals, renal EB was again higher ($P<0.01$) in the SHR than in the WKY. Neither B1 and B2 bradykinin receptor antagonists, *per se,* altered renal EB distribution in the WKY. There was no effect of B1 bradykinin receptor antagonist on EB distribution in the SHR, but the B2 bradykinin antagonist restored EB accumulation in both 4 and 8-week-old SHR, to 162 and 144 ug/g dry tissue, respectively. The values are not statistically different from control data obtained in the WKY.

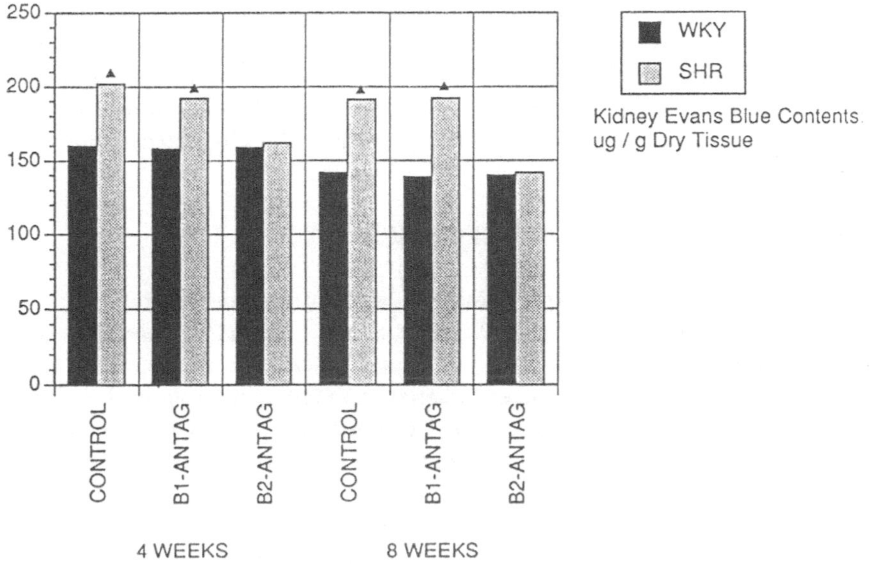

Figure 5. *Effect of B1 and B2 bradykinin antagonists (B1A and B2A) on kidney Evans blue extravasation in the WKY and the SHR at 4 (left) and 8 (right) weeks of age. Statistical significance as in Figure 3. Reproduced with permission from The Journal of Clinical Investigation.*

c) Significance of Endothelial Dysfunctions

These observations represent new information related to the pathophysiology of hypertension in the SHR: **fisrt**, they demonstrate clearly a selective renal defect in capillary permeability in the SHR strain; **second**, the selective renal sequestration of EB is apparent before the development of hypertension; **third**, B2 bradykinin receptors appear to play a role in mediating this unique permeability abnormality (41). Such findings have not been reported, to our knowledge, in this animal model, nor in human hypertension.

Why did EB preferentially accumulate in the SHR kidney? Compared with most major thoracic and splanchnic organs, the kidney has the highest blood flow per gram of tissue (16, 53). In addition, the renal microcirculation possesses a high permeability coefficient. However, it is unlikely that the renal EB sequestration observed in the SHR was merely due to these properties of the renal microcirculation. In fact the permeability characteristics of the hepatic, splenic, and gastro-intestinal submucosal microcirculation are also quite high and comparable, yet acccumulation of EB in these organs was not different in SHR and WKY groups.

Why did EB accumulate in the 4-week-old SHR kidney? The higher renal capillary permeability to albumin found in the SHR could not reasonably be related to blood pressure, since these animals were normotensive. Eventual abnormalities (physical and/or humoral) of the renal microcirculation (26), and/or alteration in the extravascular compartment (interstitial matrix, and/or cellular mass) of that organ (29), could be regarded as potential primary events leading to excess capillary extravasation of albumin. It is reasonable to propose that the renal bradykinin system, in particular the B2 receptors, initiates the selective capillary abnormalities leading to EB accumulation in the renal interstitium, weeks before hypertension develops.

4. EXPERIMENTAL DIABETES MELLITUS

a) Rationale and Description of the Experimental Model

Complications of diabetes mellitus are esentially vascular: arterial hypertension, retinal microangiopathy, accelerated atherosclerosis, coronary and renal insufficiency (1, 17, 28, 38). The endothelial cell plays a key role in the control of vascular reactivity and permeability (42). This function is achieved by the production and release of potent vaso-active autacoids (10, 50) and hormones, including the endothelium-derived relaxing factor, presumed to be nitric oxide (NO) (36). Increased capillary permeability to albumin leads to fluid and solute accumulation in the interstitial space (tissue edema), particularly in the presence of inadequate lymphatic drainage. In addition to its well known effect on the vascular smooth muscle, NO was also reported to affect capillary permeability, as well as lymphatic microcirculation (14). Because of these critical actions on microcirculation, this important autacoid represents a candidate in the development of vascular dysfunctions in diabetes mellitus (11).

Blood glucose reacts non-enzymatically with proteins to initiate a post-translational process known as glycosylation (35). This biochemical reaction leads to the production of advanced glyosylation end-products (AGEs), which are in fact elevated in conditions of sustained hyperglycemia, such as diabetes mellitus. Recent studies have shown that AGEs can react and inactivate NO, through a direct chemical reaction (10). The aim of the present study is twofold: **first**, to examine the pattern of capillary permeability (VASPE) in selected tissues obtained from streptozotocin-induced (STZ) diabetic rats; **second**, to determine the eventual contribution of NO to VASPE alterations, using aminoguanidine (AGUA), a putative inhibitor of NO quenching by AGEs.

Unanaesthetized female Sprague-Dawley rats were used in this study: one control group and five groups injected with STZ (65 mg/kg, intraperitoneally). The onset of diabetes was confirmed by measuring blood glucose, three days after STZ. In four groups of diabetic rats, VASPE was determined 3, 7, 14 and 28 days following STZ injection. In the fifth group, AGUA was administered daily (25 mg/kg) by gavage during 14 days and VASPE was then measured after this period of treatment. Capillary permeability to albumin was assessed by using Evans blue (EB), a method previously described and validated in our and other laboratories. The following organs, known to be affected in patients with diabetes mellitus, were examined: upper bronchus, lung parenchyma, heart, pancreas, duodenum, kidney, skin, and skeletal muscle (left quadriceps).

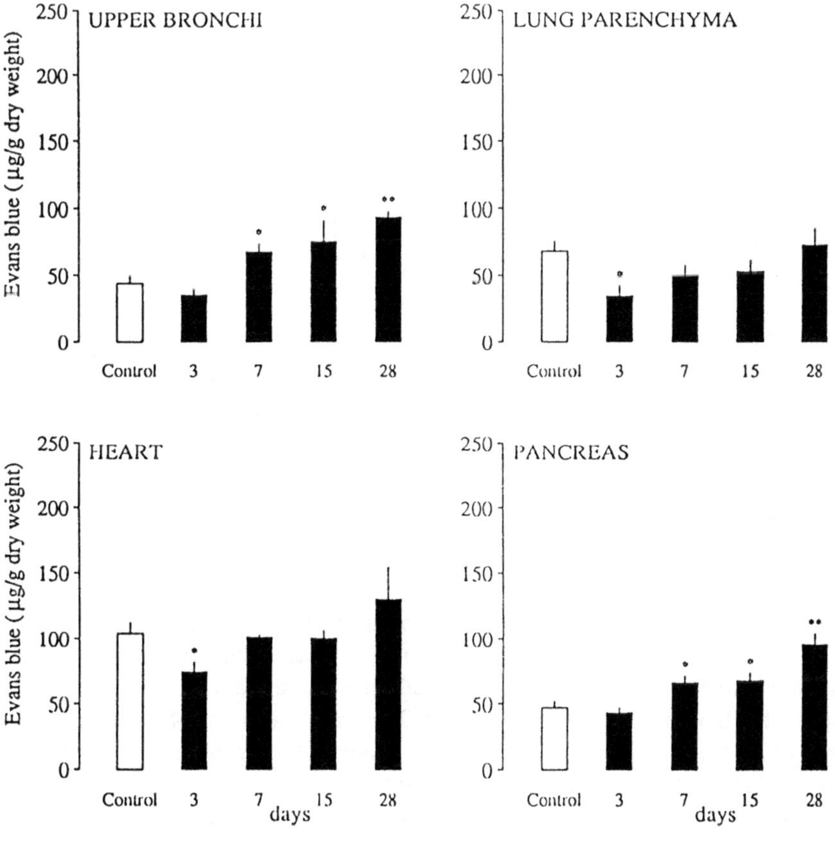

Figure 6. *Capillary permeability to Evans blue in upper bronchi, lung parenchyma, heart, and pancreas, 3, 7, 14, and 28 days following streptozotocin injection (full bars). Values are compared to control data (open bars), and statistical significance is indicated by * and ** for P values <0.05 and <0.01, respectively.*

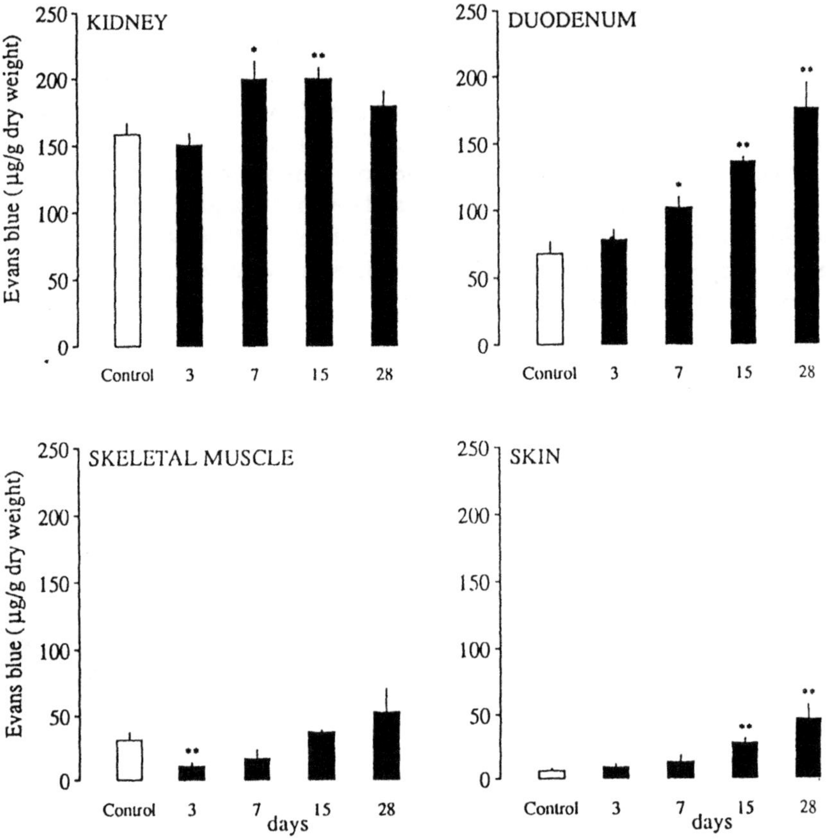

Figure 7. Capillary permeability to Evans blue in kidney, duodenum, skeletal muscle, and skin, 3, 7, 14, and 28 days following streptozotocin injection (full bars). Values are compared to control data (open bars). Statistical significance as in Figure 6.

b) Passive Endothelial Transport in Diabetes Mellitus

Changes in capillary permeability over time are illustrated in Figures 6 and 7. In lung parenchyma, heart, and skeletal muscle, there was an early (3 days) decrease of EB extravasation from 68 to 35, from 104 to 74, and from 31 to 11 ug/g dry tissue, respectively. Tissue EB contents then progressively returned to baseline values over time. In the upper bronchus, there is a significant increase in EB leakage, from 44 to 68, 75, and 94 ug/ dry tissue, 7, 14 and 28 days after STZ injection, respectively. Similar augmentation was seen as a function of time in the duodenum, and the pancreas. In the skin, the progressive increase in VASPE became significant in the second week, reaching a maximum value of 47 ug/g dry tissue, 4 weeks after STZ (a 7.8-fold increment). In the kidney, there was a significant rise in EB from 158 to 200 and 200 ug/g dry tissue, one and two weeks after STZ, but the value measured at week four was not statistically different from the control.

The effect of AGUA treatment on EB extravasation in the two-week STZ-induced diabetic rat is illustrated in Figure 8. As shown, this drug completely or almost completely (skin) restored EB extravasation back to the baseline values in organs affected by diabetes mellitus: kidney, duodenum, pancreas, upper bronchus and skin. In the three organs where an early (3 days) reduction in VASPE was noticed, the lung parenchyma, the heart, and the skeletal muscle, AGUA was associated with a significant reduction in EB extravasation. It is important to note that AGUA, *per se*, has no effect whatsoever on VASPE (data not shown).

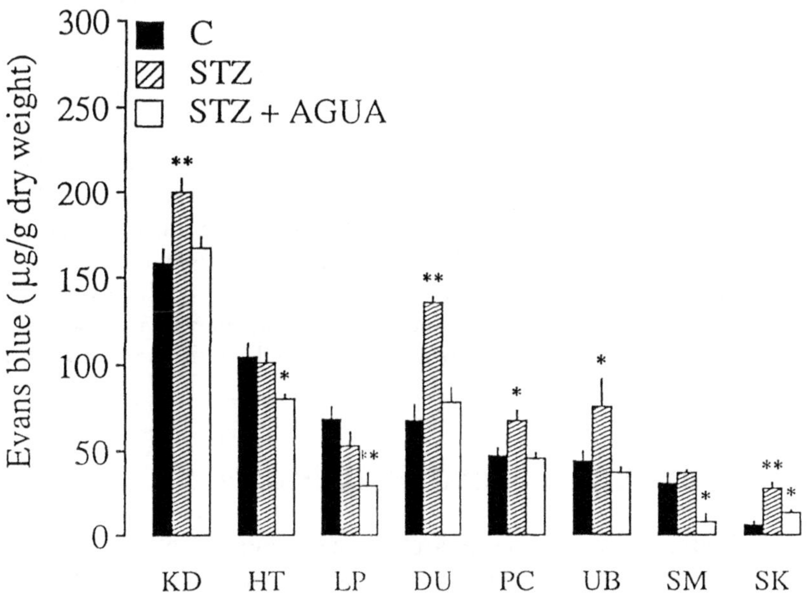

Figure 8. *Effect of 14-day tretament with aminoguanidine (AGUA: 25 mg/kg/day) on capillary permeability of the kidney (KD), heart (HT), lung parenchyma (LP), duodenum (DU), pancreas (PC), upper bronchus (UB), skeletal muscle (SM), and skin (SK). C designates control values, STZ untreated diabetic group (14 days post stertozotocin injection), and STZ + AGUA treated diabetic group with AGUA during 14 days. Statistical significance as in Figure 6.*

c) Significance of Endothelial Dysfuntions

The heterogeneous pattern of EB abnormalities observed in this model of diabetes mellitus, contrasts with the generalized transcapillary albumin escape reported in human diabetic patients (38). This phenomenon is probably the consequence of different and specific alterations in the mechanisms of vascular reactivity and permeability, responsible for the control of extra-splanchnic and splanchnic microcirculation. Under normal conditions, these vascular beds are controlled by specific neuro-humoral mechanisms. It is likely that under pathophysiological conditions, such as diabetes, similar heterogeneity of vascular damage is obtained.

The late and progressive increase in EB capillary permeability, which occurs in the splanchnic organs, as well as in the skin, would appear to be related to the basic vascular disturbances induced in this model of diabetes mellitus. In addition, these particular organs are those where the 14-day AGUA treatment completely restored the abnormal EB leakage. The apparent beneficial effect of AGUA on VASPE can either be explained by an indirect effect of this compound on vascular reactivity (pre- and/or post-capillary resistances), or by a direct influence on endothelial cell dysjuntion, the prerequisite for restoration of permeability (54). Nitric oxide is one of the most important vaso-active compounds involved in the regulation of the vascular tone. Quenching of NO by the high level of AGEs, synthetized at an advanced stage of diabetes, decreases the availability of this autacoid in the interstitial *milieu*. It is of interest that acute treatment of normal rats with L-NAME, an inhibitor of NO synthase, is associated with increased VASPE *in vivo* (unpublished data). These findings suggest that the abnormal VASPE profile documented in the present study is related to a reduction in NO availability due to increased quenching by AGEs. In fact, the apparent reversibility of altered EB extravasation obtained in all affected tissues in the 14-day diabetic rats, supports the hypothesis that NO quenching is probably implicated in the abnormal VASPE obtained.

5. EXPERIMENTAL CHRONIC RENAL FAILURE

a) Rationale and Description of the Experimental Model

Untreated chronic renal failure is a condition characterized by progressive and irreversible multi-organ failure. However, the degree of individual extrarenal organ damage does not appear to be uniform: disorganization of bone, skeletal muscle, skin, heart and blood vessel metabolism is not homogeneous during the course of adaptation to chronic renal insufficiency (9, 38, 52). External fluid and solute balance is disrupted during chronic renal failure, because of the remnant renal parenchyma incapacity to adjust fluid excretion to the needs of the *milieu intérieur*. Extrarenal adaptation to changes in the composition of this *milieu* does occur for a while, due to the contribution of other organs capable of significant exchange processes with the *milieu extérieur*, such as the digestive tract, the skin and the lungs. However, these organs are subjected to the same abnormal fluid and chemical environment as the diseased kidney, which limit their individual capacity to maintain the vascular, interstitial and cellular integrity (22, 33).

The vascular (microcirculation) mechanisms involved in the regulation of the internal fluid balance during adaptation to chronic renal failure, in particular changes in capillary permeability, have not been adequately studied. Therefore, the present study was designed to examine **first**, the pattern of capillary permeability to albumin during the course of adaptation to chronic renal failure, using the Bricker's rat model, **second**, to test the effect of captopril, an angiotensin converting enzyme inhibitor, reputed to "protect" the vascular system (including the kidney) in different models of chronic renal failure (2).

Sprague-Dawley rats were submitted to a two-step reduction of their renal mass (5/6 nephrectomy: Bricker's model). Before and after this procedure, animals were fed a normal rat chow diet. Two weeks after the second step 5/6 nephrectomy, animals were anesthetized for measurement of inulin clearance, to quantify the loss of glomerular function. At the end of the clearance periods, a bolus of Evans blue (EB) was injected for evaluation of capillary permeability in selected tissues, using an approach already described and validated in our and other laboratories. Three groups of rats were studied: a group of sham-operated animals, a group of untreated uremic rats, and a group of uremic rats which received captopril (10 mg/kg) by gavage during two weeks.

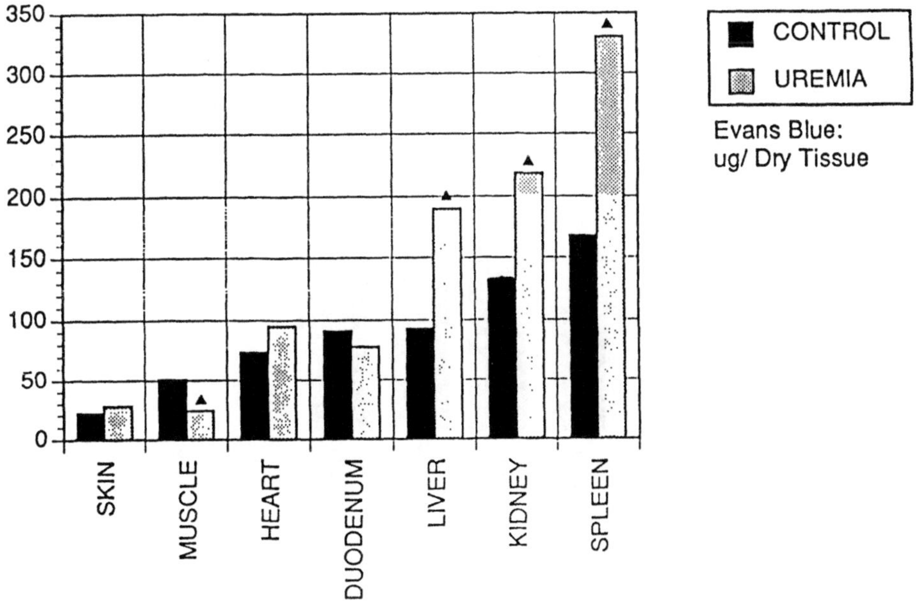

Figure 9. Evans blue dye contents in selected tissues obtained from three-week "chronic" untreated uremic rats. Statistical significance for observed differences between groups is indicated by squares (P<0.05) and triangles (P<0.01), respectively.

b) Passive Endothelial Transport in Chronic Renal Failure

The reduction in glomerular filtration obtained by the 5/6 nephrectomy procedure was approximately 80%, a finding comparable to published studies using this model. As illustrated in Figure 9, in untreated uremic rats, EB extravasation was markedly enhanced in the liver, kidney and spleen, from 95 to 198, 135 to 215, and from 168 to 345 ug/g dry tissue, respectively. On the contrary, in the skeletal muscle, capillary permeability to EB was reduced from 53 to 24 ug/g dry tissue. No significant change in permeability was observed in the skin, the heart, and duodenum.

Captopril treatment resulted in complete, partial, or no correction of EB extravasation in the organs affected by chronic uremia. As shown in Figure 10, EB contents in the liver decreased from 198 to 157 ug/g dry tissue, a value still significantly above the baseline value. In the kidney, EB content was reduced from 215 to 148 ug/g dry tissue, a value not different from the control. Captopril treatment failed to affect the abnormal uremic-induced EB extravasation in the skeletal muscle and the spleen.

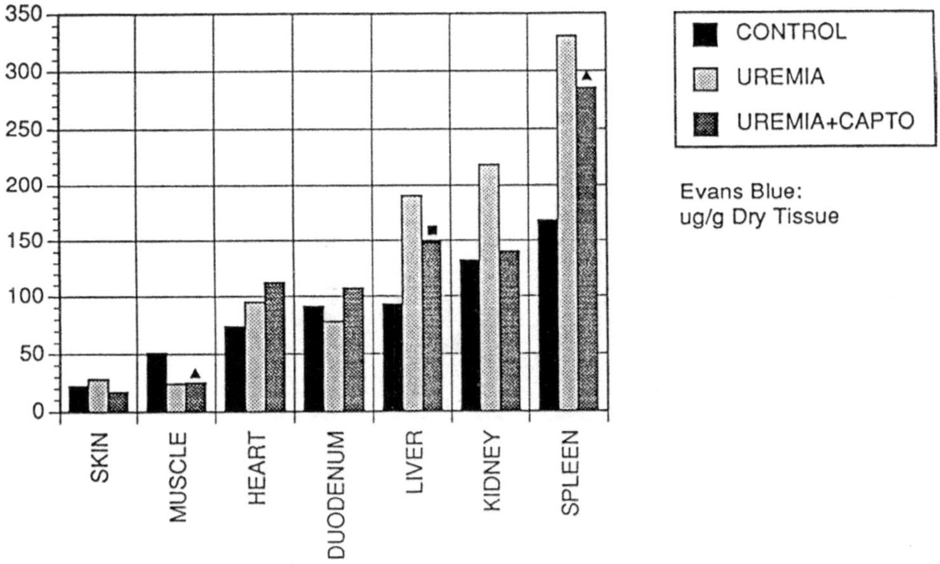

Figure 10. *Evans blue dye contents in selected tissues obtained from three-weeks "chronic" rats treated with Captopril. Data obtained from untreated rats (Figure 10) is shown for comparison. Statistical significance as in Figure 9.*

c) Significance of Endothelial Dysfunctions

In this model of chronic uremia, decreased permeability to EB was only seen in the skeletal muscle, whereas in most splanchnic organs, increased capillary permeability was observed. In a number of organs, including the skin, the heart, and the duodenum, the uremic environment failed to alter capillary permeability to Evans blue. The heterogeneous profile in permeability alterations suggests that the mechanisms involved in the regulation of pre- and post-capillary resistance, as well as the factors controling endothelial cell dysjunction, are not uniformly affected by the derangement of the *milieu intérieur* induced by chronic uremia (13).

The metabolism of several vaso-active autacoids and hormones is affected in chronic uremia. Angiotensin II, endothelins, and nitric oxide were identified as potential mediators of hemodynamic disturbances reported in models of chronic renal failure: excess production in the case of the vasoconstrictor peptides, lack of production in the case of nitric oxide, as suspected (above description) in the diabetic vasculopathy (11). In this study, we only examined the potential contribution of the renin-angiotenin system to the microcirculation disturbances, by using the converting enzyme inhibitor, captopril. Only the kidney and the liver (in part) responded to this treatment. The individual contribution of decreased angiotensin II, increased bradykinin, and/or prostaglandins induced by captopril to normalization of renal and liver capillary permeability cannot be specified, however.

6. THE CONCEPT OF TARGET ORGAN REVISITED

a) Vascular-Interstitial-Cellular Relationships

Two major physiological barriers control the traffic of fluid and solutes between the cell mass and the blood stream: the endothelium of capillaries and the plasma membrane of cells. The capillary surface exposed to the interstitial fluid compartment vary considerably from one organ to the other, and the exposure ratio determines, in part, the dynamics of fluid and solute exchanges between these two compartments. The relative perfusion of organs is also well established, but in most studies, the blood flow to individual organs is always expressed per gram of total tissue weight. Since the interstitial and cell fluid composition of tissues varies significantly, direct or indirect measurements of these volumes is required to appreciate the capillary/interstitial surface ratio, using an approach similar to that described in Figure 1 (53).

According to these data, it becomes evident that the amount of interstitium vary considerably between tissues: the skin and the digestive tract contain a large fraction of interstitial space, while the skeletal system representing a major fraction total body weight, has a smaller proportion of interstitial fluid. The capillary/interstitial volume ratio determines, in part, the dynamics of fluid and solute transfer from the microcirculation to the cellular mass, across the interstitium, and varies from 3 to 8%, the tendon and the skin having the lowest ratio, the skeletal muscle an intermediate value, and finally, segments of the digestive tract (and probably the kidney) the highest values.

The functional anatomy of the interstitial space also includes the albumin exclusion space phenomenon (6). Because of the presence of proteoglycans, the distribution of albumin in the interstitial space is heterogeneous, not only from one organ to the other, as illustrated in Figure 2, but also within the interstitium of a given organ as elegantly demonstrated in a recent paper (5). The consequence of this phenomenon on the traffic of fluid and solutes between the microcirculation and the cellular mass is obvious (25). Since fluid and solutes, as well as protein-bound substances, mainly transit within the space occupied by albumin (outside of the so-called albumin exclusion space), an organ with a high albumin exclusion space will exhibit poor conditions for the traffic of substrates and waste products between cells and the microcirculation. Therefore, it becomes potentially susceptible to initial injury and progressive damage. If, in addition to poor baseline fluid and solute traffic conditions, extravasation of albumin and tissue edema develop, following increased capillary permeability induced by a disease process, a given organ will then become much more susceptible to injury and progression of tissue damage, a condition otherwise coined target organ.

b) Significance of the Strategic Position of the Interstitium

Life is maintained by appropriate delivery of nutriments (metabolic susbstrates, oxygen), and removal of waste products (nitrogenous molecules, carbon dioxide) to and from the cellular mass of organs. The transit of these molecules takes place in the interstitium which occupies a strategic position between the cellular volume and the microcirculation. As reviewed above, profound modifications in the diffusion and convection properties of the interstitium may develop from alterations in the chemical composition of this important compartment. The heterogeneity of interstitial composition, as well as the specific microcirculation profile of each organ, therefore, become critical determinants in the extent of damage which may result from primary or secondary alteration in these parameters. Endothelial cell dysfunction associated with abnormal extravasation of plasma in different tissues, thus becomes a common denominator of the disease processes and defines, in a new, fundamental way, the concept of target organ damage.

We described three apparently different experimental models of systemic diseases, arterial hypertension (41), diabetes mellitus (11), and chronic renal failure, which have, in common, a variety of capillary permeability dysfunctions in organs reputed to be speficically affected by the original condition. This phenomenon is only apparent and may represent the consequence of basic alterations in capillary permeability and/or interstitial physiology. Morbid conditions characterized by multi-organ damage and eventual failure, could therefore be reasonably considered as endothelial cell and/or intertsitial diseases. This provocative approach in looking at pathophysiology should simplify not only our understanding at the mechanisms of mostly frequent and critical diseases, but also our goals in therapeutics.

7. PHARMACOLOGY OF PASSIVE ENDOTHELIAL TRANSPORT

a) Modulation of Pre- and Post-Capillary Reactivity

Studies on isolated resistance arteries (mesenteric) and venous preparations indicate that a variety of vaso-active autacoids and hormones exert different actions on the equivalent microcirculation sites: the pre- and post-capillary resistances. In the isolated tissue bath perfused with normal Ringer's solution, angiotensin II produces a more important contraction, for a given dose, on arterial rather than on venous tissues, suggesting the effect of this vasoconstrictor predominates on the pre-capillary segment of the microcirculation. Angiotensin-II receptor blockade, and to a certain extent (because of their additional effects on bradykinin and prostaglandins), angiotensin converting enzyme inhibitors can be considered as potential modulators of pre-capillary resistance (2). On the contrary, endothelin-1 is associated with more potent vasoconstriction on the venous segment of the microcirculation than on the arteriolar side: yet is does alter, in the same direction, pre-capillary resistance. Endothelin-1 converting enzyme inhibition, and endothelin receptor blockade can also be utilized to selectively modulate post-capillary resistance (23). The development of such pharmacological tools may be of interest in the correction of capillary permability dysfunctions, either due to enhanced plasma extravasation, in the case of predominant post-capillary net constriction (pre/post-capillary resistance ratio < 1), or due to reduced post-capillary net constriction (pre/post-capillary resistance ratio > 1). Recently, a calcium channel antagonist of the dialtiazem type, was shown to possess vasorelaxation properties, more important on the venous than on the arteriolar side of the microcirculation. Such a vascular effect, mimicking that of endothelin-1 receptor blockade, could also be of interest to modulate conditions resulting from predominant vasoconstricting effects on the post-capillary side of the microcirculation.

b) Modulation of Endothelial Cell Permeability

Since the mechanisms of endothelial cell contraction and/or tight junction organization are not entirely known at this point in time, the development of agents capable of modulating the critical physiological phenomena leading to alteration in endothelial permeability failed to cope with that of vasoactive drugs. Intracellular actin is known to be the contractile element responsible for endothelial cell contraction and ionized calcium is essential for this phenomenon to occur. In addition, intracellular tubulin also participates in the phenomenon of endothelial permeability, probably through an effect on tight junction protein assembly. Antagonists to actin and tubulin mobilization (such as colchicine and other cytoskeletal poisons) can be utilized as direct modulators of endothelial permeability and calcium channel antagonists, as indirect modulators. The complex mechanisms involved in the regulation of intracellular ionized calcium, in which the plasma membrane participates, as well as the intracellular membrane system (mitochondria, sarcomere), make it difficult, for the moment, to design specific potent modulators of endothelial cell permeability.

8. GENERAL CONCLUSIONS

The role of the microcirculation in the distribution of body fluid volumes, and the importance of vascular reactivity (pre- and post-capillary resistances), and endothelial cell permeability, on the distribution of fluid and solutes in the interstitial compartment, have been reviewed. The consequences of alterations in the physico-chemical composition of the latter compartment, on the traffic of substrates and waste products between the cellular mass and the microcirculation, emphasize the strategic nature of the interstitium. The heterogeneous characteristics of this fluid compartment, within a given tissue, as well as from one organ to the other, represent the basis for the specificity of fluid and solute dynamics between the vascular compartment and the cellular mass. The endothelium and the cell plasma membrane represent the natural barriers which control the traffic between body fluid compartments, on which a variety of autacoids and hormones interact.

The application of this concept to the pathophysiology of multi-organ systemic diseases, such as arterial hypertension, diabetes mellitus, chronic renal failure, and probably many others, is relevant since the abnormalities in fluid and solute distribution, and the resulting pattern of organ damage found in these conditions, could be related to primary or secondary dysfunctions of the capillary endothelium. The vaso-active autacoids and hormones so far suspected as potential causes for the observed abnormalities of the microcirculation in these morbid conditions can now be considered as possible candidates for pharmacological targetting.

REFERENCES

1. Alpert JS, Coffman JD, Balodimos MC, Koncz L, Soeldner JS (1972) N Engl J Med 286:454-460
2. Anderson S, Rennke HG, Brenner BM (1986) J Clin Invest 77:1993-2000
3. Aukland K, Nicolaysen G (1981) Physiol ReV 61:556-643
4. Aukland K, Reed RK (1993) Physiol Rev 73:1-78
5. Barber BJ, Nearing BD (1990) Am J Physiol 258:H556-H564
6. Bell DR, Watson PD, Renkin EM (1980) Am J Physiol 239:H532-H538
7. Bevan JA, Joyce EH (1990) Am J Physiol 259:H23-H28
8. Bevan JA, Siegel G (1991) Blood Vessels 28:552-556
9. Brenner BM, Meyer TW, Hostetter TH (1982) N Engl J Med 307:652-659
10. Bucala R, Tracey JK, Cerami A (1991) J Clin Invest 87:432-438
11. Chakir M, Plante GE (1993) J Clin Invest *in press*
12. Comper WD, Laurent TC (1978) Physiol Rev 58:255-315
13. DeWardener HE, Clarckson EM (1985) Physiol Rev 65:658-759

14. Elias RM, Eissnhoffer J, Johnston MG (1992) Am J Physiol 263:H1880-1887
15. Furchgott RF, Zawadzki JV (1987) Nature 288:373-376
16. Guyton AC, Coleman TG, Granger HJ (1982) Ann Rev Physiol 34:13-86
17. Huijberts MSP, Wolffenbuttel BHR, Struijker-Boudier HAC, Crijns FRL, Nieuwenhuijzen Kruzeman AC, Poitevin P, Levy BI (1993) J Clin Invest, *in press*
18. Jackson RL, Busch SJ, Cardin AD (1991) Physiol Rev 71:481-539
19. Koch-Weser J, Sellers EM (1976) N Engl J Med 294:311-316
20. Lai-Fook SJ, Rochester NL, Brown LV (1989) J Appl Physiol 67:606-613
21. Laposata M, Dovnarsky DK, Shin HS (1983) Blood 62:549-556
22. Leaf A (1970) Am J Med 49:291-295
23. Lehoux S, Plante GE, Sirois MG, Sirois P, D'Orleans-Juste P. (1992) Br J Pharmacol 107:996-1000
24. Lehoux S, Sirois MG, Sirois P, Plante GE (1993) J Pharmacol Exp Ther *in press*
25. Levick JR (1987) Quart J Exp Physiol 72:409-438
26. Lockette W, Otsuka Y, Carretero O (1986) Hypertension 8(Suppl.III):61-66
27. London GM, Levenson JA, London AM, Simon AC, Safar ME (1984) Kidney Int. 26:342-350
28. Lorenzi M, Cagliero E (1991) Diabetes 40:653-659
29. Lortie M, Regoli D, Rhaleb NE, Plante GE (1992) Am J Physiol 262:R72-R78
30. Lucas L, Floyer MA (1974) Clin Sci Mol Med 47:1-9
31. MacKnight ADC, Leaf A (1977) Physiol Rev 57:510-573
32. McDonald J (1988) Ann Rev Cell Biol 4:183-207
33. Manery J (1954) Physiol Rev 34:334-402
34. Moncada S, Palmer RMJ, Higgs EA (1991) Pharmacol Rev 43:109-142
35. Njorage FG, Mounier VM (1989) Prog Clin Biol Res 304:85-107
36. Palmer RMJ, Ferrige AG, Moncada S (1987) Nature 327:524-526
37. Parving MH, Gyntelberg F (1973) Circ Res 643:32-39
38. Parving MH, Rasmussen SM (1973) Scand J Lab Invest 32:81-87
39. Patterson CE, Rhoades RA, Garcia JGN (1992) J Appl Physiol 72:865-873
40. Plante GE (1989) Kidney Int 36(Suppl 27):s128-s132
41. Plante GE, Bissonnette M, Sirois MG, Regoli D, Sirois P (1992) J Clin Invest 89:2030-2034
42. Porta M, DaSelva M, Molinatti P, Molinatti GM (1987) Diabetologia 30:601-609
43. Robinson JR (1960) Physiol Rev 40:112-207
44. Rogers DF, Boschetto P, Barnes PJ (1989) J Pharmacol Meth 21:309-315
45. Ruoslahti E, Pierschbacher MD (1987) Science 238:491-497
46. Simionescu N (1983) Physiol Rev 63:1536-1640
47. Sirois MG, Jancar S, Braquet P, Plante GE, Sirois P (1988) Prostaglandins 36:631-636

48. Sirois MG, Plante GE, Braquet P, Sirois P (1990) J Lipid Mediators 2:s109-s112
49. Tarazi RC, Dustan HP, Frohlich ED (1969) Circulation 40:357-364
50. Takahashi K, Ghatei MA, Lam HC, O'Halloran DJ, Bloom SR (1990) Diabetologia 33:306-310
51. VanHoutte PM (1988) N Engl J Med 319:512-513
52. Walser M (1990) Kidney Int 37:1195-1210
53. Wigg H, DeCarlo M, Sibley L, Renkin EM (1992) Am J Physiol 263:H1222-1233
54. Wysolmerski RB, Lagunoff D (1990) Proc Nat Acad Sci 87:16-22
55. Yanagisawa M, Kurihara H, Kimura S, Tomobe Y, Kobayashi M, Mitsui Y, Yazake Y, Goto K, Masaki T (1988) Nature 332:411-415

CHAPTER 13

PHARMACOLOGY OF THE HUMAN PRECURSORS OF ENDOTHELIN: ROLE OF FUNCTIONAL RECEPTORS AND ENDOTHELIN-CONVERTING ENZYME

Pedro D'Orléans-Juste and Sabine Télémaque

Dept. of Pharmacology,
Université de Sherbrooke, Sherbrooke, Canada

INTRODUCTION

From a simple physical barrier involved in the balance of thrombogenic and thrombolytic processes, the endothelium has become an intricate endocrine and paracrine organ. Indeed, the endothelial cell secretes numerous factors which importantly modulate the vascular tone as well as thrombogenesis. Chronologically, prostacyclin (PGI_2) was the first reported autacoid generated from the endothelium (1), followed by the report on the "obligatory role of the endothelium-derived relaxing factor (EDRF)", a potent vasodilator of the underlaying smooth muscle (2). The most recently discovered endothelium-derived vasoactive factor is endothelin-1 (ET-1). This 21 amino-acid peptide remains, five years after its discovery, the most potent vasoactive peptide yet reported in the literature (3). By analogy to the renin-angiotensin system, within which the understanding of the angiotensin-converting enzyme activity has resulted in the development of important therapeutical tools, a great deal of effort is now concentrated towards the identification of the synthesis pathway for ET-1. One of the limiting steps in the synthesis of ET-1 originates from the production of an intermediate precursor, big-ET-1.

The present chapter will describe the synthesis pathway for ETs as well as pharmacological and biochemical evidences supporting the presence of an endothelin-converting enzyme (ECE). The chapter will then be concluded by some physiopathological considerations emphasizing an important role for ETs in some experimental conditions.

Synthesis Pathways of Endothelins

The genomic structure of the three separate human genes encoding ET-1, 2 and 3 has been reported, as illustrated in Figure 1; the preproendothelins-1, 2 and 3 are expressed by different genes (4). Characteristically, it is now suggested that preproendothelins are converted to their active metabolites, ET-1, 2 and 3, in a three-step enzymatic pathway.

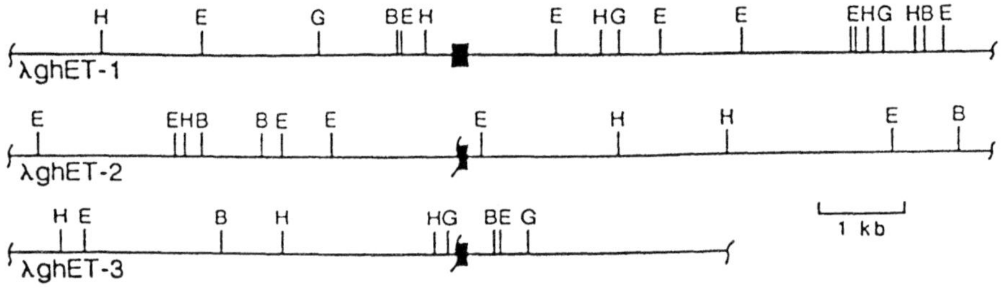

Figure 1. Partial restriction maps of three human endothelin-related clones : λghET-1, λghET-2 and λghET-3. Presumptive exons encoding mature endothelins are shown as solid boxes. B, Bam HI; E, Eco RI; G, Bg l II; H, Hind III (4).

Recent observations suggest a role for furin in the processing of the human preproendothelin of 203 amino acids (5). This precursor contains the furin consensus cleavage signal (Arg-X-X-Arg) flanking the big-ET moiety. Furin is a serine protease which is involved in the maturation of many peptides synthesized via the constitutive pathway of secretion including pro-ß-NGF, pro-von Willebrand factor and others (6). The second part of the pathway involves the processing of human big-ET-1 (38 amino acids) via a putative phosphoramidon-sensitive ECE (see Figure 2).

As illustrated in Table 1, big-ET-1 induced an array of pharmacological and toxicological effects which are all sensitive to phosphoramidon. For example, big-ET-1 has been shown to increase blood pressure with a similar potency to ET-1 in anesthetized rabbits, rats and guinea pigs (7-9). In most of these animal species,

Human big-ET-1 (1-38)	**H_2N-Cys -Ser-Cys -Ser -Ser-Leu -Met-Asp -Lys-Glu -Cys-Val** Pro-Thr -Asn -Val-**Trp -Ile -Ile -Asp-Leu -His-Cys -Phe -Tyr** Glu-His- Val- Val- Pro -Tyr-Gly- Leu-Gly-Ser-Pro -Arg -Ser HOOC
Porcine big-ET-1 (1-39)	**H_2N-Cys-Ser-Cys-Ser-Ser-Leu-Met-Asp-Lys-Glu-Cys-Val-Tyr** Glu-Pro -Thr-Asn -Val-**Trp-Ile - Ile-Asp- Leu - His -Cys- Phe** His-Ile- Val-Pro- Tyr- Gly- Leu - Gly - Ser -Pro -Ser- Arg-Ser HOOC
Porcine big-ET-1 (1-40)	**H_2N-Cys-Ser-Cys-Ser-Ser-Leu-Met-Asp-Lys-Glu-Cys-Val-Tyr** Glu -Pro -Thr - Asn-Val-**Trp- Ile- Ile -Asp- Leu-His- Cys-Phe** His-Ile-Val- Pro- Tyr-Gly- Leu-Gly- Ser-Pro- Ser-Arg-Ser-Arg HOOC
Human big-ET-2 (1-37)	**H_2N-Cys-Ser-Cys-Ser- Ser- Trp-Leu- Asp-Lys -Glu -Cys- Val** Thr- Asn - Val- **Trp -Ile - Ile -Asp -Leu -His -Cys -Phe- Tyr** Pro -Glu -Gln -Thr-Ala-Pro -Tyr -Gly-Leu -Gly -Asn -Pro -Pro HOOC
Human big-ET-3 (1-41)	**H_2N- Cys-Thr -Cys-Phe -Thr-Tyr-Lys -Asp -Lys-Glu-Cys-Val** Pro -Thr -Asn- Ile- **Trp- Ile -Ile- Asp-Leu- His-Cys- Tyr-Tyr** Glu-Gln -Thr -Val-Pro -Tyr- Gly-Leu -Ser- Asn -Tyr -Arg-Gly H_2N-Arg-Phe-Ser

Figure 2. *Primary structures of big-endothelins.*

Table 1

Phosphoramidon-sensitive effects of big-ET-1 human (1-38) or porcine (1-39)

Effect	Model	Effective concentration/dose		Ref.
		Big-endothelin-1	Endothelin-1	
Hypertension	Rat Guinea pig	1 nmol/kg 2 nmol/kg	0.5 nmol/kg 2 nmol/kg	9 13
Release of eicosanoids	Guinea pig lung Rat lung Rabbit kidney	100 nM 100 nM 100 nM	2.5 nM 5 nM 10 nM	13 19 20
Vasoconstriction	Rat mesenteric vasculature . arterial Rabbit kidney	 10 nM 100 pmol	 1 nM 5 pmol	 21 20
Bronchoconstriction	Guinea pig	4 nmol/kg	0.5 nmol/kg	8,22,23
Neurogenic contraction	Rat vas deferens	25 nM	10 nM	24
Neurotoxicity	Mouse	21 nmol/kg	3.8 nmol/kg	25
Plasma extravasation	Rat	400 pmol	400 pmol	26
Bronchoconstriction	Human bronchus	≈5 nM	5 nM	27,28
Vasoconstriction	Saphenous artery	10 nM	3 nM	29
Hypertension toxicity	Rat intrathecal administration	100 pmol	65 pmol	30
Reduction of arteriolar blood vessel diameter	Hamster cheek pouch	30 pmol	10 pmol	31

phosphoramidon was shown to markedly reduce the pressor effects of big-ET-1 without affecting the response of ET-1, suggesting that *in vivo* big-ET-1 was dynamically converted to its active metabolite, ET-1, via a phosphoramidon-sensitive ECE. This is of interest considering the fact that in all isolated tissues studied, big-ET-1 is 100-250 times less potent than ET-1 as a vasoconstrictor, either on intact or denuded vasculatures (10). Subsequent results (11) have shown that blood-borne cells, such as polymorphonuclear leukocytes, are responsible for the degradation and the production of ET-1 via a phosphoramidon-sensitive ECE. This suggests that the conversion of big- ET-1 to ET-1 may be due to an ECE which is located not only on the endothelium, but also in other cell types circulating in blood. Further, the characteristic specificity of the ECE for big-ET-1 has been established in an array of tissues, *in vivo* and *in vitro* (12-13).

Various pharmacological studies have outlined a possible involvement of a dynamic conversion of big-ET-1 to ET-1 (Table 1). Intravascular administration of big-ET-1 has been shown to markedly increase the plasma levels of immunoreactive-ET in the rabbit, rat and pig (7, 14, 15). These studies were subsequently supported by the fact that phosphoramidon markedly reduced the pressor responses to big-ET-1 in the rat (Table 1). The same inhibitor was also used to illustrate the presence of the ECE in many isolated organs and tissues. These experimental models include the pulmonary, renal and mesenteric vasculatures as well as vascular and non vascular cerebral tissues of various animal species and, more importantly, of man. The activity of the ECE was also shown in the microvasculature, in bronchial as well as genito-urinary tissues (Table 1). In all of the above-mentioned tissues, the responses to big-ET-1, but not to ET-1, were shown to be markedly inhibited by phosphoramidon.

Although the phosphoramidon-sensitive ECE has been suggested as the physiologically relevant enzyme responsible for the production of ET-1 from its intermediate precursor, one cannot yet exclude the contribution of other enzymes, such as cathepsin D-sensitive and pepstatin-sensitive enzymes (16, 17). For example, the role of a neutral endopeptidase or chymotrypsin-like enzyme which may preferentially cleave big-ET-3 *in vivo* in the rat model must be considered. In support of this hypothesis, it was shown (14) that intravascular administration of big-ET-1 induced an increase of circulating levels of immunoreactive-ET as well as hypertensive effects in rats and pigs. Although the later phenomenon is completely inhibited by phosphoramidon, the raised circulating level of immunoreactive-ET, following big-ET-1 administration, remains unaffected by the same treatment.

In addition, a recent study (18) has shown that phosphoramidon inhibits the conversion of big-ET-1 to ET-1 in isolated rat perfused lung. In that particular study, it was shown that phosphoramidon markedly reduced the big-ET-1-induced pressor action as well as immunoreactive-ET in rat lung tissues. In contrast, the increase in immunoreactive-ET levels in the lung perfusate was unaffected by phosphoramidon, yet it was markedly reduced by chymostatin, an α-chymotrypsin inhibitor.

Comparative Pharmacology of Big-Endothelins-2 and 3

Table 2

Phosphoramidon-sensitive effects of big-ET-2 (human, 1-37) and big-ET-3 (human, 1-41)

Effects	Model	Peptide	Effective concentration/ dose	Phosphoramidon sensitive	Ref .
Hypertension	Rat	Big-ET-2 Big-ET-3	1 nmol/kg 1 nmol/kg	Yes Yes	32 32
Reduction of blood flow	Hamster cheek pouch	Big-ET-3	300 pmol	No	31
Vasocon-striction	Rabbit kidney	Big-ET-2 Big-ET-3	100 pmol Inactive at 5000 pmol	Yes --	20 20
Prostacyclin release	Rabbit kidney	Big-ET-2 Big-ET-3	100 nM Inactive at 500 nM	Yes --	20 20
Neurogenic response	Rat vas deferens	Big-ET-2 Big-ET-3	500 nM Inactive at 500 nM	Yes --	24
Secretion and biosynthesis	Human renal adenocarcinoma	Big-ET-2		Yes	33

The two-above mentioned reports (14, 18) well illustrate that big-ET-1 may not only be converted by a membrane-bound phosphoramidon-sensitive ECE, but may as well be transformed to its active metabolite by a non-specific endopeptidase-like enzyme, in the rat.

Figure 2 illustrates the primary structures of human big-ET-1, 2 and 3 as well as porcine big-ET-1(1-39) and porcine big-Et-1 (1-40). Characteristically, all of these peptides contain in their N-terminal sequences the structure of the active metabolites, ET-1, 2 and 3 (represented in bold characters).

One study had shown that big-ET-1 (1-40) induced a pressor response in ganglion-blocked and anesthetized rats as well as vasoconstriction in the rat aorta only if the peptide was incubated with chymotrypsin (34). In contrast, the same peptide, in which one amino acid in C-terminal (L-Arg) had been removed, induced pressor responses in the rat without being preincubated with the same enzyme (9, 35, 36).

This was the first indication of the functional role of different enzymes in the processing of the two analogues, big-ET-1 (1-40) requiring the pretreatment of chymotrypsin, whereas big-ET-1 (1-39) could be readily converted by an endogeneous enzyme which was phosphoramidon-sensitive. This observation was of interest considering that big-ET-1 (1-40), when compared to big-ET-1 (1-39) only contained an extra amino acid in the C-terminal, which renders the peptide unrecognizable by endogeneous enzymes found in the rat *in vivo*.

This specific characteristic of the enzyme processing big-ET-1 was further substantiated in a series of studies where it was shown that big-ET-3 (1-41), when compared to human big-ET-1 (1-38), was inactive (13, 19, 20, 24). For exemple, in the anesthetized guinea pig, big-ET-1 induces phosphoramidon-sensitive pressor effects, whereas big-ET-3, at doses 10 times higher than the former peptide, remains inactive (13). Table 2 illustrates that big-ET-3 is pharmacologically inactive not only in the guinea pig *in vivo*, but also in various perfused organs and non-vascular tissues. However, three different groups (37-39) have shown a phosphoramidon-sensitive pressor effect of big-ET-3 in concious or anesthetized rats, while ET-1 and ET-3 were approximately equipotent in this animal model. It was also reported that big-ET-3 is readily converted to its active metabolite, ET-3, in cultured bovine endothelial cells (40). Hence, the specific characteristic of the ECE remains illusive. One may have to take into consideration that most of the hypertensive effects of big-ET-3 were shown in one animal model, the anesthetized rat. The rat model also shows a wide spectra of enzymatic activity, including a functional chymase-like enzyme (41) which is absent in other animal species, such as the rabbit, the guinea pig and man (42).

Interestingly, in the rat perfused lung, in which we had previously shown a phosphoramidon-sensitive release of PGI_2 induced by big-ET-1 (19), big-ET-1 was also converted by a chymostatin-sensitive enzyme (41). The same authors have suggested that the chymase or chymotrypsin was generated from activated mast cells, for the phenomenon was markedly enhanced in compound 48/80-treated lungs. In rat perfused lungs, untreated with 48/80, we have recently reported that big-ET-3 was inactive as an inducer of the release of PGI_2 (43). Whether or not big-ET-3 may be processed by a chymotrypsin-like enzymatic pathway remains to be determined.

In addition, a recent report (44) warrants some caution on the interpretation of the results obtained with the chymotrypsin-treated big-ET-1 (1-40) (34). Indeed, most of the commercially available α-chymotrypsin contains many other chymases and chymotrypsins (44) which may account for the converting properties of α-chymotrypsin. This particular aspect merits further investigation with the use of pure chymotrypsin.

Although the intermediate precursor of ET-2, big-ET-2, has now been disclosed for three years, its fate in *in vivo* and *in vitro* pharmacological models had to await two recent publications (20, 32). In those studies, big-ET-2 was shown for the first time to induce phosphoramidon-sensitive pharmacological effects, namely hypertensive responses and change in blood flow in conscious rats as well as vasoconstriction and the release of PGI_2 from the isolated perfused rabbit kidney.

Table 3

Biochemical inhibitors of the endothelin-converting enzyme

Inhibitor	Systems	Concentration range	Effect	Ref.
Phosphoramidon	Blood pressure	5 mg/kg	Inhibition of hypertensive effect of big-ET-1	8,9,18
	Vasoconstriction	50-100 μM	Inhibition of vasoconstriction	20,21
	Lung homogenate	50-100 μM	Inhibition of ET-1 production	50
	Endothelial cells	10-100 μM	Inhibition of ET-1 production	53,54,55,56
	Perfused organs	25-100 μM	Inhibition of pharmacological effects	13,19,20
	Brain homogenate	100 μM	Inhibition of ET-1 production	51,52
Suc-Ile-Ile-Trp-MCA	Lung homogenate	150 μM	Substrate inhibitor	57
EDTA	Endothelial cells	0.5 mM	Inhibition of ET-1 production	58
1,10-Phenanthroline	Endothelial cells	1 mM	Inhibition of ET-1 production	58

Big-ET-2 is a shorter peptide than human big-ET-1 (37 amino acids compared to 38), although more recent observations suggest that the previously-reported structure of big-ET-2 might be erroneous (45, 90).

In either case, it is understandable that the endothelin-converting enzyme may readily recognize big-ET-2 as well as big-ET-1 since the two peptides possess a Trp^{21}-Val^{22} bond and share a high structural homology both in the C- and N-terminal sequences. Table 2 illustrates the phosphoramidon-sensitive effects of big-ET-2 in comparison to its active metabolite, ET-2, *in vivo*, in perfused organs and in human renal adenocarcinoma.

Biochemical Evidences

As for many other enzymes, the ECE is yet to be fully characterized, both biochemically and pharmacologically. Unlike enzymes such as the neutral endopeptidase (NEP 24.11) (46) and the angiotensin-converting enzyme (ACE) (47), there is a lack of specific inhibitors for the ECE. As mentioned previously, a thermolysin inhibitor, phosphoramidon, remains the preferential tool to pharmacologically and biochemically characterize the ECE activity, the major drawback of phosphoramidon being that it is also a potent inhibitor of neutral endopeptidases which are involved in the degradation of ET-1, at least in the rat kidney (48). Table 3 resumes the various attempts to develop biochemical inhibitors of the ECE.

In most of these studies, the various inhibitors were tested against partially purified ECE preparations derived from homogenized organs, such as the kidney or the lung of various animal species. Hence, the ultimate design of a selective ECE inhibitor is hindered by the fact that the substrate used to define the enzymatic activity is limited to crude extract. Nonetheless, Ohnaka *et al.* (49) have suggested a tentative molecular weight for the ECE (120-150 kDa). More recently, it has been suggested that different enzymes are involved in the processing of big-ET-1 and 2, respectively. Indeed, Sawamura *et al.* (50) have shown that the membrane-bound ECE, which is phosphoramidon-sensitive, recognized preferentially big-ET-1; this enzyme was named M_1. Another enzymatic entity, M_2, which was found in the same tissue (porcine lung), transformes indiscriminately big-ET-1 and 2 and is sensitive to thiorphan as well as to phosphoramidon. Interestingly, M_1 and M_2 were found to be inactive against big-ET-3. It is also of interest to point out that big-ET-1, but not big-ET-3, has been found to be dynamically converted in the rat brain and human brain homogenates (51, 52).

Receptors for Endothelins

Although big-ET-1 is as potent as ET-1 as a hypertensive agent *in vivo* (9), the latter peptide is 250 times more potent than its precursor, in isolated vascular tissues in inducing contractile responses. As mentioned previously in the present chapter, to induce such potent pressor effects, big-ET-1 must be dynamically converted to ET-1 *in vivo*. Alternatively, big-ET-1 may activate ET receptors directly on the vascular smooth muscle (29). Few studies support the latter hypothesis since in all species studied, phosphoramidon markedly reduces the hypertensive effects of big-ET-1, but not those induced by ET-1 (8, 9, 37).

Interestingly, only some studies have attempted to characterize the receptors by which big-ET-1 induces its pharmacological effects. Since its discovery (59), big-ET-1 has been sparsely investigated until the first few reports using phosphoramidon. Indeed, the conversion of big-ET-1 is essential for the full expression of the ET-1-like vasoconstriction (59). Because of the lack of selective antagonist for endothelin receptors, it remained to be assessed whether big-ET-1 and ET-1 act on the same type of receptors. We reported for the first time that a selective and specific antagonist for endothelin receptors, namely BQ-123 (60), could competitively interfere with one of the pharmacological properties of big-ET-1, namely as a releaser of eicosanoids from the rat pulmonary vasculature (61).

In addition to acting as a calcium channel opener (dihydropyridine-sensitive) (3), ET-1 has been shown to interact with specific receptors. Two of those receptor types, ET_A and ET_B, were biochemically characterized and expressed in COS-7 cells (62, 63). Characteristically, the affinity order of potency for endothelins on ET_A-containing preparations is the following:

$$\text{ET-1} > \text{ET-2} > \text{ET-3}$$

In contrast, the three isomers are equipotent in ET_B-containing preparations. Table 4 describes some of the isolated preparations containing, in different proportions, one or the other type of receptor for endothelins. As illustrated in Table 4, most of the vasoconstrictive effects of endothelins are mediated via the activation of ET_A receptors, although ET_B receptor activation has recently been associated with contractions of venous blood vessels (particularly in the rabbit) and of the rabbit pulmonary artery (64). ET_B receptors have been predominantly associated with the vasodilatory properties of endothelins (65, 66). Both ET_A and ET_B receptors are of the rhodopsin type, possess seven transmembrane domains and similar molecular weight (Mr of ET_A: 48 516, Mr of ET_B: 46 901) (62, 63). To our knowledge, no attempt has been made to correlate the binding of big-ET-1 to any of those receptors expressed in COS-7 cells or other host cells. Yet, it has been suggested that big-ET-1 predominantly binds to the ET_A receptor (61, 67) in the vascular smooth muscle. Indeed, in most animal species studied (except the guinea pig), ET-1, but not big-ET-1, induced an initial transient hypotension followed by a marked and sustained pressor response. The hypotensive effect of ET-1 was clearly associated to the release of EDRF (endothelium-derived relaxing factor), via the activation of ET_B receptors. The lack of hypotensive effect of big-ET-1 *in vivo* was suggested to be due to the lack of ability of the peptide to produce sufficient ET-1 at the level where the peptide would activate the release of vasodilator substances (68). This observation may explain the inability of big-ET-1 to release EDRF *in vivo*. Yet, other vasodilator substances, such as PGI_2, have been convincingly shown to be released by the precursor, both *in vivo* (7, 13) and in various *in vitro* models (13, 19, 20). It is worthy to mention that big-ET-1 releases PGI_2 in models where the presence of ET_A receptors has been substantiated (61, 67). Although ET_A receptors have been shown to be responsible for the release of PGI_2 by big-ET-1 in the pulmonary vasculature and ET-1 in the renal vasculature, the exact location of those receptors remains to be determined.

Although BQ-123 markedly reduced the release of PGI_2 in the rat perfused lung (61) and rabbit perfused kidney (85), it is yet to be assessed whether the ET_A receptors, responsible for the release of this eicosanoid, are located on the vascular endothelium or, alternatively, on the underlying vascular smooth muscle. Interestingly, in support of the later possibility, it was recently reported that the rabbit renal artery vascular smooth muscle generated arachidonic acid cascade metabolites when stimulated with ET-1, via the selective activation of ET_A receptors (69). Thus, we suggest that ET_A-dependent release of eicosanoids counteracts the vasoconstriction induced by activation of the same receptor type, in the perfused rabbit kidney. It is worthy of notice that two recent studies have shown that BQ-123 was inactive against ET-1 inhibition of *ex vivo* platelet aggregation in the rabbit and in the dog (70, 71). Indeed, in those two animal models, ET-1 induced an indomethacin-sensitive inhibition of platelet aggregation *ex vivo*. This effect was unaltered by BQ-123 at doses which significantly reduced the hypertensive effects of ET-1. Hence, one may have to dissociate the receptors for the endothelins responsible for the generation of local eicosanoids, as opposed to the release of systemic eicosanoids which appear to be mediated by ET_B receptor activation, following intravascular administration of the peptide.

We had previously shown that big-ET-1 and ET-1 trigger the release of PGI_2 from the perfused rabbit kidney (20) and have more recently shown that ET_A receptor activation was solely responsible for the release of this prostanoid by ET-1 in the same perfused model (85). Hence, we can assume that big-ET-1, once converted to ET-1, induces vasoconstriction and the release of vasomodulatory PGI_2 via the activation of ET_A receptors (20, 61, 67).

It is also worthy of notice that in some vascular tissues (i.e. the rabbit saphenous vein (64) or the rat renal vasculature (72)), big-ET-1 may directly activate ET_B receptors, without being previously converted to ET-1 by a phosphoramidon-sensitive ECE.

Various pharmacological tools are now available to further characterize the receptors involved in the pharmacological properties of big-ET-1 and ET-1; these analogues are illustrated in Figure 3. Of these tools, BQ-123 has been the most used selective ET_A receptor antagonist with a relatively good pA_2 value of 7.2, on porcine coronary arteries. The discovery of this antagonist has been followed by a third generation antagonist, BQ-610, which has 10 times higher affinity than BQ-123 in the above-mentioned preparation (porcine coronary artery).

Table 4

Pharmacological characterization of ET_A and ET_B receptor-containing preparations

Isolated preparations	Agonist[1]				Antagonist[2]		Receptor	Effect[3]	Ref.
	ET-1	ET-2	ET-3	IRL 1620	BQ-123	IRL 1038			
Rabbit jugular vein	9.69	9.22	8.48	8.65	7.2	-	ET_A, ET_B	C	82
Rabbit pulmonary artery	9.67	-	10.09	9.24	-	-	ET_B	C	82
Rat aorta	8.35	-	6.43	>6	6.90	6.00	ET_A	C	75,82
Rat aorta	≈8.0	-	≈8.0	-	-	≈6.0	ET_B	R	66
Rat vas deferens	7.49	7.40	7.59	-	>4.77	-	ET_B[4]	C	24,82
Rat uterus	9.04	-	>7.0	-	7.76	-	ET_A	C	83
Perfused organs	**Effective concentration/dose**								
Rat mesenteric artery	25 pmol	-	1000 pmol	1000 pmol	32 nM	-	ET_A	C	84
Rat mesenteric vein	5 pmol	-	200 pmol	500 pmol	12 nM	-	ET_A	C	84
Rabbit kidney	5 pmol	10 pmol	100 pmol	>5000 pmol	100 nM	-	ET_A	C	20,85
Rabbit kidney	10 nM	10 nM	100 nM	500 nM	500 nM	-	ET_A	PGI_2[5]	20,85
Rat lung	5 nM	-	5 nM	-	500 nM	-	ET_A	PGI_2	61

1: pD_2 (-log of the ED_{50}); 2: pA_2 (-log of the molar concentration of antagonist necessary to reduce the effect of a double dose of agonist to the effect of a single dose); 3: Contraction; R: Relaxation; 4: Subtype of ET_B; 5: Release of prostacyclin.

1. Selective ET_A antagonists

BQ-123 (60)	Cyclo[D-Asp-L-Pro-D-Val-L-Leu-D-Trp]
BQ-610 (91)	 N-CO-L-Leu-D-Trp(CHO)-D-Trp

2. Selective ET_B agonists

Sarafotoxin 6C (81)	[Thr^2,Asn^4,Asp^5,Met^6,Thr^7,Glu^9,Leu^{12},Asn^{13},Glu^{17},Val^{19}]-ET-1(1-21)
BQ-3020 (73)	Ac-Leu-Met-Asp-Lys-Glu-Ala-Val-Tyr-Phe-Ala-His-Leu-Asp-Ile-Ile-Trp
IRL 1620 (74)	Suc-Asp-Glu-Glu-Ala-Val-Tyr-Phe-Ala-His-Leu-Asp-Ile-Ile-Trp
4AlaET-1 (73)	[$Ala^{1,3,11,15}$]ET-1(1-21)

3. Selective ET_B antagonist

IRL 1038 (75)	Cys-Val-Tyr-Phe-Cys-His-Leu-Asp-Ile-Ile-Trp

Figure 3. *Agonists and antagonists used for receptor classification of endothelins*

The selective ET_B agonists, such as BQ-3020 and IRL 1620, are now currently used to biochemically and pharmacologically characterize ET_B-containing preparations (73, 74). In most pharmacological assays studied until now, the two selective ET_B agonists have been shown to promote the release of EDRF from endothelium intact vascular rings (73, 75). This ET_B-mediated release of EDRF has recently been shown to be selectively inhibited by an ET_B receptor antagonist, IRL 1038. Indeed, this antagonist markedly blunted the EDRF release induced by ET-3 in the rat aortic rings (75). These above-mentioned analogues may be considered as the most recent and perhaps useful tools to be used in the further characterization of endothelin-mediated events (Table 4).

Role of Calcium Channels in the Vascular Response to Endothelin-1

It was initially suggested that ET-1 may be the endogeneous ligand for dihydropyridine-sensitive calcium channels (3). Indeed, it had been shown that ET-1

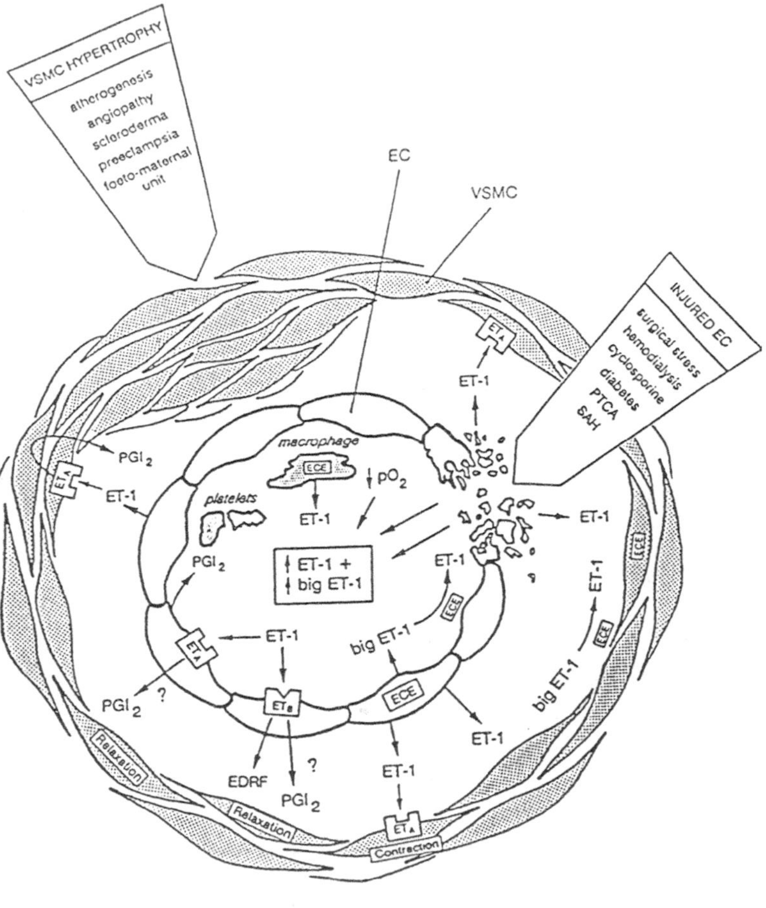

Figure 4. *Schematic diagram showing big-ET-1 conversion and ET-1 secretion from human blood vessels. Endothelin-1 is released by the endothelial cells (EC) into the circulation or abluminally toward the underlying smooth muscle layer. It is formed by the endothelium through a phosphoramidon-sensitive converting enzyme (ECE) located within the EC and vascular smooth muscle cells (VSMC) or on the surface of both layers. The precursor, big-ET-1, is also released although in larger quantities than ET-1 (ratio 2.7:1.0) and may be converted to the active form while in circulation. Release of ET-1 can activate* ET_A *and* ET_B *(and perhaps* ET_C*?) receptors on endothelial cells and vascular smooth muscle, mediating the production of prostanoids, mainly prostacyclin (*PGI_2*) and EDRF (NO). They regulate ET-1 vasoconstrictive and pressor effects and mediate the antiaggregatory properties of the peptide, through cAMP and cGMP elevation. The integrity of the endothelium would be a determining factor in the rate of synthesis and secretion of ET-1 (taken from Battistini et al. (80)).*

induced a nicardipine-sensitive contraction of the porcine coronary artery. However, subsequent studies have tampered with this concept by showing nicardipine-insensitive contractions of various isolated tissues, such as the rabbit jugular vein, the rabbit mesenteric artery and the rat aorta (76, 77).

Rather than activating directly the dihydropyridine-sensitive (DHPS) calcium channels, one is tempted to suggest an activation of receptor-operated calcium channels (ROC) by the potent vasoactive peptide, ET-1. This consideration is based on a study that has shown that the selective and competitive ET_A receptor antagonist, BQ-123, markedly reduced, with a relatively high affinity (pA_2 value of 7.2), the concentration-response curve to ET-1 in the porcine coronary artery (60). As nifedipine was shown to decrease the response of that tissue to ET-1 (3), the fact that a receptor antagonist would affect the response of the same tissue to ET-1 suggests that ET-1 may induce vasoconstriction of that particular blood vessel to subsequently trigger the opening of the DHPS calcium channel.

It is also of interest that ET-1 may interact with another type of calcium channel, namely the resting potential calcium channel (R-calcium channel) (78). Indeed, the voltage-sensitive R-calcium channel may be affected by ET-1 among other vasoactive agents (platelet-activating factor (PAF), insulin and bradykinin). The opening of this calcium channel by ET-1 may explain the biphasic calcium influx which has been reported by various groups, the initial transient calcium influx being possibly mediated directly via receptor activation or via the activation of the L-type calcium channel, whereas the slow and sustained second phase increase of calcium appears to be totally mediated via the opening of the R-type calcium channel. One of the main electrophysiological characteristics of the R-calcium type channel is that it is insensitive to dihydropyridines. In contrast, a dual L and R-calcium type channel blocker, isradipine (PN 200-10), interferes efficiently with the opening of this calcium channel by agents such as PAF and insulin (78, 79). The future use of R-type calcium channel blockers in ET-induced vasoconstrictions and vasodilations may warrant some important information on the role of calcium channels in the vascular response to this potent peptide.

Pathophysiological Considerations

As previously mentioned, because big-ET-1 is found in higher concentrations than ET-1 in human plasma, one can suggest a role for big-ET-1 as a reliable marker of vascular damage. Indeed, vascular damage is clearly associated in most cases with endothelial disruption. The lysis of endothelium (Fig. 4), as seen in surgical procedures, arterosclerosis and ischemic conditions, favors the release of endothelial-derived ET-like immunoreactivity (80). As ET-1 is a potent vasoconstrictor, any conditions that trigger an elevation of big-ET-1 and ET-1 plasma levels may have a marked influence of vascular tone. However, this is yet to be fully validated. For example, in normotensive animals, BQ-123 does not trigger a hypotensive effect *per se*, suggesting that endogeneously-released ET-1 does not influence vascular tone *in vivo* in normal conditions (86). In contrast, in experimental models of renal acute toxicity or ischemia, a selective ET_A receptor antagonist has been shown to be beneficial in inducing a partial recovery of normal renal functions (87, 88).

Even fewer studies have attempted to relate an enhanced level of big-ET-1 with vascular dysfunction. One of those reports by Vemulapalli *et al.* (89) has demonstrated that phosphoramidon reversed the increase of respiratory resistance induced by hypoxia in the guinea pig perfused lung. This interesting study suggested that hypoperfusion and hypoventilation of the lung triggered a marked increase of ET-1, which appeared to be responsible for the increase of airway resistance. Concommitantly to the reduction of pulmonary resistance, phosphoramidon also reduced the level of immunoreactive ET-1 in the lung perfusate, suggesting a marked phosphoramidon-sensitive ECE activity in this experimental condition; the ECE activity in the guinea pig pulmonary vasculature had been previously reported (13). Hence, because of the presence of marked ECE activities not only in the endothelium, but also on other cell types (blood-borne or vascular), our understanding of the synthesis pathway of big endothelins remains incomplete. In addition, both ET-1 and its precursor produce an array of pharmacological effects in the vasculature via different receptors; the complex series of events have been depicted in Figure 4. Indeed, vascular trauma induced by many conditions promotes endothelial damage and increases plasma levels of endothelins, among many other endothelial-derived autacoids. One can suggest two obvious targets of interest: the ECE activity and the receptors involved in the vascular responses to endothelins. Hence, future use of specific ECE inhibitors as well as selective receptor antagonists for endothelins (see figure 4) may allow a better understanding of the role of those potent vasoactive peptides in various vascular pathologies.

Acknowledgements

The authors would like to acknowledge H. Morin for secretarial assistance. S.T. is in receipt of a studentship from the Fonds de la recherche en santé du Québec. P.D.J. is a scholar of the Heart and Stroke Foundation of Canada. This work was financially supported by the Medical Research Council of Canada.

REFERENCES

1. Moncada S, Gryglewski R, Bunting S and Vane JR (1976) Nature 263:663-665
2. Furchgott RF and Zawadzki JV (1980) Nature 288:373-376
3. Yanagisawa M, Kurihara H, Kimura S, Tomobe Y, Kobayashi M, Mitsui Y, Yazaki Y, Goto K and Masaki T (1988) Nature 332:411-415

4. Inoue A, Yanagisawa M, Kimura S, Kasuya Y, Miyauchi T, Goto K and Masaki T (1989) Proc Natl Acad Sci 86:2863-2867
5. Laporte S, Denault J-B, D'Orléans-Juste P and Leduc R (1993) J Cardiovasc. Pharmacol, 22 (Suppl 8):S7-S10
6. Bresnahan PA, Leduc R, Thomas L, Thorner J, Gibson HL, Brake AJ, Barr PJ, Thomas G (1990) J Cell Biol 111:2851-2859
7. D'Orléans-Juste P, Lidbury PS, Warner TD and Vane JR (1990) Biochem Pharmacol 39:R21-R22
8. Fukuroda T, Noguchi K, Tsuchida S, Nishikibe M, Ikemoto F, Okada K and Yano M (1990) Biochem Biophys Res Commun 172:390-395
9. Matsumura Y, Hisaki K, Takaoka M and Morimoto S (1990) Eur J Pharmacol 185:103-106
10. Kashiwabara T, Inagaki Y, Ohta H, Iwamatsu A, Nomizu M, Morita A, Nishikori K (1989) FEBS Lett 247:73-76
11. Sessa WC, Kaw S, Hecker M and Vane JR (1991) Biochem Biophys Res Commun 174:613-618
12. Opgenorth TJ, Wu-Wong JR and Shiosaki K (1992) FASEB J 6:2653-2659
13. D'Orléans-Juste P, Télémaque S and Claing A (1991) Br J Pharmacol 104:440-444
14. Modin A, Pernow J, Lundberg JM (1991) Life Sci 49:1619-1625
15. Hemsen A, Pernow J, Lundberg JM (1991) Acta Physiol Scand 141:325-334
16. Sawamura T, Kimura S, Shinmi O, Sugita Y, Yanagisawa M, Goto K and Masaki T (1990) Biochem Biophys Res Commun 168:1230-1236
17. Matsumura Y, Ikegawa R, Takaoka M and Morimoto S (1990) Biochem Res Commun 167:203-210
18. Hisaki K, Matsumura Y, Maekawa H, Fujita K, Takaoka M and Morimoto S (1993) Jap J Pharmacol 61:135P
19. D'Orléans-Juste P, Lidbury PS, Télémaque S, Warner TD and Vane JR (1991) J Cardiovasc Pharmacol 17:S251-255
20. Télémaque S, Lemaire D, Claing A and D'Orléans-Juste P (1992) Hypertension 20:518-523
21. Hisaki K, Matsumura Y, Ikegawa R, Nishiguchi S, Hayashi Y, Takaoka M, Morimoto S (1991) Biochem Biophys Res Commun 177:1127-1132
22. Noguchi K, Noguchi Y, Hirose H, Nishikibe M, Ihara M, Ishikawa K, Yano M (1993) Eur J Pharmacol 233:47-51
23. Pons F, Touvay C, Lagente V, Mencia-Huerta JM, Braquet P (1991) Br J Pharmacol 102:791-796
24. Télémaque S and D'Orléans-Juste P (1991) N-S Arch Pharmacol 344:505-507
25. Matsuura A, Okumura H, Ashizawa N, Kobayashi F (1992) Life Sci 50:1631-1638
26. Lehoux S, Plante GE, Sirois MG, Sirois P, D'Orléans-Juste P (1992) Br J Pharmacol 107:996-1000
27. Advenier C, Sarria B, Naline E, Puybasset L, Lagente V (1990) Br J Pharmacol 100:168-172

28. Advenier C, Lagente V, Zhang Y, Naline E (1992) Br J Pharmacol 106:883-887
29. Auguet M, Delaflotte S, Chabrier PE, Braquet P (1992) Eur J Pharmacol 224:101-102
30. Poulat P, D'Orléans-Juste P, de Champlain J, Yano M, Couture R (1992) 22nd Annual Meeting of the Society for Neuroscience, Abstract #489-14
31. Lawrence E, Brain SD (1993) Eur J Pharmacol 233:243-250
32. Gardiner SM, Kemp PA and Bennett T (1992b) Br J Pharmacol 107:584-590
33. Arinami T, Ishikawa M, Inoue A, Yanagisawa M, Masaki T, Yoshida MC, Hamaguchi H (1991) Am J Hum Genet 48:990-996
34. McMahon EG, Fok KF, Moore WM, Smith CE, Siegel NR, Trapani AJ (1989) Biochem Biophys Res Commun 161:406-413
35. McMahon EG, Palomo MA, Moore WM (1991) J Cardiovasc Pharmacol 17:S29-S33
36. McMahon EG, Palomo MA, Moore WM, McDonald JF, Stern MK (1991) Proc Natl Acad Sci USA 88:703-704
37. Gardiner SM, Kemp PA, Compton AM, Bennett T (1992a) Br J Pharmacol 106:483-488
38. Matsumura Y, Fujita K, Takaoka M., Morimoto S (1993) Eur J Pharmacol 230:89-93
39. Pollock DM, Divish BJ, Milicic I, Novosad EI, Burres NS and Opgenorth T (1993) Eur J Pharmacol 231:459-464
40. Matsumura Y, Tsukahara Y, Kuninobu K, Takaoka M, Morimoto S (1992) FEBS Letters 305:86-90
41. Wypij DM, Nichols JS, Novak PJ, Lowell Stacy D, Berman J and Wiseman JS (1992) Biochem Pharmacol 43:845-853
42. Urata H, Kinoshita A, Misono KS, Bumpus FM and Husain A (1990) J Biol Chem 265:22348-22357
43. Télémaque S, Gratton J-P, Claing A and D'Orléans-Juste P (1993) J Cardiovasc Pharmacol 22(Suppl 8):S85-89
44. Ashton DS, Beddell CR, Cooper DJ, Green BN, Oliver RW, Welham KJ (1993) Biochem. Biophys. Res. Commun. 192: 75-81
45. Kimura S, Yorimitsu K, Shinmi O, Nishiyama M, Moroi K, Saito T, Inagaki Y, Sugita Y and Masaki T (1993) J Cardiovasc Pharmacol 22(Suppl 8): in press
46. Roques BP, Fournie-Zaluski MC, Soroca E, Lecomte JM, Malfroy B, Llorens C and Schwartz JC (1980) Nature 288:286-288
47. Condon ME, Petrillo EW, Ryono DE, Reid JA, Neubeck R, Puar M, Heikes JE, Sabo EF, Losee KA, Cushman DW, Ondetti MA (1982) J Med Chem 25:250-258
48. Abassi ZA, Tate JE, Golomb E and Keiser HR (1992) Hypertension 20:89-95
49. Ohnaka K, Nishikawa M, Takayanagi R, Haji M, Nawata H (1992) Biochem Biophys Res Comm 185:611-616

50. Sawamura T, Shinmi O, Kishi N, Sugita Y, Yanagisawa M, Goto K, Masaki T and Kimura S (1993) Biochimica et Biophysica Acta 1161:295-302
51. Warner TD, Budzik GP, Matsumoto T, Mitchell JA, Försterman U and Murad F (1992a) Br J Pharmacol 106:948-952
52. Warner TD, Schmidt HH, Kuk J, Mitchell JA, Murad F (1992b) Br J Pharmacol 106:505-506
53. Ohnaka K, Takayangi R, Yamauchi T, Okazaki H, Ohashi M, Umeda F, Nawata H (1990) Biochem Biophys Res Commun 168:1128-1136
54. Okada K, Miyazaki Y, Takada J, Matsuyama K, Yamaki T, Yano M (1990) Biochem Biophys Res Commun 171:1192-1198
55. Ikegawa R, Matsumura Y, Tsukahara Y, Takaoda M, Morimoto S (1990) Biochem Biophys Res Commun 171:669-675
56. Matsumura Y, Ikegawa R, Tsukahara Y, Takaoda M, Morimoto S (1990) FEBS Lett 272:166-170
57. Kundu GC, Wilson IB (1992) Life Sci 50:965-970
58. Ahn K, Beningo K, Olds G, Hupe D (1992) Proc Natl Acad Sci USA 89:8606-8610
59. Kimura S, Kasuya Y, Sawamura T, Shinimi O, Sugita Y, Yanagisawa M, Goto K and Masaki T (1989) J Cardiovasc Pharmacol 13 (suppl. 5):S5-S7
60. Ihara M, Noguchi K, Saeki T, Fukuroda T, Tsuchida S, Kimura S, Fukami T, Ishikawa K, Nishikibe M and Yano M (1992) Life Sci 50:247-255
61. D'Orléans-Juste P, Télémaque S, Claing A, Ihara M and Yano M (1992) Br J Pharmacol 105:773-775
62. Arai H, Hori S, Aramori I, Ohkubo H and Nakanishi S (1990) Nature 348:730-732
63. Sakurai T, Yanagisawa M, Takuwa Y, Miyazaki H, Kimura S, Goto K and Masaki T (1990) Nature 348:732-735
64. Moreland S, McMullen DM, Delaney CL, Lee VG and Hunt JT (1992) Biochem Biophys Res Commun 184:100-106
65. Webb DJ (1991) Trends Pharmacol Sci 12:43-46
66. Karaki H, Sudjarwo SA, Hori M, Sakata K, Urade Y, Takai M and Okada T (1993) Eur J Pharmacol 231:371-374
67. D'Orléans-Juste P, Yano M and Télémaque S (1993) J Cardiovasc Pharmacol, 22(Suppl 8):S235-S238
68. Haleen SJ, Davis LD and Keiser JA (1993) J Cardiovasc Pharmacol, 22(Suppl 8): in press
69. Doherty AM, Cody WL, He JX, DePue PL, Welch KM, Flynn MA, Reynold EE, LaDouceur DM, Davis LS and Haleen SJ (1993) J Cardiovasc Pharmacol, 22(Suppl 8): in press
70. McMurdo L, Lidbury PS, Thiemermann C and Vane JR (1993) Br J Pharmacol 109:530-534
71. Herman F, Yano M and Filep JG (1993) Eur J Pharmacol 236:143-146
72. Salvati P, Dho L, Calabresi M, Rosa B, Patrono C (1992) Eur J Pharmacol 221:267-273

73. Saeki T, Ihara M, Fukuroda T, Yamagiwa M and Yano M (1991) Biochem Biophys Res Commun 179:286-292
74. Takai M, Umemura K, Yamasaki T, Watakabe Y, Fujitani K, Oda Y, Urade Y, Inui T, Yamamura T and Okada T (1992) Biochem Biophys Res Commun 184:953-959
75. Karaki H, Sudjarwo SA, Hori M, Takai M, Urade Y and Okada T (1993) Br J Pharmacol 109:486-490
76. D'Orléans-Juste P, de Nucci G, Vane JR (1989) Eur J Pharmacol 165:289-295
77. Auguet M, Delaflotte S, Chabrier PE, Pirotzky E, Clostre F, Braquet P (1988) Biochem Biophys Res Commun 156:186-192
78. Bkaily G, Wang S, Economos E, D'Orléans-Juste P (1991) In: N Sperelakis, K Hermsmyer and H Kuriyama (Eds) Electrophysiology and ion channels of vascular smooth muscle cells and endothelial cells. Elsevier, New York, pp 199-207
79. Bkaily G, Economos D, Potvin L, Ardilouze JL, Marriott C, Corcos J, Bonneau D, Fong CN (1992) Mol Cell Biochem 117:93-106
80. Battistini B, D'Orléans-Juste P and Sirois P (1993) Lab Invest 68:600-628
81. Williams DL, Jones KL, Pettibone DJ, Lis EV, Clineschmidt BV (1991) Biochem Biophys Res Commun 175:556-561
82. Regoli D, D'Orléans-Juste P, Rouissi N, Rhaleb NE (1993) Regul Pep 45:323-340
83. Rae GA, Calixto JB, D'Orléans-Juste P (1993) Eur J Pharmacol, 240:113-119
84. D'Orléans-Juste P, Claing A, Warner TD, Yano M, Télémaque S (1993) Br J Pharmacol 110:687-692
85. Télémaque S, Gratton JP, Claing A, D'Orléans-Juste P (1993) Eur J Pharmacol 237:275-281
86. Nishikibe M, Tsuchida S, Okada M, Fukuroda T, Shimamoto K, Yano M, Ishikawa K and Ikemoto F (1993) Life Sci 52:717-724
87. Mino N, Kobayashi M, Nakajima A, Amano H, Shimamoto K, Ishikawa K, Watanabe K, Nishikibe M, Yano M and Ikemoto F (1992) Eur J Pharmacol 221:77-83
88. Fogo A, Hellings SE, Inagami T and Kon V (1992) Kidney Int 42:770-774
89. Vemulapalli S, Rivelli M, Chiu PJ, del Prado M and Hey JA (1992) J Pharmacol Exp Ther 262:1062-1069
90. Yoromitsu K, Shinmi O, Nishiyama M, Moroi K, Suzita Y, Saito T, Inagaki Y, Masaki T, Kimura S (1992) FEBS Lett 314:395-398
91. Ishikawa k, Fukami T, Nagase T, Mase T, Hayama T, Niiyama K, Fujita K, Urakawa Y, Kumagai U, Fukuroda T, Ihara M, Yano M (1993) Peptides 1992 (22 EPS) eds Schneider CH, Eberle AN (ESCOM Sci Publ BV, Leiden, The Netherlands), pp685-686

CHAPTER 14

CARBOXYL METHYLATION OF PROTEINS IN KIDNEY MEMBRANES: IMPLICATIONS IN THE REPAIR OF DAMAGED PROTEINS AND IN SIGNAL TRANSDUCTION

Dominique Boivin[*], *Denis Gingras*[*], *and Richard Béliveau*

Groupe de recherche en biothérapeutique moléculaire,
Université du Québec à Montréal,
Québec, Canada

Methylation of carboxyl groups in proteins is a widely distributed posttranslational modification that occurs on various amino acid residues [1, 2]. This reaction consists in the transfer of a methyl group from *S*-adenosyl-L-methionine (AdoMet) to the free carboxyl group of amino acid residues, resulting in the formation of a methyl ester and the neutralization of a negative charge on the protein (Fig. 1). The reverse reaction—hydrolysis of the methyl ester to form methanol—may occur either spontaneously or enzymatically, depending on the lability of the ester formed. Four types of protein carboxyl methyltransferases (PCMTs) have been identified according to the nature of the amino acid residue modified (Table 1).

Class I PCMTs are found exclusively in chemotactic bacteria where specific membrane receptor proteins are methylated on the side chain of glutamyl residues [3]. This methylation reaction is reversible and plays a role in the chemotactic response [4]. Class II PCMTs are ubiquitous and well conserved enzymes that recognize the altered aspartyl residues L-isoaspartyl [5, 6] and D-aspartyl [7].

[*] D.B. and D.G. are equal first authors

TABLE 1. **Classification of protein carboxyl methyltransferases**

	Class I L-Glutamyl methyltransferase (EC 2.1.1.80)	Class II L-Isoaspartyl/ D-aspartyl methyltransferase (EC 2.1.1.77)	Class III Isoprenyl L-cysteine C-terminal methyltransferase (EC 2.1.1.100)	Class IV Leucine C-terminal methyltransferase (non-classified)
Distribution	Bacteria	Procaryotes Eucaryotes	Eucaryotes	Eucaryotes
Subcellular localization	Cytosol	Cytosol and membrane	Membrane	Cytosol
Residue(s) methylated	L-Glutamate	L-Isoaspartate, D-aspartate	*S*-Isoprenyl-L-cysteine	C-terminal leucine
Substrates	Chemotactic receptors	Ovalbumin, calmodulin	Small G-proteins, γ-subunit of trimeric G-proteins	36 kDa protein in brain cytosol
Functions	Receptor modulation	Repair of damaged proteins	Membrane anchoring, regulation of enzyme activity?	Unknown

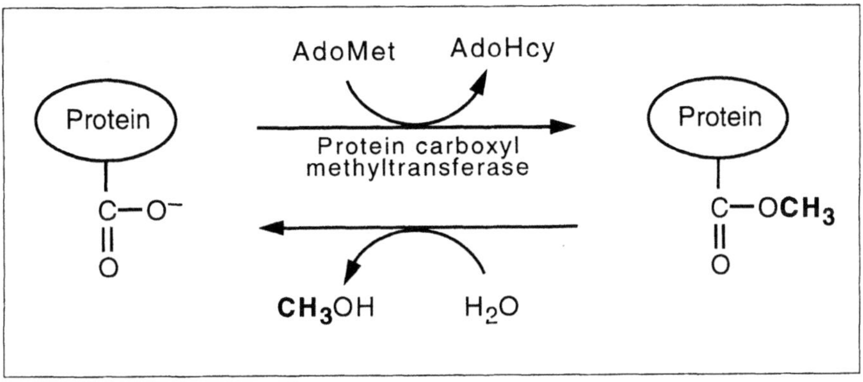

Figure 1. Reversible methyl esterification of proteins

Methylation of these abnormal amino acid residues is involved in the repair of damaged proteins that accumulate in aging cells [7, 8].

Class III PCMTs catalyze the methylation of the α-carboxyl groups of *S*-farnesylated and *S*-geranylgeranylated C-terminal cysteine residues [9–11]. C-terminal methylation is a posttranslational modification of many proteins involved in signal transduction including small and large G-proteins, rhodopsin kinase and cGMP phosphodiesterase. A fourth class of PCMTs is defined by a recently identified enzyme which methylates the C-terminal leucine residue of a 36 kDa polypeptide in brain cytosol [12]. The function of this new type of methylation remains to be established.

CLASS II PROTEIN CARBOXYL METHYLTRANSFERASE

Repair of Damaged Proteins

The atypical residues D-aspartyl and L-isoaspartyl are produced by spontaneous reactions such as deamidation of asparaginyl residues and subsequent isomerization and racemization [13, 14]. Many regulatory roles were first attributed to class II PCMTs but the very rare occurrence of these atypical residues in proteins and the lack of protein-substrate specificity are not consistent with such a function [15–16]. These unusual properties have led to the hypothesis that L-isoaspartyl/D-aspartyl PCMTs are involved in the repair of damaged proteins. The first step of the repair mechanism is the conversion of the L-isoaspartyl residue to the corresponding methyl ester, catalyzed by the methyltransferase (Fig. 2). The L-isoaspartyl α-methyl ester undergoes spontaneous demethylation to produce an unstable L-succinimidyl

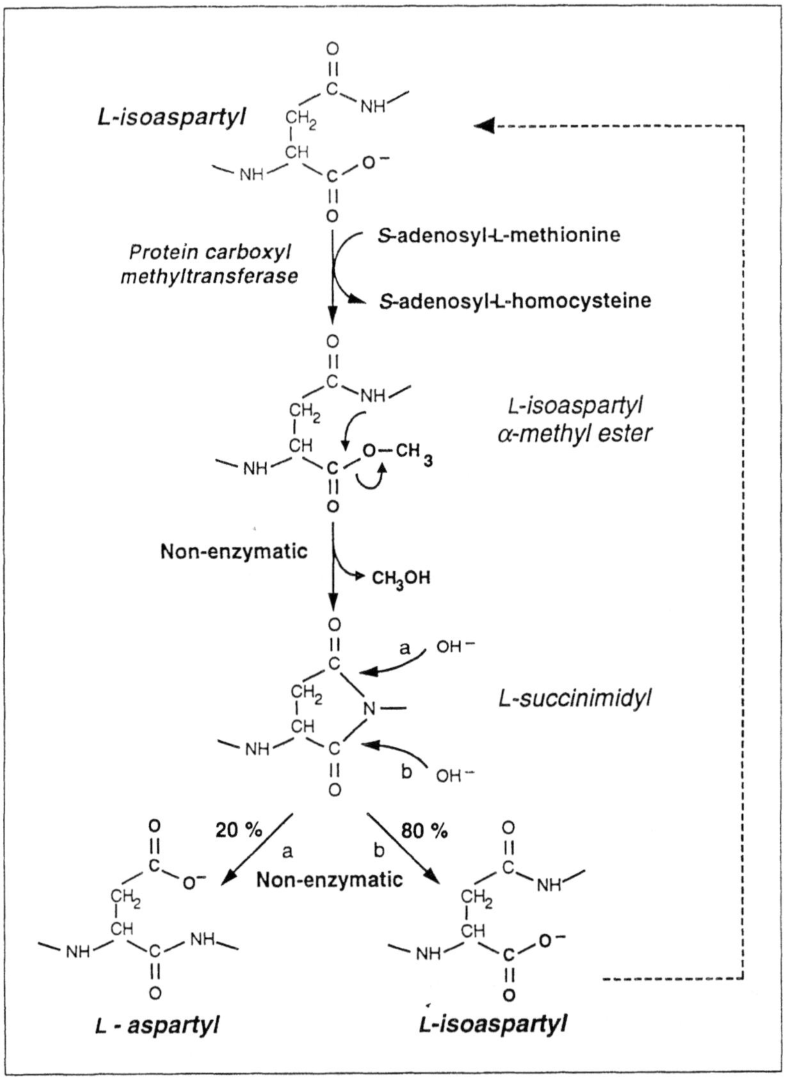

Figure 2. Conversion of l-isoaspartyl to normal L-aspartyl residues in proteins. Adapted from Lowenson and Clarke [8].

intermediate which is readily hydrolyzed to L-aspartyl (80%) and L-isoaspartyl (20%) residues. After multiple cycles of methylation, the L-isoaspartyl residues are almost completely converted to normal L-aspartyl residues. The efficacy of this repair pathway has been demonstrated *in vitro* with synthetic model peptides containing L-isoaspartyl residues [17] and with deamidation-damaged calmodulin [18].

The physiological significance of L-isoaspartyl/D-aspartyl methylation was recently investigated in a normal strain of *Escherichia coli* and in mutants deficient in PCMT [19]. The *pcm*⁻ mutants did not survive well in stationary-phase culture or under stress conditions such as heat (55°C) or exposure to an oxidative agent (H_2O_2). In HeLa cells, prolonged exposure to a temperature of 42°C caused a 70–80% increase in carboxyl methylation activity [20]. PCMT synthesis remained unaffected, suggesting that the increase in activity comes from additional altered aspartyl residues produced by the heat treatment. Together, these data suggest that methylation of altered residues plays an essential role in the repair of damaged proteins and could contribute to protect cells against injuries under normal and stress conditions.

Subcellular Distribution of L-isoaspartyl/D-aspartyl PCMTs

L-Isoaspartyl/D-aspartyl PCMT has been extensively studied, and purification of this enzyme from a variety of eucaryotic and procaryotic cells has been described [21–24]. In most tissues, the cytosol is the major source of PCMT but studies on the subcellular distribution of PCMT in brain [25] and kidney [26] have shown the existence of membrane-bound forms of the enzyme. In brain, up to 40% of the PCMT activity is found in an insoluble form [25, 27]. A membrane-bound form of PCMT, with properties similar to those of the cytosolic enzyme, has been partially purified from bovine brain membranes [28]. Extraction of this membrane-bound activity could be achieved only with detergent, suggesting a strong association of the enzyme with the membranes [28, 29].

We have been interested in the characterization of the membrane-bound form of PCMTs, using purified plasma membranes from rat kidney cortex. The plasma membrane from this tissue is differentiated into two distinct domains: a luminal brush border membrane (BBM) and an antiluminal basolateral membrane (BLM). These domains represent highly distinct subcellular fractions, as reflected by the difference in their lipid and protein compositions [30]. Addition of detergents to these membranes was necessary to measure an optimal methylation of two known substrates of class II PCMTs, ovalbumin and the synthetic substrate Thr–Ser–isoAsp-Tyr–Ser–Lys–Tyr (TS-isoD-YSKY). PCMT activity measured with TS-isoD-YSKY as the substrate is stimulated 8-fold in BBM and 2-fold in BLM by the addition of the zwitterionic detergent CHAPS, suggesting that the catalytic site of the enzyme is masked by the lipid bilayer [31]. Addition of CHAPS not only increased the rate of methylation but also resulted in a substantial extraction of the PCMT from BBM (Fig. 3). High ionic strength (1 M KCl) or EDTA-containing

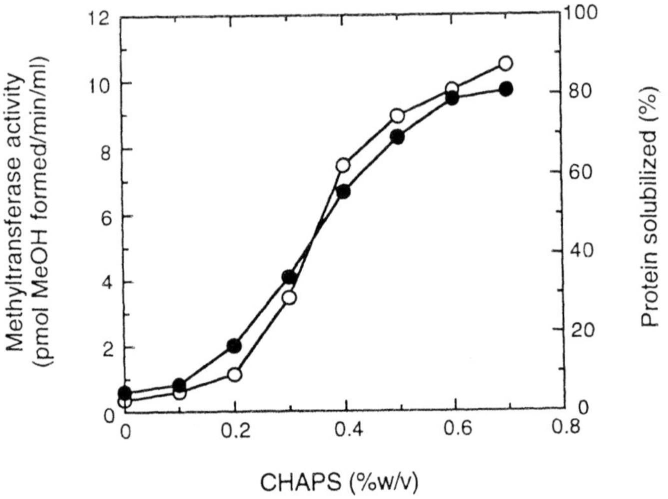

Figure 3. ***Solubilization of protein carboxyl methyltransferase from brush border membranes.*** *Membranes (10 mg protein/ml) were incubated at 4°C for 1 h in 20 mM Tris/HCl, pH 8.0, in the presence of different concentrations of CHAPS. The suspensions were centrifuged at 100 000 x g for 30 min and aliquots of the resulting supernatants were used to measure PCMT activity (○) and proteins (●). Taken from ref. 32.*

solutions were ineffective in extracting the enzyme [32], suggesting that the BBM-associated PCMT is strongly associated with the membrane.

Properties of Membrane-Bound L-isoAsp/D-Asp PCMTs from Kidney

We have recently purified and partially characterized the PCMT from kidney cortex BBM by fast protein liquid chromatography on Superdex 75 and Mono Q columns and by preparative sodium dodecyl sulfate-polyacrylamide gel electrophoresis (Fig. 4) [32]. The purified enzyme consists of a single 27.3 kDa polypeptide and recognizes exogenous substrate proteins such as ovalbumin and γ-globulins as well as synthetic peptides containing an L-isoaspartyl residue. The K_m for *S*-adenosyl-L-methionine with ovalbumin as the substrate is 1.5 μM and the purified enzyme is sensitive to inhibition by *S*-adenosyl-L-homocysteine (AdoHcy) (K_i = 0.3 μM). The purified membrane-associated enzyme shares many properties with cytosolic PCMTs such as substrate specificity and kinetic parameters. Furthermore, identical peptide fragments were obtained by limited proteolysis of the membrane-bound and cytosolic PCMTs, indicating a high degree of homology

(Fig. 5). Polyclonal antibodies raised against a synthetic peptide corresponding to the 14 amino acid residues found at the C-terminus of cytosolic PCMTs labelled a 27 kDa protein in both cytosols and membranes from kidney and brain (unpublished data).

The PCMT located in BLM showed similar catalytic properties but its activity was about 2-fold lower than that of BBM [31]. Identification of endogenous substrates located in the two plasma membranes of rat kidney cortex by acidic gel electrophoresis revealed significant differences in the methyl-accepting proteins of both membranes (Fig. 6). The methylated proteins in basolateral membrane (BLM) had sizes of 97 and 100 kDa, whereas the major brush border membranes (BBM) substrates were of 35,50 and 54 kDa [31].

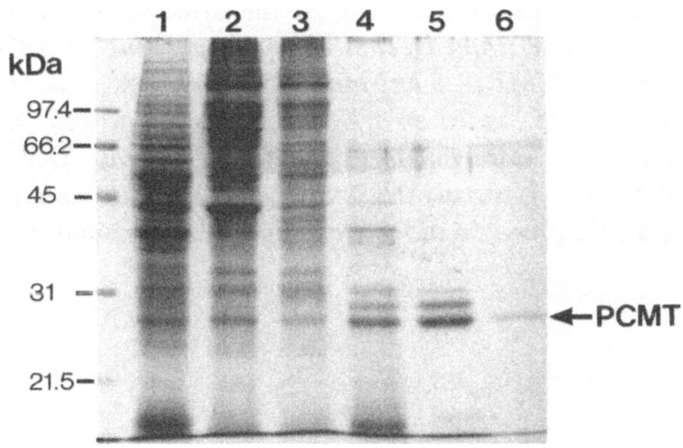

Figure 4. SDS-polyacrylamide gel electrophoresis of the fractions at different stages of purification of BBM-associated PCMT. *Samples are kidney cortex homogenate (20 μg, lane 1), brush border membranes (20 μg, lane 2), proteins extracted from BBM with 0.5% CHAPS (20 μg, lane 3), fractions obtained after chromatography on Superdex 75 (10 μg, lane 4) and Mono Q (4 μg, lane 5), and PCMT eluted from a preparative SDS-polyacrylamide gel (0.5 μg, lane 6). Taken from ref. 32.*

Fractionation of cytosols from various tissues by anion-exchange chromatography or by isoelectric focusing has revealed the presence of isoforms of PCMTs. PCMTs purified from cytosol are 26–29 kDa monomeric proteins that exist in two major isoforms, a more basic isoenzyme I (pI: 6.5) and an acidic isoenzyme II (pI: 5.5) [21, 33]. The two PCMT isoenzymes from brain and erythrocytes differ only at the C-terminus [34, 35]. The C-terminal amino acid residues of the isoenzyme I are Arg–Trp–Lys–COOH (-RWK) and the corresponding isoenzyme II sequence is

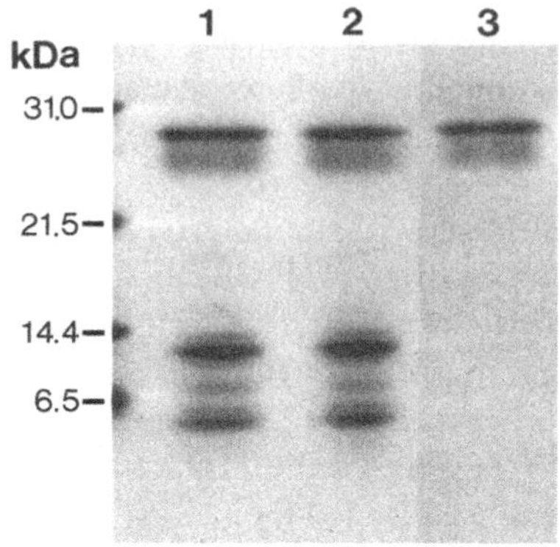

Figure 5. ***Peptide mapping of membrane-bound and cytosolic PCMTs.*** *PCMT partially purified from either the BBM or the cytosol were loaded on a 12.5% polyacrylamide gel. The stained gel slices corresponding to BBM PCMT (2.8 μg, lane 1) and cytosolic PCMT (2.8 μg, lane 2) were excised, equilibrated in stacking gel buffer for 15 min and loaded on a 15% polyacrylamide gel. The gel slices were overlayed with 10 μl of* **Staphylococcus aureus** *V8 protease (1.4 units) in 0.125 M Tris/HCl, pH 6.8, 10% glycerol, 0.1% SDS and 0.001% bromophenol blue. As a control,* **S. aureus** *V8 protease alone was applied on the same gel (lane 3). Taken from ref. 32.*

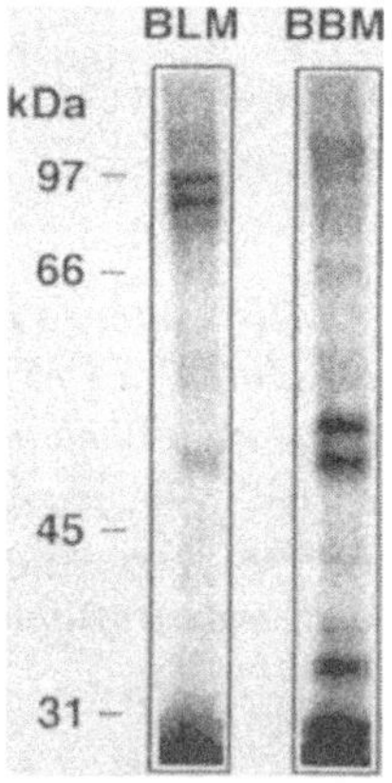

Figure 6. ***Identification of membrane-associated methyl-accepting proteins from BBM and BLM.*** *Membrane proteins (100 μg) were incubated at 37°C for 60 min in 100 mM Hepes/Tris (pH 7.5) with 5 μM [^{3}H]AdoMet (5 μCi). The methylation reaction was stopped by the addition of an equal volume of 2-fold concentrated acidic sample buffer containing 10% (w/v) 16-BAC, and the samples were applied, without heating, on 7.5% acrylamide gels. The resulting gels were exposed to autoradiographic film at −80°C for 1 month.*

Arg–Asp–Glu–Leu–COOH (-RDEL) [34, 35]. Interestingly, we observed significant differences in the behaviour of the two membrane-bound forms of PCMTs in the plasma membranes of kidney cortex upon fractionation by Mono Q anion-exchange chromatography. As shown in Fig. 7, BBM contains only a basic isoenzyme that did not bind to the column and eluted in the flow-through, while most of the BLM-associated enzyme bound to the column and eluted at a NaCl concentration of about 0.15 M. These results suggest an asymmetrical distribution of the membrane-bound forms of the enzyme in the kidney cortex cells. Since membrane and cytosolic PCMTs appear to be very similar, what is the structural determinant responsible for the membrane localization of PCMTs? One possibility is that some structural motif present in the membrane enzymes may specify their membrane attachment.

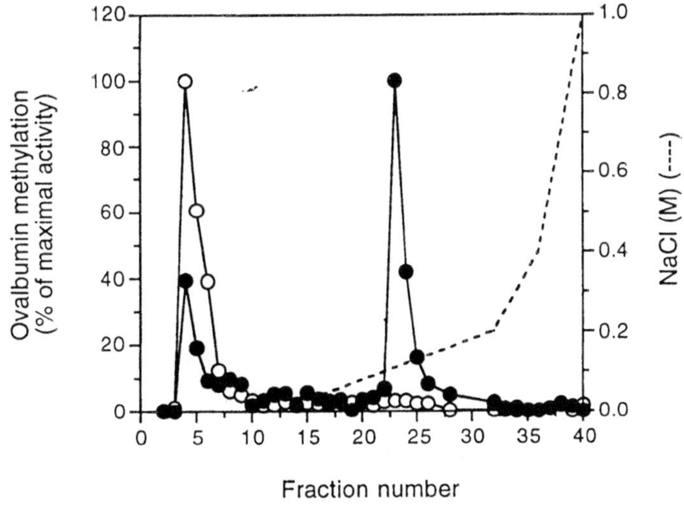

Figure 7. Mono Q anion-exchange chromatography of BBM- (○) and BLM-associated (●) PCMTs. *Solubilized membrane proteins (1 ml, 1 mg of protein) were applied to a Mono Q HR 5/5 column that had been previously equilibrated with 20 mM Tris/HCl, 0.05 % (w/v) CHAPS (pH 8.0). Proteins were eluted at a flow rate of 1.0 ml/min, using an exponential NaCl gradient, and fractions of 0.5 ml were collected. Aliquots (20 µl) of the fractions were used to measure methyltransferase activity, using ovalbumin as the substrate. Results are expressed as the percentage of activity of the most active fractions (BBM, 80 000 cpm; BLM, 35 000 cpm).*

Interestingly, the C-terminal sequence of the PCMT isoenzyme II ends in -RDEL, a sequence known to promote retention of proteins in the endoplasmic reticulum (ER) membranes [36]. It is also possible that the membrane association of the PCMTs is caused by a posttranslational modification, such as acylation, of the enzymes initially synthesized in a soluble form.

The significance of isoforms of PCMTs has not been established but recent data have suggest that these forms are generated by alternative splicing of a single PCMT gene [35, 37]. Although it is not clear whether the three forms of PCMTs detected in the kidney cortex are encoded by a single gene, it is interesting to note that 3 distinct mRNAs arising from the same PCMT gene have been detected in mouse kidney [37].

The finding that PCMTs are not restricted to a cytosolic localization but are also present in membranes provides new insight on the cellular function of these enzymes. In the polarized cells of the kidney, targeting of membrane proteins is a complex process which involves the recognition of specific signals imbedded in the protein structure [30]. The asymmetrical distribution of PCMTs between the two plasma membrane domains of these cells suggests that these enzymes play a key role in the maintenance of normal kidney cell functions. The role of the membrane-bound form of PCMTs may thus be to recognize membrane proteins that are not accessible to the cytosolic enzyme *in vivo*.

C-TERMINAL CARBOXYL METHYLTRANSFERASES

The existence of a class of methyltransferases distinct from the L-isoaspartyl/D-aspartyl protein carboxyl methyltransferases was initially inferred from studies on the structure of yeast mating peptidyl pheromones. These peptides contain at their C-termini a cysteine residue that is modified by the C_{15} isoprenoid farnesyl and a methyl ester [38, 39]. It was subsequently shown that a large number of eucaryotic proteins, including the ras family of GTP-binding proteins, possess C-terminal sequences similar to those of the yeast peptides and that these proteins may be also modified by isoprenoids and methyl groups [9]. It is now well established that a substantial group of proteins with C-terminal sequences ending in C*XXX* and C*X*C (where C is a cysteine residue and *X*, any amino acid) are modified posttranslationally by isoprenylation and α-carboxyl methylation [40] (Table 2). The processing of proteins with C*XXX* sequences has been studied in detail (Fig. 8). The C-terminal cysteine residue of the immature protein is first isoprenylated with a farnesyl or geranylgeranyl group by the action of cytosolic isoprenyltransferases [41]. The isoprenylated derivative of the protein binds loosely to cellular membranes where it undergoes the endoproteolytic removal of the C-terminal tripeptide [42] and carboxyl methylation of the newly exposed α-carboxyl group of the cysteine residue [10]. The result of this series of posttranslational modifications is the transfer of the soluble immature precursor protein to the membranes, where it becomes biologically active.

The physiological roles of isoprenylation and α-carboxyl methylation of proteins have not been elucidated in detail, but it is generally assumed that these reactions are crucial events in the association of the proteins with cellular membranes [43].

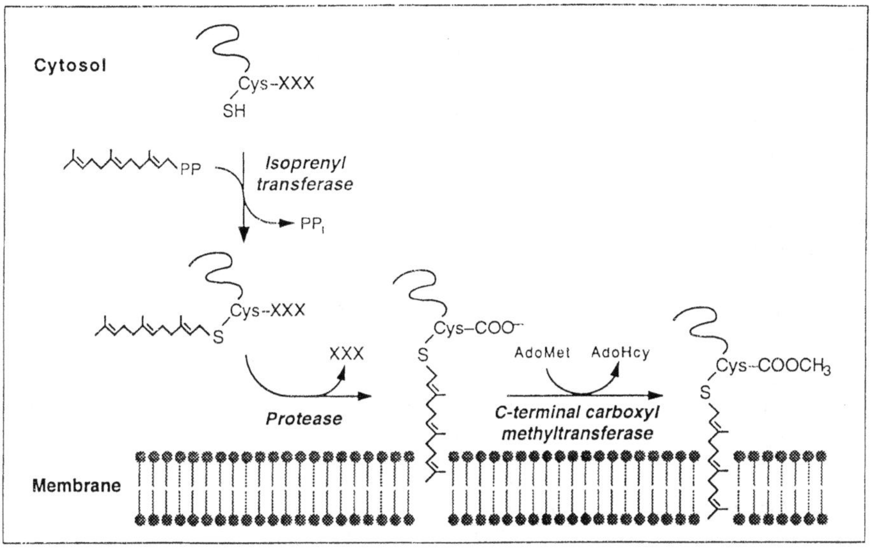

Figure 8. ***Posttranslational processing of proteins with a C-terminal CXXX sequence.***

In this regard, the isoprenylation step seems particularly important since inhibition of isoprenylation by genetic or pharmacological manipulations results in the abolition of the oncogenic potential of ras proteins [44]. The role of methylation is less clear, but also seems important since farnesylated but non-methylated ras proteins bind inefficiently to membranes [45]. Since most of the proteins identified so far with such a modified cysteine residue are implicated in signal transduction or oncogenic processes, there is great interest in elucidating more precisely the role of isoprenylation and carboxyl methylation in the function of the C*XXX*- and C*X*C-containing proteins. For this reason, intensive studies are currently carried out to characterize the enzymes involved in the various steps of these posttranslational modifications.

Subcellular Distribution of C-Terminal PCMT Activity

The use of synthetic isoprenylated substrates, such as the peptide *S*-farnesyl-LARYKC derived from the C-terminal sequence of ras from *Drosophila melanogaster* and the modified amino acid *N*-acetyl-*S*-farnesylcysteine (AFC), was a major breakthrough for the study of C-terminal protein carboxyl methyltransferase (C-terminal PCMT) activity [10, 11]. Studies on both yeast and mammalian cells

Table 2. Examples of isoprenylated and carboxyl methylated proteins

Proteins	C-terminal sequence	Modifications[a]		
Pheromones				
S. cerevisiae a-factor	D P A C V I A	F		M
Small G-proteins				
ras family				
Ha-ras	S C K C V L S	F		M
Ki-ras	K T K C V I M	F		M
N-ras	G L P C V V M	F		M
rap 1A	K K S C L L L		GG	M
rap 1B	K S S C Q L L		GG	M
rap2a	C S A C N I Q	F		
ralA	R E R C C I L		GG	M
rho family				
rhoA	K S G C L V L		GG	M
rac1	K R K C L L L		GG	
rac2	K R A C S L L		GG	
CDC42	K R K C C I F		GG	M
rab family				
rab1A	L K G G G C C		GG	
rab3A	P H Q D C A C		GG (2)	M
rab4	S A Q E C G C		GG	
rab5	R N Q C C S N		GG	
Heterotrimeric G-proteins				
γ-subunit (brain)	K F F C A I L		GG	M
γ-subunit (transducin)	K G G C V I S	F		M
Nuclear lamins				
lamin A	P Q N C S I M	F		M
lamin B	N R S C A I M	F		M
Other proteins				
GMPc phosphodiesterase				
(α-subunit)	S K S C C I Q	F		M
(β-subunit)	S S T C C I L		GG	M
Rhodopsin kinase	S G M C V L S	F		M

[a] F: farnesyl; GG: geranylgeranyl; M: methyl.

showed that the methylation of these substrates was exclusively associated with cellular membranes, with no activity in the cytosol [10, 11, 46].

More detailed fractionation of a crude membrane preparation isolated from rat liver by Percoll gradient centrifugation showed that C-terminal PCMT activity co-migrates with that of glucose-6-phosphatase, a marker enzyme for the ER, but not with that of 5'-nucleotidase, a marker enzyme for the plasma membranes (PM). This led to the proposal that C-terminal PCMTs were located in the ER [47].

We have examined the subcellular distribution of C-terminal PCMT activity in kidney cortex membranes [48]. As shown in Table 3, most of the methylation activity towards AFC is associated with membranes derived from the ER, with a 4.4-fold enrichment over the cortex homogenate. However, a high specific methylation activity was also detected in purified basolateral PM (3.1-fold enrichment) and, although to a lesser extent, in brush border PM (1.6-fold enrichment). The PM fractions are not significantly contaminated by intracellular membranes as judged by the measure of NADPH-cytochrome c reductase activity, a marker enzyme for ER [48]. These results thus suggest that, at least in the kidney, C-terminal PCMT activity is not restricted to ER membranes but is also located in PM. In our opinion, this wide distribution of C-terminal PCMT activity in the cell offers considerable flexibility to the processing of the isoprenylated proteins, considering the presence of these proteins in various subcellular compartments. For cxample, ras proteins are associated with the plasma membranes [49], rap with the Golgi apparatus [50] and lamin B with the nuclear membrane [51]. Such a co-localization of methylating enzymes and substrates also agrees with a model of regulation of protein function by reversible methylation since demethylated proteins could be methylated *in situ*.

Table 3. *N*-acetyl-*S*-farnesylcysteine carboxyl methyltransferase activities in kidney cortex subcellular fractions. Taken from ref. 48.

Fractions	Methylation of AFC: Specific activity (pmol/min/mg)	Methylation of AFC: Enrichment (*x*-fold)
Cortex homogenate	7.9 ± 0.5	1.0
Basolateral membranes	24.6 ± 0.5	3.1
Brush border membranes	13.0 ± 0.7	1.6
Intracellular membranes	34.5 ± 1.2	4.4

Properties of the Basolateral Membrane-Associated C-Terminal PCMTs

Despite the body of information available on the substrates that are carboxyl methylated in cells (Table 2), little is known about the methyltransferase itself. In *Saccharomyces cerevisiae*, carboxyl methylation of the mating pheromone α-factor and RAS proteins is catalyzed by the *STE14* gene product [52]. This gene encodes a 239 amino acid residue protein with several membrane-spanning domains. In higher eucaryotes, the molecular size of the enzyme is unknown and only some of its catalytic properties have been described. Using crude or microsomal membrane preparations as a source of enzyme, it has been suggested that the methyltransferase recognizes both geranylgeranylated and farnesylated substrates [53]. However, more detailed characterization of the enzyme at the molecular level has been hampered by its very high sensitivity to a variety of detergents [47]. For example, in BLM isolated from kidney cortex, C-terminal PCMT was found to be highly sensitive to Triton X-100, CHAPS and octylglucoside, three mild detergents widely used for the purification of membrane proteins (Fig. 9). The extent of the denaturation appears to be related to the critical micellar concentration (CMC) of the detergents which are 0.02%, 0.25% and 0.6% for Triton X-100, CHAPS and octylglucoside, respectively, at the ionic strength used in the assay.

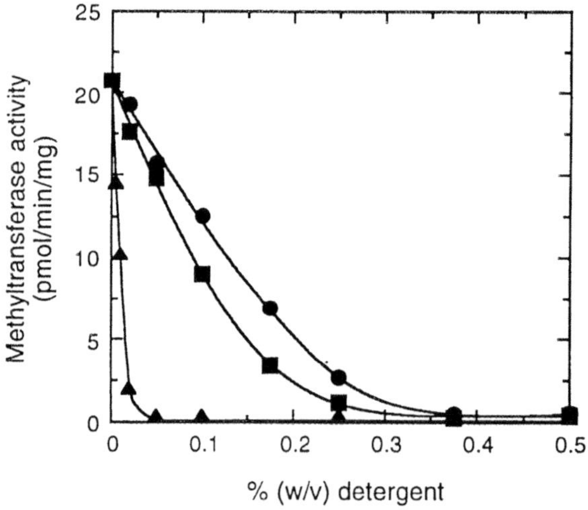

Figure 9. Effect of detergents on the activity of C-terminal PCMT. *Basolateral membranes (1mg protein/ml) were incubated with 40 μM N-acetyl farnesylcysteine, 10 μM [³H]AdoMet and 100 mM Hepes/Tris (pH 7.5) at 37°C for 20 min with n-octylglucoside (●), CHAPS (■) or Triton X-100 (▲) (at the indicated concentrations. Data are means of a representative experiment performed in triplicate.*

Because of this sensitivity to denaturation, the PCMT was studied using the radiation inactivation and fragmentation method which allows the size of enzymes to be estimated, without any solubilization steps involved [54]. BLM were irradiated with doses ranging from 0 to 10 Mrad and a semi-logarithmic plot was used to fit the data to a single exponential dependence on radiation dose (Fig. 10). A molecular size of 98 ± 14 kDa was estimated for the enzyme using an empirical equation [54]. Interestingly, this value is about 4 times that of the product of the *STE14* gene from *S. cerevisiae* (24–28 kDa). If the enzyme is conserved in mammals, this would suggest that the enzyme is a multimer. Such a conservation between yeast and mammals is suggested by the fact that the yeast farnesyltranferase, another enzyme involved in the processing of proteins with C-terminal *CXXX* sequences, is structurally and functionally similar to the mammalian enzyme [55].

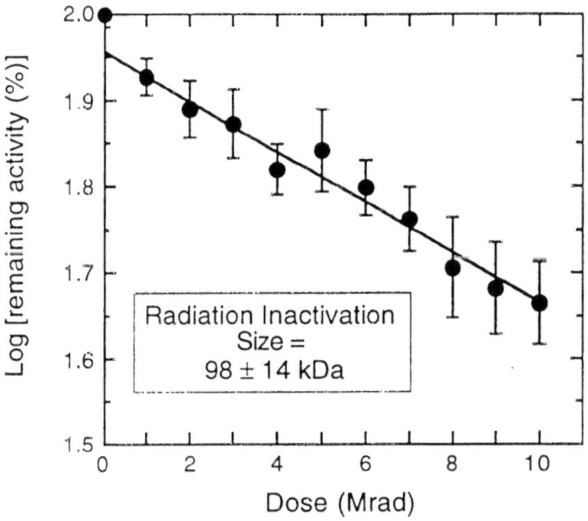

Figure 10. ***Determination of the functional size of the C-terminal PCMT.*** *Basolateral membranes were irradiated with a ^{60}Co source and incubated with 40 μM AFC, 10 μM [^{3}H]AdoMet, 100 mM Hepes/Tris (pH 7.5) at 37°C for 20 min. Data are means ± SD of five experiments performed in triplicate.*

Small G-Proteins as Substrates for C-Terminal PCMTs

Many proteins have been shown both *in vivo* and *in vitro* to be substrates for C-terminal PCMTs (Table 2). In kidney cortex, incubation of isolated intracellular and plasma membranes and cytosol with [^{3}H]AdoMet and fractionation of the proteins using SDS-PAGE resulted in the methylation of proteins within the 21–23-kDa range (Fig. 11). The incorporation of radioactivity into these proteins can be converted to volatile [^{3}H]methanol by alkaline hydrolysis, a characteristic used to distinguish carboxyl methylation from base-stable *N*-methylation [56]. Furthermore, the methylation of the 21–23-kDa proteins is competitively inhibited by AFC, suggesting that it occurs on C-terminal isoprenylated cysteine residues [48]. The similarity of the molecular mass of the carboxyl methylated membrane-associated proteins identified in kidney cortex (21–23 kDa) and small G-proteins reported to be carboxyl methylated in various tissues [57–59] strongly suggests that at least some of the carboxyl methylated proteins are small G-proteins.

Low molecular weight GTP-binding proteins (small G-proteins) are part of a superfamily of proteins that bind with high affinity and hydrolyze guanine nucleotides [60]. Members of this family share the common characteristic of switching between a GDP-bound inactive state and a GTP-bound active state. Methylation of these

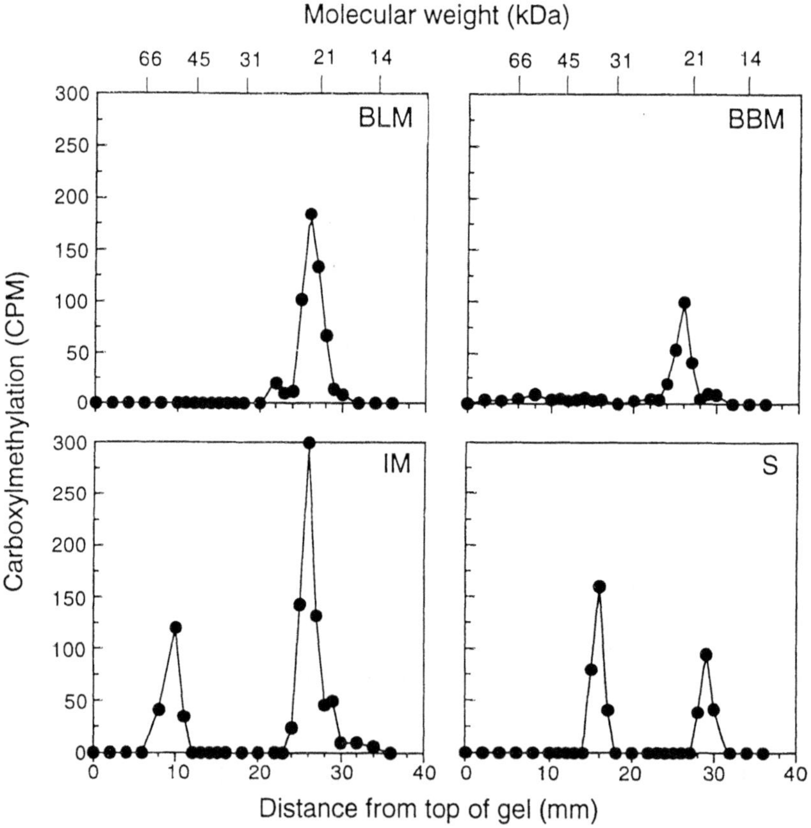

Figure 11. Identification of carboxyl methylated proteins in subcellular fractions isolated from kidney cortex. *Membrane and soluble proteins (20 μg) were incubated at 37°C for 1 h in 100 mM Hepes/Tris (pH 7.5) with 2 μM [^{3}H]AdoMet. Detection of protein methyl esters was performed as described in detail [56]. Taken from ref. 48.*

proteins seems to be influenced by their ligand-bound states since addition of the non-hydrolyzable analog of GTP, GTPγS, promoted a strong increase in the methylation of the 21–23-kDa substrates in kidney subcellular membrane fractions (Fig. 12), as well as in other tissues and cell types [58, 61].

This stimulatory effect seems restricted to the proteins in the 21–23 kDa range since GTPγS did not increase the methylation of other substrates, such as the 55-kDa intracellular membrane substrate or the soluble substrates. The stimulation induced by GTPγS could be reproduced to a similar extent by GDPβS, a non-hydrolyzable analog of GDP, but not with ATP analogs.

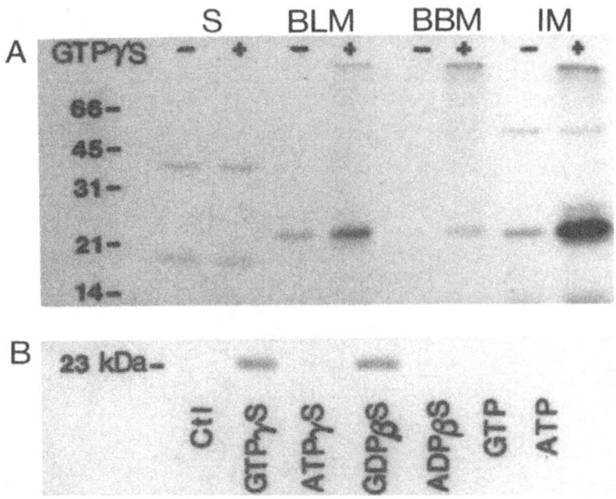

Figure 12. ***Effect of guanine nucleotides on the methylation of the subcellular fractions of the kidney cortex.*** ***A.*** *Soluble and membrane proteins were methylated as described in Fig. 11, in the presence or absence of 100 μM GTPγS.* ***B.*** *Methylation was performed in the absence (Control) or the presence of 100 μM of the indicated nucleotides. Taken from ref. 48.*

Small G-protein substrates for the membrane-bound C-terminal PCMTs are also present in the soluble fraction of cells [62]. For example, upon fractionation of a crude kidney cortex cytosol on the anion-exchanger Q-Sepharose, we observed the presence of at least two distinct peaks of proteins that were methylated by the membrane-associated enzyme (Fig. 13). These proteins are methylated to the same extent in a guanine nucleotide-dependent fashion and they migrate as 40-kDa polypeptides upon gel filtration (D. Gingras, and R. Béliveau, submitted). The identity of the two soluble proteins of the kidney cortex cytosol remains obscure, but preliminary results suggest that the more basic form (peak I) is the small G-protein CDC42Hs, a G-protein which is postulated to play a crucial role in cytosqueletal organization [63]. Moreover, the molecular size of the proteins as determined by gel filtration (40 kDa) is consistent with those observed for various small G-proteins, including the CDC42Hs protein from brain cytosol [62]. In this case, the 40 kDa size is due to the complexation of CDC42 with a GDP-dissociation inhibitor (GDI) [62].

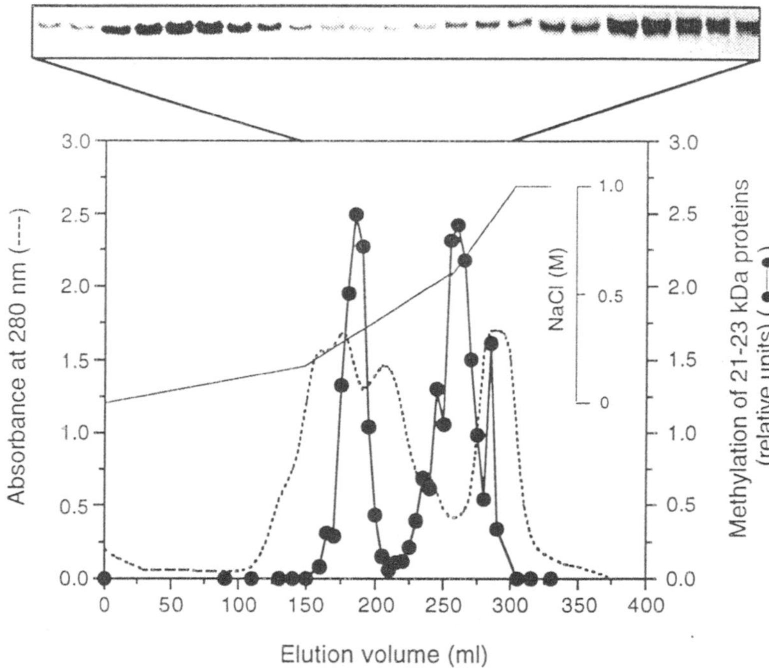

Figure 13. ***Q-Sepharose anion-exchange chromatography of kidney cortex cytosol.*** *Soluble proteins (150 ml, 1 g of protein) were loaded on a Q-Sepharose anion-exchange chromatography column previously equilibrated with Tris/HCl (pH 8.3). The column was washed with the same buffer until absorbance returned to baseline values (approx. 150 ml). After washing, bound proteins were eluted with an exponential NaCl gradient (0–1 M NaCl), and 5 ml fractions were collected. The fractions (20 μl) were incubated in 100 mM Hepes/Tris (pH 7.5) with 20 μg of kidney cortex intracellular membranes and 2 μM [^{3}H]AdoMet (3 μCi) for 30 min at 37°C and methylation of 21–23-kDa proteins visualized by SDS-PAGE and fluorography.*

PHYSIOLOGICAL AND PHARMACOLOGICAL IMPLICATIONS OF C-TERMINAL PROTEIN CARBOXYL METHYLATION

Carboxyl methylation of proteins is a widely distributed reaction that involves enzymes which are very distinct in their subcellular localization, substrate recognition, catalytic and molecular properties and thus their physiological roles. The information acquired on the class II PCMTs strongly suggests their involvement in the repair of damaged proteins from a wide variety of procaryotic and eucaryotic cells [1, 19, 20]. These enzymes may thus represent an important mechanism involved in the maintenance of normal protein function.

The physiological functions of C-terminal carboxyl methylation of eucaryotic proteins remain largely unknown. The identification of isoprenylated small G-proteins as substrates for C-terminal PCMTs [9] was a major breakthrough in the

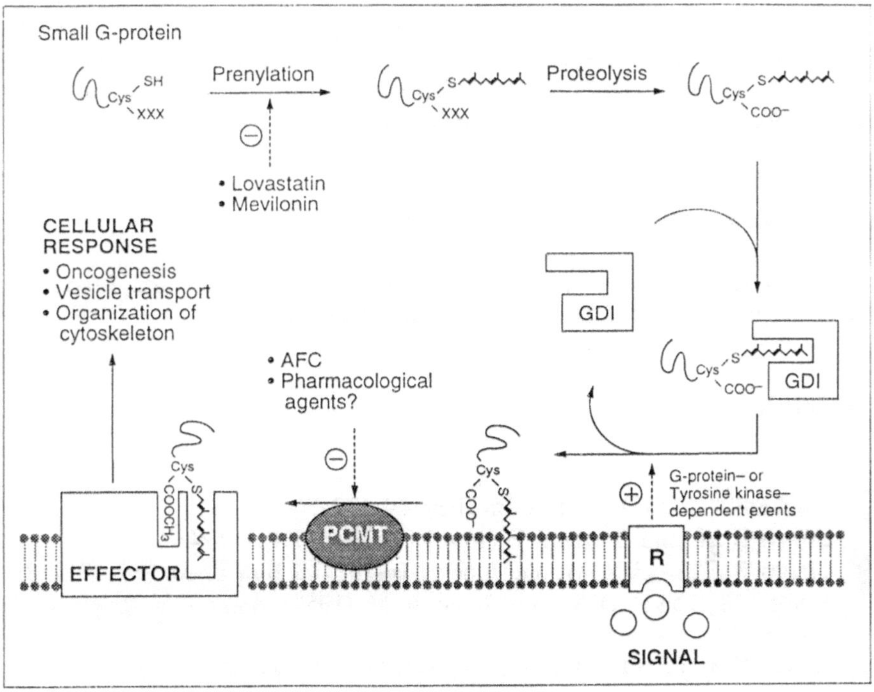

Figure 14. Possible involvement of C-terminal methylation in small G-protein-dependent cellular events.

study of carboxyl methylation of proteins because these proteins are involved in signal transduction and oncogenesis. Elucidation of the molecular mechanism as well as the role of the methylation of these proteins may thus be important for the design of pharmacological agents that are inhibitors of oncogenesis. Fig. 14 depicts the potential involvement of methylation in the cellular response to an extracellular signal. In the resting state, isoprenylated but non-methylated small G-proteins are in a soluble and inactive form, complexed with an inhibitory guanine nucleotide exchange regulatory protein which inhibits their dissociation from GDP (GDI) [64, 65]. Upon stimulation of the cell, a signal is transmitted to the G-protein (or the GDI) which results in its dissociation from the GDI and the subsequent exchange of GDP for GTP. The resulting active, GTP-bound form of the G-protein interacts loosely with the membrane where it is carboxyl methylated by the methyltransferase. The methylation of the G-protein results in its interaction with a membrane-bound effector and modulation of its activity. The methylation step is reversible [66] and thus demethylation of the protein, coupled with the hydrolysis of GTP to GDP, may bring back the protein to a GDP-bound form which would be recognized by the GDI and returned to its inactive soluble form. Although hypothetical, such a role of methylation in signal transduction processes is suggested by the findings that AFC, a competitive inhibitor of small G-protein methylation, inhibits macrophage [11] and neutrophil [67] chemotaxis as well as agonist-receptor-mediated signal transduction in platelets [68]. Thus, in addition to the inhibition of the isoprenylation step [44], inhibition of the methylation step by pharmacological agents may provide a new target for the treatment of small G-protein-dependent cancers.

Acknowledgements.

We thank the Kidney Foundation of Canada for financial support. Many thanks to Dr Vincent Vachon for his critical reading of the manuscript.

REFERENCES

1. Clarke S, (1985) Annu Rev Biochem 54:479–506
2. Barten D M, and O'Dea R F (1990) Life Sci 47:181–194
3. Simms S A, Stock A M, and Stock J B (1987) J Biol Chem 262:8537–8543
4. Stock J B, Lukat G S, and Stock A M (1991) Annu Rev Biophys Chem 20:109–136
5. Aswad D W (1984) J Biol Chem 259:10714–10721
6. Murray E D Jr, and Clarke S (1984) J Biol Chem 259:10722–10732
7. McFadden P N, and Clarke S (1982) Proc Natl Acad Sci USA 79:2460–2464
8. Lowenson J, and Clarke S (1988) Blood Cells 14:103–117
9. Clarke S, Vogel J P, Deschenes R J, and Stock J (1988) Proc Natl Acad Sci USA 85:4643–4647
10. Stephenson R C, and Clarke S (1990) J Biol Chem 265:16248–16254
11. Volker C, Miller R A, McCleary W R, Rao A, Poenie M, Backer JM, and Stock J B (1991) J Biol Chem 266:21515–21522
12. Xie H, and Clarke S (1993) J Biol Chem 268:13364–13371

13. Geiger T, and Clarke S (1987) J Biol Chem 262:785–794
14. Stephenson RC, and Clarke S (1989) J Biol Chem 264:6164–6170
15. Clarke, S., and O'Connor, C. M. (1983) Trends Biochem. Sci. 8:391–394
16. Johnson BA, and Aswad DA (1990) In Protein Methylation (WK Paik, and S Kim Eds) pp 195–208, CRC Press, Boca Raton, FL
17. Johnson BA, Murray ED Jr, Clarke S, Glass DB, and Aswad DW (1987) J Biol Chem 262:5622–5629
18. Johnson BA, Langmark EL, and Aswad DW (1987) J Biol Chem 262:12283–12287
19. Li C, and Clarke S (1992) Proc Natl Acad Sci USA 89:9885–9889
20. Ladino CA, and O'Connor CM (1992) J Cell Physiol 153:297–304
21. Aswad DW, and Deight EA (1983) J Neurochem 40:1718–1726
22. Haklai R, and Kloog Y (1987) Biochemistry 26:4200–4206
23. Gilbert JM, Fowler A, Bleibaum J, and Clarke S (1988) Biochemistry 27:5227–5233
24. Fu JC, Ding L, and Clarke S (1991) J Biol Chem 266:14562–14572
25. Diliberto EJ Jr, and Axelrod J (1976) J Neurochem 26:1159–1165
26. Gingras D, Ménard P, and Béliveau R (1991) Biochim Biophys Acta 1066:261–267
27. Kim S, Nochumson S, Chin W, and Paik WK (1978) Anal Biochem 84:415–422
28. Iqbal M, and Steenson T (1976) J Neurochem 27:605–608
29. Sellinger OZ, Kramer CM, Fischer-Bovenkerk C, and Adams CM (1987) Neurochem Int 10:155–166
30. Rodriguez-Boulan E, and Nelson WJ (1989) Science 245:718–725
31. Gingras D, Boivin D, and Béliveau R (1993) Biochem J, in press
32. Boivin D, Gingras D, and Béliveau R (1993) J Biol Chem 268:2610–2615
33. Ota IM, Gilbert JM, and Clarke S (1988) Biochem Biophys Res Commun 151:1136–1143
34. Potter SM, Johnson BA, Henschen A, and Aswad DW (1992) Biochemistry 31:6339–6347
35. MacLaren DC, Kagan RM, and Clarke S (1992) Biochem Biophys Res Commun 185:277–283
36. Pelham HRB (1990) Trends Biochem Sci 15:483–486
37. Romanik EA, Ladino CA, Killoy LC, D'Ardenne SC, and O'Connor CM (1992) Gene 118:217–222
38. Ishibashi Y, Sakagami Y, Isogai A, and Suzuki A (1984) Biochemistry 14:694–698
39. Anderegg RJ, Betz R, Carr SA, Crabb JW, and Duntze W (1988) J Biol Chem 263:18236–18240
40. Clarke S (1992) Annu Rev Biochem 61:355–386
41. Seabra MC, Reiss Y, Casey PJ, Brown MS, and Goldstein JL (1991) Cell 65:429–434
42. Ma Y-T, and Rando RR (1992) Proc Natl Acad Sci USA 89:6275–6279

43. Glomset JA, Gelb MH, and Farnsworth CC (1990) Trends Biochem Sci 15:139–142
44. Der CJ, and Cox AD (1991) Cancer Cells 3:331–340
45. Hancock JF, Cadwallader K, and Marshall CJ (1991) EMBO J 10:641–646
46. Hrycyna CA, and Clarke S (1990) Mol Cell Biol 10:5071–5076
47. Stephenson RC, and Clarke S (1992) J Biol Chem 267:13314–13319
48. Gingras D, Boivin D, and Béliveau R (1993) Am J Physiol, in press
49. Hancock JF, Magee AI, Childs JE, and Marshall CJ (1989) Cell 57:1167–1177
50. Béranger F, Goud B, Tavitian A, and Gunzburg J (1991) Proc Natl Acad Sci USA 88:1606–1610
51. Holtz D, Tanaka RA, Hartwig J, and McKeon F (1989) Cell 59:969–977
52. Hrycyna CA, Sapperstein SK, Clarke S, and Michaelis S (1991) EMBO J 10:1699–1709
53. Volker C, Lane P, Kwee C, Johnson M, and Stock J (1991) FEBS Lett 295:189–194
54. Beauregard G, and Potier M (1987) In Target-Size Analysis of Membrane Proteins (JC Venter, and CY Jung Eds) pp 61–77, Alan R Liss, New York, NY
55. Gomez R, Goodman LE, Tripathy SK, O'Rourke E, Manne V, and Tamanoi F (1993) Biochem J 289:25–31
56. Xie H, Yamane HK, Stephenson RC, Ong OC, Fung BK-K, and Clarke S (1990) Methods 1:276–282
57. Haklai R, and Kloog Y (1991) Cell Mol Neurobiol 11:415–433
58. Huzoor-Akbar Winegar DA, and Lapetina EG (1991) J Biol Chem 266:4387–4391
59. Maltese WA, Sheridan KM, Repko EM, and Erdman RA (1990) J Biol Chem 265:2148–2155
60. Bourne HR, Sanders DA, and McCormick F (1991) Nature 349:117–127
61. Backlund PS, and Aksamit RR (1988) J Biol Chem 263:15864–15867
62. Backlund PS (1992) J Biol Chem 267:18432–18439
63. Drubin DG (1991) Cell 65:1093–1096
64. Takai Y, Kaibuchi K, Kikuchi A, and Kawata M (1992) Int Rev Cytol 133:187–230
65. Regazzi R, Kikuchi A, Takai Y, and Wollheim CB (1992) J Biol Chem 267:17512–17519
66. Pérez-Sala D, Tan EW, Cañada FJ, and Rando RR (1991) Proc Natl Acad USA 88:3043–3046
67. Philips MR, Pillinger MH, Staud R, Volker C, Rosenfeld MG, Weissmann G, and Stock JB (1993) Science 259:977–980
68. Huzoor-Akbar Wang W, Kornhauser R, Volker C, and Stock JB (1993) Proc Natl Acad Sci USA 90:868–872

CHAPTER 15

TRANSPORT SYSTEMS IN KIDNEY BASOLATERAL MEMBRANE: PATHOPHYSIOLOGIC IMPLICATIONS

Animesh Sahai and Pallab K. Ganguly

Division of Cardiovascular Sciences,
St. Boniface General Hospital Research Center,
Faculty of Medicine, University of Manitoba, Winnipeg, Canada

INTRODUCTION

Body fluid volume regulation is critical to life and has been a subject of intense research for a substantial period of time. In the words of the nineteenth century physiologist, Claude Bernard, "The constancy of the internal environment is the condition of free and independent existence" (1). Research over the years has led to the appreciation of the role of the kidney in solute and water handling. Extensive and intensive investigations have provided a better understanding of the complex vascular and nephron interrelationships that allow coordination of the different components into countercurrent multiplication and exchange processes. The present overview is not an exhaustive commentary; rather, this chapter is intended to provide the reader with an appreciation of the different processes and mechanisms operating at the basolateral membrane level of the kidney. Due to the complex axial heterogeneity of the nephron, the present chapter is a generalised discussion with appropriate references being made to nephron segments where they differ. The chapter would also discuss the modifications that occur in the basolateral membrane in diabetes with special reference to some of the newer concepts generated from this laboratory.

PHYSIOLOGY OF THE BASOLATERAL MEMBRANE

Structure-Function Correlates

The kidney, with its complex and characteristic epithelial design, is recognised as a major determinant of body fluid status, and alterations in renal structural and functional integrity can effectively alter fluid disposition of the body. However, the mere arrangement of the lining cells that form the nephron is not enough to explain the directional flow of solute and water across the epithelium. In fact, certain integral features of these epithelial membranes have evolved that maintain the vectorial transport of solutes and water. First, the epithelia cell membranes are inherently disproportionate with respect to structure and function. This disparity in the properties of the apical and basolateral cell membranes is the basis for the polarization or rectification of an epithelium and defines the magnitude and orientation of the transepithelial transport processes. This fundamental doctrine of an intrinsic contrast between the apical and basolateral epithelium membranes is the basis of Ussing's double membrane model (2). Second, adjacent cells are joined by highly specialized structures known as junctional complexes. These complexes not only function in holding the cells together, but also separate the luminal and basolateral components of the cell membrane, thus creating a potential space between the cell and the basement membrane -the lateral intercellular space. Thus, solutes and water may be transported either through cells (transcellular transport) or between cells (paracellular transport), or both. However, it may be mentioned that all these paracellular transport processes ultimately rely on the transepithelial electrochemical gradients developed by cellular transport events.

The renal epithelium is essentially a three component system:- the luminal, intracellular and interstitial responsible for the overall ionic regulation of both the extracellular and intracellular fluids. The three barriers limiting these three aqueous compartments are the apical plasma membrane, the cytosol and the basolateral plasma membrane. Transport through the aqueous cytoplasmic component provides little, if any, restraint to the solute and water transport. The lipid apical or luminal plasma membrane and the lipid basolateral or contraluminal plasma membrane constitute the major impediments and provide the basis for the highly specialized directed transport. However, this represents an oversimplification of the three-compartment organization. Various investigators have suggested that other cytosolic organelles may be a decisive factor in ionic transport across the renal tubular epithelium (3,4). Likewise, the cytosolic milieu in itself may offer resistive forces to the passive movement of water and solutes (5). Kirk et. al., based on the presence of large cytoplasmic vacuoles during steady-state antidiuretic hormone-dependent flow, suggested these structures as an intermediary/parallel pathway for the flow of water (6). However, with more information being elucidated on the intricate morphological and biochemical heterogeneity of the renal tubular epithelium of the distinct nephron segments, the current opinion of transepithelial is based on the double membrane model of Koefoed-Johnsen and Ussing (2). Differences in the water and solute permeabilities of the two membranes, based on their different lipid components, the amount and prototype of integral transport proteins spanning the lipid bilayer, or both illustrate the vectorial transport of solutes and water across the membranes. Furthermore, the composition of the two membranes, in terms of lipid and transport protein composition, would impart a differential transport characteristic for each ion in question. Thus, alterations in transport properties could be achieved by qualitative (type) and/or quantitative (number and function) of the channel, enzyme, receptor protein or carrier existing in the membrane. Conversely, if the number of the carrier were to be constant (per unit area), then a quantitative increase in transport attributes could be achieved only by a corresponding augmentation of the membrane area. This principle can be further extended to imply that distinct transport characteristics in individual nephron segments, or a differential transfer of a single ion between the apical or luminal and the basolateral membrane could be possible. This is referred to as the *membrane amplification principle* (2).

Membrane amplification has been primarily realized by infoldings of the plasma membranes, thus increasing the interacting membrane surface area. At the apical side, these range from a highly specialised organisation of microvilli (brush-border membrane) to simpler infoldings called microplicae. The basolateral infoldings as well as those that comprise the cellular interdigitations, provide for the substantial increase in membrane area. These principles as well as newer insights have been provided by recent developments in quantitative micromethodologies for evaluation of morphological, physiological and biochemical differences at the single nephron or cellular level, and have guided investigators in the field of transport physiology.

Transport

In this section, some of the processes essential to kidney function have been described with particular reference to the basolateral membrane. It must be understood that due to limitations of space this is not an exhaustive account of the different mechanisms; rather, certain points have been highlighted to provide an appreciation of the complexities of the highly organised and directed transport occurring at the basolateral membrane.

Sodium Chloride Transport and the Basolateral Membrane. Salt (sodium chloride; NaCl) reabsorption occurs along the entire nephron, with the major portion being absorbed in the proximal tubule (7-10). The Na^+ and Cl^- that are absorbed from the tubular lumen via the Na-H (11-13) or Cl-base (OH^-, HCO_3^-, HCO_2^-, $C_2O_4^{2-}$) (14-16) exchangers must leave the cell across the basolateral membrane. In spite of considerable axial heterogeneity of both morphologic and functional characteristics of the tubular epithelium, certain general features can be ascribed to Na^+Cl^- exit from the cell. Na^+-K^+-ATPase, the enzymatic equivalent of the Na^+ pump, is the main mechanism which lowers the cytosolic Na^+ concentration (by pumping Na^+ out of the cell) and thus maintains the electrochemical gradient for Na^+ entry via the apical membrane (17,18). Biochemical (19), autoradiographic (20) and immunologic (21) studies have established beyond doubt the selective localisation of this ATPase to the basolateral membrane. Freeze-fracture techniques have complemented biochemical studies and established the dimeric nature of this renal protein: two α and two β subunits with each unit having a molecular weight of 10 kDa and 40 kDa respectively (22). The countertransport of 3 Na^+ and 2 K^+ against their electrochemical gradient utilising the hydrolysis of one ATP molecule indicates this pump is electrogenic. A consensus of opinions amongst different investigators shows that this enzyme, located in the basolateral membrane, is the principal factor responsible for Na^+ reabsorption (23-25). A strong correlation has been shown to exist between Na^+ transport rates and Na^+-K^+-ATPase activity in different nephron segments (24). Though, the heterogenous profile of Na^+-K^+-ATPase activity in different nephron segments could also be explained as resulting from variations in enzyme turnover rates, catalytic units or both. Mernissi and Doucet (26) showed that pump activity corresponded to the number of catalytic units in the nephron segment. This would indicate that the maximal activity of the ATPase was similar in all segments and increased activity from any one segment would be due to an increased number of functional units. Furthermore, this relationship can be extended to mean that there is a direct correlation between the Na^+-K^+-ATPase activity and the basolateral membrane area (27-29). Though these generalisations are applicable throughout the nephron-segment, two distinguishing features of the distal segment Na^+-K^+-ATPase must be highlighted. First, they have a higher activity as compared to the proximal tubule (18) and as already mentioned, these discrepancies are entirely due to the increased basolateral membrane area of the distal segments. Second, the distal ATPase is inhibited by frusemide administration while the proximal ATPase is not (30).

The mechanism of Cl^- exit from the cell is least understood with different investigators postulating different modes of transport. Primarily, four different mechanisms have been described: a) Cl^- conductance (31), b) K-Cl cotransporter (32), c) Na^+-dependent Cl-HCO_3 exchange,(33) and d) Na^+-independent Cl-HCO_3 exchange (34). Notwithstanding the importance of the other two mechanisms, only the Na-dependent and Na-independent Cl-HCO_3 exchange mechanisms have been ascribed functional characteristics in the proximal tubule (35). However, in the thick ascending limb of Henle's loop, it is the Cl^- conductance mechanism which is chiefly responsible for the Cl^- leaving the cell (36). A small but finite part is also contributed by the cotransporter (37).

Potassium Transport and the Basolateral Membrane. Potassium is the primary intracellular cation and contributes to the maintenance of electrical properties in excitable tissue as well as nonexcitable tissue (transport epithelia) (38). Intracellular K^+ homeostasis is consequential as it maintains cell volume by altering intracellular osmolality (39). It also regulates cytosolic pH by altering H^+ ion activity in the cell (40). Notwithstanding the significance of this ion, the regulatory mechanisms maintaining the intracellular K^+ constancy are poorly understood. Basolateral membrane activity does contribute a major portion to maintaining this internal environment. Experiments have shown that kidney slices in which the tubules were collapsed and in which K^+ transfer is inhibited, intracellular K^+ concentration and cell volume as well as electrical properties of the cell are maintained (41-43). Moreover, electrical measurements across the basolateral membrane coupled with simultaneous measurements of intra- and extracellular K^+ concentration have shown that the transepithelial voltage across the cell membrane cannot be entirely accounted by a passive K^+ transfer (44,45). Studies utilising inhibitors (46,47) and stimulators (48,49) have concluded that the linkage between the active Na-K pump and passive K^+ conductance serves to regulate cytosolic K^+ content (50,51), though the cellular mechanisms underlying this linkage are not entirely known (52-54). It is the Na-K pump located in the basolateral membrane which is one of the principal regulators of K^+ movement, particularly in the distal tubule and cortical collecting duct cells (55,56). Morphometry of the basolateral membrane revealed that mineralocorticoids (57), and not glucocorticoids (58), were responsible for modulating basolateral membrane area and, in accordance with the membrane amplification principle, the K^+ transport (59,60).

Acid-Base Transport and Basolateral Membrane. The kidney plays a central role in monitoring acid-base balance by excreting excess acid or alkali. Most of the luminal acidification is due to proton secretion across the apical membrane. Concurrently, a base equivalent is formed in the cell which exits across the basolateral membrane. The details of this exit are only briefly reviewed here with particular reference to the basolateral membrane and the reader is referred elsewhere for a detailed discussion of acid-base physiology (61,62).

The chief reactive ion exiting the cell across the basolateral membrane is the bicarbonate (HCO_3^-) ion. The processes mediating this transfer are: a) the electrogenic Na-HCO_3 cotransporter in the proximal tubule, and b) the chloride-base exchanger, both sodium-dependent and sodium-independent modes in the collecting ducts, mainly. The Na-3HCO_3 cotransporter exists chiefly in the proximal tubule and is electrogenic in nature (63,64). Though various investigators have shown the presence of this cotransporter (65-67), it should be understood that these studies were performed either in intact cells or vesicles, in the absence of chloride. Whereas Biagi et. al. (68) were the first to suggest this mechanism of transport which was dependent entirely on HCO_3 concentration, later on it was revealed that presence of exogenous CO_2/HCO_3 was a decisive factor too (69). In the absence of CO_2/HCO_3, the transporter functioned at one-third of its capacity (70). Furthermore, it was shown that even sulfite (SO_3^-), a structural analog of (CO_3^{2-}), could serve as a substrate for the transporter (70,71). However, Soleimani and Aronson (71) found that sulfite transport only occurred in the presence of the bicarbonate ion.

The second mode of transport across the basolateral membrane, as part of the processes that acidify the cell, is the Cl-base exchange. Edelman et.al. (72) were the first to identify such a transporter, which was later shown to be Na-coupled (73).

This mode of HCO_3 transport was inhibited in the absence of chloride, suggesting that this mechanism is linked to both Na and Cl. Perfusion studies later revealed that the majority of the HCO_3 transport is dependent on the Na-dependent Cl-HCO_3 exchange (probably the Na$(HCO_3)_2$-Cl) (73), while a smaller contribution is also by simple Cl-HCO_3 exchange. A Na-independent Cl-HCO_3 exchange in the proximal tubules has also been described (75). A caveat may be entered here as studies with basolateral membrane vesicles have revealed only Cl-HCO_3 exchange, while little evidence exists for the Na-coupled Cl-HCO_3 exchange (76,77). As part of the renal acidification mechanisms, ammonium (NH_4^+) transport across the basolateral membrane also is a major phenomenon. It does so by substituting for K^+ on the Na^+-K^+-ATPase (78,79)and by a passive conductance mechanism through K^+ channels driven by the negative cell voltage (80). This NH_4^+ ion then acts as a source for proton transport for luminal acidification (78).

Finally, Na/H antiporter has been described in the basolateral membrane (81), though conflicting evidence exists (82). The role of this mode of transport across the basolateral membrane awaits further investigation (83).

Organic Anion and Cation Transport and Basolateral Membrane. Organic anion and cation transport represents a composite of numerous processes employed by the kidney for excreting exogenous xenobiotics and endogenous metabolites. The voluminous literature of these aspects of basolateral transport precludes a detailed account and only a brief mention of the varied transport mechanisms are presented. Though various substances have been investigated in great detail, N^1-methylnicotinamide (NMN) and tetraethyl-ammonium (TEA) are considered as examples of organic cations, while para-aminohippurate (PAH) is symbolic of organic anions.

The basic event in the transport mechanics across the basolateral membrane is the presence of a transepithelial electrochemical potential difference generated by the Na^+-K^+-ATPase of the basolateral membrane. By coupling an anion transport (e.g. dicarboxylate) to a downhill cation influx (Na ion/anion cotransport) this secondary active process of organic anion transfer can be accomplished (84-88). This cotransport of Na ions with a dicarboxylate has been described as electrogenic with three Na^+ ions being exchanged for one dicarboxylate anion carrying two negative charges (87,88). Similarly, the tricarboxylate ion may participate in this process. By virtue of this property, this transport mode, located in the basolateral membrane, is capable of accepting all citric acid cycle intermediates (86). The intracellular accumulation of such a dicarboxylate anion may then participate in the "uphill" transfer of another organic anion, a process referred to as tertiary active. The PAH transporter is such a mechanism, whereby a dicarboxylate (possibly 2-oxoglutarate) translocates with PAH and other hydrophobic ions (89,90). This process also includes mechanisms for the secretion of xenobiotics across the basolateral membrane (90). Though this is not directly coupled to Na, this process requires the presence of Na/anion secondary active cotransport (91-93). Similarly, a sulfate transporter has also been identified in the basolateral membrane and is shared by other anions such as thiosulfate, arsenate, sulfate-monoesters, etc (94-97). Like PAH, urate translocation has also been described at the basolateral membrane, and a urate-Cl exchanger identified (98). Though a Na-urate cotransport is yet to be identified, urate transport has also been explained as occurring by simple diffusion. The role of the enzyme uricase in mediating this transport mechanism is also suggested (99,100).

Organic cation transport is similar to organic anion transport in that it is also dependent on the presence of a polarized epithelial cell membrane - the basolateral membrane. The presence of a transepithelial electrochemical voltage generated by the Na^+-K^+-ATPase is responsible for mediating this transport (101,102). This process is temperature-dependent, and inhibited by anoxia and cellular poisons and other organic cations (103-105). Changes in intracellular voltage can also modify basolateral transport. Few studies have dealt with organic cation transport thoroughly and more evidence is forthcoming on both intracellular organic ion content and the precise mechanisms mediating these transport processes.

Amino Acid Transport and the Basolateral Membrane. Amino acid transport across the basolateral membrane is coupled to Na-translocating mechanisms located in the basolateral membrane (106). Moreover, the basolateral membrane is instrumental in regulating amino acid transfer along its gradient out from the epithelial cell (107,108). The transfer of amino acids and the role of basolateral membrane is of considerable clinical significance and alterations of these active processes/backflux mechanisms may explain the pathophysiology of congenital aminoacidurias.

Na-dependent active mechanisms have been identified for uptake of glycine (109) and glutamine (110). Neutral amino acids are transported in a Na-independent fashion, mediated probably via a facilitated diffusion process (111-113). It cannot be overstressed that glutamine transport mechanisms are essential, as they provide a means for accumulating this amino acid in high concentration as part of the acid-base regulation (114,115). At present, about seven distinct Na-coupled amino acid transporters have been identified in the basolateral membrane. Future investigations may reveal more transporters and contribute more to our knowledge of congenital aminoacidurias.

Other Transport Processes and the Basolateral Membrane. The preceding has been an overview of some of the major transporters across the basolateral membrane. There are other translocating mechanisms, too numerous to mention, situated in the basolateral membrane. However, this section would be incomplete without reference to two lesser known but nevertheless important transporters - phosphate and hexose transporters.

Transport of phosphate across the basolateral membrane is along its gradient though the precise transporter remains elusive. Both a Na-dependent (116) and a Na-independent (117) process have been described. The Na-independent path is linked to the anion exchange process, most likely bicarbonate (117), and is probably the major transporter, while the Na-dependent mechanism is a low-capacity system

(116). Furthermore, a voltage-dependent Na-independent process has also been described though little is known about it (118).

Hexose transport across the basolateral membrane is necessary for the reabsorption of hexoses. Studies with vesicles have shown that this transport across the basolateral membrane, for hexose, is a concentration-dependent and Na-independent process, in contrast to the apical membrane (119). Further, certain sugars e.g. L-glucose are transported only across the basolateral membrane and not the apical membrane (120). A closer look at the differential characteristics of the hexose transport system at the basolateral and apical membrane reveal differences in inhibitor action, substrate specificity, etc., and the transport at the basolateral membrane is more consistent with that of the erythrocyte membrane (121).

CALCIUM TRANSPORT ACROSS THE KIDNEY BASOLATERAL MEMBRANE

Consistent with the important role of Ca^{2+} in biology, active transport of this divalent cation has been demonstrated in eukaryotic cells (122). Calcium has only been recently recognised as a universal cell messenger. It was the classical work by Ringer in 1882 (122), on the role of Ca^{2+} in muscle contraction, which showed that Ca^{2+} was fundamental to cell signalling mechanisms. Subsequently, the developments in the field of Ca^{2+} ionophores, the discovery of calmodulin and other intracellular Ca^{2+} binding proteins, the progress in studies on membrane transport of Ca^{2+} and the development of precise techniques for measuring intracellular Ca^{2+} concentration have contributed to the continued expansion of interest in the role of Ca^{2+}. Paralleling these developments, there has been an autocatalytic growth in the understanding of mechanisms that are involved in the regulation of cytosolic Ca^{2+} concentration. Our laboratory has been involved in understanding the regulatory mechanisms underlying Ca^{2+} transport in the kidney across the renal basolateral membrane. Efforts have primarily been directed towards identifying the cellular and subcellular defects in these transport mechanisms in diabetes in an attempt to explain the pathophysiological mechanisms of diabetic nephropathy.

Considering the importance of calcium as an intracellular messenger, it is conceivable that mechanisms exist to maintain an optimal cytosolic Ca^{2+} concentration. Calcium homeostasis is maintained by the integrated actions of the intestine, bone and kidneys. Interchange of Ca^{2+} between skeleton and blood, its absorption from the intestine, and renal reabsorption of filtered Ca^{2+} are all involved, but the relative contribution of each is still controversial. Although the gut and skeleton do play significant roles, Nordin and Peacock (123) and Nordin et al. (124) have shown that the kidney is the chief regulator of calcium.

In the blood, 35 to 45% of total calcium is bound to plasma proteins, and only the remaining fraction passes freely across the glomerular basement membrane. In the healthy man, this free fraction of calcium approximates 11,000 mg, while only 0.5% to 1.0% of the filtered Ca^{2+} appears in the urine. It is obvious from this that there are effective mechanism(s) for tubular reabsorption of Ca^{2+}. Data from micropuncture and microperfusion studies (125-127) also indicate active Ca^{2+} reabsorption along the tubule. Both micropuncture studies and the evaluation of urine samples collected following transient ureteral occlusion reveal tubular fluid/ultrafiltrable plasma ratios of Ca^{2+} well below 1.0 (123,128-130), despite an electric potential difference across the tubule that should lead to the appearance of tubular fluid/ultrafiltrable plasma ratios above 1.0. These findings suggest that a very effective active transport mechanism exists for Ca^{2+} in the renal tubule.

Under basic conditions, nearly all the Ca^{2+} in the glomerular filtrate is reabsorbed along the nephron. Though the maximal load of the filtered Ca^{2+} is absorbed along the proximal portion of the tubule, reabsorption is considerable along the distal segments of the nephron (131). Both active and passive modes of transport have been postulated to exist for tubular reabsorption of Ca^{2+}, though the cellular events responsible for active Ca^{2+} transport by the renal tubule remain undefined.

It is important to understand a few essential concepts in the area of cellular Ca^{2+} homeostasis. Most of the Ca^{2+} is reabsorbed from the tubular lumen into the cell across the brush-border membrane along an electrochemical gradient by a facilitated diffusion transport mechanism which is powered by an electrogenic sodium transport system (132,133). In other words, Ca^{2+} reabsorption into the cell is a passive mechanism. Ca^{2+} is maintained at low concentrations in the cell with the background concentration of free Ca^{2+} in most cytosol oscillating between 0.1 and 0.2 μM (134-136). However, as important as Ca^{2+} is in cell signalling and other biochemical processes, the situation of Ca^{2+} overload could compromise cellular function and integrity (137). Indeed, flooding of the cytosol with Ca^{2+} is a characteristic and ominous event in cell pathology (137). This, understandably, would compel the presence of strategies designed to eject Ca^{2+} from the cells, thereby countervailing its destruction. Moreover, the electrochemical gradient at the basolateral side is reversed, requiring energy-dependent mechanisms for the extrusion of Ca^{2+} (138,139) and it is by concerted actions of the importing system (Ca^{2+} channel) and the exporting systems (Ca^{2+} pump and Na^+-Ca^{2+} exchanger) that an optimal cytosolic Ca^{2+} concentration is maintained. The Na^+-Ca^{2+} exchanger is a large-capacity, low-affinity electrogenic antiporter (140). However, kinetic studies have shown that it is in equilibrium at cytosolic free Ca^{2+} levels and does not contribute greatly to Ca^{2+} extrusion (141). The Ca^{2+} pump on the other hand is a low capacity system with a high affinity for Ca^{2+} (139). Its high affinity enables it to interact with Ca^{2+} even at the very low background intracellular concentration of resting cells. It therefore functions continuously, satisfying the demands for the fine tuning of intracellular Ca^{2+}.

Renal Ca^{2+}-ATPase

Cytoplasmic free Ca^{2+} concentration (0.1 to 0.2 μM) is several orders of magnitude lower than in the extracellular milieu. Though the diffusion of Ca^{2+} into the cell is passive, the electrochemical gradient at the basolateral side is reversed, requiring energy-dependent mechanisms for Ca^{2+} exit (138). In transporting epithelia, pump and exchanger systems participate in the vectorial translocation of calcium for maintaining intracellular Ca^{2+} homeostasis (138,141). Thus, membrane-bound Ca^{2+} transporting enzymes have been identified in cortex homogenates (142), suspensions of cortical tubules (143) and microsomes (144-147) from rat, rabbit and hog kidneys. Strong corroborative evidence for a function of Ca^{2+}-ATPase in renal tubular Ca^{2+} transport has been obtained from two studies on plasma tubules. First, separation of vesicles from the luminal and basolateral aspect of the cell has shown that these enzymes are located exclusively in the basolateral membrane (145-147), where they extrude Ca^{2+} into the peritubular space against steep gradients. Second, Ca^{2+} uptake by (inside-out) basolateral membrane vesicles (corresponding in vivo to Ca^{2+} extrusion from the cell), depends on, and is supported by, ATP found in the cytosol and in concentrations optimal for Ca^{2+}-dependent ATPase activity (132).

General Properties. Kinne-Saffran and Kinne (144) were the first to describe the presence of a Ca^{2+}-stimulated ATPase in the renal basolateral membrane. Studies showed that the enzyme was present along the entire nephron and was activated by millimolar concentrations of either Ca^{2+} or Mg^{2+}, neither of which was necessary for activation by the other cation (148). The physiological role of this low-affinity enzyme in transmembrane calcium transport is not clear, but is activated by Ca^{2+} concentrations in the millimolar range, whereas the cytosolic free Ca^{2+} concentration in kidney cells is of a lower magnitude (136). It is conceivable that the high capacity of the enzyme compensates for its low affinity (144), or that Ca^{2+} may be "compartmentalized" at the inner surface of the membrane at a higher concentration than in the rest of the cytosol. The exact status of this low-affinity Ca^{2+}-ATPase in Ca^{2+} transport still remains to be determined.

A more acceptable hypothesis is the presence of a high-affinity Ca^{2+}-ATPase for extruding Ca^{2+} out from the renal tubular cells, activated by normal intracellular Ca^{2+} concentrations. In rat renal cortex basolateral membranes and rabbit renal tubules, a Mg^{2+}-dependent, high-affinity, low capacity calmodulin-sensitive Ca^{2+}-ATPase has been identified (140,146,149-152). Moreover, the relationship between the K_m for Ca^{2+} of this ($Ca^{2+}+Mg^{2+}$) ATPase (0.68 μM) and that of the ATP-dependent Ca^{2+} uptake (0.5 μM) in vesicle preparations lends further credence to the existence and involvement of this enzyme in transmembrane Ca^{2+} transport (132,146).

Ca^{2+} + Mg^{2+} ATPase and Ca^{2+} Transport

The general mechanism of the Ca^{2+} pump follows the pattern of that of all other P-type ion pumps (153,154). The cycle begins with the Ca^{2+} dependent transfer of the terminal phosphate of ATP to an aspartic acid residue in the pump (155,156). The phosphorylated intermediate has a high Ca^{2+} affinity. The pump bearing the phosphorylated intermediate (E_1-P in the conventional formalism) undergoes a conformation transition promoted by Mg^{2+}: the new conformer is called E_2-P (157). The matter of the affinity for ATP is more complex. The hydrolysis to E_2 and P_i is assumed to be accelerated by the binding of ATP to a site on the pump that has less affinity (158,159) than the site that binds ATP to initiate phosphorylation (158,159). In the last step of the reaction cycle, the dephosphorylated E_2 conformer reverts back to the initial E_1 conformation. Analogous to other P-type ion pumps, vanadate (a pentacoordinate, trigonal bipyramid stereo analogue of phosphate), also inhibits the Ca^{2+}-ATPase at very low concentrations (160). It is generally assumed that vanadate inhibits by binding to the aspartic acid residue. Not only has it been shown to act as a non-competitive inhibitor of the binding of ATP to the high affinity site (161), but also as a mixed, partially competitive inhibitor at the low affinity site (161). Vanadate interacts with the E_2 conformer of the ATPase, blocking the last step of the reaction (i.e. E_2--E_1 transformation), thereby stabilizing the enzyme in the E_2 conformation.

One aspect of the reaction cycle which is still open to question is the step during which translocation of Ca^{2+} across the hydrophobic barrier of the membrane takes place. Two steps in the reaction cycle have been tentatively identified. The first refers to a conformational change of the pump protein, i.e. E_1-P to E_2-P, and the second, from conformer E_2 to E_1. In either case, the Ca^{2+} binding site, located on the interior of the membrane before the translocation step (E_1 conformation), will be located on the external side of the plasma membrane at the end of the translocation step, i.e. the Ca^{2+} binding site on the E_2 conformer of the pump faces the exterior of the cell (162). Apart from the dispute regarding the site of Ca^{2+} transfer, considerable debate also persists in the number of Ca^{2+} atoms translocated during the cycle. The observation of a Hill coefficient greater than one for the activation of ATPase, saturation of the pump and its rate of Ca^{2+} transfer, in erythrocytes is proportional to the square of cytosolic Ca^{2+}, indicating two high-affinity Ca^{2+} binding sites (163,164). It is possible that the sigmoidal kinetics of this pump could be attributed to other factors, e.g. the Ca^{2+}-dependent binding of the activator calmodulin (165). Even if two Ca^{2+} binding sites existed, whether the translocation step involves one to two Ca^{2+} atoms is still an open issue. Kinetic studies under optimal conditions (166) or in purified preparations (167) favor a 1:1 stoichiometry. Others have found that Ca^{2+}: ATP stoichiometry depends on the Ca^{2+} concentration, approaching two above 500 μM (168,169).

Kidney Basolateral Membrane in Diabetes

The action of insulin, like other hormone systems, involves the initial binding of the hormone to a specific receptor on the cell membrane, which would initiate a series of biochemical events. The potential role of Ca^{2+} in mediating insulin action is under current study (170). A potential role for Ca^{2+} in insulin-mediated cellular responses is suggested by observations of direct modulation, by insulin, of high affinity $Ca^{2+}+Mg^{2+}$ATPase activity in adipocyte plasma membranes (171). If there is some regulatory role for this enzyme in mediating insulin action, it is reasonable to assume that in situations of insulin resistance, insulin's effect on $Ca^{2+}+Mg^{2+}$ATPase may be different from that found under normal conditions. Levy et al. (170) examined the direct effect of insulin on the high affinity $Ca^{2+}+Mg^{2+}$ATPase of kidney basolateral membranes in streptozotocin-induced diabetic rats. The enzyme activity in basolateral membranes from diabetic rats was higher at all Ca^{2+} concentrations tested due to a higher maximum velocity of the enzyme from diabetic rats. The enzyme was inhibited by trifluperazine and no difference was noted in the calmodulin content in membranes from diabetic and control rats. Furthermore, insulin significantly increased the activity of the $Ca^{2+}+Mg^{2+}$ATPase in the kidney basolateral membranes, though there was a lack of such response from the membranes of diabetic rats. These findings not only demonstrate a defect in the ability of insulin to regulate the $Ca^{2+}+Mg^{2+}$ATPase activity in diabetic rats, but also inculpate an inherent defect in enzyme activity as responsible for the impaired insulin action in diabetic rats. It may also be pointed out that in the presence of an increased Ca^{2+} pump in diabetes, rats are resistant to the development of congestive heart failure unless superimposed with hypertension (172). However, a causal relationship has not yet been established. Thus, it is clear from the foregoing discussion that an impaired intracellular Ca^{2+} homeostasis represents a crucial factor in the pathogenesis of altered insulin effects in target tissues.

Our Data

Our results showed a direct stimulatory effect of atrial natriuretic peptide (ANP) on this pump, suggesting that ANP may have significant bearings on Ca^{2+} transport. In spite of the fact that a considerable body of evidence suggests that ANP participates in intracellular Ca^{2+} homeostasis by altering membrane-bound activities in other tissues such as smooth muscle (173), this is the first direct demonstration that ANP may be involved in intracellular Ca^{2+} transport in the kidney by altering the Ca^{2+} pump activity (174). Furthermore, ANP is known to decrease intracellular Ca^{2+} concentration (175) and it is possible that ANP may be stimulating Ca^{2+} efflux mechanisms (Ca^{2+} pump) in order to maintain optimal levels of intracellular Ca^{2+}.

ANP has been shown to mediate its actions by stimulating particulate guanylate cyclase in responsive tissues (176). There is coexistence of the ANP receptor and guanylate cyclase activities on a single polypeptide chain, suggesting that this transmembrane protein contains both the information for cell recognition and its translation into a second messenger, viz. cGMP (177). Wong et al. (178) have shown that cGMP is an excellent marker for renal biological actions of ANP. Therefore, as a consequence of its involvement in direct stimulation of the guanylate cyclase system in the kidney basolateral membrane and its participation in Ca^{2+} homeostasis via its stimulatory action on the Ca^{2+} pump, the possibility exists that the Ca^{2+} pump and guanylate cyclase system are coupled together (179). Examination of the data revealed that isotopic assay of $Ca^{2+}+Mg^{2+}$ATPase, after stimulating the guanylate cyclase system by GTP, resulted in a decreased activity (180). This would suggest an inhibitory role for the guanylate cyclase system on the Ca^{2+} pump. Corroborative evidence of this distinctive effect of guanylate cyclase on the Ca^{2+} pump came from experiments using non-hydrolysable GTP analogues. The substitution of GTP with a non-hydrolysable analogue, GTP-γ-s, was incapable of altering Ca^{2+} pump activity. This suggests that the utilisation of the substrate (GTP) by guanylate cyclase is imperative for the modulation of the Ca^{2+} pump activity. Further evidence of the inhibitory character of guanylate cyclase on $Ca^{2+}+Mg^{2+}$ATPase came from our studies employing different concentrations of GTP to activate the guanylate cyclase system. Our results demonstrated a dose-dependent inhibition of $Ca^{2+}+Mg^{2+}$ATPase by GTP. This would be interpreted to mean that augmentation of the guanylate cyclase system, by increasing substrate (GTP) utilisation, showed a corresponding step-wise inhibition of $Ca^{2+}+Mg^{2+}$ATPase. It can be postulated that a reciprocal modulation exists between the activities of guanylate cyclase and the Ca^{2+} pump, in which alterations at the catalytic site of guanylate cyclase are able to modulate Ca^{2+} pump activity. This inhibitory effect of guanylate cyclase $Ca^{2+}+Mg^{2+}$ATPase was specific to the kidney cortex. Similar studies in the liver plasma membranes showed that there was minimal inhibition by GTP (181).

A logical extension of this observation was to determine the mechanism of this inhibitory effect of the guanylate cyclase system on $Ca^{2+}+Mg^{2+}$ATPase. It is theoretically possible that this could be a post-receptor, i.e. cGMP, mediated effect. However, our data showed no change in $Ca^{2+}+Mg^{2+}$ATPase activity when cGMP was included in the incubation media, thus invalidating the original hypothesis. This inability of cGMP to alter the $Ca^{2+}+Mg^{2+}$ATPase activity would indicate that the inhibitory effect of guanylate cyclase on $Ca^{2+}+Mg^{2+}$ATPase is not mediated via cGMP; that is, it is not a post-receptor phenomenon. The possibility that $Ca^{2+}+Mg^{2+}$ATPase and guanylate cyclase are coupled together is therefore raised (180,181). Accordingly, we propose the following working model as a prospective mechanism of the coupling of these two enzymes. It is suggested, according to our model, that guanylate cyclase is able to induce conformational changes in the $Ca^{2+}+Mg^{2+}$ATPase with decreased expression of the pump activity. The validity of this proposed model was examined by modulating the guanylate cyclase using various stimulators and inhibitors. Sodium azide, a well-known powerful stimulator of guanylate cyclase (182), was added to the incubation media and the $Ca^{2+}+Mg^{2+}$ATPase assayed. The results clearly showed that $Ca^{2+}+Mg^{2+}$ATPase was depressed due to an increase in the guanylate cyclase activity. Our studies with guanylate cyclase antibody further lend credence to this model. This antibody cross-reacted with the guanylate cyclase enzyme in the kidney BLM and was able to nearly block its action. Radioisotopic assay of $Ca^{2+}+Mg^{2+}$ATPase after inhibiting guanylate cyclase within this antibody showed an elevation of the $Ca^{2+}+Mg^{2+}$ATPase activity. Inclusion of the antibody in the incubation medium was able to block the guanylate cyclase and thereby remove the inhibitory influence of guanylate cyclase on $Ca^{2+}+Mg^{2+}$ATPase, leading to increased expression of the latter enzyme (180).

It can be argued that depression in $Ca^{2+}+Mg^{2+}$ATPase could arise possibly as a result of competitive substrate utilisation. In other words, since the incubating media contains both ATP and GTP, the preferential utilisation of one instead of the other could lead to a decrease in enzyme activity. Accordingly, another ATPase, was tested for possible effects due to preferential substrate utilisation. No inhibitory effect of GTP on $Na^{+}+K^{+}$ATPase was noted, indicating that the GTP inhibitory effect was exclusive to the $Ca^{2+}+Mg^{2+}$ATPase. These data provide conclusive proof that these enzyme systems, viz. guanylate cyclase and the $Ca^{2+}+Mg^{2+}$ATPase are coupled together and a reciprocal modulation exists between the two (179). Further analysis of this proposed model shows that ANP is acting on two sites to modulate the $Ca^{2+}+Mg^{2+}$ATPase. As also mentioned previously, ANP is directly stimulating the enzyme (174) while it also, paradoxically, has an inhibitory influence on the enzyme via the guanylate cyclase system. A reflection on this paradoxical situation would suggest that probably the $Ca^{2+}+Mg^{2+}$ATPase is maintained at an optimal level by these two radically opposed influences. As a consequence, the cytosolic Ca^{2+} concentration is regulated.

The fact that cGMP may be involved in the removal of Ca^{2+} from the cell is suggested by the findings that intracellular Ca^{2+} is depleted more rapidly in the presence of agents that elevate cytosolic cGMP (183). In keeping with an effect of ANP on Ca^{2+} exit, ANP reduces intracellular Ca^{2+} concentrations in rat and aortic smooth muscle cells and rat glomerular mesangial cells (173,184). The available evidence thus favors the view that ANP, through stimulation of guanylate cyclase (185-187), modulates plasma membrane Ca^{2+} ATPase activity (174), thus regulating removal of Ca^{2+} from the cell. A lack of response of this pump activity to ANP may lead to intracellular Ca^{2+} overload. An increased intracellular Ca^{2+} concentration may decrease oxidative phosphorylation in mitochondria (16). As a result, the capacity of the renal cell to generate ATP becomes attenuated, which in turn compromises renal cellular function or precipitates nephropathy (137). Furthermore, increased intracellular Ca^{2+} may also result in other effects. In the presence of phospholipid, raised cytosolic Ca^{2+} either by itself or alongwith diacylglycerol may activate protein kinase C which has been implicated in the transduction of extracellular signals from growth factors, and consequently activating cellular mechanisms for growth and proliferation (renal hypertrophy) (188). Increased Ca^{2+} may also potentiate renal hypertrophy by switching on the transcription of proto-oncogenes such as myc and fos which appear to contribute to the onset of DNA synthesis (189). Additionally, Ca^{2+} may augment the formation of cytochrome P-450 arachidonic metabolites, thereby elevating the blood pressure. Altered Ca^{2+} concentration may contribute to the insulin resistance encountered in the non-insulin-dependent diabetes mellitus (190). Besides, Ca^{2+} may act as a modulator of the tubuloglomerular feedback loop system via the mesangial cells, and thus modify the cellular response in pathological situations (190 a).

Other Changes in Diabetes

There are other changes which have been documented as part of the spectrum in diabetes. Considering the relevance of Na^+-K^+-ATPase in the renal work and renal Na^+ transport, changes have been characterised in its activity in diabetes. Both increased (191,192) and unchanged (193) activity has been shown depending on the nephron segment examined and type of diabetes, i.e. insulin-dependent or non-insulin-dependent. Wald and Popovtzer showed increased activity after one week of induction of diabetes in all segments except the distal convoluted tubule (194). Again, insulin-dependent animals showed an increase in basal Na^+-K^+-ATPase activity (195), while non-insulin-dependent animals had unchanged activity (193). Though it has been suggested that probably the renal hypertrophy and attendant increase in renal function is the basic phenomenon, the prevalent view is that notwithstanding the regional variations, there is an increase in Na^+-K^+-ATPase activity obtained from two to four week old diabetic rats (196). The renal ouabain-inhibitable Na^+-K^+-ATPase activity is increased in diabetes, while insulin therapy restores this activity to normal (196,197). Another interesting phenomenon associated with Na^+-K^+-ATPase activity in diabetes is its response to cholera and pertussis toxin. Control cases show an increase in activity while this effect is abolished in diabetes (199). It is postulated that the stimulatory effects might be related to persistent stimulation of adenylate cyclase (in the case of cholera toxin) or an unmasking of inhibitory effect of Gi on adenylate cyclase (in the case of pertussis toxin). This would suggest a defective post-receptor mechanism underlying the elevated Na^+-K^+-ATPase in diabetes.

CONCLUSIONS

From the previous discussion, it is apparent that numerous transport processes exist in the kidney basolateral membrane. It is important to identify each and every aspect of these processes in order to understand the overall kidney function. Though various studies have tried to distinguish the various biochemical abnormalities in pathophysiologic processes, most of the studies have been limited to identifying whole organ defects. The majority of the work has involved recognizing the comprehensive ion defect by measuring serum levels or quantifying transport after ion loading. Very little has been accomplished in terms of identifying the fundamental transporter defect and how they affect overall ionic balance. Interestingly, Ca^{2+} transport in the kidney has gained considerable attention in the recent past. Future studies need to be directed towards establishing the different modes of transport in the basolateral membrane and identifying the pathophysiological mechanisms underlying the altered states.

Acknowledgements

The support of Dr.P.K.Ganguly by Manitoba Health Research Council and St. Boniface General Hospital Research Foundation is gratefully acknowledged. AS is a Fellow of the Canadian Diabetes Association and PKG is a Scholar of the Heart and Stroke Foundation, Canada.

REFERENCES

1. Bernard C (1885) Leçons sur les Phénomènes de la vie Communs aux Animaux et aux Végétaux. Vol 1, J.B. Bailliere et Fils, Paris
2. Koefoed-Johnsen V and Ussing (1958) Acta Physiol Scand 42:298-310
3. Mollgard K and Rostgaard J (1981) Transport Across Epithelia. Munksgaard, Copenhagen
4. Bergeron M and Thiery G (1981) Biol Cell 42:43-52
5. Schafer JA and Andreoli TE (1972) J Clin Invest 51:1264-69
6. Kirk KL, Schafer JA and DiBona DB (1984) J Membr Biol 79:65-74
7. Giebisch G, Klose RM and Windhager EE (1964) Am J Physiol 206:687-92
8. Giebasch G, and Windhager EE (1964) Am J Med 36:643-648
9. Kawamura S, Imai M, Seldin DW and Kokko JP (1975) J Clin Invest 55:1269-1276
10. Chantrelle BM, Cogan MG and Rector FC,Jr. (1985) Miner Electr Metab 11:209-214
11. Cassano G, Stieger B and Murer H (1984) Pflügers Arch 400:309-317
12. Murer H, Hopfer U and Kinne R (1976) Biochem J 154:597-603
13. Baum M (1977) Am J Physiol 233:307-312
14. Burnham C, Munzesheimer C, Rabon E and Sach G (1982) Biochim Biophys Acta 685:260-266
15. Chen PY, Illsley NP and Verkman AS (1988) Am J Physiol 254:114-120
16. Ives HE, Chen PY and Verkman AS (1986) Biochim Biophys Acta 863:91-100
17. Heidrich HG, Kinne R, Kinne-Saffran E and Hannig K (1972) J Cell Biol 54:232-245
18. Schmidt U, Schmidt H, Funck B and Dubach UC (1974) Ann N Y Acad Sci 242:489-500
19. Schmidt U and Dubach UC (1971) Pflügers Arch 330:265-274
20. Shaver JLF and Stirling C (1978) J Cell Biol 76:278-283
21. Kyte J (1976) J Cell Biol 68:287-292
22. Maunsbach AB, Skriver E, Deguchi N and Jorgensen PL (1980) Acta Histochem Cytochem 13:103-112
23. Epstein FH and Silva P (1974) Ann N Y Acad Sci 242:519-523
24. Garg LC, Knepper MA and Burg MB (1981) Am J Physiol 240:536-542

25. Doucet A and Katz AI (1980) 238:380-386
26. Mernissi A and Doucet A (1984) Am J Physiol 247:158-167
27. Horster M (1978) Am J Physiol 235:387-393
28. Aperia A, Larsson L and Zetterstrom R (1981) Am J Physiol 241:356-362
29. Aperia A and Larsson L (1984) Acta Physiol Scand 121:133-41
30. Schmidt U and Dubach UC (1971) Nephron 7:447-458
31. Gogelein H and Greger R (1986) Pflügers Arch 407:S142
32. Sasaki S, Ishibashi K, Yoshiyama N and Shigaii I (1988) J Clin Invest 81:194-199
33. Kondo Y and Fromter E (1988) Pflügers Arch 411:R100
34. Grassl SM, Holohan PD and Ross CR (1987) J Biol Chem 262:2682-2687
35. Ullrich KJ, Capasso G, Rumrich G, (1977) Pflügers Arch 368:245-253
36. Greger R (1985) Physiol Rev 65:760-797
37. Greger R and Schlatter E (1983) Pflügers Arch 396:325-332
38. Ussing HH (1960) Handbuch der Experimentallen Pharmacologie. Springer-Verlag, Hiedelburg, pp 1-195
39. Whittembury G and Grantham JJ (1976) Kidney Int 9:103-109
40. Adler S and Fraley DS (1977) Kidney Int 11:433-439
41. Mudge GH (1953) Am J Physiol 173:511-517
42. Whittembury G (1965) J Gen Physiol 44:689-96
43. Proverbio F, Robinson JWL and Whittembury G (1970) Biochim Biophys Acta 211:327-333
44. Biagi B, Kubota T, Sohtell M and Giebisch G (1981) Am J Physiol 240:200-206
45. Cemerikic D, Wilcox CS and Giebisch G (1982) J Membr Biol 69:159-163
46. Katz AI (1982) Am J Physiol 242:207-213
47. Jacobson MP, Rodriguez HJ, Hogan WC and Klahr S (1980) Am J Physiol 239:281-286
48. Koeppen BM, Biagi BM and Giebisch GH (1983) Am J Physiol 244:35-47
49. Koeppen BM and Giebisch GH (1984) In: Regulation and Development of Membrane Transport Processes, Raven Press, New York
50. Shultz SG (1981) Am J Physiol 241:579-590
51. Shultz SG (1982) Semin Nephrol 2:343-372
52. Lang F, Messner G, Wang W and Oberleithner H (1983) Klin Wochenschr 61:1029-1037
53. Schwarz W and Passow H (1983) Ann Rev Physiol 45:359-374
54. Lewis SA and deMoura JLC (1982) Nature 297:685-687
55. Hayslett JP and Binder HJ (1982) Am J Physiol 243:103-110
56. Katz AI, Doucet A and Morel F (1979) Am J Physiol 237:114-119
57. Stanton B, Janzen A, Klein-Robbenhaar G, Wade J, Giebisch G and DeFronzo R (1982) Kidney Int 23:267 (Abstr)
58. Stanton B, Janzen A, Klein-Robbenhaar G, DeFronzo R, Giebisch G and Wade J (1985) J Clin Invest 75:1327-1334

59. Rastegar A, Biemesderfer D, Kashgariah M and Hayslett JP (1980) Kidney Int 18:293-300
60. Wade JB, O'Neil RG, Pryor JL and Boulpaep EL (1979) J Cell Biol 81:439-444
61. Preisig PA and Alpern RJ (1989) Am J Physiol 256:751-765
62. Alpern RJ (1990) Physiol Rev 70:79-97
63. Burckhardt BC, Sato K and Frömter E (1984) Pflügers Arch 401:43-51
64. Alpern RJ (1985) J Gen Physiol 86:613-620
65. Biagi B, Kubota T, Sohtell M and Giebisch G (1981) Am J Physiol 240:200-207
66. Boron WF and Boulpaep EL (1983) J Gen Physiol (1983) 81:53-94
67. Alpern RJ and Chambers M (1987) J Gen Physiol 89:581-598
68. Biagi B and Sohtell M (1986) Am J Physiol 250:267-272
69. Krapf R, Alpern RJ, Rector FS,Jr, and Berry CA (1987) J Gen Physiol 90:833-853
70. Akiba T, Warnock DG (1987) Clin Res 35:633A
71. Soleimani M and Aronson PS (1988) Kidney Int 33:407 (Abstr)
72. Edelman M, Bouthier M and Anagnostopoulos T (1981) J Membr Biol 62:7-13
73. Guggino WB, London R, Boulpaep EL and Giebisch G (1983) J Membr Biol 71:227-240
74. Preisig PA and Alpern RJ (1988) J Clin Invest 28:1445-1452
75. Kurtz I (1989) J Clin Invest 83:616-622
76. Sasaki S, Shiigai T and Takeuchi J (1985) Am J Physiol 249:417-423
77. Grassl SM, Karniski LP and Aronson PS (1985) Kidney Int 27:282 (Abstr)
78. Garvin JL, Burg MB and Knepper MA (1985) Am J Physiol 249:785-788
79. Kurtz I and Balaban RS (1986) Am J Physiol 250:497-502
80. Soltoff SP and Mandel LJ (1984) J Gen Physiol 84:623-642
81. Boron WF and Boulpaep EL (1983) J Gen Physiol 81:29-52
82. Ives HE, Yee VJ and Warnock DG (1983) J Biol Chem 258:13513-13519
83. Sabolic I and Burckhardt G (1983) Biochim Biophys Acta 734:210-216
84. Herndon RF and Freeman S (1958) Am J Physiol 192:369-372
85. Fritzsch G, Rumrich G, Fasold H and Ullrich KJ (1984) Pflügers Arch 400:250-256
86. Ullrich KJ, Fasold H, Rumrich G and Klöss S (1984) Pflügers Arch 400:241-249
87. Burckhardt G (1984) Pflügers Arch 401:254-261
88. Sheikh MI, Kragh-Hansen U, Jorgensen KE and Roigaard-Petersen H (1982) Biochem J 208:377-382
89. Miller JV and Shiekh MI (1983) Physiol Rev 34:315-359
90. Pritchard JB (1987) Biochim Biophys Acta 906:295-308
91. Chung ST, Park YS and Hong SK (1970) Am J Physiol 219:30-33
92. Dantzler WH and Bentley SK (1976) Am J Physiol 230:256-262

93. Maxild J, Miller JV and Shiekh MI (1980) Biochim Biophys Acta 601:490-499
94. Ullrich KJ, Rumrich G and Klöss S Pflügers Arch 404:293-299
95. Ullrich KJ, Rumrich G and Klöss S Pflügers Arch 404:300-306
96. Ullrich KJ, Rumrich G and Klöss S Pflügers Arch 404:307-310
97. Ullrich KJ, Rumrich G and Klöss S Pflügers Arch 404:311-318
98. Kahn AM, Shelat H and Weinman EJ (1985) Am J Physiol 249:654-661
99. Abramson RG, King VF, Reif MC, Leal-Pinto E and Baruch SB (1982) Am J Physiol 242:158-170
100. Abramson RG and Lipkowitz MS (1985) Am J Physiol 248:574-584
101. Farah A, Frazer M and Porter E (1959) J Pharmacol Exp Ther 126:202-208
102. Farah A and Rennick S (1956) J Pharmacol Exp Ther 117:478-485
103. McKinney TD (1983) J Pharmacol Exp Ther 224:302-307
104. Schali C and Roch-Ramel F (1982) J Pharmacol Exp Ther 223:811-818
105. Schali C, Schild L, Overney J and Roch-Ramel F (1983) Am J Physiol 245:238-243
106. Schafer JA and Barfurs DW (1980) Am J Physiol 238:335-346
107. Kinne R (1985) Ann N Y Acad Sci 456:39-47
108. Schafer JA and Williams JC (1985) Ann Rev Physiol 47:103-125
109. Barfuss DW, Mays JM and Schafer JA (1980) Am J Physiol 238:324-333
110. Kuwahara M, Sasaki S, Shiigai T and Takeuchi J (1986) Kidney Int 30:340-347
111. Evers J, Murer H and Kinne R (1976) Biochim Biophys Acta 426:598-615
112. Kragh-Hansen U and Sheikh MI (1984) J Physiol 354:55-67
113. Slack EN, Liang CT and Sacktor B (1977) Biochem Biophys Res Commun 77:891-897
114. Silverman M, Vinay P, Shinobu L, Gougoux A and Lemieux G (1981) Kidney Int 20:359-365
115. Reynolds RA, Wald H and Segal S (1982) Biosci Rep 2:883-890
116. Ullrich KJ and Murer H (1982) Philos Trans R Soc London B Biol Sci 299:549-558
117. Schwab SJ, Klahr S and Hammerman MR (1984) Am J Physiol 246:663-669
118. Hammerman MR and Schwab SJ (1984) In: Bronner F and Peterlik M (eds) Epithelial Calcium and Phosphate Transport:Molecular and Cellular Aspects, Liss, New York, pp 325-330
119. Kinne R, Murer H, Kinne-Saffran E, Thees M and Sachs G (1975) J Membr Biol 21:375-395
120. Silverman M and Turner RJ (1982) Am J Physiol 242:711-720
121. Kleinzeller A, McAvoy EM and McKibbin RD (1980) Biochim Biophys Acta 600:513-529
122. Ringer S (1882) J Physiol Lond 3:380-393
123. Nordin BEC and Peacock M (1969) Lancet 2:1280-1283

124. Nordin BEC (1972) In: Talmage, RV Munson PL (eds) Calcium, Parathyroid Hormone and the Calcitonins, Excerpta Medica, Amsterdam, pp 263-272
125. Lassiter WE (1963) Am J Physiol 204:771-775
126. Frick A (1963) Arch Ges Physiol 286:109-117
127. Duarte CG and Watson JF (1967) Am J Physiol 212:1355-1360
128. Wesson LG, Jr. and Lauler DP (1959) Proc Soc Exp Biol Med 101:235-236
129. Howard PJ (1959) Am J Physiol 197:337-341
130. Grollman AP (1963) Am J Physiol 205:697-701
131. Suki W (1979) Am J Physiol 237:1-6
132. Gmaj P, Murer H and Kinne R (1979) Biochem J 178:549-557
133. Somermeyer MG, Knauss TC, Weinberg JM and Humes HD (1983) Biochem J 214:37-46
134. Blaustein MP (1976) Fed Proc 35:2574-2578
135. Lee CO, Taylor A and Windhager EE (1980) Nature 287:859-861
136. Murphy E and Mandel LJ (1982) Am J Physiol 242:124-128
137. Nayler WG, Poole-Wilson PA and Williams A (1979) J Mol Cell Cardiol 11:683-706
138. Kinne-Saffran E and Kinne R (1974) J Memb Biol 17:263-274
139. Moore L, Fitzpatrick DF, Chen TS and Landon EJ (1974) Biochim Biophys Acta 345:405-418
140. Van Heeswijk MPE, Geertsen JAM and Van Os CH (1984) J Memb Biol 79:19-31
141. Ramachandran C and Brunette MG (1989) Biochem J 257:259-264
142. Parkinson DK and Radde IC (1971) Biochem Biophys Acta 242:238-246
143. Tsukamoto Y, Suki WN, Liang CT and Sacktor B (1986) J Biol Chem 261:2718-2724
144. Vieyra A, Nachbin L, De Bios-Abad E, Goldfield M, Meyer-Fernandes JR and De Moraes L (1986) J Biol Chem 261:4247-4255
145. Rorive G and Kleinzeller A (1972) Biochim Biophys Acta 274:226-239
146. Gmaj P, Murer H and Carafoli E (1982) FEBS Lett 144:226-230
147. DeSmedt H, Parys JB, Borghgraefr R and Wuytack F (1983) Biochim Biophys Acta 728:409-418
148. Katz AI and Doucet A (1980) Int J Biochem 12:125-129
149. Doucet A and Katz AI (1982) Am J Physiol 242:346-352
150. DeSmedt H, Parys JB, Borghgraef R and Wuytack F (1981) FEBS Lett 131:60-62
151. DeSmedt H, Parys JB, Wuytack F and Borghgraef R (1984) Biochim. Biophys. Acta 776:122-132
152. Gmaj P, Zurini M, Murer H and Carafoli E (1983) Biochim Biophys Acta 778:481-488
153. Pedersen PL and Carafoli E (1987) Trends Biochem Sci 12:146-150
154. Pedersen PL and Carafoli E (1987) Trends Biochem Sci 12:186-189
155. Katz S and Blostein R (1975) Biochim Biophys Acta 389:314-324

156. Knauf PA, Proverbio F and Hoffmann JF (1974) J Gen Physiol 63:324-326
157. Garrahan PJ and Rega AF (1978) Biochim Biophys Acta 513:59-65
158. Muallem S and Karlish SJD (1979) Nature Lond 277:238-240
159. Richards DE, Rega AF and Garrahan PJ (1978) Biochim Biophys Acta 511:194-201
160. Bond GH and Hudgins P (1980) Biochim Biophys Acta 600:781-790
161. Barrabin H, Garrahan PJ and Rega AF (1980) Biochim Biophys Acta 600:796-804
162. Schatzmann HJ (1982) In: Carafoli E (ed) Membrane Transport of Calcium. Academic, London pp 41-108
163. Ferreira HG and Lew YL (1976) Nature 259:47-49
164. Lew VL, Tsien RY, Miner C and Bookchin RM (1982) Nature 298:478-480
165. Larsen FL, Katz S and Roufogalis GD (1981) Biochem J 200:185-191
166. Schatzmann HJ (1973) J Physiol 235:551-569
167. Niggli V, Adunyah ES, Penniston JT and Carafoli E (1981) J Biol Chem 256:395-401
168. Sarkadi B, Szasz I, Garloczi A and Gardos G (1977) Biochim Biophys Acta 464:93-107
169. Sarkadi B (1980) Biochim Biophys Acta 604:159-190
170. Levy J, Grunberger G, Karl I and Gavin III JR (1990) Metabolism 39(1):25-33
171. Pershadsingh HA and McDonald JM (1980) J Biol Chem 255(9):4087-4093
172. Sahai A and Ganguly PK (1991) Am Heart J 122: 164-170
173. Meyer-Lehnert H, Caramelo C, Tsai P and Schrier RW (1988) J Clin Invest 82:1407-1414
174. Sahai A and Ganguly PK (1990) Biochem Biophys Res Commun 169:537-544
175. Appel RG, Wang J, Simonson MS and Dunn MJ (1986) Am J Physiol 251:1036-1042
176. Chinkers M, Garbers DL, Chang M-S, Lowe DG, Chin H, Goeddel DV and Schulz S (1989) Nature Lond 338:78-83
177. Leitman DC and Murad F (1987) Endocrinol Metab Clin North Am 16:79-105
178. Wong KR, Xie M-H, Shi L-B, Liu F-Y, Huang C-L, Gardner DG and Cogan MG (1988) Am J Physiol 255:1220-1224
179. Weiser S, Sahai A and Ganguly PK (1991) In: Proceedings of Canadian Federation of Biological Societies, 309, 1991
180. Sahai A, Weiser SJ and Ganguly PK (1993) Angiology (in press)
181. Sahai A and Ganguly PK (1992) J Mol Cell Cardiol 24:S205
182. DeRubertis FR and Craven PA (1976) J Biol Chem 251:4651-4658
183. Kobayashi S, Kanaide H and Nakamura M (1985) Science Wash DC 229:553-556
184. Chartier L and Schiffrin EL (1987) Am J Physiol 252:485-491
185. Rapoport RM, Draznin MB and Murad F (1982) Proc Natl Acad Sci USA 79:6470-6474

186. Zeidel ML, Silva P, Brenner BM and Seifter JL (1987) Am J Physiol 252:551-559
187. O'Grady SM, DeJonge HR, Vaandrager AB and Field M (1985) Am J Physiol 249:531-534
188. Nishizuka Y (1986) Science 233:305-312
189. Berridge MJ (1986) Cancer Surv 5:413-430
190. Sacerdoti D, Escalante B, Abraham NG, McGiff GC, Levere RD and Schwartzman ML (1989) Science 243:388-390
190a. Brenner BM, Ballerman BJ, Gunning ME and Zeidel ML (1990) Physiol Rev 70:665-699
191. Wald H and Popovtzer MM (1984) Pflügers Arch 401:97-100
192. Ku DD and Meezan E (1984) J Pharmacol Exp Ther 229:664-670
193. Levy J, Avioli LV, Roberts ML and Gavin JR, III (1986) Biochem Biophys Res Commun 139:1313-1319
194. Wald H, Scherzer P and Popovtzer MM (1986) Am J Physiol 251:164-170
195. Rasch R (1986) Scand J Lab Invest (1986) 46:59-62
196. Ku DD, Sellers BM and Meezan E (1986) Endocrinol 119:672-679
197. Ku DD, Roberts RB, and Sellers BM (1987) Endocrinology 120:2166-2173
198. Chen CC (1992) Eur J Pharmacol 225:275-279

CHAPTER 16

EFFECT OF INSULIN AND GLUCAGON ON HEPATOCELLULAR ALANINE UPTAKE IN NORMAL AND STREPTOZOCIN-INDUCED DIABETIC RATS

*Anwar B. Bikhazi**,
George E. Haddad, Habib M. Alloush,
Fadia H. Uthman, Najla A. Fakruddin and Muna El-Kasti

Department of Physiology, Faculty of Medicine
American University of Beirut, Beirut, Lebanon

ABSTRACT

Hepatocytes from normal and streptozocin-treated diabetic rats were dispersed by collagenase perfusion. Cells were suspended in Ca-Mg free phosphate buffer containing oxamic acid and treated with insulin, glucagon and/or cycloheximide prior to measurement of ^{14}C-alanine transport. Samples were withdrawn at specified intervals, centrifuged, and hepatocellular uptake was monitored by measurement of alanine intracellularly and extracellularly. Alanine uptake in insulin-treated normal cells increased (86%) compared to controls. However, in cycloheximide-treated cells, uptake decreased by 25%. Cycloheximide, in the presence of insulin, blocked increased uptake induced by insulin treatment. Diabetic cells showed a 25% increase in uptake compared to normal cells. Insulin treatment enhanced uptake (67%),

---

*To whom inquiries should be directed

while cycloheximide decreased it by 13% compared to diabetic controls. No change in uptake was observed when diabetic cells were treated with insulin and cycloheximide simultaneously. Insulin enhances alanine transport in normal and diabetic hepatocytes contrary to the effect of cycloheximide. However, the data disagrees with the "down-regulation" concept. The increase in uptake exhibited by diabetic cells may be explained by an increase in glucagon plasma levels which has been shown to be a potent stimulator of the alanine carrier transport receptors.

INTRODUCTION

Insulin acts on the target organ, the liver, to inhibit gluconeogenesis. Thus, insulinopenic diabetes is characterized by an increase in gluconeogenesis which occurs in the face of diminished circulating levels of alanine and other glucogenic amino acids. This has strongly suggested that liver extraction of amino acids is enhanced in diabetes. The increase in the transport of amino acids into the hepatocytes of diabetic animals will result in an increased supply of glycogenic precursors and, consequently, will further increase the already uncontrollable hyperglycaemia associated with insulin deficiency. It was Van Slyke and Meyer in 1913 (1) who first recognized that amino acid uptake by the liver occurs against a concentration gradient (carrier-mediated transport).

Glucagon is an important catabolic agent which can be considered to be the dominant hormone in the fasting state. It acts to deplete stored glucose and fatty acids in the liver and adipose tissue and increases the supply of energy for oxidation. Glucagon has a rapid hyperglycaemic effect. It is a powerful hepatic glycogenolytic agent and has a potent effect on gluconeogenesis. It also has a lipolytic action, in that it stimulates lipase activity in both liver and adipose tissue. Thus, it stimulates fatty acid uptake and oxidation in liver and muscles. Glucagon is a potent protein catabolic agent and it increases the rate of uptake and deamination of amino acids in the liver. However, it decreases the rate of uptake of amino acids in muscle cells. The net outcome of the lipolytic and proteolytic effect of glucagon is to supply more fatty and animo acids for the gluconeogenic pathway.

Amino Acid Transport Systems

A survey of the studies done on the amino acid transport in and out of cells reveals a variety of transport systems whose reactivity to substrates seems, in general, to be rather similar among animal tissues, despite few differences observed in specialized tissues. The transport of amino acids by animal cells is divided among four main systems that have overlapping substrate reactivities. These are known as System A, System ASC, System L, and System N.

System A. System A is most reactive with amino acids having short, polar or linear side chains, such as alanine, glycine, and the non-metabolized analog, α-aminoisobutyric acid, and its N-methylated derivative, 2-methylamine isobutyric acid. It is Na-dependent and is greatly inhibited by pH values <7.0 and by metabolic inhibitors (2-4).

System ASC. A second transport system was later characterized as a portion of the Na-dependent uptake of alanine, serine and cysteine, and other small neutral amino acids that was not blocked by an excess of 2-methylamine isobutyric acid, an indicator of System A mediation (4). This was termed ASC for its restrictiveness to alanine, serine, and cysteine. System ASC is found to be ubiquitous in tissues of higher animals (5-7).

System L. This system was observed to have reactivity towards amino acids with branches or rings on the side chain (8). However, all neutral amino acids were shown to be transported via this system to a greater or lesser extent (9). The most striking property differentiating System L from other transport systems is its Na-independence (10, 11) especially where the analog 2-amino-bicyclo-(2,2,1)-heptane-2-carboxylic acid was used because of its specificity in the Ehrlich cell (12).

Christensen and Handlogten (13) described a glycine specific system in isolated rat hepatocytes previously found in pigeon erythrocytes (7). The glycine system is distinguished from System A by its lack of adaptive and hormonal control.

System N. Recently, a sodium-dependent transport system restricted in reactivity to glutamine, aspargine, histidine and few synthetic analogs was described in cultured rat hepatocytes (14,15). It is called System N, for nitrogen containing side chains of each of its substrates. System N differs from System A by its insensitivity to hormones such as insulin and glucagon and from System ASC by its increased activity following amino acid starvation (14). It is worthy to mention that System N is found to be unique to the liver tissue (16).

Regulation of Neutral Amino Acid Transport by the Liver Adaptive Regulation

Adaptive Regulation. The adaptive regulation links the transport activity to the availability of relevant amino acids in the cell environment. This phenomenon was originally described by Gazzola et al. (17) in chick embryo heart cells when they observed that incubation of certain tissues in an amino acid-free medium for 3 to 5 hours led to increased neutral amino acid uptake. This response has been referred to as adaptive control or starvation-induced transport enhancement which has been proven to have a wide range of distribution in a variety of cells and tissues of avian and mammalian origin.

Hormonal Regulation. Extensive studies during the last two decades have provided abundant descriptive information on the effects of hormones on the transport of amino acids in cells and tissues; they have shown that the uptake of neutral amino acids by liver tissue is responsive to a variety of hormones (18,19). In most instances, there is a correlation between the stimulation of amino acid uptake by a hormone in its specific target tissue and the increase in the anabolic activities of the cell, such as macromolecular synthesis, energy production or cell division. Conversely, transport is either unaffected or inhibited when the effect of the hormone is catabolic, as in the case of glucocorticoid inhibition of amino acid uptake in muscle (20,21).

From numerous published studies on the hormonal regulation of amino acid uptake, on a wide variety of cells and tissues, it was generally agreed, with few exceptions, that System A is the hormone-sensitive system. For the liver, two lines of evidence suggest that System A is the only amino acid transport system that responds to hormones. First, the transport system affected by hormones is Na-dependent, thus ruling out the involvement of System L which is Na-independent (22,23). Second, the hepatic transport ASC, N (14), and L systems (23-25), as measured by the test substrates cysteine, and glutamine, appear to be non-responsive to hormones.

The effect of glucagon on hepatic amino acid transport has been studied extensively in perfused rat liver (26,27), rat liver slices (28,29), freshly isolated rat hepatocytes (30), and primary cultures of liver parenchymal cells (31). Glucagon was shown to enhance amino acid uptake in the liver by inducing the activity of System A (14). Kinetically, the glucagon-induced uptake appears to be the result of an increase in the maximal velocity ($\dot{V}_{max}$) with almost no change in K_m (32,33). The mechanism by which glucagon stimulates amino acid transport is biphasic. For the initial 15-30 minutes, the enhancement in amino acid uptake is independent of *de novo* protein synthesis, whereas cycloheximide completely abolishes any further increase by the hormone (34).

Current evidence indicates that insulin does not penetrate the cell membrane. It attaches itself to specific receptors, which are molecules with a high degree of affinity for binding insulin, and the complex of hormone to receptor activates the cell. For many peptide hormones, the hormone receptor complex activates a membrane bound enzyme which releases cAMP, and possibly Ca, known as "second messenger" that is thought to transmit the hormonal signal into the cell interior. For insulin, the identity of the intracellular messenger is unknown (35).

A stimulatory effect of insulin on amino acid uptake has been reported for a variety of biological preparations, indicating that insulin regulation affects predominantly or entirely the activity of transport System A. Insulin's effect on System A occurs at physiological levels of the hormone, beginning at about 0.1 nM and attaining a maximum increase at 0.1 μM (36,37). The time needed to observe the stimulatory effect of insulin was reported to vary between 30 and 120 minutes.

Other possible mechanisms for the hormone-induced stimulation of amino acid transport include the activation of inactive carriers or the recruitment of cryptic membrane transport carriers. Glucagon is known to enhance transport activity by causing an increased conversion of the inactive carriers to their active forms (38). Moreover, the treatment of rat adipocytes (39) and rat diaphragm cells (40) with insulin resulted in an increase in the number of glucose transport carriers in the plasma membrane fraction, as assayed by the binding of cytochalasin B, and a corresponding decrease in the number of transport carriers in the microsomal fraction.

There are three proposed explanations for the mechanism of stimulation of amino acid uptake by glucagon: (i) *De novo* Synthesis of Transport Carriers: In the studies of Edmondson and Lumeng (34), the initial 15-30 minutes period of stimulation by glucagon was not altered by treatment with cycloheximide, whereas the increase in transport activity after 30 minutes was abolished by this inhibitor. Similar results have been obtained after treatment of the cells with cAMP rather than glucagon (41). These results suggest the existence of two separate phases of hormone stimulation of System A hepatocytes, an initial period independent of *de novo* macromolecular synthesis and a second phase dependent on protein synthesis. Also, decay suppression by inhibitors of protein synthesis, which have a stabilizing effect even if added after decay has commenced, have been interpreted as indicating that *de novo* synthesis of proteins plays an obligatory role during the second phase (42). (ii) Alterations in Membrane Potential: An increase in the magnitude of the transmembrane electrochemical gradient could produce an increase in transport by System A as a result of the dependence of this system on the flow of Na^+ ions. A hormone could hyperpolarize the cell membrane by increasing the membrane permeability to K^+ ions, by decreasing permeability to Na^+ ions or by increasing the activity of the electrogenic Na^+-K^+ ATPase (43). Friedmann and Dambach (44) observed an increase in membrane potential in rat liver from 39.0 to 47.2 mV after glucagon treatment and similar results have been reported by Graf and Peterson (45). Addition of insulin resulted in an antagonistic effect, such that glucagon stimulation of the membrane potential was abolished. Edmondson and Lumeng (34) suggested that the glucagon induced System A uptake is the result of a concomitant increase in membrane potential based on the observed temporal similarities between glucagon stimulated transport and increased transport caused by an artificially created increase in the membrane potential. (iii) Activation of Inactivated Transport Carriers: glucagon is postulated to enhance transport activity by causing an increased conversion of the inactive carrier to its active from (31). In theory, this second step could be either dependent or independent of *de novo* protein synthesis, although most of the induction of transport by glucagon is cycloheximide sensitive (34).

MATERIALS AND METHODS

Various methods have been described for the isolation of parenchymal cells from mammalian livers. Some investigations resulted in a low yield (5-10%) due to forcing the liver through cheese-cloth, shaking slices with glass beads (46), or by mincing with a tissue press (47). However, Anderson (48) made the first attempt at mass isolation of whole cells from rat liver by perfusion through the dorsal aorta for 3-4 minutes. But this method, together with previous ones, had proven to be inappropriate since it was dependent on mechanical treatment that would damage cells by abrasion of the plasma membrane, resulting in breaking of microvilli or tearing of the plasma membrane at the junctional complex between adjacent cells. Furthermore, the cells did not respire sufficiently when suspended in typical extracellular medium, and hence they were not reliable for research study. The production of intact and viable cells was the main concern of Howard et al. (49) who introduced the first enzymatic technique that employed the preparation of liver cells by shaking liver slices in a buffered medium containing two enzymes, hyaluronidase and collagenase. The liver cells obtained respired in the presence of calcium and were intact and viable after the utilization of the trypan blue test. The disadvantage of this method was the poor yield of cells, being only 5% of the original liver cells. Later, Berry and Friend (50) developed a new technique which employed *in situ* liver perfusion for 45 minutes with a solution containing the same enzymes used by Howard et al. (49). The liver was filtered through a nylon mesh filter that resulted in the separation of about 50% of the liver cells, retaining their structural integrity. All problems of oxygen diffusion associated with the surviving slice technique were overcome by the use of isolated cell preparations.

Ingebresten and Wagle (51) modified this technique by perfusing the liver *in vitro*, with enzymes containing perfusion solution. This procedure requires 40-50 minutes from hepatectomy to cell isolation with collagenase (20 mg per 100 ml) for 10-15 minutes. Approximately, 90-95% of the cells isolated by this method excluded the vital stain of trypan blue when checked under the microscope. Afterwards, other modifications or improvements have been introduced in the perfusion system, such as addition of nutrients, chelating agents, and heparin (52).

Preparation of Calcium-Free Perfusion Solution (Krebs-Henseleit)

Stock solutions of 0.154 M sodium chloride, potassium chloride, potassium dihydrogen phosphate, magnesium sulfate and sodium bicarbonate were prepared. Then, 400 ml of the 0.154 M sodium chloride solution, 20 ml of the 0.154 M potassium chloride solution, 5 ml of the 0.154 M magnesium sulfate solution, 105 ml of the 0.154 M sodium bicarbonate solution, and 15 ml of distilled water were mixed together. This solution then contained 0.034 M NaCl, 0.0017 M KCl, 0.00043 M KH_2PO_4, 0.00043 M $MgSO_4$, and 0.0089 M $NaHCO_3$. To 550 ml of the above calcium-free solution, 20 ml of 0.16 M sodium pyruvate solution, 35 ml of 0.10 M sodium fumarate solution, 20 ml of 0.16 M sodium glutamate solution and 25 ml of 0.30 M D-glucose solution were added. This solution then contained 0.00192 M of sodium pyruvate, 0.0021 M of sodium fumarate, 0.00194 M of sodium glutamate, and 0.0042 M of D-glucose. The prepared solution was gassed with 95% O_2/5% CO_2 for 15 minutes. The pH of the final solution was then adjusted to 7.4 by the addition of either 1 N HCl or 1N NaOH.

Preparation of Calcium and Magnesium Free Phosphate Buffer Saline (CMF-PBS)

About 1.7 gm of potassium phosphate dibasic were dissolved in a 0.14 M solution of sodium chloride to produce a 0.01 M solution. The solution was oxygenated, as in the case of the perfusion solution. The pH was adjusted to 7.4 by the addition of either a 1 M solution of potassium dihydrogen phosphate or 1 N sodium hydroxide.

Preparation of Calcium and Magnesium Free Phosphate Buffer Saline Containing 0.5 mM Oxamic Acid (Amino-Oxyacetate)

A 0.1 mM solution of oxamic acid was prepared by dissolving 0.4453 gm of oxamic acid in 50 ml of CMF-PBS. From this stock 0.251 ml were added to 50 ml of buffer solution to have a final concentration of 0.5 mM oxamic acid.

Preparation of Calcium and Magnesium Free Phosphate Buffer Saline Containing 0.1 μM Insulin or 0.1 μM Glucagon

A 50 mg sample of insulin was dissolved in 20.7 ml of buffer. From this stock, 12.4 μl was pipetted and added to 50 ml of (CMF-PBS) solution to produce a final concentration of 0.1 μM insulin. Similarly, a 1 mg sample of glucagon was dissolved in 5 ml of buffer. From this stock, 87.5 μl was withdrawn and added to 50 ml of (CMF-PBS) solution to get a final concentration of 0.1 μM glucagon.

Preparation of Calcium and Magnesium Free Phosphate Buffer Saline Containing 0.1 mM Cycloheximide

A 0.1 M solution of cycloheximide was prepared by dissolving 0.5628 gm of cycloheximide in 20.4 ml of buffer. From this stock, 50 μl was added to 50 ml of buffer solution to have a final concentration of 0.1 mM cycloheximide.

Preparation of Rat Hepatocyte Suspensions

Suspensions of isolated liver parenchymal cells were prepared by modification of the enzymatic perfusion technique, established by Ingebresten and Wagle (51). Briefly, male Sprague-Dawley rats weighing 200-250 g and fed *ad libitum* were anesthetized by intraperitoneal injection of 80 mg/kg of 5 sec butyl-5-ethyl-2-thiobarbituric acid sodium salt. The rat's liver was initially perfused with 200 ml Ca-free Krebs-Henseleit solution containing 1000 U.S.P. units of heparin sodium via the portal vein at a flow rate of 30-40 ml/min. Perfusion at 100 ml/min was then started at 37^{0}C with 250 ml of Ca-free Krebs-Henseleit solution containing 0.4 mg/ml collagenase and 1000 U.S.P. units of heparin sodium. All perfusion media were gassed with 95% O_2/5% CO_2 throughout the experiment. After 30 min of perfusion, the fibrous tunic was scraped and the hepatocytes dispersed by gentle motion. The cells were filtered through a course 250 μM nylon-mesh filter, washed and suspended in 30 ml Ca and Mg free phosphate buffered saline. The viability of the hepatocytes in suspension was then estimated by the trypan blue exclusion test (53-56).

Determination of Extracellular Alanine in a Ca and Mg-Free Phosphate Buffer Saline Containing Dispersed Rat Hepatocytes

A certain volume of the hepatocyte suspension was pipetted into a 110 ml CMF-PBS to attain around 10^4 cells/ml. From this suspension, 50 ml was pipetted and placed in a water-jacketed beaker which was maintained at 37^{0}C throughout the experiment. The cell suspension was continuously stirred at 120 rpm to prevent agglutination of the cells. Approximately 4 nmoles of L-(1-^{14}C)-alanine, with a specific activity of 52 mCi/mmol, was added into the 50 ml CMF-PBS containing the hepatocytes. At 5, 10, 15, 20, 30, 40, 50 and 60 minutes intervals, after the addition of radioactive alanine, 1 ml aliquots of the cell suspension were withdrawn into 1.5 ml plastic centrifuge tubes. The cells were then sedimented by centrifugation for 30 sec with a microcentrifuge at 10000 xg. Then, 0.5 ml aliquots of the supernatant were withdrawn and placed in a liquid scintillation vial containing 5 ml of Bray's solution.

The same procedure, described above, was used for the determination of extracellular alanine in the CMF-PBS solution containing hepatocytes while applying all the modification (such as addition of glucagon, cycloheximide etc.....).

Description of the Cell Treatments

Controls. Hepatocytes for controls were suspended in 50 ml of CMF-PBS solution. The cells were treated with 0.5 mM oxamic acid for 30 min before the addition of labelled alanine.

Treatment with Insulin. The cells were treated similarly as for controls, except, here, the cells were treated with 0.1 μM insulin for 90 min before the addition of alanine.

Treatment with Glucagon. The cells in suspension were treated the same as controls. However, 15 min prior to alanine addition, the cells were treated with 0.1 μM glucagon.

Treatment with Cycloheximide. The same treatment, mentioned above under "controls", was followed, except, here, the cell suspension was treated with 0.1 mM cycloheximide for 30 min.

Treatment with Insulin or Glucagon and Cycloheximide. The same procedure was followed as for control experiments. Moreover, the cells were treated with 0.1 μM insulin or 0.1 μM glucagon and 0.1 mM cycloheximide simultaneously for 90 min or 30 min before alanine addition, respectively.

Treatment with Insulin or Glucagon Followed by Cycloheximide. The cells for control experiments were treated with 0.1 μM insulin for 90 min or 0.1 μM glucagon for 15 min. After alanine addition, 3 samples, at 5 min intervals, were withdrawn for uptake measurement. Then, the cells were treated with 0.1 mM cycloheximide for 30 minutes, after which the experiments were conducted as described above.

Determination of Mean Hepatocyte Count and Diameter

The mean hepatocyte count and diameter were determined by the Coulter Counter, model A. The counter was precalibrated with a 200 μm aperture glass tube dipped in isotonic sodium chloride solution containing 18.5 μm diameter polystyrene Divinyl Benzene Latex. Half a ml of the hepatocyte suspension was placed in 200 ml CMF-PBS solution and read for a cell count at thresholds of 2, 5, 10, 15, 20, 30, 40, 50, 60, 80 and 100 respectively. The mean hepatocyte count and diameter (20 μm) were estimated as by the procedure as reported by Bikhazi et al. (53).

RESULTS AND DISCUSSION

Effect of Insulin

The data reported in Table I and Figure 1 are a resume of the steady state percentage uptake of alanine by normal rat hepatocytes for different experimental protocols. Treating normal hepatocytes with insulin for 90 min resulted in an 86% increase in alanine uptake, compared to controls. However, 30 min cellular treatment with cycloheximide resulted in a 25% decrease in alanine retention, compared to controls. Since cycloheximide is a potent endogenous protein-synthesis

Table 1. Summary of the Results on Steady State Uptake of Alanine by Normal and Diabetic Hepatocytes Treated with Insulin and/or Cycloheximide

Treatment	pM Alanine Uptake per 10^6 cells ± SEM	% Uptake per 10^6 cells ± SEM
→.Control (Normal)	1267 ± 53[a]	16.2 ± 0.7[A]
.Control (Diabetic)	1605 ± 48[a']	20.5 ± 0.6[A']
→.90 min treated normal cells with 0.1 μM insulin	2355 ± 93[b]	30.0 ± 1.2[B]
.90 min treated diabetic cells with 0.1 μM insulin	2684 ± 87[b']	34.2 ± 1.1[B']
→.30 min treated normal cells with 0.1 mM cycloheximide	946 ± 61[c]	12.1 ± 0.7[C]
.30 min treated diabetic cells with 0.1 mM cycloheximide	1400 ± 59[c']	17.9 ± 0.8[C']
→.90 min treated normal cells with insulin + cycloheximide	1260 ± 74[d]	16.1 ± 0.9[D]
.90 min treated diabetic cells with insulin + cycloheximide	1136 ± 47[d']	14.5 ± 0.6[D']
→.90 min treated diabetic cells with insulin	2656 ± 77[e]	33.9 ± 1.0[E]
followed by 30 min treatment with cycloheximide	2560 ± 75[e']	32.7 ± 1.0[E']

% difference between: (a,b) 86%; (a,c) 25%; (a,a') 27%; (a',b') 67%; (a',d') 29%; (A,A') 27%; (C,C') 48%.
Combinations (a,d); (a',c'); (a',e'); (e,e'); (B,B'); (D,D'); not significant

inhibitor (34), the decrease in alanine uptake may be due to the suppression of the synthesis of new carrier-protein(s), and/or, possibly, altering the already existing carriers which are responsible for System A activity. This is also indicated in the results of normal cells treated with insulin and cycloheximide, which resulted in a blockage of the increase in alanine uptake caused by insulin treatment alone (Table 1). It has been reported (36) that a lag period of 30 min is needed for the onset of

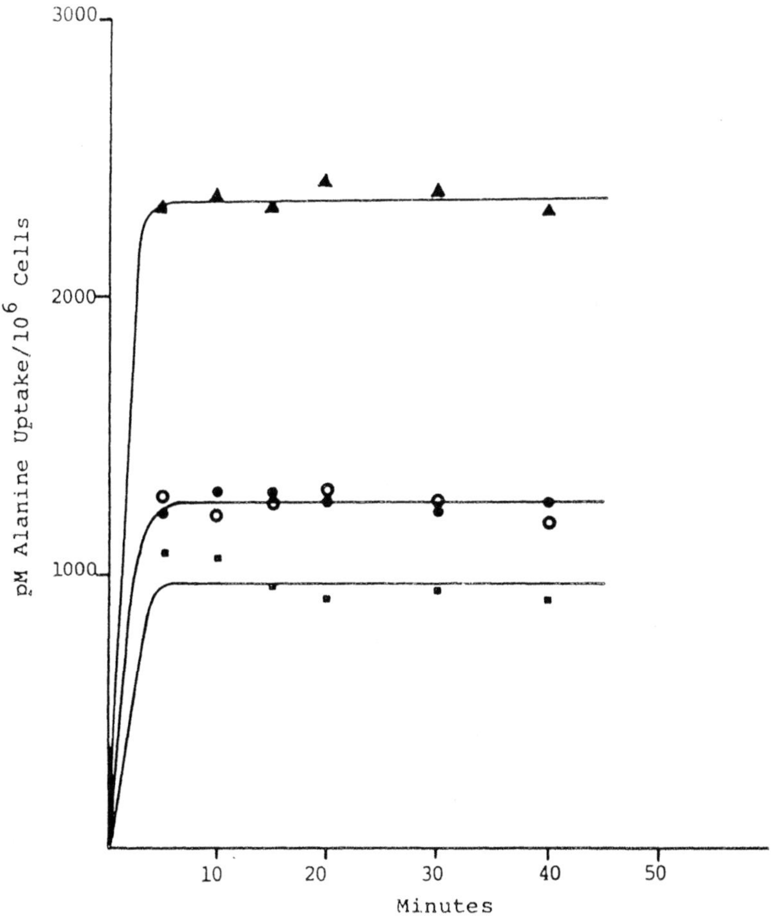

Figure 1. *Time-dependent uptake of alanine per 10^6 hepatocytes. Key: ● normal cells; ▲ insulin-treated normal cells; ■ cycloheximide-treated normal cells; ○ insulin-and cycloheximide-treated normal cells. Steady state is observed after 5 minutes.*

Table 2. Summary of the Results on Steady State Uptake of Alanine by Normal and Diabetic Hepatocytes Treated with Glucagon and/or Cycloheximide

Treatment	pM Alanine Uptake per 10^6 cells ± SEM	% Uptake per 10^6 cells ± SEM
→.Control (Normal)	1267 ± 53[a]	16.2 ± 0.7[A]
.Control (Diabetic)	1605 ± 53[a']	20.5 ± 0.6[A']
→.15 min pretreated normal cells with 0.1 μM glucagon	2808 ± 73[b]	35.8 ± 0.9[B]
.15 min pretreated diabetic cells with 0.1 μM glucagon	3292 ± 52[b']	42.0 ± 0.6[B']
→.30 min pretreated normal cells with 0.1 mM cycloheximide	946 ± 61[c]	12.1 ± 0.7[C]
.30 min pretreated diabetic cells with 0.1 mM cycloheximide	1400 ± 58[c']	17.8 ± 0.7[C']
→.30 min pretreated normal cells with cycloheximide and glucagon	1223 ± 59[d]	15.6 ± 0.8[D]
.30 min pretreated diabetic cells with cycloheximide and glucagon	1663 ± 57[d']	21.2 ± 0.7[D']
→.15 min pretreated normal cells with glucagon	2814 ± 81[e]	35.9 ± 1.0[E]
followed by 30 min treatment with cycloheximide	2748 ± 67[f]	35.0 ± 0.9[F]
→.15 min pretreated diabetic cells with glucagon	3302 ± 54[e']	42.1 ± 0.7[E']
followed by 30 min treatment with cycloheximide	3233 ± 57[f']	41.3 ± 0.7[F']

% difference between: (a,b) 122%; (a,c) 26%; (a,e) 122%; (a',b') 105%; (a',e') 105%; (A,A') 27%; (B,B') 17%; (C,C') 47%; (D,D') 36%, (E, E') 17%; (F,F') 18%

Combinations (a,d); (b,e); (e,f); (a',c'); (a',d'); (b',e'); (e',f'); not significant insulin action, resulting in an increase of alanine $\dot{V}_{max}$. Therefore, it appears that the uptake of alanine through the rat hepatocytes is dependent upon endogenous protein synthesis. Thus, insulin increases amino acid transport by inducing the synthesis of new carrier-protein(s) responsible for amino acid uptake (57, 58).

The results compiled in Table 2 summarises that diabetic rat hepatocytes showed a 27% increase in alanine uptake, as compared to their normal control counterparts. This is explained by the fact that one of the major manifestations of diabetes is the possible increase in glucagon plasma levels. Since it is well known that glucagon is

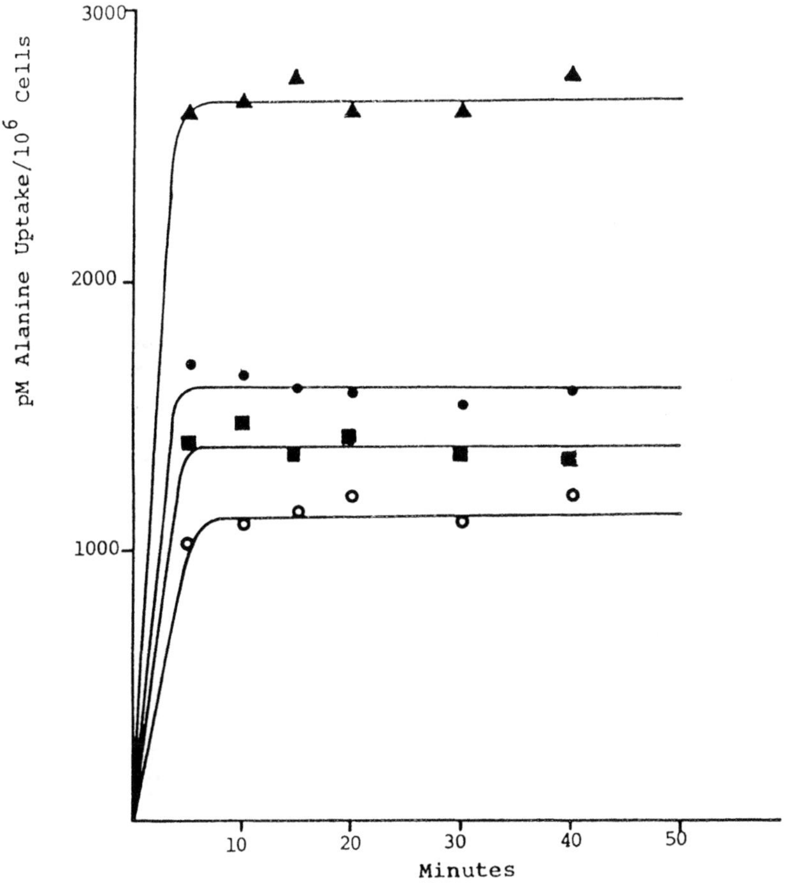

Figure 2. Time-dependent uptake of alanine per 10^6 hepatocytes. Key: • diabetic cells; ▲ insulin-treated diabetic cells; ■ cycloheximide-treated diabetic cells; ○ insulin- and cycloheximide-treated diabetic cells. Steady state is observed after 5 minutes.

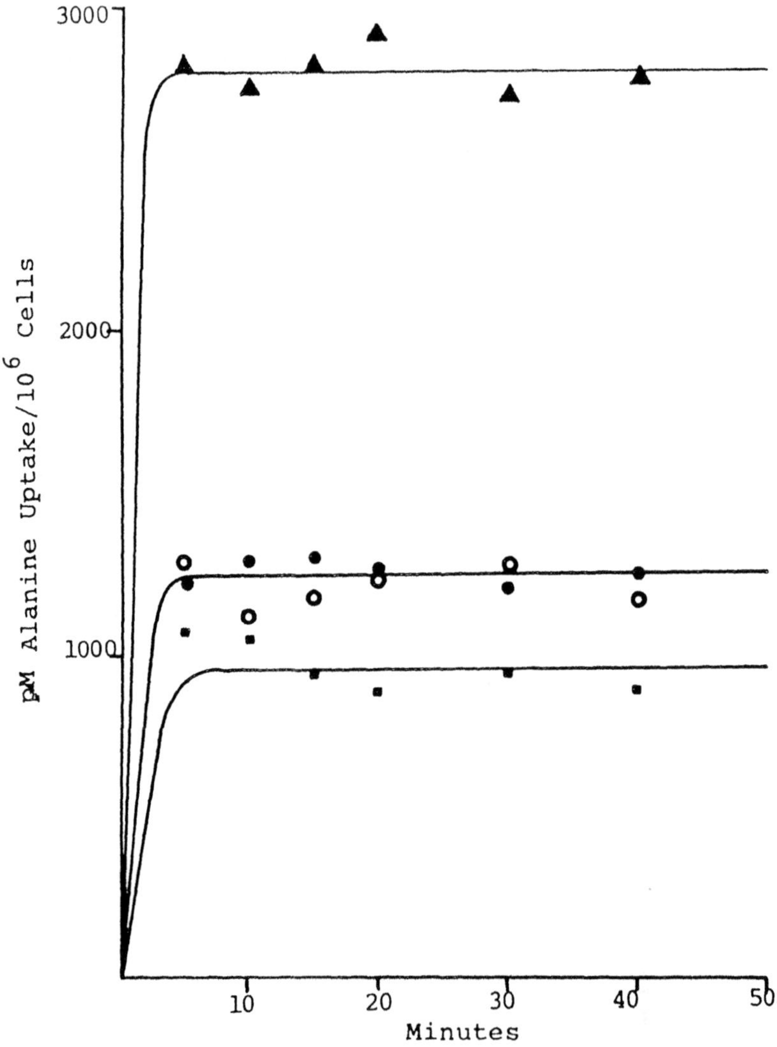

Figure 3. *Time-dependent uptake of alanine per 10^6 hepatocytes. Key:* • *normal cells;* ▲ *glucagon-treated normal cells;* ■ *cycloheximide-treated normal cells;* ○ *glucagon- and cycloheximide-treated normal cells. Steady state is observed after 5 minutes.*

a potent stimulator of System A activity in liver tissue (32, 33), it is expected that hepatocytes isolated from diabetic rats will show an increase in transport activity of System A. Thus, the increase in alanine uptake by diabetic cells treated with insulin (Table 2 and Figure 2) indicates that the transport system was not stimulated maximally in the diabetic state and that the effect of glucagon and insulin on transport activity is additive, as observed by Kelley and coworkers (8). However, 30 min of cycloheximide treatment of the hepatocytes caused a 13% reduction in alanine uptake, as compared to the diabetic controls (Table 2, Figure 2). Also, as observed in normal hepatocytes, 90 min of insulin treatment resulted in an increase of alanine transport by 67% and was also observed to be dependent on protein synthesis since the simultaneous presence of cycloheximide with insulin in the medium blocked the stimulatory effect of insulin (29% decrease). Thus, insulin can be considered to be an inducing agent in the synthesis of carriers or other proteins involved in alanine transport across the cell membrane. However, the addition of cycloheximide after 90 min of insulin treatment caused an insignificant change in alanine uptake. This shows that cycloheximide, which is a protein synthesis inhibitor, does not interfere with the already available carrier-proteins produced as a result of insulin presence when cells were exposed for 30 min with cycloheximide. Longer periods of cycloheximide treatment (90 min) may result in a reduction of these transport sites (Table 1).

The effect of insulin on alanine uptake by normal and diabetic rat hepatocytes is compared in Table 1. Diabetic hepatocytes exhibited a diminished responsiveness to insulin with regard to the stimulation of amino acid uptake since there was an 86% increase in alanine transport through normal hepatocytes after insulin treatment. However, their diabetic counterparts showed only a 67% increase in alanine uptake. This disagrees with the general concept known as "down-regulation", which states that there is an inverse relationship between insulin binding, hence its activity, and the level of endogenous insulin levels.

The data summarized in Table 1 compares the effect of 30 min of cycloheximide treatment on alanine uptake by normal and diabetic rat hepatocytes. Normal hepatocytes showed a 25% decrease in alanine transport as compared to the control. However, alanine uptake by diabetic hepatocytes showed only a 13% decrease as compared to the diabetic control. Therefore, one can conclude that there are more amino acid transport sites available in the cells due to the effect of diabetes and consequently, support the analysis that diabetic cells are richer in these carriers.

Effect of Glucagon

Glucagon is considered to be one of the hormones responsible for the control of glycogenolysis (59). Its role in the control of diabetes and uptake of glyconeogenetic substances is still under considerable debate (60).

Table 2 and Figures 3 and 4 report the data on the role of glucagon in normal and diabetic hepatocellular uptake of alanine. Normal and diabetic cells which have

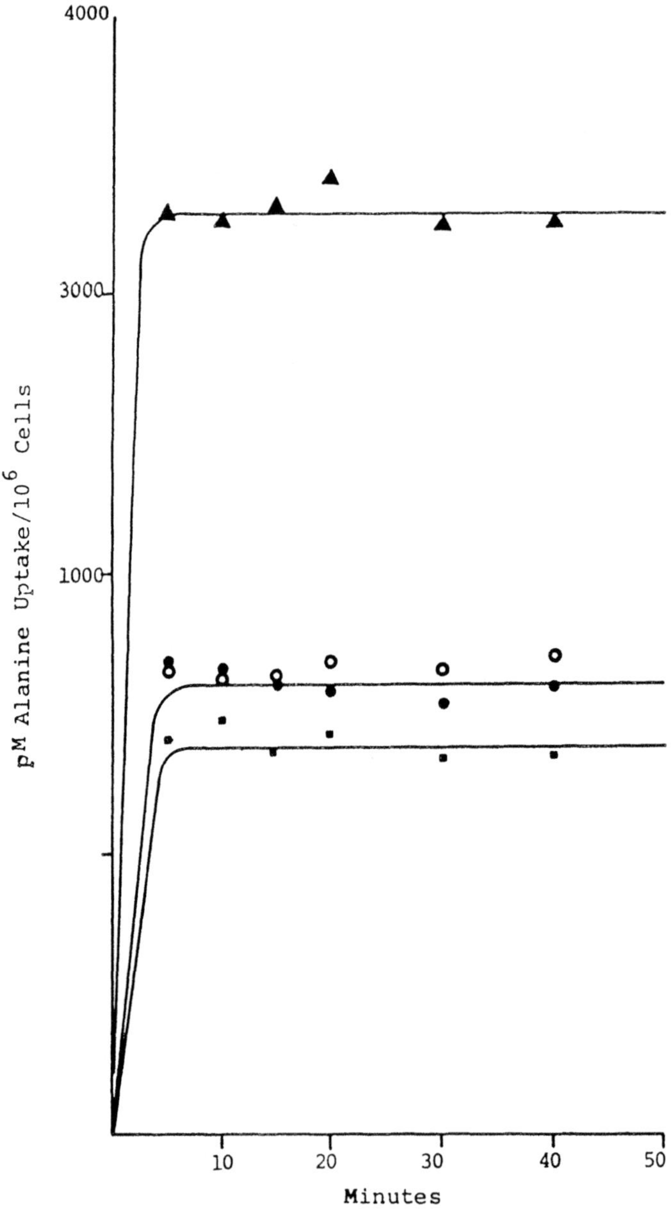

Figure 4. *Time-dependent uptake of alanine per 10^6 hepatocytes. Key:* • *diabetic cells;* ▲ *glucagon-treated diabetic cells;* ■ *cycloheximide-treated diabetic cells;* ○ *glucagon- and cycloheximide-treated diabetic cells.*

been pretreated for 15 min with 0.1 μM glucagon show a $\simeq$ 2.2 and 2.0 fold increase in the uptake of alanine respectively (approximately 121% and 105% respective increase). However, cells treated for 30 min with 0.1 mM cycloheximide show approximately a 25% and 15% decrease in alanine uptake, respectively. These data suggest that glucagon stimulates the production of new carrier proteins which are involved in System A activity for alanine uptake. Cycloheximide, on the contrary, inhibits the production of System A carrier, as well as reduces the number of available System A carriers for analine transport which are already existing on the hepatocellular membrane.

In order to further explain the role of cycloheximide and glucagon on alanine transport and uptake by liver cells, hepatocytes were incubated for 30 min in a buffer containing, simultaneously, 0.1 μM glucagon and 0.1 mM cycloheximide. The results, as shown in Table II, support the finding that glucagon did not stimulate the production of System A carriers for alanine transport in the presence of cycloheximide. In order to validate the data, the normal and diabetic hepatocytes were incubated with 0.1 μM glucagon for 15 min and the results show that the uptake of alanine is almost the same as indicated above for both types of cells (Figures 3 and 4). The same cells were treated, afterwards, with 0.1 mM cycloheximide for 30 min and showed the same increase in the uptake of alanine (2.2 and 2.0 fold as compared to controls), suggesting that after glucagon stimulated the production of alanine-transport-carriers, cycloheximide did not produce any drastic effect on the increase of alanine transport. This explanation contradicts, somehow, the previous explanation that 30 min of incubation with 0.1 mM cycloheximide resulted in a 25% and a 13% decrease of alanine uptake by normal and diabetic cells. However, a plausible explanation of this effect is that cycloheximide, after glucagon treatment, did not drastically alter the number of carriers for alanine transport (61).

In conclusion, glucagon has been shown to be a very potent stimulator of alanine carrier transport receptors and, therefore, liver cells from diabetic animals carry, on their membranes, more carrier proteins than their normal counterparts (62).

Acknowledgement

The authors wish to acknowledge the valuable contribution of Miss S. Mahseredjian, both in typing and editing of this manuscript.

REFERENCES

1. Van Slyke DD and Meyer GA (1913) J Biol Chem 16:197-212
2. Christensen HN and Handlogten ME (1977) J Biol Chem 37:193-211
3. Garcia-Sancho J, Sanchez A and Christensen HN (1977) Biochem Biophys Acta 404:295-312
4. Christensen HN, Liang M and Archer EG (1967) J Biol Chem 242:5237-5246

5. Wise WC (1976) J Cell Physiol 87:199-212
6. Thomas EL and Christensen HN (1971) J Biol Chem 246:1682-1688
7. Evanson E and Christensen HN (1967) J Biol Chem 242:5386-5696
8. Kelley DH, Shull JD and Potter VR (1980) J Cell Physiol 103:159-186
9. Oxender DL and Christensen HN (1963) J Biol Chem 236:3686-3699
10. Christensen HN Fed Proc (1973) 32:19-28
11. Inui Y and Christensen HN J Gen Physiol (1966) 50:203-224
12. Christensen HN, Handlogten ME, Lam I, Tager HS and Zand R (1969) J Biol Chem 244:1510-1520
13. Christensen HN and Handlogten ME (1981) Biochem Biophys Res Commun 98:102-107
14. Kilberg MS, Handlogten ME and Christensen HN (1980) J Biol Chem 255:4011-4019
15. Handlogten ME, Kilberg MS and Christensen HN (1982) J Biol Chem 257:345-348
16. Kilberg MS (1982) J Membr Biol 69:1-12
17. Gazzola GC and Franchi R (1972) Biochim Biophys Acta 266:407-421
18. Kilberg MS, Christensen HN and Handlogten ME (1979) Biochem Biophys Res Commun 88:744-751
19. Riggs TR (1970) In Biochemical Action of Hormones - Litwack G ed pp 157-208 Academic Press New York
20. Wool IG and Weinsshelbaum EI (1960) Am J Physiol 198:360-362
21. Kostyo JO and Schmidt JE (1963) Am J Physiol 204:1031-1038
22. Le Cam A and Freychet P (1976) Biochem Biophys Res Commun 72:893-901
23. Harrison LI and Christensen HN (1971) Biochem Biophys Res Commun 43:49-125
24. Adamson LF and Ingbar SH (1967) Endocrinology 81:1372-1378
25. Adamson LF and Ingbar SH (1967) J Biol Chem 242:2646-2652
26. Mallete LE, Exton JH and Park CR (1969) J Biol Chem 244:5724-5728
27. Chambers JW, George RH and Bass AD (1968) Endocrinology 83:1185-1192
28. Tews JK, Woodcock NA and Harper AE (1970) J Biol Chem 245:3026-3032
29. Tews JK, Woodcock Colosi NA and Harper AE (1975) Life Sciences 16:739-760
30. Kletzien RF, Parisa MW, Becker JE and Potter VR (1975) Nature 256:46-47
31. Pariza MW, Butcher FR, Kletzien RF, Becker JE and Potter VR (1976) Proc Natl Acad Sci USA 73:4511-4515
32. Fehlmann M, LeCam A and Freychet P (1979) J Biol Chem 254:10431-10439
33. Kelley DS, Campbell HA and Potter VR (1982) J Cell Physiol 12:67-75
34. Edmondson JW and Lumeng L (1980) Biochem Biophys Res Commun 96:61-68
35. Freychet P (1976) Diabetoglia 12:83-100
36. Kletzien RF, Pariza MW, Becker JE, Potter VR and Butcher FR (1976) J Biol Chem 251:3014-3020
37. Le Cam A and Freychet P (1978) Diabetoglia 15:117-123
38. Pariza MW, Kletzien RF, Butcher FR et al (1976) Adv Enz Req 14:103-115

39. Cushman SW and Wardzala LJ (1980) J Biol Chem 255:4758-4762
40. Wardzala LJ and JeanRenaud B (1981) J Biol Chem 256:7090-7093
41. McGivan JD, Samsell JC and Lacey JH (1981) Biochim Biophys Acta 644:295-304
42. Guidotti GG, Franchi-Gazzola R and Gazzola GC (1974) Biochim Biophys Acta 356:219-230
43. Zierler K and Rogus EM (1981) Fed Proc 40:119-120
44. Friedmann N and Dambach G (1980) Biochim Biophys Acta 596:180-185
45. Graf J and Petersen OH (1978) J Physiol 284:105-126
46. Le Cam A, Guidotti A and Freychet P (1976) Exp Cell Res 98:383-395
47. Schneider WE and Potter VR (1943) J Biol Chem 149:217
48. Scott JF Simpson E and Bucher NL (1951) Cancer Res 11:240
49. Howard RB (1967) J Cell Biol 35:675
50. Berry MN and Friends DS (1969) J Cell Biol 43:506-520
51. Ingebresten WR and Wagle SR (1972) Biochim Biophys Res Commun 47:403
52. Seglen PO (1973) Exp Cell Res 82:391-398
53. Bikhazi AB, Abbud EM, Argulian SC and Nassar CF (1980) Pflügers Archiv 86:245-250
54. Bikhazi AB, Nubani HM and Coe L (1983) J Pharm Sci 72(2):116-118
55. Bikhazi AB, Baasiri GM, Boulos NZ and Khuri RN (1983) J Pharm Sci 72(3):296-298
56. Bikhazi AB and Turyaki LH (1984) J Pharm Sci 73(5):628-630
57. Picket JB (1992) Muscle-Nerv 1517:761-767
58. Exton JH (1990) Diabetes 40:521-526
59. Selkurt E (1984) Physiology Little Brown
60. LeFebre P and Under R (1972) in Glucagon: Molecular Physiology Clinical and Therapeutic Implications Pergamon Press
61. Dunbar JC, Schultz S, Houser F and Walker J (1989) Horm Res 31/5-6: 244-249
62. Couet C, Fukagawa NK, Mathews DE, Bier DM and Young VR (1990) Am J Physiol 258:E78-E85

CHAPTER 17

THE MODULATION OF MEMBRANE ION MOVEMENTS BY CHOLESTEROL

Kanzhi Liu & Grant N. Pierce

Division of Cardiovascular Sciences
St. Boniface General Hospital Research Centre
Univ. of Manitoba, Faculty of Medicine
Winnipeg, Manitoba, Canada

THE STRUCTURE AND CHARACTERISTICS OF CHOLESTEROL IN MEMBRANE

Chemical Structure of Cholesterol

Cholesterol structure can be broken down into three important regions: 1) a relatively rigid steroid nucleus composed of four fused rings, referred to as the A,B,C and D rings. The formation of A ring can alter its conformation in the crystal state permitting more flexibility for the whole molecule (1); 2) in ring A, an alcoholic hydroxyl group is covalently attached at carbon 3; and 3) a freely movable, branched hydrocarbon side chain attached to carbon 17 of the D ring (Figure 1). It is important to point out that all three of the features of cholesterol are important for characteristic cholesterol-like performance.

Cholesterol is a weak amphiphile due to the hydrophobic rings, the hydrocarbon side chain and the polar 3ß-hydroxyl group. Being an amphiphile, cholesterol is able to fit into a phospholipid bilayer with its polar hydroxyl group protruding into the

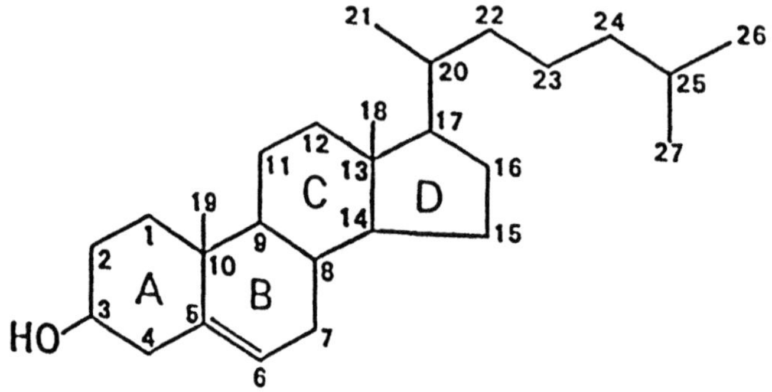

Figure 1. The chemical structure of cholesterol.

polar surface region, and the hydrophobic steroid rings oriented parallel to, and buried in, the hydrocarbon chains of the phospholipids. Cholesterol creates a physical state which is intermediate between the liquid crystal and gel phases in the surrounding lipids of the membrane, possibly by a close fit contact between the steroid and specific lipid chains (2-5). From its specific conformation in the membrane, cholesterol confers to the membrane a state of "intermediate fluidity", enabling motion of hydrocarbon chains in the gel phase while restricting motion in the liquid crystalline phase. Cell membranes usually work best when the lipid bilayer membrane is in the liquid crystalline state (6).

Distribution and Movement of Cholesterol in the Cell

Distribution of Cholesterol in Cell Membranes. The content of cholesterol in eukaryotes is entirely different among the various membranes of the cell (7). By using the enzyme cholesterol oxidase, which can oxidize all plasma membrane cholesterol without affecting intracellular cholesterol, plasma membrane cholesterol has been studied as a percentage of total cellular cholesterol. In fibroblasts and Chinese hamster ovary cells, more than 90% of total cell cholesterol was found in the plasma membrane (6). Studies in different cells also showed that 80% of total cellular cholesterol is in the plasma membrane of hepatocytes (7) and 82% of unesterified cholesterol was associated with the plasma membrane in rat ovarian granulosa cells (8). It has been noted that the distribution of the cholesterol among the membranes of mammalian cells is very uneven. Higher concentrations of cholesterol are found in the plasma membranes and secretory vesicles, moderate concentrations in the Golgi apparatus and secondary lysosomes, and lower concentrations of cholesterol in the endoplasmic reticulum and mitochondria (9).

More recently, another method has been developed that allows a determination of cholesterol content in biological membranes without isolating the membranes (10). Filipin, an antibiotic that binds specifically to membrane cholesterol, was used to treat the cell membrane and then viewed by freeze-fracture electron microscopy. The density of large pits, which bind filipin, is representative of the cholesterol concentration in the membrane (10). Using this method, cholesterol concentration in the basolateral membranes was found to be higher than in brush-border membranes in the rat intestinal mucosa (11,12). Furthermore, the cholesterol concentration in the plasma membrane of the same colonic cells is different according to the region (13). In the colonic epithelium of the guinea pig, the cholesterol concentration in the plasma membrane in the proximal colon was higher than in the distal colon. Therefore, there is a greater fluidity in the membranes of the absorbing cells in the distal colon which can be correlated to the higher absorption rates of short chain fatty acids characteristic of this region (13).

In the cardiomyocyte, filipin-cholesterol complexes were observed mainly in the sarcolemma where they occurred in patches or, sometimes, they densely populated the entire cell surface (14). Clusters of filipin-induced lesions can be seen in plasma membrane, which indicates the presence of cholesterol-rich domains surrounded by regions of much lower cholesterol content (15). Some can be seen in the sarcoplasmic reticulum and in the mitochondria, but they were very rare in the sarcoplasmic reticulum (16). A non-homogeneous distribution of cholesterol in SR was observed. The filipin-induced deformations occurred more commonly in free SR at or near the Z-region of the sarcomere than in other parts of the free SR or the junctional SR (17). Few existed in the nuclear envelope in cardiac muscle. By studying filipin-sterol complexes in the nuclear membrane, a heterogeneous distribution of cholesterol was observed (16). The outer nuclear membrane was more homogeneously distributed than the inner membrane (16).

There have been suggestions that cholesterol is not uniformly distributed between the inner and outer leaflets of a membrane as well. In the erythrocyte, the external surface of the membrane was significantly richer in cholesterol than the cytoplasmic surface (18). This observation was further confirmed by several other studies in red cell ghosts and in erythrocytes where there is a very slight enrichment of cholesterol in the outer membrane leaflet (19,20). This unequal distribution of cholesterol may be due to the difference in phospholipid composition between the two halves of the erythrocyte membranes (21). Since cholesterol had a preferential affinity for sphingolipid and phosphatidylcholine, and the phospholipids having the higher affinity for cholesterol are concentrated at the outer membrane leaflet (22,23), it is reasonable to hypothesize that this side of the membrane should contain more cholesterol than the cytoplasmic leaflet (5,22).

Cholesterol Movement in the Cell. Membrane cholesterol may come from two sources: plasma lipoproteins and cellular biosynthesis. Low density lipoprotein, a major carrier of cholesterol in the blood, enters cells through both receptor mediated (24) and non-receptor-mediated pathways (25). Mammalian cells, cultured in the presence of serum lipoproteins, acquire cholesterol necessary for growth from the uptake and lysosomal hydrolysis of LDL.

Newly synthesized cholesterol is made in the endoplasmic reticulum of cells but at short time intervals it can be detected in the Golgi apparatus, lysosomes, mitochondria and plasma membranes using radioactively labelled sterol precursors (26). Therefore, cholesterol appears to be transported very quickly from its site of synthesis to the other membranes of the cell. Several studies have traced intracellular cholesterol movement in various cell types. Transport of newly synthesized cholesterol to the cell surface seems to be vesicular and energy and temperature dependent (27). The transport of newly synthesized cholesterol from the endoplasmic reticulum to the plasma membrane ceased at 15 ^{0}C, whereas cholesterol synthesis continued in Chinese hamster ovary cells (27). When synthesis occured at 15^{0}C, the newly synthesized cholesterol was found to accumulate in the endoplasmic reticulum and in a low density, lipid-rich vesicular fraction (27). Also in Chinese hamster ovary cells, the transfer of newly synthesized cholesterol from the endoplasmic reticulum to the plasma membrane occured with a half-time of 10 minutes (28). Human skin fibroblasts and Chinese hamster ovary cells require a half-time of about 1 hour to transfer cholesterol between plasma membrane and internal membranes (29). In rat ovarian granulosa cells, newly synthesized cholesterol can reach the plasma membrane in less than 20 minutes (t1/2) (8). Cholesterol movement to the plasma membrane appears to be regulated by gonadotropins in these cells. Cholesterol movement was much slower (t1/2 approximately 2 h) in cells stimulated by FSH and androstenedione (8).

It has been proposed that the movement of cholesterol in the cell may be through several mechanisms: a) diffusion through the cytoplasm; b) protein-mediated transport, and c) vesicular transport (5). The movement of intracellular cholesterol might be unidirectional since radio-labelled cholesterol introduced into the plasma membrane did not subsequently move to the cell interior (29). One exception is that zymosterol, the precursor of cholesterol, can move rapidly and in both directions between the rough endoplasmic reticulum (a sterol-rich intracellular membrane bearing nascent cholesterol) and the plasma membrane (30). It is possible that the cholesterol transfer occurs via a pool of sterol molecules in the aqueous medium since cholesterol can readily transfer between structures such as liposomes, lipoproteins, and erythrocytes (31). However, studies in vitro have demonstrated that when membranes, lipoproteins or vesicles containing little sterol are incubated with similar structures containing a high concentration of sterol, there is a net transfer of sterol from the latter to the former so that the concentration differences are reduced (32). The large differences in the cholesterol content of mammalian cell membranes would suggest that either this transfer process does not occur intracellularly in vivo, or that it occurs too slow to equalize the concentration differences, or that its effects are outweighed by other factors. Therefore, it is unlikely that cholesterol transfer occurs by simple diffusion within the cell.

Another possible way of cholesterol transport is via an active transport in which newly synthesized cholesterol is transported against its concentration gradient to the plasma membrane and then diffuses back down the gradient to the cell interior. However, this possibility is excluded by experiments showing that for many hours at 37^{0}C, exogenous [^{14}C]-cholesterol inserted into the plasma membrane did not mix with newly synthesized [^{3}H] cholesterol (29).

The existence of sterol carrier proteins is suggested by the observation that the rate of movement of cholesterol between mitochondria and microsomal vesicles could be increased by rat liver supernatant (33). Several small soluble, cytoplasmic sterol-binding proteins have been identified in cells: a) a sterol-carrier protein (34); b) the "fatty acid binding protein", which is identical to the sterol-carrier protein described by Dempsey et al. (35); c) ligandin (36). There are several roles of these sterol-carrier proteins in the regulation of intracellular cholesterol movement such as: a) they promote the transfer of intermediates in the final stages of cholesterol biosynthesis between microsomes (37); b) they can increase the rate of transfer of cholesterol to microsomal vesicles containing acyl-CoA:cholesterol acyltransferase (34,38); c) they may be involved in the movement of cholesterol both from plasma membrane lipid droplets to mitochondria and from the outer to the inner mitochondria membrane (39,40); d) in some pathologic conditions such as hepatoma, the low abundance of the high molecular weight form of sterol carrier protein-2 from hepatoma peroxisomes may be related to the accumulation of cholesterol in the cells (41). Sterol carrier proteins may be involved in cholesterol transport between membranes not only in steroidogenic cells but also in non-steroidogenic cells (42).

The transfer of newly synthesized cholesterol to the plasma membrane may also be mediated by membrane vesicles. Various types of vesicles were present in cells such as those that appear during endocytosis or as part of the transport of proteins from the rough endoplasmic reticulum via the Golgi apparatus to the cell surface for insertion in the plasma membrane (43,44). There are some differences between vesicles involved in cholesterol transport to the plasma membrane and those involved in bulk protein transport to the cell surface (44). However, this type of membrane flux is probably too slow to account for the transfer of newly-made cholesterol to the plasma membrane (28).

The cellular mechanisms of cholesterol efflux via high density lipoprotein (HDL) receptor-mediated transport have been studied in cultured human fibroblasts and bovine aortic endothelial cells (45). HDL3 induced a rapid movement of [3H] cholesterol from a preplasma membrane compartment to the plasma membrane preceding [3H] cholesterol efflux. It was suggested that receptor binding of HDL3 could facilitate removal of cellular cholesterol from specific intracellular pools by initiation of translocation of intracellular cholesterol to the plasma membrane (45).

Pathological Alterations in Distribution & Movement of Cholesterol in Membranes. Abnormalities of membrane cholesterol have been observed in some diseases. In liver disease, an increased cholesterol/phospholipid ratio in erythrocyte membrane resulted from abnormalities in the metabolism and composition of plasma lipoproteins. Ultimately, this produced a decreased membrane fluidity (46). In obese children with either abnormal and normal plasma cholesterol levels, the fluidity of erythrocyte membranes decreased significantly which may indicate that the erythrocyte membrane responds very quickly to a modification of plasma cholesterol levels (47). In rats fed an atherogenic diet (cholesterol supplemented), the membrane fluidity of liver microsomes was also decreased (48). In the erythrocyte plasma membrane from insulin-dependent diabetes mellitus patients, the total cholesterol and phospholipid content was lowered, and the cholesterol-to-phospholipid molar ratio was significantly decreased as well (49). In the chronically diabetic rat, the cholesterol content of sarcolemmal membranes from heart was significantly increased (50). This was membrane specific because cardiac mitochondria (51) and SR (52) membrane cholesterol content was unaffected during chronic diabetes.

In neoplastic disease, a decreased membrane fluidity was observed (53). More rigid membranes produced by the decreased membrane fluidity, accompanied by increases of cholesterol, was observed in solid tumours like hepatomas (53). The synthesis of cholesterol was deregulated and control was lost early after exposure to carcinogenic events and much earlier than the appearance of malignant growth in the malignant cells (54). The ratio of cholesterol/phospholipid increases whereas the total lipid content and the degree of fatty acid unsaturation decreases with increasing growth rate of a tumour (55). In hepatomas, the regulation of cholesterol metabolism has been altered (56). For example, the content of free cholesterol in hepatomas was about twice as much as that in control liver. Esterified cholesterol increased about 2-5 times in hepatomas (56). More striking, the rates of in vivo hepatoma cholesterol synthesis were 1.5-4 times higher than those in the control liver (56). The activity of the LDL receptor was also found to decrease in renal carcinoma (57) and to increase in leukemia (58). LDL receptor expression in experimental murine tumours may be related to the in vivo uptake of lipoprotein (59). The authors suggested that the relative high in vivo uptake of LDL was determined by the elevated LDL-receptor expression in the tumours (59).

The effects of ischemia and anoxia-induced alterations in cholesterol content of sarcolemmal membranes in rat heart cells has been studied (60). Cholesterol content in cultured heart cells was significantly decreased after 1 hour anoxia. This change in sarcolemmal cholesterol content preceded the release of cytoplasmic proteins from the cells. Therefore, the authors concluded that anoxia in cardiac cells destroyed sarcolemmal structure and function secondary to the loss of cholesterol (55). An effect of ischemia on membrane cholesterol content has also been observed. The cholesterol content of sarcolemma decreased and that of mitochondria rose in the ischemic myocardium (61,62). It was suggested that cholesterol is redistributed within the ischemic heart cell. Longer durations of ischemia (15-60 minutes) in isolated perfused rat heart caused a progressive loss of cholesterol from sarcolemma and sarcoplasmic reticulum, concomitant with a significant increase in mitochondrial cholesterol content (63). Following these compositional changes, a marked increase in sarcolemmal and mitochondrial microviscosity were noticed. In contrast, others reported no change in cardiac sarcolemmal membrane cholesterol content during an in vitro ischemic challenge (64). In reperfused ischemic canine myocardium, the normally random distribution of free membranous particles underwent a dramatic change to patches of aggregation. This may have been due to the lateral movement of intramembranous proteins to lipid phases which were nonhomogeneously distributed over the membrane area (65). The mechanism may involve sarcolemmal cholesterol depletion, which can introduce semicrystalline phases, forcing protein to move to patchy areas with normal fluidity.

A decrease in membrane cholesterol content has also been observed in hereditary cardiomyopathy in hamsters (66). In contrast, others report that in genetic cardiomyopathy (67) and muscular dystrophy in hamsters, the contents of cholesterol from the heart homogenate, mitochondria and microsomal vesicles (68) are increased. However, although pretreatment of the dystrophic hamster for 20 days with verapamil prevented cardiac necrosis, it had no effect on membrane cholesterol (68).

In erythrocyte membranes from guinea pigs fed diets supplemented with 1% cholesterol, the cholesterol content increases up to 2-fold whereas the phospholipid content and the fatty acid composition of these membranes remains essentially unchanged (69). In cultured rabbit aortic smooth muscle cells (SMCs) and isolated SMC plasma membrane microsomes, incubation with medium containing LDL and cholesterol enriched phospholipid liposomes resulted in an increased cholesterol mass without affecting PL mass (70).

Niemann-Pick disease type C (NPC) is characterized by substantial intracellular accumulation of unesterified cholesterol. In the cultured NPC fibroblasts, the transport of LDL-derived [3H] cholesterol to the plasma membrane is slower than control (71). However, the rate of appearance of [3H] acetate-derived, endogenously synthesized [3H] cholesterol at the plasma membrane is the same for normal and NPC cells (71).

The Effect of Cholesterol on the Biophysical Properties of Membrane

Cholesterol Effects on Cellular Membrane Fluidity. Lipid fluidity is defined as the motional freedom of lipid molecules or lipid soluble probes within a membrane bilayer. Free cholesterol is thought to participate in the regulation of fluidity in mammalian plasma membranes in general. The molar ratio of free cholesterol to phospholipid (FC/PL) in the membrane correlates with lipid microviscosity (72). There is ample evidence that cholesterol decreases the fluidity of the plasma membrane (69-79). For example, enrichment of membrane cholesterol resulted in an increase in the local viscosity of the membranes compared to the controls in guinea pig erythrocyte (69). It is also believed that cholesterol is able to condense and rigidify the plasma membrane by restricting the random motion and mean cross-sectional area occupied by the neighbouring phospholipid acyl chain (73,74). Very recently, more data have been available to support the view that increases in membrane cholesterol content are associated with decreased membrane fluidity. This relationship has been identified in arterial smooth muscle cells (70), leukaemia cells (75), erythrocytes (46,76), rat intestinal microvillus (72,77), liver surface membranes (78) and fibroblasts (79).

Cholesterol Effects on Permeability of the Cell Membrane. Cholesterol affects the permeability of both artificial phospholipid membranes and biological membranes (80-91). Pure lipid bilayer membranes are characterized by an extremely low permeability to diffusible ions, which is principally mediated by channels or pores. However, an alteration of the passive permeability produced by the amount of membrane cholesterol has been observed. Low cholesterol content in the membrane can increase membrane permeability whereas higher cholesterol concentrations can strongly suppress the permeability (81). The effect of cholesterol on membrane permeability may be due to the effect of cholesterol on phospholipid hydrocarbon chain ordering in membranes (82,83).

In artificial phospholipid membranes, the permeability of many solutes such as glycerol, glucose and erythritol, decreases considerably upon incorporation of cholesterol into the membrane (84,85). In biological membranes, cholesterol also appears to affect permeability. Enrichment of membrane cholesterol in rat liver mitochondria decreases the passive proton permeability (86). In erythrocytes from cholesterol-fed guinea pigs, the permeability to both hydrophilic and amphiphilic non-electrolytes and to both active and the passive Na^+ efflux was decreased (87). The possible mechanism may be a tighter packing of the lipid components which in turn decreases the mobility of the permeants within the membrane matrix. A recent study also demonstrated that enrichment of cholesterol in renal cortical brush-border membranes can decrease membrane fluidity and stimulate Na^+-dependent D-glucose uptake by increasing the affinity of the carrier (88). Other studies have shown that in human erythrocytes, partial removal of membrane cholesterol led to an increase in the permeability of the membrane to glycerol (89), although the permeability to water was not changed (90). However, some results conflict with the above observations. For example, oxidation of membrane cholesterol did not produce alterations in the permeability of red blood cell membranes to cations (91). Furthermore, in control and ischemic hearts, loss of myocardial membrane cholesterol did not affect cell permeability (63). It could be that membrane permeability is unaffected until some threshold for membrane cholesterol content is passed.

THE EFFECT OF CHOLESTEROL ON MEMBRANE PHYSIOLOGICAL FUNCTIONS

Membrane-Bound Proteins

Cholesterol may affect the activity of a membrane-bound enzyme via an indirect route, a special molecular ordering effect on the membrane, or bind directly to the enzyme, and thereby alter its behaviour. The effect upon a membrane protein by cholesterol is dependent upon the particular membrane protein. Cholesterol may inhibit, stimulate, or have no effect at all (82, 92-99).

As an example of the former possibility, when human leukaemic cells were incubated with cholesterol, Na^+, K^+-ATPase activity was reduced significantly compared to that of untreated cells (92). Similar inhibitory effects of cholesterol have been observed on alkaline phosphatase of rat intestinal microvillus membrane (93), UDP-glucuronosyltransferase in the lipid bilayer (94) and bovine rhodopsin (95). Cholesterol may also stimulate membrane protein function. For example, in Novikoff hepatoma cells, a five- to sixfold increase in gap junction assembly between reaggregated cells and a 50% increase in cellular cholesterol content were observed with 20 μM added cholesterol (96). A similar effect has been observed on carrier mediated lactate transport (97) and sodium coupled glucose uptake in renal proximal tubular cells (98). However, some membrane proteins seem to be insensitive to cholesterol alterations in the membrane. It has been observed that the susceptibility of the band 3 protein of the human erythrocyte membrane to proteolytic digestion at either surface of the membrane was not altered when the membrane cholesterol was increased by 65-103% (99). Sucrase and lactase activities of the rat intestinal microvillus are also not affected by alterations in the membrane cholesterol levels as well (72).

Membrane Receptors and Cell Growth

Membrane receptors and cell growth are also subject to control by membrane cholesterol level. In liver cirrhosis patients, impairments in insulin receptor processing was associated with an increase in the erythrocyte membrane cholesterol to phospholipid molar ratio (100). Furthermore, similar changes in insulin receptor processing were observed when the molar ratio of cholesterol to phospholipid in normal erythrocytes was modified in vitro by incubation with cholesterol-rich liposomes (100). Another study has also shown that low membrane cholesterol decreased the number of insulin receptors in erythrocytes without altering the affinity (101). The possible mechanism may be that membrane cholesterol affects insulin receptor behaviour through an alteration of membrane fluidity, depending on the phase state of the membrane (101). The author further demonstrated that the increased cholesterol content in the erythrocyte membrane was responsible for the increased receptor number and insulin binding to the erythrocytes from patients suffering from alcoholic liver cirrhosis (101). The requirement for membrane cholesterol has also been observed in the acetylcholine receptor (102), ß-adrenergic receptor density (103), adenylate cyclase activity stimulated by receptors (104) and the glycophospholipid-anchored membrane receptor for 5-methylterahydrofolate (105).

Recent studies have demonstrated that cholesterol plays a central role in the control of membrane biosynthesis and cell growth. For example, myeloma cells (106) and Mycoplasma (107) have an absolute growth requirement for exogenous sterol. Most animal tissues are capable of synthesizing cholesterol de novo from acetate, and, perhaps because of this, there are relatively few mammalian cells which express an absolute requirement for an exogenous source of cholesterol in vitro. For example, rat and mouse myeloma cells which had been cultivated in serum-free medium containing cholesterol for more than 6 months still required cholesterol in vitro for growth in serum -free medium. This requirement for cholesterol could be replaced by human low density lipoprotein but not by mevalonic acid lactone (106). Similarly, in Mycoplasma, sterol starvation inhibited both lipid and protein synthesis, which suggested an important link between sterols and cell growth (107). Furthermore, cholesterol prolonged the survival of adult rat hepatocytes in a serum-free primary culture (108). The effect appeared to be due to stabilization of the plasma membrane. Cholesterol (at low levels) also stimulated unsaturated phospholipid synthesis, which was followed by RNA and protein synthesis in the same cells (109).

MODULATION OF ION MOVEMENT BY MEMBRANE CHOLESTEROL

Na^+-K^+ ATPase

Na^+-K^+ ATPase or the Na^+ pump, which is present in the plasma membrane of all cells of the body, is responsible for pumping sodium ions out of the cell in exchange for extracellular potassium ions against their respective concentration gradients. This is an energy consuming process. Therefore, this pump is responsible for maintaining the sodium and potassium concentration differences across the cell membrane as well as for establishing a negative electrical potential inside the cells (110). Na^+-K^+ ATPase activity is electrogenic because it creates an electrical potential across the cell membrane as it pumps 3 Na^+ ions to the outside of the cell for every 2 K^+ ions pumped to the interior. These properties place this enzyme in a vital role in a number of cellular processes, including maintaining normal cell volume, establishing electrical potentials along the plasma membrane and contributing to sodium cotransport systems.

Since Na^+-K^+ ATPase is an important membrane transport system which plays an essential role in cellular function and is affected by changes in membrane composition, extensive studies have been carried out to investigate the effect of membrane cholesterol on the function of Na^+-K^+ ATPase. Recent study has demonstrated that Na^+-K^+ ATPase activity from cardiac sarcolemmal membranes strongly required cholesterol (111). This sterol requirement was highly specific for cholesterol in that several cholesterol analogues with minor structural changes were unable to support Na^+-K^+ ATPase activity. These data suggested the importance of cholesterol structure in the interaction with the Na^+ pump and was suggestive of a direct interaction of cholesterol with the Na^+-K^+ ATPase in the membranes. Others have shown that the stimulatory effects of cholesterol were dependent upon the initial membrane cholesterol content. For example, when the membrane cholesterol content is lower than that found in the native membranes of bovine kidney, cholesterol could stimulate the activity of Na^+-K^+ ATPase (112). This effect had structural specificity in that some cholesterol analogues such as ergosterol and lanosterol were not as effective as cholesterol at stimulating the activity of Na^+-K^+ ATPase (112). It is very possible that cholesterol may interact directly with Na^+-K^+ ATPase as an essential positive effector rather than influence the motional order of the membrane (112,113). In other studies, depletion of membrane cholesterol had a biphasic effect in human erythrocytes. Depletion by 5-25% increased Na^+-K^+ ATPase activity by a mean of 16.1%, whereas depletion by 35-50% decreased Na^+-K^+ ATPase activity by a mean of 14.8% (120). Similar biphasic responses of Na^+-K^+ ATPase to cholesterol depletion were also found in rabbit renal medullar membrane vesicles (112) and human red blood cells (115). It was assumed that the effect of cholesterol depletion was mediated through a redistribution of the phospholipids located in the microenvironment of the protein which have been shown to play a role in the activation of the Na^+-K^+ ATPase (115). The biphasic response of Na^+-K^+ ATPase to cholesterol depletion supports the idea of an 'optimum' membrane cholesterol for maximal activity of a membrane transport system (113). In certain pathological conditions, Na^+-K^+ ATPase activity is stimulated with increasing cholesterol content. For example, in sarcolemmal membranes from hearts of rats fed a 2% cholesterol diet for 12 weeks, Na^+-K^+ ATPase activity was increased as was the sarcolemmal cholesterol/phospholipid molar ratio (116). A similar result was found in sarcolemma from genetically myopathic hamsters which display increased membrane cholesterol (118) and enhanced Na^+-K^+ ATPase activity (117).

However, several other studies dispute these findings in varying degrees. In the purified Na^+-K^+ ATPase, the enzyme activity appears to be relatively insensitive to cholesterol content (119) or could be depressed in artificial membranes by cholesterol (120). Several groups have studied cholesterol modulation of the Na^+-K^+ ATPase in membrane vesicles and intact cells. Early work noted that cholesterol inhibits delipidized and soluble Na^+-K^+ ATPase activity from rabbit kidney outer medulla (121). The Na^+-K^+ ATPase activity is only partially inhibited by cholesterol in the presence of unsaturated phospholipids, while complete inhibition can be achieved in the presence of saturated phospholipid which may indicate the interaction of cholesterol with phospholipids of different chain length and unsaturation (121). Subsequently, other experiments have shown an inhibitory effect of high membrane cholesterol content on Na^+-K^+ ATPase from human red blood cells membranes (113), rabbit erythrocyte membranes (122,126), normal clear bovine and human lens (124), canine ventricular sarcolemmal vesicles (125,126), rabbit myocardium after cholesterol rich diet (127), rat liver membranes (128) and kidney basolateral membranes (112).

Several mechanisms have been proposed for the inhibitory effect on the Na^+-K^+ ATPase by membrane cholesterol. The rigid sterol structure of cholesterol may lead to an increase in the anisotropic motional ordering of the lipid bilayer of the membrane, which in turn may lead to an increase in the ordering of the conformation of the Na^+-K^+ ATPase (85,113). Therefore, the capability of the Na^+-K^+ ATPase to undergo conformational change will be reduced, and as a result, would inhibit its function. A second possibility may involve a direct interaction between cholesterol and the Na^+-K^+ ATPase which will modulate its activity (6). Direct binding of cholesterol to various membrane proteins (such as the band 3 protein from human erythrocytes) has been reported (129). A third mechanism may be the lipid peroxidation that may occur after the enrichment of cholesterol content in the cell membrane (122). Membrane lipid peroxidation would certainly explain the decreased Na^+-K^+ ATPase activity (122).

Calcium Pumps

Calcium pumps appear to be ubiquitous components of eukaryotic cells and their major function is to translocate Ca^{2+} across membranes against a Ca^{2+} gradient. Biochemically, calcium pumps are expressed as a Ca^{2+}-ATPase since they require energy supplied from ATP hydrolysis. The best characterized Ca^{2+}-ATPases are located in the sarcoplasmic reticulum and plasma membrane.

Sarcoplasmic Reticulum Ca^{2+}-ATPase. The Ca^{2+}-ATPase is responsible for sequestering Ca^{2+} which has been released in muscle cells to stimulate myofilament contraction. Two moles of Ca^{2+} are translocated for each mole of ATP hydrolysed (130). Amongst all the proteins in the sarcoplasmic reticulum (SR), 80% are ATPase proteins responsible for calcium uptake (131). This enzyme requires a hydrophobic environment or fluid membrane to sustain its activity (132). It has been proposed that about 30 phospholipid molecules interact directly with the hydrophobic surface of the ATPase spanning the bilayer, and form the minimal lipid environment required to maintain full Ca^{2+}-ATPase activity (133).

Earlier study suggested that cholesterol does not affect the activity of sarcoplasmic reticulum calcium-ATPase from the rabbit muscle because the protein in the Ca^{2+}-ATPase specifically excludes cholesterol from the immediate vicinity of the protein (134). If cholesterol replaces phospholipids in the annulus, a potent inhibition of activity of Ca^{2+}-ATPase was observed and this inhibition can be abolished by sufficient phospholipid to reform a complete annulus (134). Therefore, the authors proposed that cholesterol may be inhibitory because it does not meet the requirement of the protein for specific chemical structures in the annulus (134). More studies agreed that cholesterol is normally excluded from direct contact with Ca^{2+}-ATPase enzyme (135-137) and excluded from other membrane enzymes as well (138). Using a reconstituted system, another study showed that a high membrane cholesterol content in proteoliposomes (cholesterol/phospholipid molar ratio = 0.3) reduced the Ca^{2+}-ATPase activity (139). However, this decline in Ca^{2+}-ATPase activity induced by cholesterol is quite small which agrees reasonably well with results derived from a reconstitution study undertaken by Warren et al (134). In rabbit skeletal muscle, the SR Ca^{2+}-ATPase was notably unaffected by cholesterol in reconstituted phosphatidylcholine : phosphatidylserine vesicles (140).

In reconstitution experiments in which phosphatidylethanolamine was used as a dominant lipid component in the membrane, cholesterol enhanced the Ca-transport function of the Ca^{2+}-ATPase (139). Therefore, under certain circumstances, cholesterol may stimulate Ca^{2+}-ATPase in the reconstituted proteoliposomes via, possibly, direct binding of cholesterol to Ca^{2+}-ATPase (139). Nevertheless, some studies have disagreed with the above conclusion that Ca^{2+}-ATPase was not sensitive to cholesterol. Warren et al showed that Ca^{2+}-ATPase activity is decreased in direct proportion to the cholesterol content of the membrane. They proposed the effect of cholesterol on Ca^{2+}-ATPase is consistent with an effect on the fluidity of the SR membrane (141,142). Recovery of Ca^{2+}-ATPase activity was achieved upon the removal of cholesterol from the SR membrane, which suggested that cholesterol does reversibly inactivate the enzyme (141). Similar qualitative results have been interpreted quite differently by other investigators. In bovine arterial smooth muscle cells, increased cholesterol content of the cells through liposome fusion resulted in a reduction of SR Ca^{2+}-ATPase activity (143). Since the alteration of membrane cholesterol content was paralleled by a decrease in the Ca^{2+}-ATPase activity of SR (142), and the inhibition was accompanied by a loss in enzyme sensitivity to calmodulin (144), they proposed that the change of Ca^{2+}-ATPase activity was caused by a direct effect of cholesterol on the enzyme (143). Other studies have had similar conclusions that the inhibitory effect of cholesterol on the Ca^{2+}-ATPase was due to the direct interaction of cholesterol on the enzyme (145) and not a change of membrane fluidity induced by cholesterol (146,147).

The effect on Ca^{2+}-ATPase activity of an atherogenic (1% cholesterol) diet fed to swine (148) or rabbits (150) has been studied. In swine fed an atherogenic diet, total Ca^{2+} uptake was decreased as was the altered cholesterol/phospholipid ratio without having effect on Ca^{2+}-ATPase activity (148). In rabbits fed an atherogenic diet for 12 weeks, the rate of SR calcium binding was unchanged but calcium uptake was increased (150).

Plasma Membrane Ca^{2+}-ATPase. This enzyme acts to pump Ca^{2+} out of the cell to regulate the intracellular $[Ca^{2+}]$. Two calcium ions are transported across the membrane for each ATP hydrolysed. Membrane fluidity might be an important determinant of Ca^{2+}-ATPase activity. A 50% increase in Ca^{2+}-ATPase activity was observed concomitant with increased membrane fluidity (151).

There has been a great controversy for several years concerning an indirect versus a direct effect of cholesterol upon the muscle Ca^{2+}-ATPase. In a cardiac sarcolemmal preparation, cholesterol has an inhibitory effect upon the Ca^{2+}-ATPase activity (125,144,145). Figure 2 displays the effect of cholesterol enrichment on the sarcolemmal Ca^{2+}-pump (125). The Ca^{2+} accumulating ability of the Ca^{2+}-ATPase is inhibited 40% by cholesterol enrichment at all incubation times (125). The same inhibitory effect has been observed in sarcolemmal Ca^{2+}-ATPase activity in rats fed 2% cholesterol diet for 24 weeks (116). Not only can the incorporation of cholesterol into sarcolemma inhibit Ca^{2+}-ATPase activity, depletion of cholesterol can induce a much more pronounced, stimulatory effect (144). This may indicate that there is a critical concentration of cholesterol in the sarcolemma which normally modulates ATPase activity (144). Since the slight changes in E_a (Arrhennius activation energy) and transition temperatures that cholesterol produced did not have an important effect upon the fluidity of the membranes, the authors suggested that cholesterol may have a direct effect of cholesterol upon the sarcolemmal Ca^{2+}-ATPase (144). The authors also suggested that there is a possibility of a slow exchange of cholesterol between the tightly bound lipid surrounding the Ca^{2+}-ATPase and the bulk lipid of the

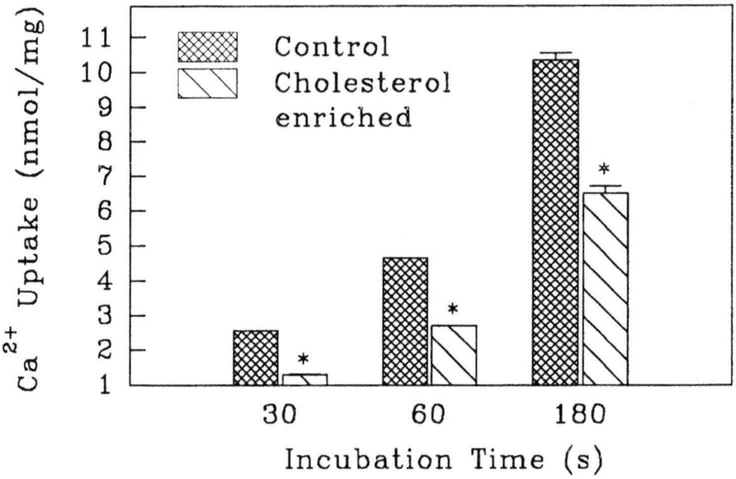

*Figure 2. ATP-dependent Ca^{2+} uptake in control and cholesterol-enriched sarcolemma vesicles. (n=4). *: $p<0.05$ versus control values. Data adapted from reference 125.*

sarcolemma (144). Another study demonstrated that cholesterol enrichment of the sarcolemmal membrane was associated with a loss of calmodulin sensitivity of Ca^{2+} ATPase (145).

Ca^{2+} uptake via the sarcolemmal Ca^{2+} pump was also inhibited if the endogenous membrane cholesterol was oxidized with cholesterol oxidase (149). Cholesterol oxidase inhibited ATP-dependent Ca^{2+} uptake by 22%, 36%, and 52% at 15, 30, and 60 seconds of reaction time, respectively (149). However, it was not clear if the inhibitory effect was a direct action on the Ca^{2+} pump or if it was primarily a reflection of a non-specific increase in membrane permeability.

The Calcium Channel

The Ca^{2+} channel is a protein which facilitates the selective movement of Ca^{2+} across a membrane down its concentration gradient. There are at least two types of Ca^{2+} channels: one is potential dependent and the other is receptor-operated (152). The first report of the effect of membrane cholesterol on the calcium channel was carried out in human red blood cells. Calcium influx was strongly influenced by the cholesterol content of the plasma membrane (153). Calcium influx through the calcium channel increased when the membrane was enriched with cholesterol using liposomes, and conversely, cholesterol depletion decreased influx (153,154).

Arterial smooth muscle cells are the primary cell type which are enriched with unesterified cholesterol in an atherosclerosis plaque (155). The entry of extracelluar calcium is a major regulatory factor in a number of vascular smooth muscle functions. Vascular smooth muscle Ca^{2+} content may also be important in plaque development. Therefore, it is important to understand the interactions of membrane cholesterol and intracellular calcium in arterial smooth muscle cells. Calcium channel activity in arterial smooth muscle cells was increased upon the enrichment of membrane cholesterol by exposure to either cholesterol-rich liposomes or LDL (low density lipoprotein) (70,140,143,156,157). For example, in cultured rabbit aortic smooth muscle cells, cholesterol enrichment of cell membranes was associated with a significant increase in the unstimulated calcium influx rate which is sensitive to channel blockers (70). A similar observation was reported for intact carotid arterial segments after perfusion with cholesterol-rich liposomes (157). The cholesterol-induced calcium influx was reversible after exposure to medium containing 10% serum (no liposome) for 4 days (70). Since lipid dynamics in the arterial smooth muscle cells membrane can be altered by cholesterol enrichment, the authors proposed that cholesterol enrichment in smooth muscle cell membranes unmasked a new, or otherwise silent calcium channel (present, but inactive under normal conditions) with pharmacological characteristics similar to L-type channels, i.e. dihydropyridine sensitivity (70).

The mechanism by which cholesterol-rich liposomes and cholesterol enrichment increase calcium currents in vascular smooth muscle cells was investigated by using whole-cell patch clamp recording (140). Enrichment of membrane free cholesterol in arterial smooth muscle cells resulted in a gradual increase in the L-type current over 20 hours and a plateau ($73 \pm 7\%$ increase over basal) between 20 and 32 hours (140). The authors demonstrated that the effect of cholesterol enrichment was not due to either an increase in L-type channel expression (i.e. increased synthesis, decreased degradation) or availability (e.g., recruitment), since there was no increase in the maximum L-type current stimulated by (+)-PN-202-791 in cholesterol-enriched cells (140). Instead, a shift in the inactivation potential and a decrease in sensitivity to dihydropyridine antagonists (140), suggested that cholesterol enrichment may be associated with a change in the functional properties of the L-type calcium channel. Therefore, altered L-type channel function might be a result of a change in membrane physical properties such as fluidity and thickness which are modulated by cholesterol (140).

The effects of LDL on calcium transients in isolated rabbit cardiomyocytes have been examined (159). Intracellular [Ca^{2+}] was measured spectrofluorometrically using the Ca^{2+} indicator dye fura-2. As shown in Figure 3, a slow ($\geq$ 30 min) but significant increase (2-fold) in the cellular Ca^{2+} transient was observed after incubation of cardiomyocytes with $\geq$ 1 mg of LDL cholesterol/ml perfusion medium (159). The effects correlated well in a time-dependent manner with LDL-associated cholesterol deposition in the cardiomyocyte. Lowering the extracellular Ca^{2+} concentration removed the stimulatory effect of LDL on the Ca^{2+} transient (Fig. 4). Therefore, the mechanism of LDL inducing the calcium transient increment was

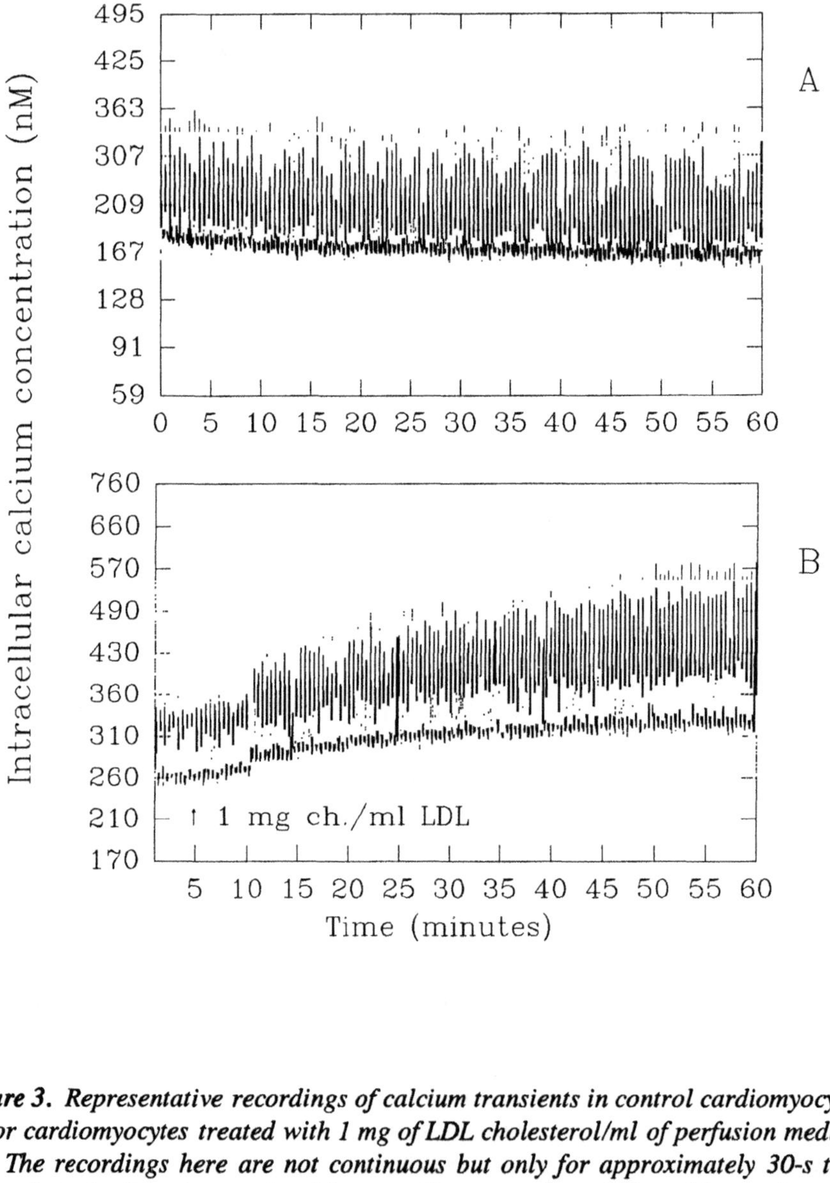

Figure 3. *Representative recordings of calcium transients in control cardiomyocytes (A) or cardiomyocytes treated with 1 mg of LDL cholesterol/ml of perfusion medium (B). The recordings here are not continuous but only for approximately 30-s then stopped. At the 5th min, another 30-s recording was taken, followed by another 30-s recording at 10th min, etc. Note the stability in the* Ca^{2+} *transient recording in the control cell (A) over the entire measurement period and the increase in the size of the* Ca^{2+} *transient induced in another cell after treatment with LDL (B). Data adapted from reference 159.*

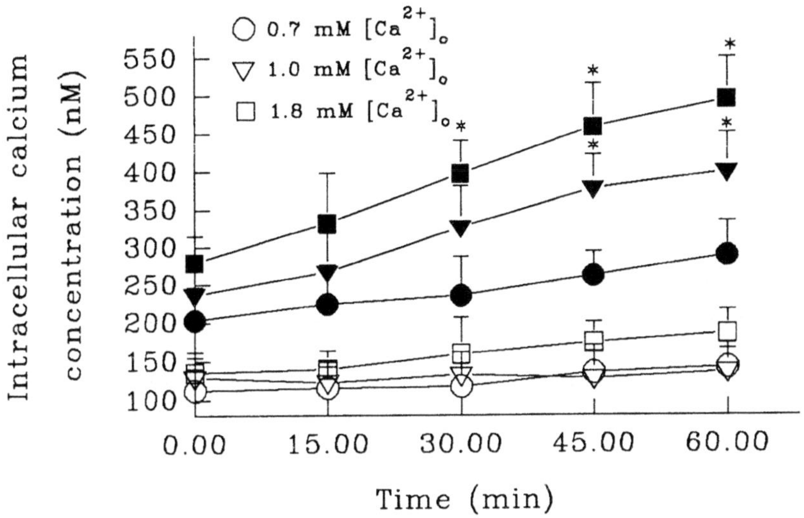

*Figure 4. Influence of extracellular calcium concentrations on the effects of LDL enhancing calcium transients in rabbit ventricular myocytes. Value of systolic (filled symbols) and diastolic (open symbols) Ca^{2+} concentration represent the mean ± S.E. from four experiments using different cells. *:$P < 0.05$ versus same extracellular Ca^{2+} group at 0 time point. Data adapted from reference 159.*

suggested to be due to a stimulation of a trans-sarcolemmal Ca^{2+} transport pathway.

A general insensitivity to various drugs was found after cardiomyocytes were treated with LDL (159). Ryanodine, a plant alkaloid used to probe the role of Ca^{2+} release from the SR, could strongly depress the Ca^{2+} transient in control but not LDL-treated cardiomyocytes. Similar results were observed when other drugs, such as nicardipine (calcium channel antagonist) or dichlorobenzamil (inhibitor of Na^+/Ca^{2+} exchange) or BAYK 8644 (calcium channel stimulator), were applied to control and LDL-treated cardiomyocytes. Cholesterol enrichment of the membrane may slow or inhibit drug interactions with membrane receptors (160). Significantly, these results suggested a potential for hypercholesterolemic patients to possess a general insensitivity to drug therapy (159).

Oxidized LDL (oxLDL) can also induce a significant increment in intracellular calcium in the isolated rabbit cardiomyocytes during relatively short incubation times (≥ 16 min) (161). Systolic Ca^{2+} transients were significantly increased after treatment of cardiomyocytes with 100 μg of oxLDL cholesterol/ml. Diastolic $[Ca^{2+}]$ was only affected if the [oxLDL] increased ten-fold (161). The effects of 4 μM nicardipine on the Ca^{2+} transient were more potent in oxLDL-treated cells than control cells, although the calcium transient from both groups could be blocked (161). Lowering the extracellular [Ca^{2+}] removed the stimulatory effects of oxLDL. These data suggest that the calcium channel characteristics were modified by oxLDL.

Na^+-H^+ exchange

The plasma membrane Na^+-H^+ exchanger is a membrane-bound glycoprotein which exists in a wide range of cell types. It catalyses the coupled transmembrane exchange of extracellular sodium (Na^+) for intracellular H^+ without consuming any ATP (162). The main physiological roles of the Na^+-H^+ exchanger are regulating intracellular pH, controlling cell growth and proliferation, and adjusting cell volume (163,164). In some pathological states, including essential hypertension, ischemia/reperfusion, diabetic cardiomyopathy and diabetic nephropathy, Na^+-H^+ exchange also plays an important role (165-168).

Earlier studies have shown that the fluidity of rat colonic brush-border membranes can influence Na^+-H^+ exchange activity. As discussed previously, fluidity changes commonly occur following alterations in membrane cholesterol (169,170). When cultured human lymphoblasts were incubated with liposomes of varying cholesterol concentration, alterations of Na^+-H^+ exchange were observed (171). Enrichment of lymphoblast membrane cholesterol resulted in an inhibition of antiporter activity, whereas depletion of membrane cholesterol enhanced the activity of Na^+-H^+ exchange (171). The mechanism of the cholesterol effect on Na^+-H^+ exchange remains unclear. One recent study on the effect of cholesterol on human platelets found cholesterol enrichment resulted in an increase in phospholipase A_2 (172). They proposed that membrane-modulated phospholipase A_2 may act as a molecular-switch for the regulation of Na^+-H^+ exchange in cholesterol-enriched platelets (172).

Na^+-Ca^{2+} exchange

Another pathway contributing to trans-plasma membrane Ca^{2+} flux is Na^+-Ca^{2+} exchange. It has been found in various cells including epithelia, nerves and muscles. The Na^+-Ca^{2+} exchanger was first hypothesized by Reuter and Seitz 20 years ago (173), and was then identified in isolated cardiac sarcolemmal membranes by Reeves and Sutko (174). This is a carrier-mediated process coupling the flux of Na^+ in one direction to the movement of Ca^{2+} in the opposite direction (175). The stoichiometry of Na^+-Ca^{2+} exchange is three Na^+ for each Ca^{2+} transported, making it electrogenic and sensitive to membrane potential (176). One of putative roles of Na^+-Ca^{2+} exchange is to maintain the intracellular concentration of Ca^{2+} in cells at a resting level (177).

All studies concerning the influence of membrane cholesterol on Na^+-Ca^{2+} exchange have used cardiac sarcolemmal vesicles (111,125,149,179,180). Cholesterol enrichment of sarcolemmal vesicles resulted in a significant increase in the initial rate of Na^+-Ca^{2+} exchange (125). There is a very good general correlation between the cholesterol content of sarcolemmal membranes and Na^+-Ca^{2+} exchange (Fig. 5). It was suggested that a cholesterol-rich annulus surrounding the Na^+-Ca^{2+} exchange protein may modulate its function (125). This was confirmed later in a study investigating the thermotropic transition points of Na^+-Ca^{2+} exchange in proteoliposomes reconstituted with varying cholesterol concentrations (181).

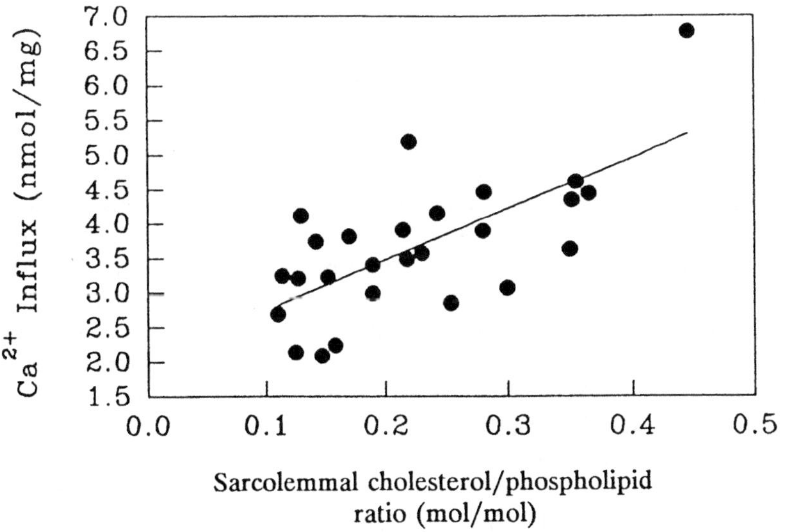

Figure 5. *Na^+_i-dependent Ca^{2+} influx rate as a function of membrane cholesterol/phospholipid ratio. Sarcolemmal membrane vesicles were incubated overnight with liposomes of various cholesterol /phospholipid ratios. Ca^{2+} uptake occurred for 1.5 s at $[Ca^{2+}]$ = 10 μM. Data adapted from reference 125.*

In situ oxidation of membrane cholesterol with cholesterol oxidase resulted in an inhibition of Na^+-Ca^{2+} exchange (125,149). This inhibition was particularly prominent (67%) during the early linear part of the reaction (0.5 seconds) (149). Oxysterols are distributed quite differently in the membrane. They are less polar than cholesterol and occupy a higher molecular area in the membrane (178). The effect on Na^+-Ca^{2+} exchange produced by oxysterols may due to the disturbance of cholesterol-rich domains in the membrane by the oxidation reaction (149). This would further support the previous conclusions that a cholesterol rich annulus may be associated with the Na^+-Ca^{2+} exchange protein (125). If large amounts of cholesterol were oxidized in the membrane, Na^+-Ca^{2+} exchange could be completely eliminated. However, this was primarily due to an elimination of any effective permeability barrier in the membrane (149).

In a study of reconstituted phosphatidylserine/phosphatidylcholine vesicles, no exchange was observed unless significant amounts (20% by weight) of cholesterol were also present (179). It is interesting to note that this amount of cholesterol is similar to endogenous levels of sarcolemmal cholesterol (179). This observation was confirmed by another study (180). This sterol requirement for Na^+-Ca^{2+} exchange is highly specific for cholesterol. Several cholesterol analogues with minor structural modifications were unable to substitute for cholesterol in supporting exchange activity (111).

SUMMARY

Cholesterol is one of the major components in the cell membrane. Being an amphiphile, cholesterol creates an intermediate physical state between the liquid crystal and gel phases in the surrounding lipids of the membrane. The distribution of cholesterol among the various membranes of the cell is entirely different. The majority of cholesterol is located in the plasma membrane. In some cells (like colonic epithelial cells), cholesterol distribution in the plasma membrane varies according to their function. Even in the same membrane, there is an uneven distribution between the inner and outer leaflets. Several mechanisms may account for the intracellular movement of cholesterol among membranes, such as diffusion, vesicular transport and sterol carrier proteins. Under certain pathological conditions, changes in the ratio of cholesterol-to-phospholipids have been observed. These alterations could lead to an decrease in membrane fluidity. A redistribution of membrane cholesterol is another common alteration during pathological conditions like ischemia.

Cholesterol affects two major biophysical characteristics of membrane: fluidity and permeability. Cholesterol decreases the fluidity of the plasma membrane, condensing and rigidifying the plasma membrane. In both artificial phospholipid membranes and biological membranes, the permeabilities to both hydrophillic and amphiphilic non-electrolytes and some passive ion movements are decreased upon the enrichment of membrane cholesterol. The effects of cholesterol on the activity of membrane-bound protein is variable depending upon the particular membrane protein examined. Membrane cholesterol also has a controlling effect on membrane receptors and cell growth.

Cholesterol has been shown to regulate membrane ion movements. In some cases, some ion transport proteins show an absolute dependence on cholesterol for activity whereas some may not. Cholesterol is necessary for maintaining Na^+-K^+ ATPase activity but both stimulatory and inhibitory effects have been described. Cholesterol appears to exert a stimulatory effect on Na^+-Ca^{2+} exchange and L-type Ca^{2+} channel function but has an inhibitory effect on Ca^{2+} pumps and Na^+-H^+

exchange. Overall, cholesterol deposition will enhance the Ca^{2+} transient in cardiomyocytes. Oxidation of membrane cholesterol will also result in significant changes in Na^{+}-Ca^{2+} exchange, Ca^{2+} pump activity and passive Ca^{2+} permeability characteristics. The Ca^{2+} transient of cardiomyocytes is also strikingly affected after exposing the cell to oxidized LDL.

High circulating cholesterol levels in human plasma are a relatively common pathological condition in industrialized countries today. It is now clear that cell function is altered by changes in membrane cholesterol content. It is becoming more and more likely that cholesterol , in addition to its well known role in atherogenic plaque development, may also exert a direct effect on cell function throughout the body which may, in itself, represent an important pathological process.

Acknowledgment

This work was supported by the Medical Research Council of Canada. K.Z. Liu was a Trainee of the Manitoba Health Research Council. G.N. Pierce was a Scientist of the Medical Research Council of Canada.

REFERENCES

1. Duax WL, Griffin JF, Rohere DC and Weeks CM (1980) Lipids 15: 783-792
2. Small DM, Bourges M and Dervichian DG (1966) Biochim Biophys Acta 125: 563-580
3. Rand RP and Luzatti V (1968) Biophys J 8:125-137
4. Chapman D (1982) In: Biological Membranes, Vol.4, Academic Press, New York.
5. Lange Y (1986) In: Scanu AM and Spector AA, Marcel Dekker (eds) Biochemistry and Biology of Plasma Lipoproteins. INC. New York.
6. Yeagle PL (1989) FASEB J 3:1833-1842
7. Lange Y and Ramos B (1983) J Biol Chem 258:15130-15134
8. Lange Y, Schmit VM and Schreiber JR (1988) Endocrinology 123:81-86
9. Green C (1977) In: Goodwin TW, (ed) Biochemistry of Lipid II University Park Press, Baltimore, pp 101-152
10. Miller RG (1984) Cell Biol Int Rep 8:519-535
11. Chapelle S and Gilles-Baillien M (1983) Biochim Biophys Acta 753:269-271
12. Luciano L, Reale E, Konitz H, Boseck U and Boseck S (1989) J Histochem Cytochem 37:1421-1425
13. Luciano L, Konitz H and Reale E (1989) Cell Tissue Res 258:339-37
14. Sommer JR, Dolber PC and Taylor I (1982) J Ultras Res 80:98-103
15. Severs NJ (1981) Europ J Cell Biol 25:289-299
16. Alroy J, Merk FB, Goyal V and Ucci A (1981) Biochim Biophys Acta 649:239-243

17. Severs NJ (1982) Cell Tissue Res 224:613-624
18. Fisher KA (1976) Proc Natl Acad Sic USA 73:173-177
19. Lange Y, Culter HB and Steck TL (1980) J Biol Chem 255:9331-9337
20. Lange Y and Slayton JM (1982) J Lipid Res 23:1121-1127
21. Op den Kamp JAF (1979) Annu Rev Biochem 48:47-71
22. Demel RA, Jansen JWCM, Van Dijick PWM and Van Deenen LLM (1977) Biochim Biophys Acta 465:1-10
23. Slotter JP and Bierman EL (1988) Biochem J 250:653-658
24. Brown MS, Kovanen PT and Goldstein JL (1981) Science 212:628-635
25. Attie AD, Pittman RC and Steinberg D (1982) Hepatology 2:269-281
26. Lange Y and Matthies HJG (1984) J Biol Chem 259:14624-14630
27. Kaplan MR and Simoni RD (1985) J Cell Biol 101:446-453
28. DeGrella RF and Simoni RD (1982) J Biol Chem 257:14256-14262
29. Huang CH (1977) Lipids 12:348-356
30. Lange Y, Echevarria F and Steck TL (1991) J Biol Chem 266:21439-21443
31. McLean LR & Phillips MC (1981) Biochemistry 20:2893-2900
32. Cooper RA, Arner EC, Wiley JS & Shattil SJ (1975) J Clin Invest 55:115-126
33. Bell FP (1975) Biochim Biophys Acta 398:18-27
34. Gavey KL, Noland BJ & Scallen TJ (1981) J Biol Chem 256:2993-2999
35. Dempsey ME, McCoy KE, Baker HN, Vafiadou-Dimitriadou A, Lorsbach T and Howard JB (1981) J Biol Chem 256:1867-1873
36. Tiltaman AJ (1984) J Pathol 142:61-66
37. Ishibashi T and Bloch K (1981) J Biol Chem 256:12962-12967
38. Trzaskos JM and Gaylor JL (1983) Biochim Biophys Acta 751:52-65
39. Chanderbhan R, Noland BJ, Scallen TJ and Vahouny GV (1982) J Biol Chem 257:8928-8934
40. Conneely OM, Headon DR, Olson CD, Ungar F and Dempsey ME (1984) Proc Natl Acad Sci USA 81:2970-2974
41. Lyons HT, Kharroubi A, Wolins N, Tenner S, Chanderbhan RF, Fiskum G and Donaldson RP (1991) Arch Biochem Biophys 285:238-245
42. van Haren L, Teerds KJ, Ossendorp BC, van Heusden GP, Orly J, Stocco DM, Wirtz KW and Rommerts FF (1992) Biochim Biophys Acta 1124:288-296
43. Steinman RM, Mellman IS, Muller WA and Cohn ZA (1983) J Cell Biol 96: 1-27
44. Voelker DR (1991) Microbiol Rev 55:543-560
45. Slotte JP, Oram JF and Bierman EL (1987) J Biol Chem 262:12904-12907
46. Cooper RA (1977) N Engl J Med 297:371-377
47. Ferretti G, Curatola G, Bertoli E, Catassi C and Giorgi PL (1991) Biochem Med Metab Biol 46:1-9
48. Muriana FJ, Ruiz-Gutierrez V and Vazquez CM (1992) Biochimie 74:551-556
49. Baldini P, Sandra I, Lambert-Gardini S, Spinedi A and Luly P (1989) Diabetes 38:825-831
50. Pierce GN and Dhalla NS (1983) Am J Physiol C241-C247
51. Pierce GN and Dhalla NS (1985) Can J Cardiol 1:48-54

52. Ganguly PK, Pierce GN, Dhalla KS and Dhalla NS (1983) Am J Physiol 244:E528-E535
53. Shinitzky M and Inbar M (1976) Biochim Biophys Acta 433:133-149
54. Coleman PS and Lavietes BB (1981) Crit Rev Biochem 11:341-393
55. Bartoli GM, Bartoli S, Galeotti T and Bertoli E (1980) Biochim Biophys Acta 620:205-211
56. Erickson SK, Cooper AD, Barnard GF, Havel CM, Watson JA, Feingold KR, Moser AH, Hughes-Fulford M and Siperstein MD (1988) Biochim Biophys Acta 960:131-138
57. Clayman RV, Bilhartz LE, Buja LM, Spady DK and Dietschy JM (1986) Cancer Res 46:2958-2963
58. Vitols S, Gahrton G, Bjorkholm M and Peterson C (1985) The Lancent 23:1150-1153
59. Lombardi P, Norata G, Maggi FM, Canti G, Franco P, Nicolin A and Catapano AL (1989) Biochim Biophys Acta 1003:301-306
60. Altona JC and Laarse van der A (1982) Cardiovasc Res 16:138-143
61. Laarse van der A (1987) In: Stam H and Vusse van der GJ (eds) Lipid metabolism in the normoxic and ischaemic heart. CIP-Kurztitelaufnahme der Deutschen Bibliothek, pp 137-145
62. Rouslin W, Macgee J, Gupte S, Wesselman A and Epps DE (1982) Am J Physiol 242:H254-H259
63. Venter H, Genade S, Mouton R, Huisamen B, Harper IS and Lochner A (1991) J Mol Cell Cardiol 23:1271-1286
64. Bersohn MM, Philipson KD and Fukushima JY (1982) Am J Physiol 242:C288-C295
65. Ashraf M and Halverson CA (1977) Am J Pathol 88:583-588
66. Owens K and Weglicki WB, Sonnenblick EH and Gertz EW (1972) J Mol Cell Cardiol 4:229-236
67. Szollar L, Pucsok J, Szelenyi I and Sos J (1973) In: Bajusz E and Rona G (eds) Recent Advances in Studies on Cardiac Structure and Metabolism. Vol. 2. University Park Press, Baltimore, pp 313-320
68. Slack BE, Boegman RJ, Downie JW and Jasmin G (1980) J Mol Cell Cardiol 12:179-185
69. Kroes J, Ostwald M and Keith A (1972) Biochim Biophys Acta 274:71-74
70. Gleason MM, Medow MS and Tulenko TN (1991) Circ Res 69:216-227
71. Liscum L, Ruggiero RM and Faust JR (1989) J Cell Biol 108:1625-36
72. Brasitus TA, Dahiya R, Dudeja PK and Bissonnette BM (1988) J Biol Chem 263:8592-8597
73. Myant NB (1981) In: Myant NB (ed) Steroids in Biological Membranes. London, William Heimnemann Medical Books Ltd. pp 161-396
74. Shinitzky M and Inbar M (1976) Biochim Biophys Acta 433:133-149
75. Wainberg RH, Walden TLJr, Stebler BA and Yatvin MB (1991) Prostaglandins 41:501-513

76. Cantin B, Brun LD, Gngne C, Murthy MR, Lupien PJ and Julien P (1992) Biochim Biophys Acta 1139:25-31
77. Schwarz SM, Bostwick HE and Medow MM (1988) Am J Physiol 254:G687-G694
78. Davis RA, Kern F, Showalter R, Sutherland E, Sinensky M and Simon FR (1978) Proc Natl Sci USA 75:4130-4134
79. Incerpi S, Jefferson JR, Wood WG, Ball WJ and Schroeder F (1992) Arch Biochem Biophys 298:35-42
80. Sommer JR, Dolber PC and Taylor I (1982) J Ultrastruct Res 80:98-103
81. Corvera E, Mouritsen OG, Singer MA and Zuckermann MJ (1992) Biochim Biophys Acta 1107:261-270
82. Yeagle PL (1985) Biochim Biophys Acta 822: 267-287
83. Yeagle PL (1992) In: Gold P, Grover S and Roncari DAK (eds), Cholesterol and Coronary Heart Disease. Parthenon Publishing Group Ltd. Park Ridge, New Jersey.
84. de Gier J, Mandersloot JG and van Deenen LL Van (1968) Biochim Biophys Acta 150:666-675
85. Demel RA, Bruckdorfer KR and Deenen LLMv (1972) Biochim Biophys Acta 255:321-330
86. Baggetto LG, Clottes E and Vial C (1992) Cancer Res 52:4935-4941
87. Kroes J and Ostwald R (1971) Biochim Biophys Acta 249:647-650
88. Bruckdorfer KR, Demel RA, de Gier J and van Deenen LLM (1969) Biochim Biophys Acta 183:334-345
89. Shaafi RI, Gary Bobo CM and Solomon Ak (1969) Biochim Biophys Acta 173:141-142
90. Imai Y, Scoble JE, McIntyre N and Owen JS (1992) J Lipid Res 33:473-483
91. Cardenas AM, Cortes MP, Fernandez E and Pena W (1992) Toxicology 72:145-151
92. Diomede L, Bianchi R, Modest EJ, Piovani B, Bubba F and Salmona M (1992) Biochem Pharmacol 43:803-807
93. Brasitus TA, Dahiya R, Dudeja PK and Bissonnette BM (1988) J Biol Chem 263:8592-8597
94. Rotenberg M and Zakim D (1991) J Biol Chem 266:4159-4161
95. Mitchell DC, Straume M, Miller JL and Litman BJ (1990) Biochemistry 822: 9143-9149
96. Meyer R, Malewicz B, Baumann WJ and Johnson RG (1990) J Cell Sci 96: 231-238
97. Grunze M, Forst B and Deuticke B (1980) Biochim Biophys Acta 600:860-868
98. Vrtovsnik F, el-Yandouzi EH, Le-Grimellec C and Friedlander G (1992) Kindey Int 41:983-991
99. Kirby CJ and Green C (1980) Biochim Biophys Acta 598:422-425
100. Peterson SW, Angelico M, Masella R, Foster K, Gandin C and Cantafora A (1992) Ital J Gastroenterol 24:65-71
101. Meahara K (1991) Igaku Zasshi 82:586-602

102. Criado M, Eibl H and Barrantes FJ (1982) Biochemistry 21:3622-3629
103. O'Connor SW, Scarpace PJ and Abrass IB (1984) Biochim Biophys Acta 778:497-502
104. Whetton AD, Gorden LM and Houslay MD (1983) Biochem J 210:437-449
105. Rothberg KG, Ying YS, Kamen BA and Anderson RG (1990) J Cell Biol 111:2931-2938
106. Li JL, Li YJ, Chao YJ, Lin LX, Ouyang MH, Peng YB and Chang WS (1991) Cytobios 68:15-22
107. Dahl JS, Dahl CE and Block K (1981) J Biol Chem 256:87-91
108. Miyazaki M, Bai L and Namba M (1991) Res Exp Med Berl 191:77-83
109. Dahl JS and Dahl CE (1983) Proc Natl Acad Sci USA 80:692-696
110. Wallick ET, Lane KL and Schwartz A (1979) Annu Rev Physiol 41:397-411
111. Vemuri R and Philipson KD (1989) J Biol Chem 264:8680-8685
112. Yeagle PL, Young J and Rice D (1988) Biochemistry 27:6449-6452
113. Yeagle PL (1983) Biochim Biophys Acta 727:39-44
114. Lucio FJ, Hendry BM and Ellory JC (1991) Exp Physiol 76:437-443
115. Giraud F, Claret M, Bruckdorfer R and Chailley B (1981) Biochim Biophys Acta 647:249-258
116. Moffat MP and Dhalla NS (1985) Can J Cardiol 1:194-200
117. Sulakhe PV, Dhalla NS (1973) Exp Mol Pathol 18:100-111
118. Slack BE, Boegman RJ, Downie JW and Jasmin G (1980) J Mol Cell Cardiol 12:179-185
119. De Pont JHHM, Peters WHM & Bonting SL (1983) Current Topics in Membranes and Transport 19:163-166
120. Anner BM (1985) Biochim Biophys Acta 822:319-334
121. Kimelberg HK and Papahadjopoulos D (1974) J Biol Chem 249:1071-1080
122. Uysal M (1986) Int J Vitam Nutr Res 56:307-310
123. Kamber E and Kopeikina-Tsiboukidou L (1986) Z Naturforsch C 41: 301-309
124. Baghieri S and Garner MH (1992) Curr Eye Res 11:459-467
125. Kutryk MJB and Pierce GN (1988) J Biol Chem 263:13167-13172
126. Grupp I, Lee WB, Lee CO, Pecker MS and Schwartz A (1985) J Physiol Lond 360:149-160
127. Kutscherskij E, Gunther J and Methley E (1983) Experientia 40:812-815
128. Field FJ, Albright E and Mathur SN (1986) Gastroenterology 91:297-304
129. Klappauf E and Schubert D (1977) FEBS Lett 80:423-425
130. Hasselblach W and Makinose M (1962) Biochem Biophys Res Commun 7:132-136
131. Warren GB, Toon PA, Birdsall NJM, Lee AG and Metcalfe JC (1974) Proc Natl Acad Sci USA 71:622-626
132. Hidalgo C, Thomas DD and Ikemoto N (1978) J Biol Chem 253:6879-6887
133. Warren GB, Toon PA, Birdsall NJM, Lee AG and Metcalfe JC (1974) Biochemistry 13:5501-5507
134. Warren GB, Houslay MD, Metcalfe JC and Birdsall NJM (1975) Nature 255:684-687

135. Johannson A, Keithly CA, Smith BA and Metcalfe JC (1981) Biochem J 196:505-511
136. Simmonds AC, East JM, Jones OT, Rooney EK, McWhirter J and Lee AG (1982) Biochim Biophys Acta 693:398-406
137. Almeida LM, Vaz WLC, Zachariasse KA and Madeira VMC (1984) Biochemistry 23:4714-4720
138. Houslay MD and Stanley KK (1982) In: Dynamics of Biological Membranes, Publisher: Chichester. Wiley, New York
139. Cheng KH, Lepock JR, Hui SW and Yeagle PL (1986) J Biol Chem 261: 5081-5087
140. Sen L, Bialecki RA, Smith E, Smith TW and Colucci WS (1992) Circ Res 71:1008-1014
141. Madden TD, Chapman D and Quinn PJ (1979) Nature 279:538-541
142. Madden TD, King MD and Quinn PJ (1981) Biochim Biophys Acta 641: 265-269
143. Zhou Q, Jimi S, Smith TL and Kummerow FA (1991) Biochim Biophys Acta 1085:1-6
144. Ortega A and Mas Oliva J (1984) Biochim Biophys Acta 773:231-236
145. Ortega A and Mas Oliva J (1986) Biochim Biophys Res Commun 139: 868-874
146. Abeywardena MY, McMurchie EJ, Russell GR, Sawyer WH and Charnock JS (1984) Biochim Biophys Acta 776:48-59
147. East JM, Jones OT, Simmonds AC and Lee AG (1984) J Biol Chem 259:8070-8071
148. Jacobson MS, Ambudkar IS, Young EP, Naseem SM, Heald FP and Shamoo AE (1985) Atherosclerosis 55:81-91
149. Kutryk MJB, Maddaford TG, Ramjiawan B and Pierce GN (1991) Circ Res 68:18-26
150. Gunther J and Will H (1979) Acta Biol Med Germ 38:1467-1472
151. Strittmatter WJ, Hirata F and Axelrod J (1979) Biochim Biophys Res Commun 88:147-153
152. Triggle DJ and Swamy VC (1983) Circ Res 52:1-17
153. Locher R, Neyses L, Stimpel M, Kuffer B and Vetter W (1984) Biochem Biophys Res Commun 124:822-828
154. Rosier F, M'Zali H and Giraud F (1986) Biochim Biophys Acta 863:253-263
155. Shio H, Haley NJ and Fowler S (1978) Lab Invest 39:390-397
156. Bialecki R, Tulenko TN and Colucci WS (1991) J Clin Invest 88:1894-1900
157. Sachinidis A, Locker R and Vetter W (1990) Klin Wochenschr 68:(suppl 20), 65-67
158. Bialecki RA and Tulenko TN (1989) Am J Physiol 257:C306-C314
159. Liu KZ and Pierce GN (1993) J Biol Chem 268:3767-3775
160. Mason RP, Moisey DM and Shajenko L (1992) Mol Pharmacol 41:315-321
161. Liu KZ, Massaeli H and Pierce GN (1993) J Biol Chem (in press)
162. Mahnensmith RL and Aronson PS (1985) Circ. Res. 57:773-788
163. Roos A and Boron WF (1983) Physiol Rev 61:296-434

164. Parker JC (1983) Am J Physiol 244:C324-C330
165. Livne A, Balfe JW, Veitch R, Marquez-julio A, Grinstein S and Rothstein A (1987) Lancet 1:533-1536
166. Meng H, Lonsberry BB and Pierce GN (1991) J Pharmacol Exp Ther 258:772-777
167. Ng LL, Simmons D, Frighi V, Garrido MC, Bomford J and Hockaday TDR (1990) Diabetologia 33:371-377
168. Pierce GN, Ramjiawan B, Dhalla NS and Ferrari R (1990) Am J Physiol 258:H255-H261
169. Brasitus TA, Dudeja PK, Worman HJ and Foster ES (1986) Biochim Biophys Acta 855:16-24
170. Dudeja PK, Foster ES and Brasitus TA (1987) Biochim Biophys Acta 905:485-493
171. Poli De Figueiredo CE, Ng LL, Davis JE, Lucio-cazana FJ, Ellory JC and Hendry BM (1991) Am J Physiol 261:C1138-C1142
172. Kochhar N and Kaul D (1992) FEBS Lett 299:19-22
173. Reuter H and Seitz N (1968) J Physiol Lond 195:451-470
174. Reeves JP and Sutko JL (1979) Proc Ntal Acad Sci USA 76:590-594
175. Reuter H (1982) In: Martonosi AN (ed) Membranes and transport. Vol. 1. N.Y., Plenum Press, pp 623-631
176. Bridge JHB and Bassingthwaighte JB (1983) Science 219:178-180
177. Crespo LM, Grantham CJ and Cannell MB (1990) Nature 345:618-621
178. Theunissen JJH, Jackson RL, Kempen HJM and Demel RA (1986) Biochim Biophys Acta 860:66-74
179. Vemuri R and Philipson KD (1987) Biochim Biophys Acta 937:258-268
180. Dalla-serra M, Pederzolli C, Antolini R, Cusinato F, Luciani S and Menestrina G (1991) Cardioscience 2:193-197
181. Luciani S, Bova S, Cargnelli G, Cusinato F and Debetto P (1991) In: Blaustein MP, DiPolo R and Reeves JP (eds) Sodium-calcium exchange, Proceedings of the second international conference. Ann N York Acad Sci 639:156-165

CHAPTER 18

HUMAN INTESTINAL BRUSH BORDER MEMBRANE HYDROLASES

Daniel Ménard and Jean-François Beaulieu

Groupe de recherche en biologie du développement,
Département d'anatomie et de biologie cellulaire,
Faculté de médecine, Université de Sherbrooke,
Sherbrooke, Québec, Canada J1H 5N4

The apical brush border of enterocytes is designed to expand the luminal membrane surface available for hydrolysis and transport. In 1961, Crane et al described the brush border of intestinal epithelial cells as a functionally unique subcellular organelle, as well as digestive-absorptive surface (1). Since then, the microvillus membrane, which represents the apical domain of enterocytes, has been shown to possess an array of membrane-anchored hydrolases as illustrated in table 1.

Brush-border membrane disease was defined over 25 years ago (2) in order to clarify relationships among a growing list of congenital and/or acquired gastrointestinal disturbances characterized by intolerance to specific dietary components. Accordingly, it was defined as an absence or a greatly diminished activity of a specific functional protein element of the brush border membrane.

Over the years, a great deal of attention has been focussed on establishing the basic data, and the proper techniques and experimental models in order to elucidate the mechanisms that control specific gene expression in human gut. In this review, we cover the most recent observations related to the regulatory mechanism of the

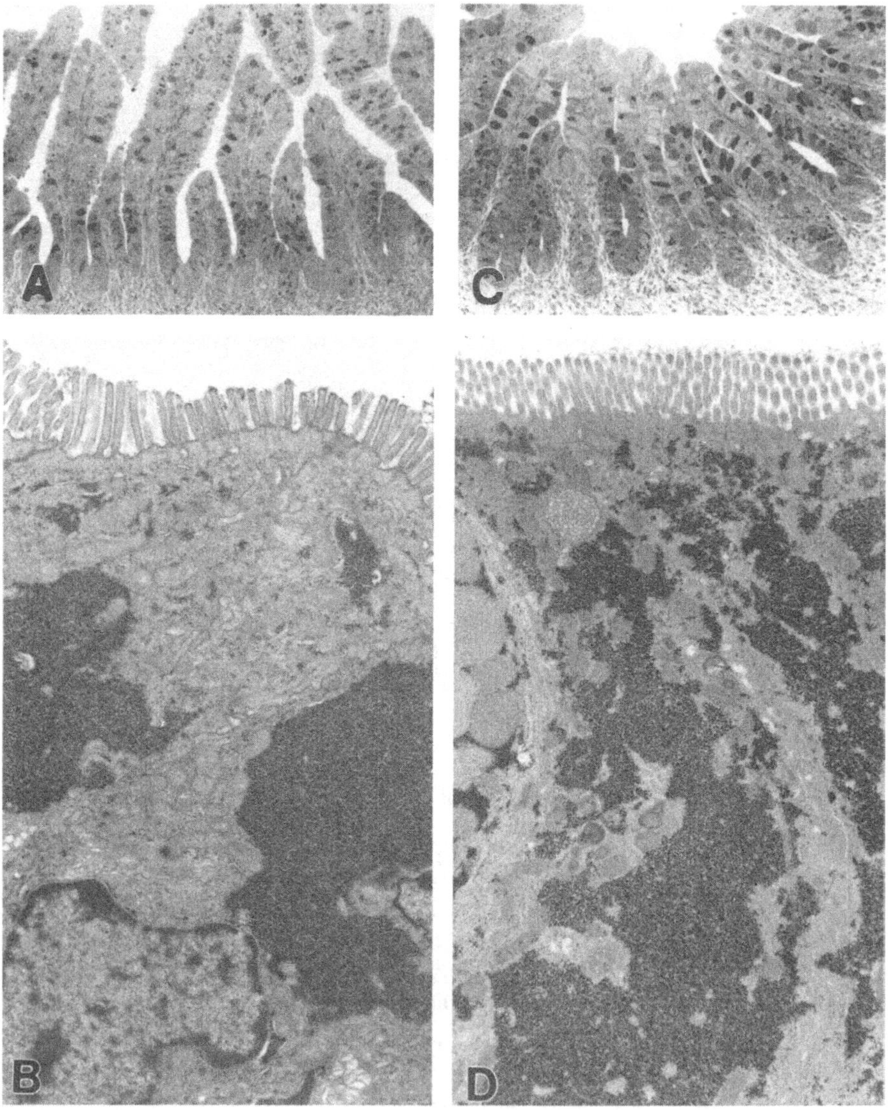

Figure 1. *One μ-thick epoxy sections of the jejunum (A) and colon (C) in a 16-week-old fetus stained with a mixture of 1% methylene blue, 1% azure II, and 1% borax. Villi and developing crypts are found in both segments. At the ultrastructural level, villus epithelial cells in the jejunum (B) and colon (D) exhibit a well developed brush border as well as supranuclear accumulation of glycogen.*

Table 1. Microvillar membrane enzymes

Sucrase-isomaltase (3.2.1.48-10)
Lactase-phlorizin hydrolase (3.2.1.23-62)
Maltase-glucoamylase (3.2.1.20)
Trehalase (3.2.1.28)
Aminopeptidase A (3.4.11.7)
Aminopeptidase N (3.4.11.2)
Enteropeptidase (3.2.21.9)
γ-glutamyltranspeptidase (2.3.2.2)
Dipeptidyl peptidase IV (3.4.14.5)
Endopeptidase 24.11 (3.4.24.11)
Angiotensin-converting enzyme (3.4.15.1)
Alkaline phosphatase (3.1.3.1)

development of different brush border hydrolases in human small intestine and colon, the regulation of expression of brush border hydrolases in the adult gut, the pathophysiology of expression of brush border membrane enzymes in congenital diseases and colon cancer, and the use of a colonic adenocarcinoma cell line (Caco-2) as a tool to study various aspects of human intestinal brush border enzyme expression.

HYDROLASE EXPRESSION DURING FETAL DEVELOPMENT

Small Intestine

Between 8 and 10 weeks of gestation, the luminal surface of the entire length of the small intestine is covered by stratified epithelium (3-5). At 9 weeks of gestation, villi are present in the duodenum and jejunum, forming in the proximal ileum, and still absent from the distal ileum. However, by 11 weeks the entire small intestinal mucosa is lined by villi, most of which are covered by a simple columnar epithelium mainly composed of well differentiated /enterocytes (Figs 1A, B). Disaccharidase activities, namely sucrase-isomaltase, maltase, and trehalase, are all present at 10-13 weeks of gestation at levels comparable to those in mature intestine (5-7). The cranio caudal gradient of these α-glucosidase activities, characteristic of the adult small intestine, is already established by 15 weeks of gestation. It is interesting to note that the sucrase-isomaltase complex is present only as the high molecular weight pro-sucrase-isomaltase form between 12 and 30 weeks, whereas, from 30 weeks until birth the two subunits are also present, as in normal adult intestine (8-10). The enzymatic activity of lactase, like the other disaccharidases, is also present at 10 weeks of gestation. However, its activity is very low up to 34 weeks, representing only 30% of the activity found in full-term babies. A marked increase of lactase activity occurs towards the end of gestation, bringing the levels at term to two to four times the values found in normal infants 2 to 11 months of age (11).

Alkaline phosphatase activity is also detectable in the jejunum of 11-week old fetuses and rises four-fold by 23 weeks of gestation, but is still considerably lower than the levels reported for mature jejunum (12). The exact developmental profile from 23 weeks to birth is still undefined. A recent study pointed out the difficulty in studying alkaline phosphatase development (13). Indeed, the tissular distribution of its activity differs strikingly from all other brush border enzymes studied, as only 50% of this activity is of epithelial origin in the proximal intestinal epithelium itself, compared to 90- 99% of total disaccharidase and γ-glutamyltranspeptidase activities.

Enteropeptidase activity appears late in gestation and cannot be detected before the 26th week (11). Between the 26th and the 40th week of gestation, the activity rises and approximates 17% of that found in children between 1 and 4 years of age. The activities of the other peptidases are generally found as early as the other brush border enzymes during fetal development (14, 15), but exhibit different development patterns. For example, γ-glutamyltranspeptidase activity is higher in fetuses than adults, dipeptidylpeptidase IV activity is at adult level in the youngest studied fetuses, and aminopeptidase A activity is significantly lower than in adults. The physiological significance of the different patterns remains to be elucidated.

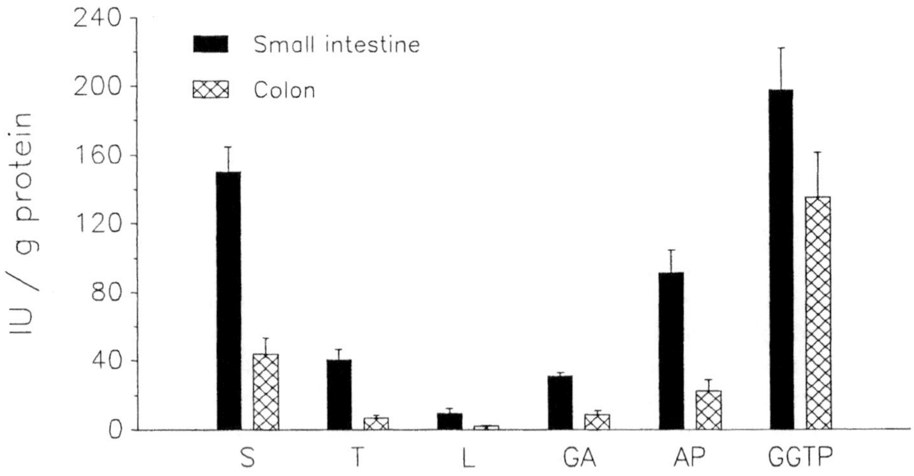

Figure 2. Specific activity of brush border enzymes in isolated intestinal and colonic epithelial cells at 15-18 weeks of gestation. S: sucrase; T: trehalase; L: lactase; GA: glucoamylase; AP: alkaline phosphatase; GGTP: gamma glutamyltranspeptidase. Adapted from Ménard and Pothier (ref. 13).

Colon

The human fetal colon undergoes substantial morphogenetic changes before attaining its full development. Between 9 and 16 weeks of gestation, one striking event is the formation of villi lined by a simple columnar epithelium (Figs 1C, D) similar to that of the small intestine (5, 16). However, the presence of villi in the colon is transitory since these structures will disappear between 30 weeks of gestation and birth (17).

As early as 13 weeks of gestation, all studied intestinal brush border hydrolases are also present in fetal colon (13) and figure 2 illustrates in a comparative manner, enzymatic levels in both segments at 15-18 weeks of gestation.

Overall, the intestinal enzymic levels are always higher than those in the colon. Of particular interest during colonic development is the presence of sucrase/isomaltase complex, which is the classical marker of enterocytic differentiation and the presence of low but definite lactase activity in isolated colonic epithelial cells at all fetal age groups studied. Before 24 weeks of gestation, the colonic sucrase-isomaltase complex is also expressed only as the high molecular weight prosucrase-isomaltase (9, 18).

The developmental patterns for colonic digestive enzymes are described with a limited number of fetuses (19) and are not well defined. A more recent study reported the variations of hydrolase activities within the colon of five preterm (28-30 weeks of gestational ages) and three near-term infants (36-38 weeks) (17). The levels of sucrase, aminopeptidase, and glucoamylase showed no significant differences along the colon in the preterm newborns. However, in near-term babies, a drastic decrease of sucrase, a slight decrease of glucoamylase, and a significant enhancement of aminopeptidase levels were noticed only in the proximal colon. Although preliminary, these observations illustrate regional differences in enzymic levels along the developing colon and stress the need for more investigations regarding the development of brush border hydrolases in the colon. This is particularly important, in view of the fact that normal development leads to the general disappearance of most of these enzymic activities, although some of them are still expressed in the adult colon (see Table 3). Nevertheless, they continue to be highly synthesized in the small intestinal enterocytes. Furthermore, expression of digestive enzymes are described to occur during the process of carcinogenesis in the human colon as discussed later on.

REGULATORY MECHANISM OF BRUSH BORDER HYDROLASES EXPRESSION DURING FETAL DEVELOPMENT

Until recently, the regulatory mechanism of the developmental program of the human intestinal hydrolases was completely unknown and our understanding of it originated mainly from studies on rat and mouse (20,22). Hormones (glucocorticoids, thyroxine, insulin) and growth factors (epidermal growth factor) were found to be important modulators of the intestinal brush border enzymic expression in small rodents; but whether they also exerted specific modulatory influences on human fetal intestine remained an open question. The establishment of a serum-free organ culture technique allowing in vitro morphological and physiological maintenance of human fetal small intestine (23) and colon (24) for more than 5 days offered, at last, the mean to answer this basic question (25). The addition of hydrocortisone to the medium of cultured small intestinal explants at representative fetal serum levels induced significant increases of lactase and alkaline phosphatase activities without affecting sucrase, trehalase and glucoamylase activities (26). This specific effect on intestinal lactase activity was confirmed later on using dexamethasone, a synthetic glucocorticoid (27). Interestingly, hydrocortisone did not affect any brush border enzymatic activities in cultured colon explants (28) suggesting a different regulatory mechanism for the small and large intestine during human development. Epidermal growth factor (EGF) was also shown to induce small intestinal lactase activity (29) like hydrocortisone, but there was no synergistic action between the two. However, the colonic lactase activity remained unaffected by EGF. On the other hand, EGF did significantly decrease tissue sucrase activity in both segments, suggesting that this growth factor affects the synthesis and/or the degradation of the enzyme. Therefore, EGF would influence the processing of some digestive enzymes in both the small intestine and the colon. Obviously more investigations are needed in order to delineate specific hormones and growth factors involved in the modulation of the enzymic development of the human gut.

Table 2. Statistical analysis of the effects of organ culture alone or with hydrocortisone on levels of lactase activity and LPHmRNA

Condition 1	Condition 2	P value* Lactase	Condition 2 vs 1 [fold effect] LPH mRNA
No organ culture	Organ culture	0.455	0.002 [1.9X]
No organ culture	Organ culture + HC	0.023 [2.1X]	0.004 [2.3X]
Organ culture	Organ culture + HC	0.007 [1.8X]	0.198

Fetal jejunum was analyzed without treatment or after 5 days of organ culture with or without 50 ng/ml hydrocortisone (HC). *Conditions 1 and 2 were compared in a paired, two-tailed Student's t-test with n= 10. In cases where the two conditions differed significantly, the magnitude of the effect is shown in square brackets (condition 2 results divided by condition 1 results). From Villa et al, ref.31, with permission.

The recent availability of sucrase and lactase cDNAs permitted the investigation of the primary control level of gene expression of these two major integral glycoproteins of the human fetal small intestinal brush border membrane. The correlation between the sucrase-isomaltase activities and mRNA levels during the spontaneous appearance of pro- sucrase-isomaltase in the small intestine and colon, between 9 and 21 weeks of gestation, strongly suggests that the biosynthesis of sucrase-isomaltase is most likely to be controlled at the level of transcription (30). The relatively high level of lactase activity seen at 37 weeks of gestation, compared to 14-20 weeks, is accompanied by a corresponding high level of lactase-phlorizin hydrolase (LPH) mRNA; suggesting that the developmental regulation of lactase during the fetal period is, at least in part, at the level of mRNA accumulation (31). However, culturing of human fetal jejunum for 5 days induced, by itself, a 2-fold increase in LPHmRNA without any significant change in brush border lactase activity. The fact that the enzymatic activity did not increase in parallel implied an additional translational or posttranslational regulatory component. In presence of 50 ng/ml hydrocortisone, during 5 days of organ culture, jejunal lactase activity doubled, in agreement with previous studies (27, 29). However, this increase was not accompanied by an increase of LPHmRNA beyond that seen during culture without hormone. Table 2 summarizes the statistical analysis.

These data, therefore, suggest that hydrocortisone modulates lactase expression at a step beyond mRNA accumulation, possibly post-translationally. It is noteworthy that no consensus binding site for glucocorticoid hormone receptor was found upstream of the human LPH gene, at least within the 1 Kb segment sequenced (32). These observations could, therefore, lead to the following suggestion for the interpretation of human LPH development: between 14-20 weeks of gestation there is a negative transcriptional regulation, possibly mediated by a factor which is either not present or is inactivated during culture. During gestation, the factor declines leading to the accumulation of LPHmRNA as observed at 37 weeks of gestation. Between 35 and 40 weeks, the doubling of the fetal serum hydrocortisone levels modulate LPH activity expression at a post-transcriptional level. As discussed below, a post-transcriptional control of lactase expression is likely to be one of the regulatory mechanisms during adult life.

BRUSH BORDER HYDROLASE EXPRESSION IN ADULT GUT

As detailled in the preceeding section, the human intestinal mucosa exhibits, at birth, all brush border hydrolase activities at levels comparable to those found in the adult, (table 3) with few exceptions.

The crypt-villus axis represents the functional unit of the intestinal mucosa and it develops very early during the ontogeny of the gut (5,7). The villi are lined primarily by mature absorptive cells and mucus-secreting goblet cells, and the crypts lined by stem cells, proliferating undifferentiated cells and differentiated secretory cells which include Paneth, goblet, and endocrine cells. Intestinal epithelial cell proliferation, migration and differentiation are susceptible to various influences along the crypt-villus axis, including compositional changes in the basement membrane and differential expression of cellular receptors for these components (33-35). Therefore, differentiation of each cell type takes place while the cells move up to the villus tip (36). It has been long recognized that it is the mature villi enterocytes that exhibit all brush border hydrolases and these enzymes have, therefore, been used as specific markers for the enterocytic differentiation studies.

Sucrase-isomaltase is one of the best characterized marker enzymes of the mature absorptive cells in adult small intestine. Its structure, biosynthesis, complex intracellular processing, and transport to the microvillus membrane have been investigated in great detail (see reviews 37, 38). The recent development of a panel of monoclonal antibodies specific for sucrase-isomaltase shed new lights on the differentiation program and the level of control of SI expression in intestinal epithelial cells (39, 40). Two markedly different patterns of immunofluorescence staining were observed; one specific for the luminal membrane of the absorptive villus cells and one in which the crypt cells were also stained (Fig. 3A).

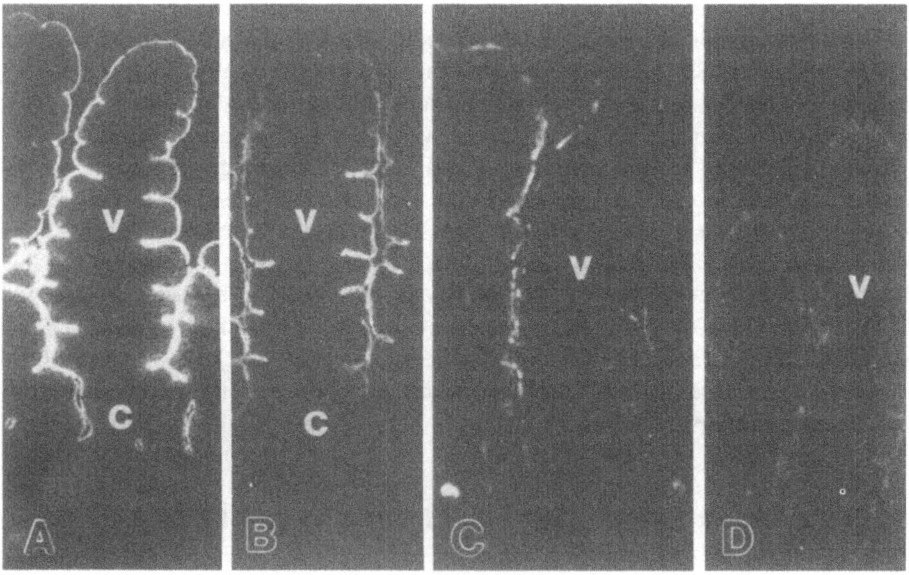

Figure 3. Expression of sucrase-isomaltase (A) and lactase (B-D) in the human adult jejunal mucosa. Sucrase-isomaltase is found at the luminal surface of both villus (v) and crypt (c) enterocytes (A), while lactase expression is restricted to villus cells (B). (C,D) Specimens from presumptive hypolactasia subjects as indicated by lactase immunodectection in the villus epithelium according to a patchy staining pattern (C) or in very low level (D) [Beaulieu and Vachon, unpublished data].

Interestingly, immunofluorescence staining also demonstrated the presence of sucrase-isomaltase in the apical region of normal colonic crypt cells but not on surface epithelium. The lack of staining of the surface colonic epithelium containing mature absorptive cells contrasts with the small intestine and suggests different regulatory mechanism(s) of sucrase-isomaltase expression in these differentiated cells. On the other hand, the presence of immunologically related sucrase-isomaltase forms in proliferative crypt cells in both small intestine and colon clearly supports the hypothesis that undifferentiated crypt cells could normally express brush border membrane markers. In fact, as shown in Table 3, two classes of brush border markers can be delineated: one specific for the differentiated enterocytes such as maltase-glucoamylase and lactase (Figure 3B) and one expressed both in undifferentiated crypt cells and differentiated villus cells such as sucrase-isomaltase, aminopeptidase A and dipeptidyl peptidase IV. Again, the markers expressed in the intestinal crypt cells appear to also be expressed in colonic crypt cells (Table 3), suggesting more similarities than anticipated between these two crypt cell populations.

The primary site of control of gene expression is a key issue in our understanding of the mechanism involved in the regulation of intestinal brush border membrane functions. As discussed in the preceeding section, sucrase-isomaltase appears to be regulated at transcriptional level during fetal development (30) and investigations suggest a complex developmental regulation of lactase both at transcriptional and post-transcriptional levels (31). The new evidences that sucrase-isomaltase can be expressed in intestinal and colonic crypt cells, that this expression is a transient phenomenon restricted to proliferative migrating cells in the colon, and that sucrase-isomaltase processing and its conversion into a distinct conformational species are dependent on the state of enterocytic differentiation (39, 40) lead to the suggestion of a primary regulation of sucrase-isomaltase at the posttranslational level in contrast to a transcriptional level reported in fetal development. The extent to which either of these mechanisms, alone or in combination, regulate sucrase-isomaltase expression in adult intestine remains to be explored. Furthermore, the cellular and/or extracellular factors involved in the activation of sucrase-isomaltase gene transcription are yet to be defined although the first step has been made identifying regulatory elements of the sucrase-isomaltase gene (41, 42). Congenital diseases, such as sucrase-isomaltase deficiency and adult-type hypolactasia as well as colon cancer and human colon cancer cell lines offer unique opportunities to get insights about regulation of brush border hydrolase gene expression in the human.

BRUSH BORDER MEMBRANE DISEASES

In theory, we could expect to find a brush border membrane disease for each enzyme indicated by the activities in table 1. However, they have not all been seen and only a limited number have been described: sucrose-isomaltose maldigestion (43), congenital lactose maldigestion (44, 45), adult lactose maldigestion (46, 47), trehalose maldigestion (48), and enteropeptidase deficiency (49). We will consider two types of congenital abnormalities of the human small intestine which are at present best characterized namely, the congenital sucrase-isomaltase deficiency and the adult-type hypolactasia.

Congenital Sucrase-Isomaltase Deficiency (CSID)

This autosomal recessive disease in humans leads to osmotic diarrhea. CSID is characterized by undetectable sucrase activity, while the isomaltase activity can vary from absent to practically normal. The molecular basis of CISD is still unknown, but recent observations clearly suggest that CSID is not due to a single molecular defect. In 1973, Dubs et al reported the presence of cross-reacting material in enzyme-deficient intestine by use of indirect immunofluorescence (50). In the following year (51), the complete absence of sucrase-isomaltase proteins in purified brush border membranes from two siblings diagnosed as sucrase-isomaltase deficient was reported and confirmed with other patients (52-54).

These apparently conflicting results were recently unified by the hypothesis that different mutations may lead to CSID (55, 56). Up to now, studies at the subcellular and protein level with monoclonal antibodies against sucrase-isomaltase revealed at least five phenotypes. Phenotype 1 is characterized by the intracellular accumulation in the endoplasmic reticulum of the sucrase-isomaltase protein (56,57). In phenotype II, there is a normal transport of sucrase-isomaltase to the Golgi apparatus but it is degraded prior to complex glycosylation possibly in the cis-Golgi apparatus (55, 58). In phenotype III, sucrase-isomaltase is transported to the brush border and cleaved into subunits. However, the catalytic activity of the sucrase subunit only is affected, leading to a patchy distribution and partial missorting to the basolateral membrane occurs (55). The phenotype IV is characterized by the synthesis of a mannose-rich pro-sucrase-isomaltase of an apparent molecular weight practically indistinguishable from normal controls. However, this pro-sucrase-isomaltase does not undergo terminal glycosylation in the Golgi apparatus and is missorted to the basolateral membrane and/or accumulated in the endoplasmic reticulum (56). In phenotype V, the pro-sucrase-isomaltase is intracellularly cleaved leading to the lost of the sucrase subunit which is not transported to the brush border membrane and is most likely degraded intracellularly. On the other hand, the isomaltase subunit is transported and correctly sorted to the brush border membrane providing evidence for the presence of putative sorting signals of the sucrase-isomaltase complex on the isomaltase subunit (56). A phenotype in which no immunoreactive protein is detectable has not yet been described.

Overall, these studies suggest that different mutations in the sucrase-isomaltase gene lead to the synthesis of transport-incompetent or functionally altered enzyme leading to congenital sucrose intolerance. Analysis of mutant phenotypes at the genetic level will provide important clues for determinants involved in the synthesis, sorting event and transport of the sucrase-isomaltase complex to the human brush border membrane. Furthermore, they do provide interesting evidence for posttranslational regulation of sucrase-isomaltase expression in adult intestinal epithelium. The extent to which the different phenotypes described for sucrase-isomaltase deficient subjects would also be reflected at the colonic crypt cell through expression of sucrase-isomaltase, remains an open question.

Adult-Type Hypolactasia

In a great majority of the human population, lactase activity decreases during childhood leading to adult-type hypolactasia; the most frequent genetically based syndrome in man (59). Therefore, the term deficiency or disease is only justified for ethnic groups with high adult lactase levels such as Caucasians and many other milk-drinking populations of the world. This autosomal recessive disease is characterized by a reduced level of lactase activity. Two different phenotypes have been identified (60-62). Phenotype I is characterized by a reduced synthesis of precursor LPH while phenotype II is associated with ample precursor synthesis but reduced conversion of the protein into its mature molecular form. Recent investigations were aimed at elucidating the mechanisms that control LPH gene expression in both deficient and control subjects. A study of the levels of LPH-specific mRNA reported poor correlation with enzyme activity in control and hypolactasic subjects and, therefore, a posttranscriptional control of human LPH gene expression was postulated (63). A second investigation suggested that in the majority of the studied hypolactasic subjects, pre-translational mechanisms accounted for the predominate regulatory control of LPH expression in the proximal intestine (62). This study did, however, report one individual with adult hypolactasia who demonstrated synthesis of the LPH protein at a rate exceeding that of the control but there was a marked reduction of its conversion into its mature brush border form, suggesting a posttranslational biosynthetic defect. Finally, a study designed to get insights into the mechanism leading to the well-defined differences in lactase activity levels found between racial groups (oriental, black and white patients) strongly suggested basic regulation at the level of gene transcription (64). These data reinforce the hypothesis that adult hypolactasia may arise as a result of at least two distinct mechanisms of regulatory control. The extent to which these mechanisms, alone or in combination, regulate the programmed decline of intestinal LPH in defined ethnic groups remains to be explored.

Genetic polymorphism in regard to adult hypolactasia has been strengthened by the immunohistological analysis of the expression of LPH protein in adults with hypolactasia (65). The existence of two patterns of expression of the lactase protein were characterized by either absent mucosal staining with anti-LPH monoclonal antibodies or by a mosaic pattern showing mosaic expression of the protein by individual enterocytes scattered over the villi. Such patterns are illustrated in figures 3C and D. These observations have yet to be correlated with the two phenotypes already described and it would not be surprising if more phenotypes would be characterized in future investigations.

Table 3 Expression of brush border hydrolases in the small intestine and the normal and neoplastic colorectal epithelium

Hydrolase (antibody)	Jejunum		Colon			Ref.
	C[1]	V[2]	Normal	Adenoma	Carcinoma	
Sucrase-isomaltase						
(HBB-2/614)			-		7/27[3]	66
(HSI-1,-5)	-	++	-	-	-	
(HSI-3,-9,-12,-14)	++	++	+	10/16	4/61	39,40
(HBB-2/614)			+	21/23	22/57	69
(8A9)	++	++	+			70
(polyclonal)			-		24/30	68
(HBB-2/614)			+		19/46	72
(HSI-14)	++	++				Fig.3A
Lactase						
(HBB-1/909)			-		-	66
(HLA-1)	-	++	-	-	-	71
(HBB-1/909)			1/4		2/29	72
(MLAC-)	-	++				Fig.3B
Maltase-Glucoamylase						
(HBB-2/143)			-		-	66
(HMA-1 to 7)	-	++	-	-	-	71
(HBB-2/143)			1/4		2/29	72
Aminopeptidase N						
(HBB-3/153)			-		7/27	66
(HAPN-1,-2,-3)	++	++	+	20/23	17/35	71
(F23)			+		28/39	72
Dipeptidylpeptidase IV						
(HBB-3/775)			-		7/27	66
(4H3)	++	++	+			70
(DPPIV-1)	++	++				71
(HBB-3/775)			+		34/46	72

Brush border hydrolases detected by immunohistochemistry: -, not detected; +, detected; ++, detected according to a strong staining pattern; [1] C: crypt epithelium; [2] V: villus epithelium; [3] number of positive/ total number of specimens examined.

EXPRESSION OF BRUSH BORDER HYDROLASES IN COLON CANCER

Several studies have suggested that the process of carcinogenesis in the human large intestine is often accompanied by abnormal expression of the digestive enzymes characteristic of the small intestinal brush border membrane. Their potential usefulness as markers for the identification and characterization of colon cancer was first proposed by Zweibaum et al. (18,66) who demonstrated the expression of sucrase-isomaltase, aminopeptidase N and dipeptidylpeptidase IV in some human colon cancer cell lines and primary colorectal tumors. As mentionned above, small intestinal hydrolases occur transiently in the fetal colon between 10 and 30 weeks (13,19). It was thus suggested that expression of brush border membrane enzymes in colonic neoplasia may represents the resurgence to a fetal pattern of gene expression.

Investigations on human adult colonic mucosa, adenomas and adenocarcinomas with various mono- and polyclonal antibodies for the expression of small intestinal hydrolases (Table 3) have raised, however, some conflicting data concerning their occurence. For instance, sucrase-isomaltase was initially found to be absent from the normal colonic epithelium (66,67,68), but was later demonstrated in normal colonocytes by means of both immunocytochemistry (40,69,70) and biochemistry (40). Similarly, dipeptidylpeptidase IV and aminopeptidase N were also recently shown to be consistently expressed in normal colonic mucosal specimens (70,71). Contradictory observations were also reported relative to the expression of sucrase-isomaltase in the peritumoral mucosa (66,68,69). A likely explanation for these differences may be the existence of distinct immunoreactive forms of the enzymes. Indeed, as illustrated for sucrase-isomaltase, it was possible, by using a panel of specific monoclonal antibodies, to distinguish various forms of the molecule (Table 3), namely a fully active enzymatic form detected exclusively in differentiated enterocytes of the small intestinal villus and a poorly active form detected in the crypts of the normal colonic mucosa. Both were immunologically related, although biochemically distinct, to the sucrase-isomaltase present in the small intestinal crypt (39,40). Thus with respect to these studies demonstrating immunoreactive sucrase-isomaltase in the normal colon, the expression of this hydrolase in colon cancer, which appears heterogeneous among (Table 3) and within tumors (40,66,69-71), could be seen as a deregulated manifestation associated with neoplasia rather than a true neoexpression of an oncofetal marker.

Lactase and maltase-glucoamylase, two other brush border hydrolases expressed by epithelial cells of both the small intestine and colon during the fetal period (5), were also investigated as potential markers of human colon cancers (66,71,72). They were consistently found to be absent from colorectal tumors, irrespective of whether they were detected with antibodies to conformationally sensitive or insensitive epitopes (71). Interestingly, the same antibodies failed also to reveal the presence of these enzymes in the normal adult colon as well as in the small intestinal crypts (Table 3).

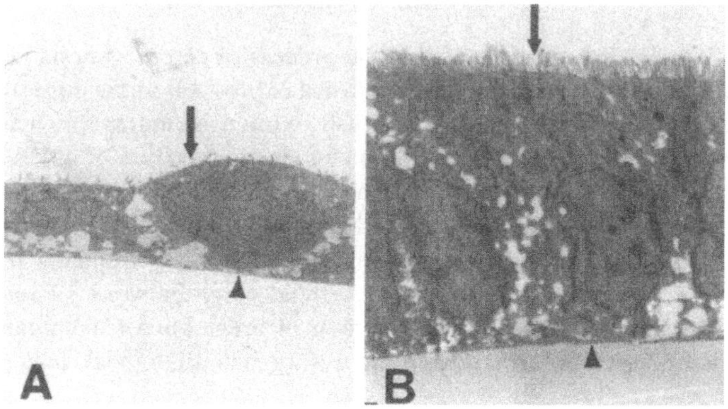

Figure 4. *Morphological differentiation of Caco-2 cells in culture. (A) At 1 day postconfluence, the cells exhibit a relatively flat morphology with prominent nuclei. (B) At 30 days of postconfluent culture, the cells have a typical enterocyte-like morphology with a well-organized brush border. Arrows and arrowheads denote the luminal and the growth surfaces respectively [Beaulieu and Vachon, unpublished data].*

In a recent study, Quaroni et al (71) re-examined the expression of aminopeptidase N in normal human colonic mucosa and in colonic tumors. Contrary to previous observations (66), they found the enzyme to be consistently expressed in the normal colonic mucosa by both immunofluorescence and immunoblotting. However the normal colonic mucosa by both immunofluorescence and immunoblotting. However, in contrast to its localization in the small intestine where the enzyme is confined to the microvillus border of the enterocytes (71,73), aminopeptidase N was exclusively detected at the base of epithelial cells in most specimens of colon obtained from individuals free of intestinal tumors; conversely, in large-intestinal mucosa distal to tumors, adenomas and hyperplastic polyps, from members of hereditary colon cancer syndrome families, and in colonic tumors, the enzyme was frequently localized at the luminal membrane of the epithelium. These findings suggest that aminopeptidase N is subject to differential ways of sorting in enterocytes and colonocytes, a process which may be altered at an early stage of colon carcinogenesis (71).

These studies thus provide new evidence for the usefulness of intestinal hydrolases as potential biomarkers for prognostic and/or diagnosis of colorectal cancer. Among them, sucrase-isomaltase, aminopeptidase N and dipeptidypeptidase IV, which are expressed in the normal adult colonic mucosa, may be of interest as differentiation markers to investigate abnormal cellular properties associated with the progression of colonic mucosal neoplasia. Their altered patterns of expression, in both quantity and cellular localization as well as their patchy/heterogeneous distribution within most tumors remain, nevertheless, to be further investigated on a much larger number of patients as well as on *in vitro* models. For instance, up to now, no clear correlation has been found between the peculiar expression of these markers and the degree of invasiveness or histological differentiation of the tumors. Moreover, cellular and molecular mechanisms involved remain poorly understood.

THE CACO-2 CELL LINE AS AN EXPERIMENTAL MODEL

The use of the Caco-2 cell line has provided an inportant tool to study various aspects of human intestinal brush border enzyme expression, namely the biosynthesis, processing, intracellular transport and regulation (39,40,73-76). The cell line was initially established from a colonic adenocarcinoma by Fogh et al. (78). Under standard culture conditions, Caco-2 cells form a confluent monolayer which over a period of 20-40 days of postconfluent culture gradually acquire full morphological polarity (Fig. 4) and express relatively high levels of brush border hydrolases [sucrase-isomaltase, lactase, dipeptidylpeptidase IV and aminopeptidase N] (73,76,79-81). Among all intestinal cell lines established so far, including the one hundred and more derived from human colon carcinomas, the Caco-2 cell line is the only one able to undertake a spontaneous and complete intestinal-like program of differentiation

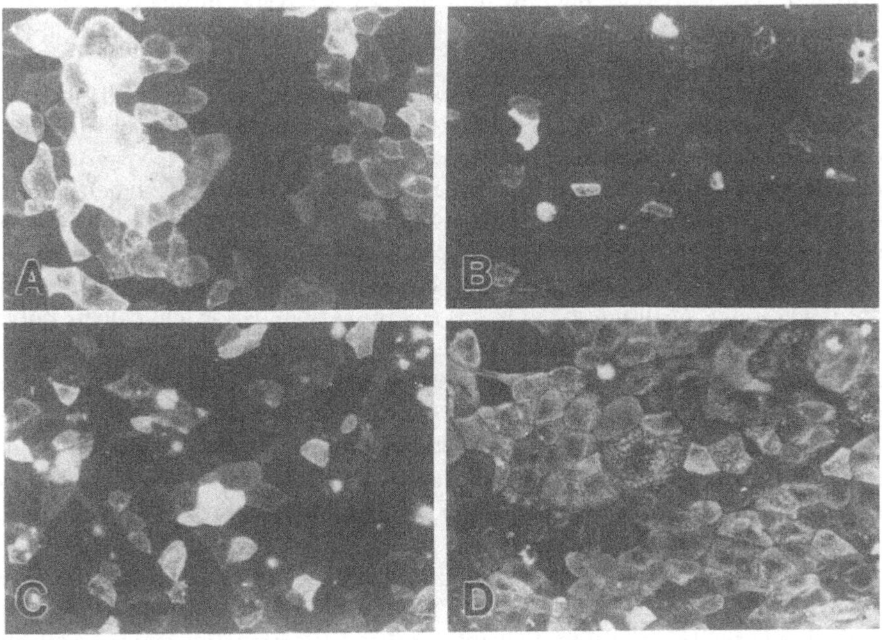

Figure 5. *Mosaic pattern of expression of sucrase-isomaltase (A), lactase (B), aminopeptidase N (C) in 12-day postconfluent Caco-2 cell monolayers as determined by indirect immunofluorescence. In the contrary, dipeptidylpeptidase IV is detected in most cells (D) in the same conditions [Beaulieu and Vachon, unpublished data].*

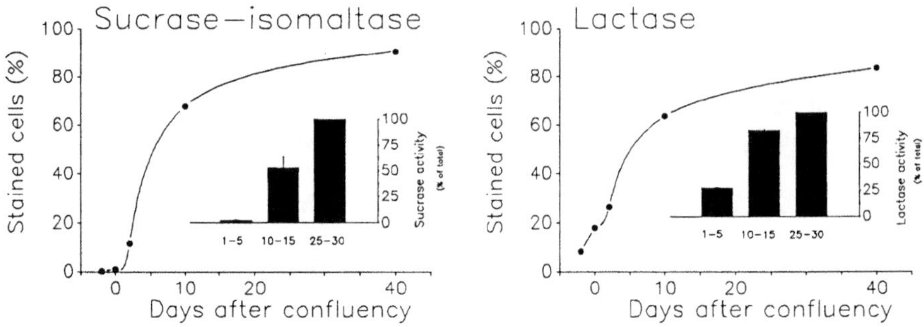

Figure 6. *Sucrase-isomaltase and lactase expression during the differentiation of Caco-2 cells. Estimation of the proportion of immunoreactive cells reveals that the heterogeneous pattern of expression of brush border hydrolases is a transient phenomenon since, in older cultures, most of the cells (≥ 90%) express these two differentiation markers. Moreover, it is worth emphasizing that for both markers, percentages of stained cells closely parrallel enzyme activities from homogenates (inserts) [Beaulieu and Vachon, unpublished data].*

(67,82,83). The increasing interest, over the last ten years, in this cell line as an experimental model to study most aspects of intestinal epithelial cell biology *in vitro* is, therefore, not surprising.

Whether Caco-2 cells grown *in vitro* represent mature small intestinal enterocytes, colonic crypt cells or fetal differentiated colonic cells may still appear controversial when considering individual features. However, there are several indications that the closest analogy which can be found is with the fetal human colon (84). Indeed, as metioned earlier, the hydrolases expressed by Caco-2 cells are also detected at high levels in the human fetal colon at mid-pregnancy (13,19,66,76,80). The form of pro-sucrase-isomaltase sythesized by the cells is clearly distinct from those found in the small intestine (39,73) and in the adult colon and colonic polyps (40), but similar to that identified in the fetal colon (85; Beaulieu et al. unpublished data). Moreover, Na^+-dependent glucose transport in Caco-2 cells shares properties with that normally found in brush border membranes from 16-22 week-old human fetal colons (80); morphologically, post-confluent Caco-2 cells also exhibit similar features to their normal fetal counterparts such as well developed brush borders and junctional complexes, and accumulation of large amounts of glycogen in the supranuclear region (5,81,84).

The potential value of Caco-2 cells to study various aspects of human intestinal cell physiology is now well documented (39,40,73-75,80,86-94). The cell line also represents an excellent model to study intestinal cell differentiation and its regulatory mechanisms. This is well exemplified by recents studies on intestinal brush border hydrolase expression, namely for sucrase-isomaltase which provided evidence for both transcriptional and post-translational control sites (39,76,95), and modulatory

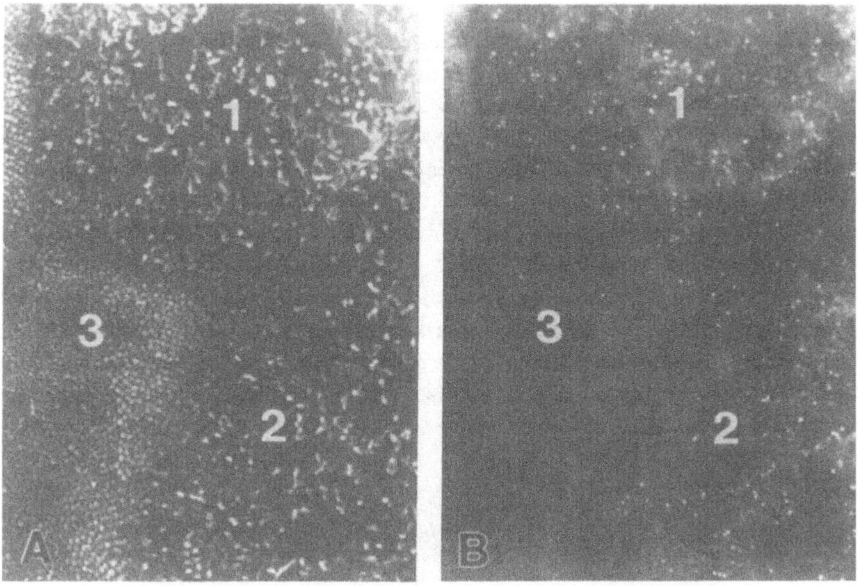

Figure 7. Scanning electrom micrograph (A) and corresponding backscattered electron imaging micrograph (B) of 12-day postconfluent Caco-2 monolayer after colloidal gold immunolabeling for the detection of lactase. As illustrated, cells with a well-developped brush border (A, cell #3) are frequently found devoided of labeling (B) while, at the opposite, immunogold particles can be seen at the level of poorly organized apical cell surfaces of other cells with a poorly-organized brush border (A,B, cells #1 and 2) [Beaulieu and Vachon, unpublished data].

influences by glucocorticoids (30) and growth factors (76,96).

Studies of brush border hydrolases in Caco-2 cells at the cellular level revealed another interesting property of these cells: a remarkable heterogeneity in their differentiated characteristics. As illustrated in Figure 5, immunofluorescent staining of postconfluent Caco-2 cells with specific monoclonal antibodies reveals consistantly that the expression of sucrase-isomaltase (Fig. 5A), lactase (Fig. 5B) and aminopeptidase N (Fig. 5C) are highly heterogeneous among differentiated cells; while in contrast, the same cells are homogeneous for dipeptidylpeptidase IV (Fig. 5D). The cloning of the original cell line Caco-2 and the subsequent analysis of sucrase-isomaltase and aminopeptidase N revealed that this mosaic pattern of differentiation is likely an intrinsic property of the Caco-2 cell line since all the subclones studied remained heterogeneous for the expression of these two markers (76). The question of whether the mosaic appearence of brush border hydrolases reflects solely individual differentiation-related gene expression or, rather, is a manifestation of an overall enterocytic differentiation process was then investigated. Surprisingly, the morphological differentiation of Caco-2 cells, as evaluated by brush border ultrastructural organization, was also found to proceed heterogeneously (81). Moreover, comparative studies examining the morphological and functional differentiation of Caco-2 cells at various culture stages revealed that this heterogeneity is only transient since after 20-30 days of post-confluent culture, the vast majority of the cells were homogeneously polarized and expressed sucrase-isomaltase (81). Similarly, it was recently shown that the other brush border hydrolases, lactase and aminopeptidase N, were also expressed according to a transient mosaic pattern in these cells (97; Fig. 6). However, unexpectedly, there was no relationship with brush border organization and brush border enzyme expression at the cellular level as assessed by backscattering EM (Fig. 7). Indeed, well-polarized cells and their less polarized neighbours were found indistinctly positive or negative for the differentiation marker. Furthermore, no relationship was observed within brush border hydrolase expression in cells at mid-stage of differentiation (Fig. 8).

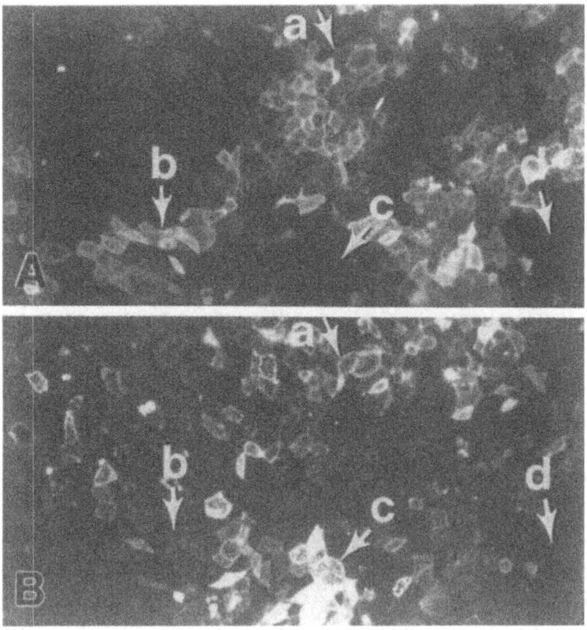

Figure 8. Double immunofluorescent staining for the detection of sucrase-isomaltase (A) and lactase (B) at the apical surface of 12-day postconfluent Caco-2 cells. Positive (a,b) and negative (c,d) cells for sucrase-isomaltase can be found positive (a,c)) or negative (b,d) for lactase expression [Beaulieu and Vachon, unpublished data].

These observations indicated that the morphological and functional differentiation of Caco-2 cells progresses concomitantly according to a transient mosaic pattern which is uncoupled at the cellular level (81,97). Therefore, in contrast to the *in vivo* situation where gene expression appears highly coordinated, such as along the crypt-villus axis (see above sections), the expression of the same genes in Caco-2 cells is clearly regulated but appears totally uncoordinated, suggesting that, an overall control mechanism is lacking in these cells.

The nature and significance of the uncoupled heterogeneity in Caco-2 cell morphological and functional differentiation remain unclear. Interestingly, studies on intestinal epithelial cytodifferentiation with chimeric and transgenic mice have shown the emergence of mosaic patterns in the expression of specific gene markers along the crypt-villus unit, suggesting that this may reflect the differential influences of factors involved in the establishment and maintenance of the enterocytic phenotype (98). This is also reminiscent of the mosaic pattern of lactase expression demonstrated recently with human adult-type hypolactasia (65) or for various intestinal hydrolases in some differentiated colorectal tumors (see the previous section). The Caco-2 cell line, thus, represents a promising model to further analyze, *in vitro*, brush border hydrolases at the molecular and cellular levels and to investigate the mechanisms which trigger, modulate and coordinate their expression.

Acknowledgments

This work was supported by the Medical Research Council of Canada (J.-F. B., D. M.) and the Cancer Research Society (J.-F. B.). J.-F. Beaulieu is a "Chercheur boursier du Fonds de la recherche en santé du Québec".

The authors wish to thank Drs. A. Bilodeau, C. Poulin, M. Morin, F. Jacot, and L. Lemieux of the Département de la santé communautaire, and Dr. J. Poisson of the Département de chirurgie générale, Centre Hospitalier Universitaire de Sherbrooke for their excellent cooperation in providing tissue specimens for this investigation.

REFERENCES

1. Crane RK, Miller D, Bihler I (1961) In Membrane Transport and Metabolism (Kleinzeller A, Kotyk A, eds) Academic Press, New York, pp 439-450
2. Crane RK (1966) Gastroenterology 50: 254-262
3. Colony Moxey P, Trier JS (1978) Anat Rec 191: 269-86
4. Lacroix B, Kedinger M, Simon-Assman P, Haffen K (1984) Gut 25: 925-30
5. Ménard D (1989) In Human Gastrointestinal Development (Lebenthal E, ed.) New York, Raven Press, pp 123-150
6. Aurrichio S, Rubino A, Murset G (1965) Pediatrics 35: 544-54
7. Antonowicz I, Chang SK, Grand RJ (1974) Gastroenterology 67: 51-58

8. Skovbjerg IT (1982) Pediatr Res 16: 948-949
9. Triadou N, Zweibaum A (1985) Pediatr Res 19: 136-138.
10. Beaulieu JF, Jutras S, Kusakabe M, Perreault NC (1993) Biochem Biophys Res Comm 192: 1086-1092
11. Antonowicz I, Lebenthal E (1977) Gastroenterology 72: 1299-1303
12. Dahlqvist A, Lindberg T (1966) Biol Neonate 9: 24-32
13. Ménard D, Pothier P (1987) J Pediatr Gastroenterol Nutr 6: 509-516
14. Grand RJ, Watkins JB, Torti FM (1976) Gastroenterology 70: 790-810
15. Aurrichio S, Stallato A, De Vizia B (1981) Pediatr Res 15: 991-995
16. Arsenault P, Ménard D (1989) Biol Neonate 55: 137-142
17. Raul F, Lacroix B, Aprahamian M (1986) Early Hum Dev 13: 225-234
18. Zweibaum A, Triadou N, Kedinger M, Augeron C, Robine-Léon S, Pinto M, Rousset M, Haffen K (1983) Int J Cancer 32: 407-412
19. Lacroix B, Kedinger M, Simon-Assmann P, Rousset M, Zweibaum A., Haffen K (1984) Early Hum Dev 9: 95-103
20. Koldovsky O (1981) In Gastrointestinal Development and Perinatal Nutrition (Lebenthal E, ed) Raven Press, New York, pp 77-82
21. Henning SJ (1985) Ann Rev Physiol 47: 231-245
22. Ménard D, Calvert R (1990) In Growth of the Gastrointestinal Tract (Morisset J, Solomon T, eds) CRC Press, Boca Raton, pp 147-162
23. Ménard D, Arsenault P (1985) Gastroenterology 88: 691-700
24. Ménard D, Arsenault P (1987) Anat Embryol 176: 441-448
25. Ménard D (1989) In Adaptation and Development of Gastrointestinal Function (Smith MW, Sepulveda FV, eds) Manchester University Press, Manchester, pp 79-91
26. Arsenault P, Ménard D (1985) J Pediatr Gastroenterol Nutr 4: 893-901
27. Simon-Assmann P, Lacroix B, Kedinger M, Haffen K (1986) Early Human Dev 13: 65-74
28. Ménard D, Corriveau L, Arsenault P (1990) J Pediatr Gastroenterol Nutr 10: 13-20
29. Ménard D, Arsenault P, Pothier P (1988) Gastroenterology 94: 656-663
30. Sebastio G, Hunziker W, O'Neill B, Malo C, Ménard D, Aurrichio S, Semenza G (1987) Biochem Biophys Res Comm 149: 830-839
31. Villa M, Ménard D, Semenza G, Mantei N (1992) FEBS Lett 301: 202-206
32. Boll W, Wagner P, Mantei N (1991) Am J Human Genet 48: 889-902
33. Beaulieu JF, Vachon PH, Chartrand S (1991) Anat Embryol 183: 363-369
34. Beaulieu, JF (1992) J Cell Sci 102: 427-436
35. Beaulieu JF, Millane G, Calvert R (1992) Develop Dynam 193: 325-331
36. Leblond CP (1981) Am J Anat 160: 114-158
37. Semenza G (1981) In Carbohydrate Metabolism and its Disorders (Randle PJ, Steiner DF, Whelan WJ, eds) Academic Press, London, pp 691-706
38. Semenza G (1986) Ann Rev Cell Biol 2: 255-314
39. Beaulieu JF, Nichols B, Quaroni A (1989) J Biol Chem 264: 20000-20011

40. Beaulieu JF, Weiser MM, Herrera L, Quaroni A (1990) Gastroenterology 98: 1467-1477
41. Traber PG, Yul WW GD, Judge TA (1992) Am J Physiol 262: G123-G130
42. Chantret I, Lacosa M, Chevalier G, Ruf J, Islam I, Mantei N, Edwards Y, Swallow D, Rousset M (1992) Biochem J 285: 915-923
43. Aurrichio S, Prader A, Mürset G, Witt G (1961) Helv Paediatr Acta 16: 483
44. Holzel A, Schwartz V, Sutcliffe KW (1959) Lancet 1: 1126-1128
45. Freiburghaus AV, Schmitz J, Schindler M (1976) N Engl J Med 294: 1030-1032
46. Aurrichio S, Rubino A, Landolt M, Semenza G, Prader A (1963) Lancet 2: 324-326
47. Dahlqvist A, Hammond JB, Crane RK, Durphy J, Littman A (1963) Gastroenterology 45: 488-491
48. Bergoz R (1971) Gastroenterology 60: 909-912
49. Hadorn G, Tarlow M, Lloyd JK, Wolff OH (1969) Lancet 1: 812-813
50. Dubs RB, Steinmann B, Gitzelmann R (1973) Helv Paediatr Acta 28: 187-198
51. Preiser H, Ménard D, Crane RK, Cerda J (1974) Biochim Biophys Acta 363: 279-282
52. Crane RK, Ménard D, Preiser H, Cerda J (1976) In Membranes and Disease (Bolis L, Hoffman JF, Leaf A, eds) Raven Press, New York, pp 229-241
53. Gray GM, Conklin KA, Townley RRW (1976) N Engl J Med 294: 750-753
54. Schmitz J, Bresson JL, Triadou N, Bataille J, Rey J (1980) Gastroenterol Clin Biol 4: 251-256
55. Naim HY, Roth J, Sterchi EE, Lentze M, Milla P, Schmitz J, Hauri HP (1988) J Clin Invest 82: 667-679
56. Fransen JAM, Hauri HP, Ginsel LA, Naim HY (1991) J Cell Biol 115: 45-57
57. Lloyd ML, Olsen WA (1987) N Engl J Med 316: 438-442
58. Hauri HP, Roth J, Sterchi E, Lentze M (1985) Proc Natl Acad Sci USA 82: 4423-4427
59. Semenza G, Aurrichio S (1989) In the metabolic basis of inherited disease (Scriver CR, Beaudet AL, Sly WS, Valle D, eds) Mc Graw-Hill, New York, pp 2975-2997
60. Sterchi EE, Mills PR, Fransen JAM, Gauri HP, Lentze MJ (1990) J Clin Invest 86: 1329-1337
61. Witt J, Lloyd M, Lorenzsonn V, Korsmo H, Olsen W (1990) J Clin Invest 86: 1338-1342
62. Lloyd M, Vevissen G, Fiser M, Olsen W, Goodsperd D, Genini M, Boll W, Semenza G, Mantei N (1992) J Clin Invest 89: 524-529
63. Sebastio G, Villa M, Sartorio R, Guzzetta V, Poggi V, Aurrichio S, Boll W, Mantei N, Semenza G (1989) Am J Hum Genet 45: 489-497
64. Escher JC, de Koning ND, van Engen CGJ, Arora S, Buller HA, Montgomery RK, Grand RJ (1992) J Clin Invest 89: 480-483

65. Maiuri L, Raia V, Potter J, Swallow D, Ho MW, Fiocca R, Finzi G, Cornaggia M, Capella C, Quaroni A, Aurrichio S (1991) Gastroenterology 100: 359-369
66. Zweibaum A, Hauri HP, Sterchi E, Chantret I, Haffen K, Bamat J, Sordat B (1984) Int J Cancer 34: 591-598
67. Zweibaum A, Chantret I (1989) In Adaptation and development of gastrointestinal function (Smith MW, Sepulveda FV, eds). New-York, Manchester: Manchester University Press, 1989: 103-112
68. Wiltz O, O'Hara CJ, Steele GD, Mercurio AM (1983) Gastroenterology 100: 1266-1278
69. Czernichow B, Simon-Assmann P, Kedinger M, Arnold C, Parache M, Marescaux J, Zweibaum A, Haffen K (1988) Int J Cancer 44: 238-244
70. Gorvel JP, Ferrero A, Chambraud L, Rigal A, Bonicel J, Maroux S (1991) Gastroenterology 101: 618-625
71. Quaroni A, Nichols BL, Quaroni E, Hurst K, Herrera L, Weiser MM, Hamilton SR (1992) Int J Cancer 51:404-411
72. Real FX, Xu M, Vila MR, Debolos C (1992) Int J Cancer 51:173-181
73. Hauri H-P, Sterchi EE, Bienz D, Fransen JAM, Marxer A (1985) J Cell Biol 101: 838-851
74. Rousset M, Chantret I, Darmoul D, Trugnan G, Sapin C, Green F, Swallow D, Zweibaum A (1989) J Cell Physiol 141: 627-635
75. Matsumoto H, Erickson RH, Gum JR, Yoshioka M, Gum E, Kim YS. Biosynthesis of alkaline phosphatase during differentiation of the human colon cancer cell line Caco-2. Gastroenterology 1990; 98: 1199-1207
76. Beaulieu J-F, Quaroni A. (1991) Biochem J 280: 599-608
77. Gilbert T, Le Bivic A, Quaroni A, Rodriguez-Boulan E (1991) J Cell Biol 113: 275-288
78. Fogh J, Fogh JM, Orfeo T (1977) J Natl Cancer Inst 59: 221-226
79. Pinto M, Robine-Léon S, Appay MD, Kedinger M, Triadou N, Dussaulx E, Lacroix B, Simon-Assmann P, Haffen K, Fogh J, Zweibaum A (1983) Biol Cell 47: 323-330
80. Blais A, Bissonette P, Berteloot A (1987) J Membrane Biol 99: 113-125
81. Vachon PH, Beaulieu J-F (1992) Gastroenterology 103: 414-423
82. Chantret I, Barbat A, Dussaulx E, Brattain MG, Zweibaum A (1988) Cancer Res 48: 1936-1942
83. Rutzky LP, Moyer MP (1990) In: Colon Cancer Cells (Moyer MP and Poste GH eds.), pp. 155-202, Academic Press, SanDiego
84. Rousset M (1986) Biochimie 68: 1035-1040
85. Triadou N, Zweibaum A (1985) Pediatr Res 19: 136-138
86. Grasset E, Pinto M, Dussaulx E, Zweibaum A, Desjeux J-F (1984) Am J Physiol 247: C260-C267
87. Hidalgo IJ, Raub TJ, Borchardt RT (1989) Gastroenterology 96: 736-749
88. Hughson EJ, Hopkins CR (1989) J Cell Biol 110: 337-348

89. Dix CJ, Hassan IF, Obray HY, Shah R, Wilson G(1990) Gastroenterology 98: 1272-1279
90. Quick TC, Ong DE. Vitamin A metabolism in the human intestinal Caco-2 cell line. Biochemistry 1990; 29: 11 116-11 123
91. Rousset M, Trugnan G, Brun J-L, Zweibaum A (1986) FEBS Lett 208: 34-38
92. Daniele B, Quaroni A (1990) Gastroenterology 99: 1675-1687
93. Traber MG, Kayden HJ, Rindler MJ (1987) J. Lipid Res. 28: 1350-1363
94. Vachon PH, Durand J, Beaulieu J-F (1993) Anat Rec 236: 567-576
95. Darmoul D, Lacasa M, Baricault L, Marguet D, Sapin C, Trotot P, Barbat A, Trugnan G (1992) J biol Chem 267: 4824-4833
96. Cross HS, Quaroni A (1991) Am J Physiol 261: C1173-C1183
97. Vachon PH, Beaulieu J-F (1993) Gastroenterology 104: 1064a
98. Gordon JI, Schmidt GH, Roth KA (1992) FASEB J 6: 3039-3050

CHAPTER 19

CYTOSOLIC Ca^{2+} HOMEOSTASIS IN HUMAN HeLa CARCINOMA CELLS

Rémy Sauvé, Rui Wang and Line Garneau

Research Group on Membrane Transport
Department of Physiology,
University of Montreal, Canada

INTRODUCTION

It is now well established that the stimulation of a wide variety of membrane receptors in electrically non-excitable cells causes a biphasic elevation in cytosolic Ca^{2+} concentration characterized by a rapid initial Ca^{2+} rise, followed by a long-lasting entry of Ca^{2+} from the extracellular medium (1, 2). There is currently considerable evidence that the initial transient increase in Ca^{2+} concentration initiated following membrane receptor stimulation reflects the breakdown by phospholipase C of the membrane inositol lipid phosphatidylinositol-4,5-diphosphate, with the concomitant release of the putative Ca^{2+} mobilizing second messenger inositol 1,4,5-triphosphate ($InsP_3$) (2, 3, 4). Moreover, cytosolic Ca^{2+} measurements carried out in the presence of mitochondrial inhibitory agents have revealed that the initial Ca^{2+} rise evoked by agonists that operate through the $InsP_3$ system is independent of external Ca^{2+}. This was regarded as an indication that $InsP_3$-sensitive and potentially $InsP_3$-insensitive internal Ca^{2+} stores constitute the main source of Ca^{2+} ions in this case (2, 5). There was also additional evidence that the site of Ca^{2+} release in non-excitable cells is the endoplasmic reticulum (ER) or a specialized subcompartment of the ER (5, 6, 7).

The molecular mechanism that underlies the secondary phase of the receptor-evoked Ca^{2+} response in non-excitable cells remains still poorly circumscribed. It is generally agreed that part of the cytosolic Ca^{2+} elevation observed following membrane receptor stimulation is related to an entry of Ca^{2+} from the extracellular medium since a sustained Ca^{2+} increase was never observed in external Ca^{2+} free conditions (2). Several Ca^{2+} permeable pathways have been identified already. Stretch or shear stress-activated channels, as well as receptor and/or second messenger-operated Ca^{2+} channels, have been reported in a wide variety of cellular preparations (see for instance (8, 9)). Inositol 1, 3, 4, 5-tetrakisphosphate has been often described as an important agent mediating Ca^{2+} entry following membrane receptor stimulation (1). It has also been suggested that $InsP_3$ is responsible for the agonist-evoked Ca^{2+} influx in certain cell types, either directly by acting on $InsP_3$ receptors located at the plasma membrane level, or indirectly by communication of $InsP_3$-sensitive Ca^{2+} pools with the plasma membrane (2, 3). In T cells for instance, the $InsP_3$ receptor appears to be responsible for the entry of Ca^{2+} that initiates proliferative responses (1, 2, 3). There is also increasing evidence that the agonist-stimulated Ca^{2+} entry in non-excitable cells involves a pathway controlled by the fullness state of the intracellular Ca^{2+} stores. According to this mechanism, termed capacitative model, the Ca^{2+} influx across the plasma membrane is secondarily activated by emptying the intracellular Ca^{2+} stores (10, 11, 12). Such a model has been confirmed in mast cells using both electrophysiological and fluorescent Ca^{2+} indicator techniques (13). Despite an obvious need for more information on the nature of the Ca^{2+} entry mechanisms in non-excitable cells, it is clear now that the receptor-evoked Ca^{2+} influx in this case is not mediated by classical voltage-operated-Ca^{2+}-channels (VOC). In most non-excitable cells for instance, the Ca^{2+} influx linked to receptor stimulation is not sensitive to well known VOC agonist and/or antagonist drugs. In addition, depolarizing membrane potentials caused a decrease, not an increase in intracellular Ca^{2+} concentration. This effect was attributed to a decreased Ca^{2+} entry due to the lower electrochemical driving force acting on external Ca^{2+} ions under these conditions.

The increase in cytosolic Ca^{2+} initiated in response to membrane receptor stimulation cannot be always described in terms of an initial Ca^{2+} rise followed by a plateau phase where the Ca^{2+} level remains at a more or less constant value. For instance, in HeLa cells, an epithelial cell line derived from an epidermoid carcinoma of the cervix, the cytosolic Ca^{2+} response evoked following H_1 receptor stimulation by histamine is often characterized by the appearance of periodic Ca^{2+} spikes occurring at low frequency (1-6/min) (14, 15). Ca^{2+} oscillations have been reported in a large variety of non-excitable cells stimulated with agonists known to be linked to the phosphatidylinositol signaling pathway (8, 16). One important feature of agonist-evoked Ca^{2+} oscillations concerns the generation of Ca^{2+} spikes in the absence of external Ca^{2+}. This observation indicates quite clearly that the underlying mechanism is not related to the cyclic activation of a Ca^{2+} influx pathway as reported in many studies on excitable cells. However, the presence of external Ca^{2+} appeared essential to sustain the appearance of periodic Ca^{2+} spikes. Therefore, agonist-evoked Ca^{2+} oscillations appear as a general phenomenon likely to arise from the interaction of intracellular Ca^{2+} pools involved in intracellular Ca^{2+} homeostasis.

Ca^{2+} OSCILLATIONS: GENERAL CONSIDERATIONS

The fundamental mechanism underlying cytosolic Ca^{2+} oscillations in non-excitable cells remains still a matter of debates. Four different classes of models have so far been proposed namely: (a) models in which the cyclic release of a second messenger, such as $InsP_3$, drives the Ca^{2+} oscillations (17, 18); (b) models where an $InsP_3$-sensitive Ca^{2+} pool provides the primer Ca^{2+} which overloads and triggers a Ca^{2+} release from an $InsP_3$-insensitive store (2, 16, 19); (c) models in which Ca^{2+} oscillations are generated through feedback activation and inhibition by Ca^{2+} of the $InsP_3$ dependent Ca^{2+} release process (20, 21, 22, 23) and, finally, (d) models based on the cooperative effect of $InsP_3$ releasing Ca^{2+} from internal stores and Ca^{2+} stimulating phospholipase C (PLC) (24, 25). All these mechanisms can probably coexist at the same time, but are expressed at different degrees depending upon the cell type. One of the key features common to all these models is the requirement of a positive feedback where Ca^{2+} amplifies its own release. However, there are major discrepencies concerning the role in the Ca^{2+} spike generation process of the $InsP_3$ production machinery and the nature of the Ca^{2+} pools involved in Ca^{2+} oscillations.

Ca^{2+} OSCILLATIONS IN HeLa CELLS AND THE CYCLIC PRODUCTION OF $InsP_3$

Many reports have shown that the frequency of the receptor-evoked Ca^{2+} oscillatory process is a function of the agonist concentration in the external medium (17, 26, 27). Moreover, the shape and temporal behaviour of Ca^{2+} oscillations in some cell types, appeared to vary according to the agonist used (see for instance 18). These observations led to the proposal that each agonist which operates through the phosphoinositide pathway is capable of activating the mechanisms underlying the generation of Ca^{2+} spikes in a unique and specific manner. A model in which Ca^{2+} oscillations arise from the cyclic production of $InsP_3$ can easily take into account such a behaviour. Central to this model is the feedback control of the $InsP_3$ production by protein kinase C through an inhibition of phospholipase C (17, 18). It is indeed well established that the hydrolysis of phosphatidylinositol-4,5-diphosphate leads not only to the formation of $InsP_3$ but also to the production of diacylglycerol, a lipid soluble agent that activates protein kinase C (1, 2). Furthermore, there is strong evidence that protein kinase C can down regulate phospholipase C activity (28, 29). Finally, several studies have indicated that the activities of protein kinase C and of phospholipase C are sensitive to cytosolic Ca^{2+} (24, 30). It follows that the combined action of cytosolic Ca^{2+} and diacylglycerol on protein kinase C is likely to result in a reduced $InsP_3$ production, and thus in a decreased Ca^{2+} mobilization from $InsP_3$-sensitive Ca^{2+} pools. Because Ca^{2+} sequestration in $InsP_3$-sensitive Ca^{2+} pools would be optimized under such conditions, there should be an uncompensated decrease in cytosolic Ca^{2+} concentration to basal level, leading to a reduced protein kinase C activity and a regeneration of the $InsP_3$-production system. It has been proposed that an $InsP_3$ production controlled by protein kinase C would constitute a reasonable model for Ca^{2+} oscillations in HeLa cells, since the histamine-induced inositol phosphate formation in these cells appeared dependent on cytosolic Ca^{2+} and could be inhibited by pretreatment with the protein kinase C activator PMA (31, 32). Furthermore, because the $InsP_3$ level is only transiently elevated in histamine-stimulated HeLa cells (< 10 s), it was suggested that the generation of Ca^{2+} spikes in this case arises from periodic pulses of $InsP_3$ (32). However, it has to be mentioned that no significant increase in phospholipase C activity was observed within the Ca^{2+} concentration range 100 nM to 1 μM (32). A control of phospholipase C activity by cytosolic Ca^{2+} during Ca^{2+} oscillations thus appears unlikely.

Because an estimation of the $InsP_3$ concentration at the single cell level is currently out of reach, the validity of this model cannot be assessed directly. However, it is possible to verify whether or not Ca^{2+} oscillations can occur under conditions where the $InsP_3$ concentration is held constant. This problem was approached experimentally by recording Ca^{2+} fluctuations in whole-cell patch-clamp experiments using patch pipettes containing a non-metabolizable $InsP_3$ analogue. Ca^{2+} fluctuations then appear as current variations coming from the activation of Ca^{2+}-dependent ionic channels. For instance, fluctuations in cytosolic Ca^{2+} have been measured in HeLa cells through the activation of a Ca^{2+}-dependent K^+ channel ($K(Ca^{2+})$) of small conductance (40 pS in 200 mM KCl) (15). Patch excised single channel recordings have shown, in this particular case, that the $K(Ca^{2+})$ channel in these cells is voltage insensitive within the voltage range -100 to 0 mV, and responds to submicromolar variations in cytosolic Ca^{2+} concentration (33). In addition, there was no indications of a second class of $K(Ca^{2+})$ channels with different biophysical properties (33). A K^+ selective channel characterized by a strong inward rectifying current/voltage curve was identified, but there was no evidence of cytosolic Ca^{2+} being involved in channel regulation in this case. These observations led to the conclusion that a $K(Ca^{2+})$ channel with voltage independent gating properties would constitute an excellent endogenous Ca^{2+} probe well suited to monitor Ca^{2+} fluctuations occurring at the plasma membrane inner phase. Clearly, this experimental approach can provide information otherwise not accessible through microspectrofluorometric measurements where an average Ca^{2+} concentration is measured over the entire cell.

An example of a current trace measured in whole-cell configuration with a patch electrode containing the poorly metabolizable $InsP_3$ analogue 3-F-$InsP_3$ (D-myo-inositol 1,4,5-triphosphate,3-Deoxy-3-Fluoro-) is presented in Figure 1. Cyclic variations in K^+ current are observed indicating the occurrence of repetitive cytosolic

Ca^{2+} pulses. This conclusion also derives from results coming from cell-attached patch-clamp experiments where pulses of single channel activity were recorded following H_1 receptor stimulation. Furthermore, the single channel current/voltage relationship measured in cell-attached during a burst of single channel activity was

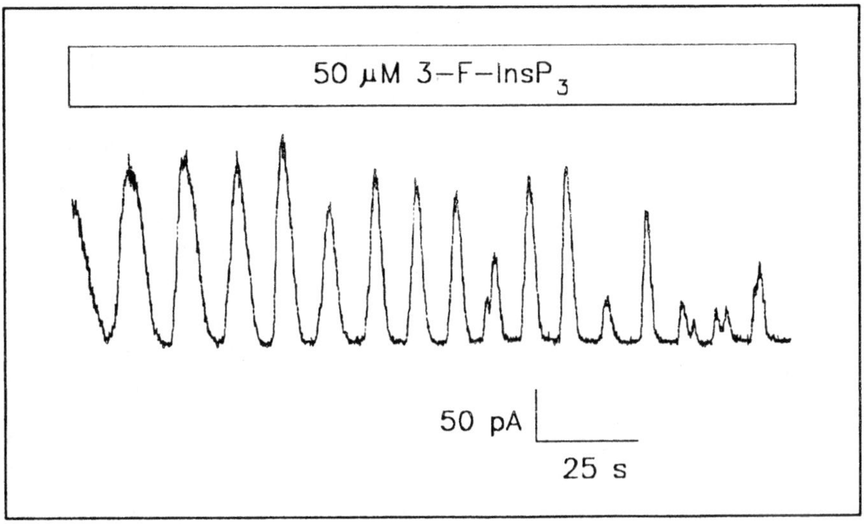

Figure 1. *Whole-cell patch-clamp recording of repetitive K(Ca^{2+}) current pulses measured with a patch pipettte containing 50 μM 3-F-$InsP_3$ in a K^+ solution which consisted of (in mM) 140 K^+-glutamate, 10 NaCl, 1 $MgCl_2$, 1 Na-ATP, 0.6 Na-GTP buffered at pH 7.3 with 4 KOH and 10 HEPES. The external bathing solution was an Earle medium having the standard composition (in mM): 121.3 NaCl, 6.0 $NaHCO_3$, 1.0 NaH_2PO_4, 5.4 KCl, 0.8 $MgSO_4$, 1.8 $CaCl_2$ and 5.5 glucose buffered at 7.3 with 25 HEPES plus 10 mM NaOH. The potential was maintained to 0 mV throughout. All experiments were carried out at room temperature.*

found to be identical to that obtained for the K(Ca^{2+}) channel in excised patch-clamp experiments. These observations were regarded as direct evidence that the current variations illustrated in Figure 1 reflect temporal changes in cytosolic Ca^{2+} concentration. The current recording presented in Figure 1 would thus support the proposal of Ca^{2+} oscillations taking place in HeLa cells without repetitive pulses of $InsP_3$. Similar observations were reported in a large variety of non-excitable cells including pancreatic acinar cells and golden hamster eggs (34, 35). For instance, whole-cell recordings carried out on single pancreatic acinar cells with patch electrodes containing the non-metabolizable $InsP_3$ analogue $InsPS_3$ quite clearly demonstrated that cyclic Ca^{2+}-dependent Cl^- currents could be generated at constant intracellular $InsPS_3$ level (34). These observations were interpreted as indications that $InsP_3$ fluctuations in pancreatic acinar cells are not essential to initiate a Ca^{2+} oscillatory activity. However, these experiments do not totally rule out the possibility that endogenous $InsP_3$ can act locally to cause a transient Ca^{2+} release from Ca^{2+} pools located at the plasma membrane inner face.

Ca^{2+} OSCILLATIONS IN HeLa CELLS AND Ca^{2+}-INDUCED-Ca^{2+}-RELEASE

Considerable attention has been so far given to the so-called two-pool model where the $InsP_3$-sensitive pools do not participate in the oscillatory activity, but act indirectly by supplying Ca^{2+} to Ca^{2+} reservoirs with Ca^{2+}-induced-Ca^{2+}-release (CICR) properties. CICR was first described by Endo et al. (1970) (36) and characterized later in details by Fabiato (1985) in skinned canine cardiac Purkinje cells (37). The CICR process in excitable cells corresponds essentially to a release of Ca^{2+} from the sarcoplasmic reticulum (SR) in response to a rapid rise in cytosolic Ca^{2+} concentration. The molecular structure responsible for CICR in the SR has been recently identified as a Ca^{2+} permeable channel, the activity of which is sensitive to cytosolic Ca^{2+} and ATP (38). One key prediction of models based on CICR is that an increase in intracellular Ca^{2+} can lead, under the proper conditions, to Ca^{2+} oscillations without $InsP_3$ production. Evidence that Ca^{2+} oscillations can be initiated without $InsP_3$ production has been reported in several studies. For instance, it was observed in pancreatic acinar cells that the inhibitor of Ca^{2+}-ATPase in endoplasmic reticulum membrane, thapsigargin (TG), initiates Ca^{2+} oscillations similar to those evoked following receptor stimulation but without $InsP_3$ production (39, 40). It was proposed that TG caused an influx of Ca^{2+} that fueled in turn a CICR mechanism from an $InsP_3$-insensitive Ca^{2+} pool (40, 41). The influx of Ca^{2+} triggered by TG would arise from the depletion of the $InsP_3$-sensitive Ca^{2+} pools in accordance with the capacitative model proposed by Putney (12). Similar observations were reported in hepatocytes using the monohydroxylated bile acid taurolithocholate (TLC-S), which is known for its ability in mobilizing Ca^{2+} from the endoplasmic reticulum in these cells (42).

The contribution of a CICR process to Ca^{2+} oscillations was also confirmed using agents known to modify the Ca^{2+}-dependent-Ca^{2+} release channel in muscle preparations. For instance, caffeine, a methylxanthine reported to activate the Ca^{2+}-dependent-Ca^{2+} release channel in sarcoplasmic reticulum (43), was found to either potentiate at a low concentration (1 mM) the $InsP_3$-evoked Ca^{2+} fluctuations or to directly induce a Ca^{2+} spiking activity under certain conditions (41, 44, 45). In contrast, an inhibitory action on Ca^{2+} oscillations was observed in acinar and chromaffin cells with ryanodine, a plant alkaloid that binds and at the proper concentration either blocks or maintains the SR Ca^{2+} release channel in an open state of lower conductance (41, 46). It has to be mentioned that caffeine, at high concentrations (20 mM), was found also to block the Ca^{2+} oscillations induced by $InsP_3$ in several preparations (45, 47). This effect was originally attributed to a discharge of the $InsP_3$-insensitive Ca^{2+} pools and thus to a suppression of the Ca^{2+} oscillation process (47). However, the conclusions derived on the basis of results obtained with these agents remain limited to the SR Ca^{2+} release channel. Other

Ca^{2+}-release channels may have different pharmacological properties. For instance, a study by Schmid et al. (1990) (48) has shown that the endoplasmic reticulum membrane of pancreatic acinar cells contains an $InsP_3$- and ryanodine-insensitive Ca^{2+} channel that can be activated by caffeine and blocked by 10 μM ruthenium red. A potential contribution of this channel to cytosolic Ca^{2+} oscillations cannot currently be ruled out.

Direct evidence to support the idea that Ca^{2+} oscillations can be generated in some cell types due to a CICR process in $InsP_3$-insensitive Ca^{2+} pools was finally obtained from experiments in which an action of $InsP_3$ on Ca^{2+} mobilization was prevented using the $InsP_3$ receptor antagonist, heparin. In pancreatic acinar cells for instance, periodic Ca^{2+}-dependent Cl^- current pulses were successfully triggered by Ca^{2+} infusion in the presence of heparin, whereas heparin blocked the same Cl^- currents following $InsP_3$ application in whole-cell patch-clamp experiments (45, 49). These experiments provided direct support to a model in which Ca^{2+} oscillations are not mediated by $InsP_3$-gated Ca^{2+} channels. Such a behaviour would be in accordance with the two-pool model presented previously, if one assumes that the production of $InsP_3$ in whole-cell experiments is not significantly affected by sudden changes in cytosolic Ca^{2+} concentration.

Ca^{2+}-REGULATION OF THE $InsP_3$-INDUCED Ca^{2+} RELEASE PROCESS: IMPLICATION FOR Ca^{2+} OSCILLATIONS IN HeLa CELLS

In many cases, the applicability of the two-pool model can be questioned, suggesting that other mechanisms may also account, at least in part, for the generation of receptor-evoked Ca^{2+} spikes. In HeLa cells, for instance, we observed that TG causes a release of Ca^{2+} from intracellular pools and stimulates the entry of Ca^{2+} without actually generating intracellular Ca^{2+} spikes (50). In fact, our results provided evidence that the action of TG resulted in a depletion of the $InsP_3$-sensitive Ca^{2+} pools, but an action of TG on $InsP_3$-insensitive Ca^{2+} reservoirs could not be ruled out entirely. Finally, TG was found to irreversibly inhibit the Ca^{2+} oscillations resulting from the stimulation of H_1 receptors by exogenous histamine. On the basis of these observations, it was concluded: 1) that the entry of Ca^{2+} in these cells is mainly controlled by the emptiness of the Ca^{2+} pools, as predicted by the capacitative model of Putney (12), and 2) that a model in which TG initiates Ca^{2+} oscillations by generating a Ca^{2+} influx which, in turn, provides Ca^{2+} to an $InsP_3$-insensitive Ca^{2+} pool with CICR properties is not applicable in this case. However, such a model could still be valid if the action of TG is not strictly limited to the $InsP_3$-sensitive Ca^{2+} reservoirs but also extends to $InsP_3$-insensitive Ca^{2+} pools (see for instance (51, 52)). Under these conditions, TG would affect not only the $InsP_3$-sensitive Ca^{2+} stores, but also the $InsP_3$-insensitive Ca^{2+} pools involved in CICR. In both cases, the generation of Ca^{2+} spikes would be inhibited by TG due to its action on the Ca^{2+} uptake process that normally insures the refilling of the Ca^{2+} pools involved in Ca^{2+} oscillations.

Another aspect of the Ca^{2+} response in HeLa cells that provides evidence against Ca^{2+} oscillations being due to an SR type CICR mechanism, concerns the effect of the CICR promotor caffeine. Caffeine is known to cause in muscle cell preparations a release of Ca^{2+} from the SR by increasing the affinity of the Ca^{2+}-release channel to cytosolic Ca^{2+} (43). The results obtained so far on HeLa cells indicate: 1) that caffeine does not mobilize Ca^{2+} in resting cells or in cells stimulated by histamine or TG and 2) that caffeine increases the Ca^{2+} uptake in $InsP_3$-sensitive pools in conditions where the H_1 receptor is stimulated by exogenous histamine (50). The latter conclusion was derived from fura-2 experiments performed in Ca^{2+} free conditions, where the release of Ca^{2+} in response to a second application of histamine appeared substantially higher in cells that were exposed to caffeine during the first histamine application, than in cells initially stimulated by histamine in the absence of caffeine. In fact, there is now increasing evidence that the release of Ca^{2+} from $InsP_3$-sensitive Ca^{2+} stores is inhibited by caffeine at high concentrations (22, 53, 54). For instance, a recent study by Brown et al. (1992) has provided data consistent with the view that the opening of the $InsP_3$-sensitive Ca^{2+} channel from rat cerebellum is inhibited by caffeine at concentrations ranging from 10 to 50 mM (53). Our results would be compatible with such a finding, since a partial inhibition of $InsP_3$-dependent Ca^{2+} release during receptor stimulation should lead to an increased Ca^{2+} uptake in the $InsP_3$-sensitive Ca^{2+} pool, thus creating conditions for a more important liberation of Ca^{2+} in response to a subsequent receptor-evoked $InsP_3$ production. In addition to an effect on Ca^{2+} uptake in $InsP_3$-sensitive Ca^{2+} pools, caffeine was also found to affect Ca^{2+} oscillations in HeLa cells. A strong inhibition by caffeine (10 mM) of the histamine evoked Ca^{2+} oscillatory process was observed in fura-2 and cell-attached patch-clamp experiments where single $K(Ca^{2+})$ channels were used to monitor Ca^{2+} fluctuations occurring at the plasma membrane interface (see Figure 2a). A similar inhibitory effect was recorded in whole-cell patch-clamp experiments where 3-F-$InsP_3$ was added directly to the patch electrode (Figure 2b). The latter experiment was performed to rule out any indirect effect of caffeine on the $InsP_3$ production machinery, as reported in mouse pancreatic acinar cell where caffeine reduced the agonist-evoked $InsP_3$ production (55). In addition, whole-cell experiments in which the production of cAMP was promoted using forskoline (10 μM) failed to inhibit the appearance of $K(Ca^{2+})$ current pulses evoked by 50 μM 3-F-$InsP_3$ in the patch pipette (data not shown). The well documented action of caffeine on phosphodiesterases cannot explain therefore the inhibitory effect illustrated in Figures 2a and 2b. These observations would be also in agreement with the biochemical analysis of Tilly et al. (32), who reported a lack of cross-talk between the $InsP_3$- and cAMP-generating pathways in HeLa cells. Finally, the current traces in Figure 2b show that the removal of caffeine (10 mM) from the bathing medium is correlated with the reappearance of current pulses characterized by an initial current spike of higher peak amplitude. Such a phenomenon was systematically observed in whole-cell and fura-2 experiments, thus providing further support to an action of caffeine leading to more Ca^{2+} being sequestered in the Ca^{2+} pools which sustain the periodic generation of

Ca^{2+} spikes. Together, the inhibition of Ca^{2+} oscillations by caffeine, its promotion of a greater Ca^{2+} uptake in $InsP_3$-sensitive Ca^{2+} pools and the absence of Ca^{2+} spiking activity with TG, support a model in which Ca^{2+} oscillations are closely linked to Ca^{2+} mobilization from the $InsP_3$-Ca^{2+} stores.

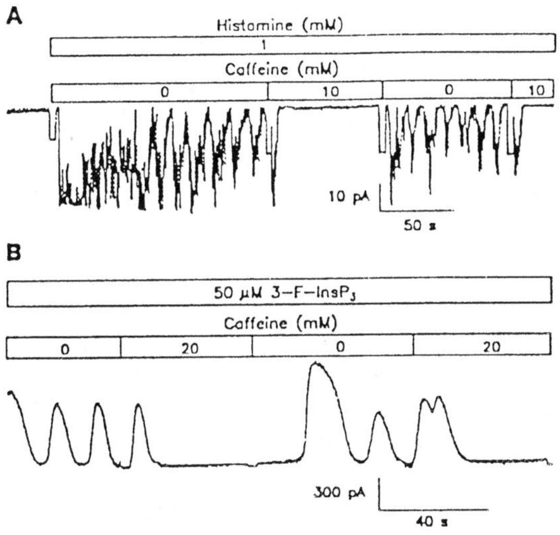

Figure 2. *Effect of caffeine on Ca^{2+} oscillations in HeLa cells.* ***(A)****: Inhibitory effect of caffeine on histamine evoked-periodic activation of single $K(Ca^{2+})$ channels. Patch-clamp current record obtained in cell-attached with a pipette containing (in mM): 200 KCl, 0.5 $MgSO_4$, 1 K_2-EGTA and 0.91 $CaCl_2$ for a free Ca^{2+} concentration of 1 μM. The pH was adjusted to 7.3 with 10 mM KOH and 25 HEPES for a total K^+ concentration of 212 mM. The external medium was a standard Earle solution.* ***(B)****: Inhibition by caffeine of $K(Ca^{2+})$ current pulses evoked by 50 μM 3-F-$InsP_3$ in the patch pipette. Recording conditions as described in fig. 1.*

Evidence against the potential contribution of a caffeine-sensitive CICR process to the Ca^{2+} spike generation in HeLa cells was also obtained using the plant alkaloid ryanodine. Figure 3 presents the result of a whole-cell experiment performed with a patch pipette containing both the $InsP_3$ analogue 3-F-$InsP_3$ and ryanodine (100 μM), in the presence of ryanodine (100 μM) in the bathing solution. The resulting current trace shows clearly that ryanodine failed to affect the appearance of periodic $K(Ca^{2+})$ current pulses in the presence of 3-F-$InsP_3$ in the patch electrode. This finding is not compatible with a model for Ca^{2+} oscillations in which a SR type CICR process would play a key role. However, it must be

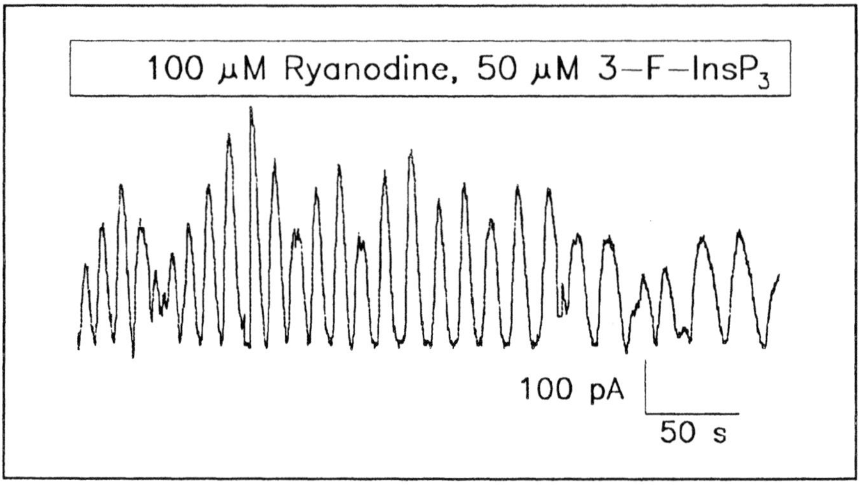

***Figure 3.** Absence of inhibitory effects of ryanodine on $K(Ca^{2+})$ current pulses measured in whole-cell configuration with a patch pipette containing 50 μM F-$InsP_3$ plus 100 μM ryanodine. Experimental conditions as described in fig.1 with 100 μM ryanodine added to the external medium.*

stressed that the pharmacological properties of the Ca^{2+}-release channel in HeLa cells need not to be identical to those prevailing for the Ca^{2+} channel of the SR. As mentioned previously, Ca^{2+}-release channels that are activated by caffeine but remain insensitive to ryanodine have been reported already in endoplasmic reticulum membranes (48). Such Ca^{2+} conducting pathways may be present in HeLa cells, thus providing an explanation for the effects of caffeine and ryanodine measured in patch-clamp experiments. Furthermore, different ryanodine receptors have already been identified (see for instance (1)). One receptor regulated by transforming growth factor beta (TGF-ß), appeared insensitive to caffeine while being still capable of inducing a ryanodine-dependent Ca^{2+} release in mink lung epithelial cells (56). There seems therefore to be a diversity of Ca^{2+}-release channels characterized by different pharmacological properties. However, the evidence gathered so far does not support the presence of a ryanodine-sensitive Ca^{2+}-release channel in HeLa cells. In fact, on the basis of fura-2 experiments carried out with TG, it was concluded that more than 85% of the sequesterd Ca^{2+} in these cells can be mobilized through receptor stimulation. There should be therefore a limited contribution coming from Ca^{2+} stores with CICR properties not related to the $InsP_3$ receptor.

Finally, the contribution of $InsP_3$-sensitive Ca^{2+} pools to the Ca^{2+} oscillatory activity in HeLa cells was recently investigated using thimerosal, a thiol alkylating agent known to mobilize intracellular Ca^{2+} in leucocytes (57), platelets (58) and pancreatic acinar cells (59). Bootman et al. (1992) showed that thimerosal causes a dose-dependent sensitization of $InsP_3$-evoked Ca^{2+} mobilization in HeLa cells

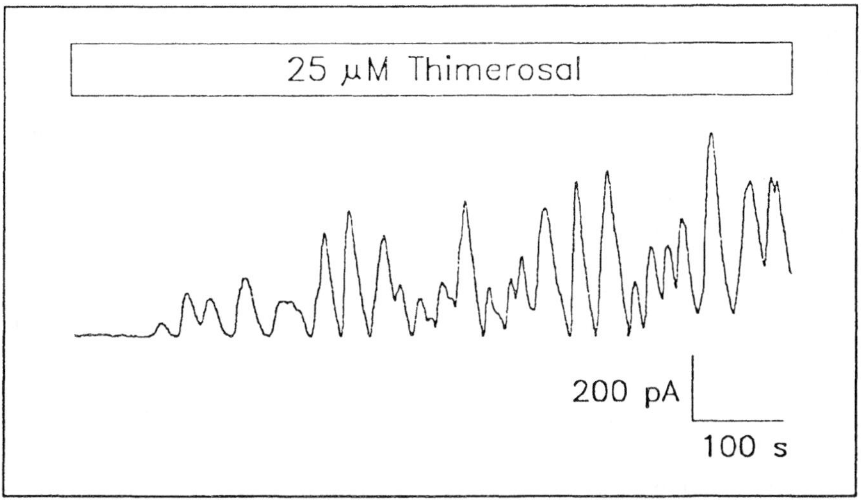

***Figure 4.** $K(Ca^{2+})$ current pulses induced by bath application of thimerosal in the absence of $InsP_3$ in the patch pipette. Whole-cell experiment carried out as described in fig.1.*

without stimulating the production of $InsP_3$ (60). In fact, thimerosal appeared to inhibit inositol phosphate production in this case. In addition, the results obtained with fura-2 loaded cells provided clear evidence that low concentrations of thimerosal lead to Ca^{2+} oscillations similar in shape to those evoked by histamine. These observations were confirmed subsequently in our laboratories in whole-cell current recordings such as the one presented in Figure 4, where thimerosal (25 μM) caused the appearance of repetitive $K(Ca^{2+})$ channel current spikes despite the absence of $InsP_3$ in the patch electrode. Such a behaviour contrasts with the increase in cytosolic Ca^{2+} initiated by TG which failed to induce a Ca^{2+} oscillatory activity. It was proposed that the effect of thimerosal on HeLa cells was due to the sensitization of the $InsP_3$ receptor to resting levels of $InsP_3$ (60). Within this framework, the generation of Ca^{2+} oscillations in HeLa cells would be closely related to the Ca^{2+} mobilization properties of the $InsP_3$ receptor.

What model would then be the most likely to account for the generation of repetitive Ca^{2+} spikes in HeLa cells? Several reports have shown that the release of Ca^{2+} from $InsP_3$-sensitive Ca^{2+} pools is a complex process controlled to a large extent by cytosolic and luminal Ca^{2+} (61). For instance, substantial evidence has been provided indicating that the $InsP_3$-gated Ca^{2+} channel behaves in some cell types as a Ca^{2+}-activated channel at low concentrations (300 nM), while being inhibited at higher Ca^{2+} concentration values (21, 22, 23, 62, 63). In the presence of $InsP_3$, the $InsP_3$ receptor seems therefore to correspond to a Ca^{2+}-release channel, but with a sensitivi/ty to cytosolic Ca^{2+} greater than that reported for the ryanodine receptor (1). Recent results obtained on *Xenopus oocytes* that lack ryanodine receptors have further confirmed that Ca^{2+} waves can be generated by injecting the non-metabolizable analogue $InsPS_3$, whereas Ca^{2+} injection produced local Ca^{2+} increases without wave formation (1). In addition, a recent study by Nunn and Taylor (1992) (61) has established that the release of Ca^{2+} by $InsP_3$ is dependent on the luminal Ca^{2+} concentration in the $InsP_3$-sensitive Ca^{2+} stores. A potential effect of luminal Ca^{2+} on the binding properties of $InsP_3$ to its receptor may be partly responsible for the all-or-non Ca^{2+} mobilization in histamine stimulated HeLa cells reported by Bootman et al. (1992) (64). In fact, heterogeneous $InsP_3$-sensitive Ca^{2+} stores have been implicated in the phenomenon of quantal Ca^{2+} release observed in a variety of cellular types, including HeLa cells. This particular process may indeed be accounted for by a diversity in luminal Ca^{2+} content leading to $InsP_3$ receptors with variable $InsP_3$ binding properties, and/or by an intrinsic receptor heterogeneity. Despite the uncertainties concerning the exact role of luminal Ca^{2+} in the regulation of the $InsP_3$-evoked Ca^{2+} release process, a control by the cytosolic Ca^{2+} of the $InsP_3$-gated Ca^{2+} channel is probably sufficient to lead to Ca^{2+} oscillations in conditions where the $InsP_3$ concentration is constant as observed experimentally. Furthermore, a control by luminal and cytosolic Ca^{2+} of the $InsP_3$-dependent Ca^{2+} release process will likely provide all the required feedback mechanims needed to generate periodic Ca^{2+} spikes. This mechanism would explain the inhibitory action of caffeine and the absence of ryanodine effect on the generation of Ca^{2+} spikes. Finally, a Ca^{2+} oscillatory process based on a dynamic control of the $InsP_3$ receptor is expected to respond to a sensitization of the $InsP_3$ receptor by the generation at low $InsP_3$ levels of Ca^{2+} oscillations. Such a behaviour is in accordance with the results reported to date.

CONCLUDING REMARKS

On the basis of the observations gathered so far, we would like to suggest that the Ca^{2+} oscillations in HeLa cells do not involve a distinct $InsP_3$-insensitive Ca^{2+} store capable of CICR, but that the generation of Ca^{2+} spikes in this case is linked to the control of the Ca^{2+} release process from $InsP_3$-sensitive pools by Ca^{2+}. Our findings would thus be compatible with a revised version of the two-pool model in which the $InsP_3$-sensitive pools are responsible for Ca^{2+} influx and the CICR process that generates Ca^{2+} spikes (64). The physiological significance of Ca^{2+} oscillations is still unclear. It is often proposed that Ca^{2+} oscillations constitute a novel mechanism by which information is coded in response to extracellular stimuli. Despite no direct evidence supporting this hypothesis, an analysis of the cellular mechanisms underlying the appearance of repetitive pulses of cytosolic Ca^{2+} clearly remains essential to better understand the role of Ca^{2+} as a second messenger.

Acknowledgements

Dr. Rémy Sauvé is a senior fellow from the Fonds de la Recherche en Santé du Québec. Dr. Rui Wang is a post-doctoral fellow from the Medical Research Council of Canada. This work was supported by a grant from the Medical Research Council of Canada.

REFERENCES

1. Berridge MJ (1988) Proc R Soc Lond B 234:359-378
2. Berridge MJ, Irvine RF (1989) Nature (London) 341:197-204
3. Streb H, Irvine RF, Berridge MJ, Schulz I (1983) Nature (London) 306:67-69
4. Streb H, Bayerdorffer E, Haase W, Irvine RF, Schulz I (1984) J Membrane Biol 81:241-253
5. Dehlinger-Kremer M, Zeuzem S, Schulz I (1991) J Membrane Biol 119:85-100
6. Bond M, Vadasz G, Somlyo AV, Somlyo AP (1987) J Biol Chem 262:15630-15636
7. Bayerdorffer E, Streb H, Eckhardt L, Hasse W, Schulz I (1984) J Membrane Biol 81:69-82
8. Rink TJ, Hallam TJ (1989) Cell Calcium 10:385-395
9. Himmel HM, Whorton AR, Strauss HC (1993) Hypertension 21:112-127
10. Merritt JE, Rink TJ (1987) J Biol Chem 262:17362-17369
11. Takemura H, Putney JW (1989) Biochem J 258:409-412
12. Putney JW (1990) Cell Calcium 11:611-624
13. Hoth M, Penner R (1992) Nature (London) 355:353-356
14. Sauvé R, Simoneau C, Parent L, Monette R, Roy G (1987) J Membrane Biol 96:199-208
15. Sauvé R, Diarra A, Chahine M, Simoneau C, Morier N, Roy G (1991) Cell Calcium 12:165-176
16. Berridge MJ (1991) Cell Calcium 12:63-72
17. Woods NM, Cuthberson HSR, Cobbold PH (1987) Cell Calcium 8:79-100
18. Cobbold PH, Sanchez-Bueno A, Dixon CJ (1991) Cell Calcium 12:87-95
19. Berridge MJ, Galione A (1988) FASEB J 2:3074-3082
20. Payne R, Flores TM, Fein A (1990) Neuron 4:547-555
21. Parker I, Ivora I (1990) Proc Natl Acad Sci USA 87:260-264
22. Parker I, Ivora I (1991) J Physiol (London) 433:229-240
23. Finch EA, Turner TJ, Goldin SM (1991) Science 252:443-446
24. Meyer T, Stryer L (1988) Proc Natl Acad Sci USA 85:5051-5055
25. Harootunian AT, Kao JPY, Paranjape S, Adams SR, Potter BVL, Tsein RY (1991) Cell Calcium 12:153-164
26. Jacob R (1991) Cell Calcium 12:127-134

27. Rooney TA, Renard DC, Sass EJ, Thomas AP (1991) J Biol Chem 266:12272-12282
28. Brock TA, Rittenhouse SE, Powers CW, Ekstein LS, Gimbrome MA, Alexander RW (1985) J Biol Chem 260:14158-14262
29. Orellana SA, Solski PA, Brown JM (1985) J Biol Chem 260:5236-5239
30. Taylor CW, Berridge MJ, Cooke AM, Potter BVL (1989) Biochem J 259:645-656
31. Tilly BC, Lambrechts AC, Tertoolen LGJ, de Laat SW, Moolenaar WH (1990) FEBS LETT 265:80-84
32. Tilly BC, Tertoolen LGJ, Lambrechts AC, Remorie R, de Laat SW, Moolenaar WH (1990) Biochem J 266:235-243
33. Sauvé R, Simoneau C, Monette R, Roy G (1986) J Membrane Biol 92:269-282
34. Wakui M, Potter BVL, Petersen OH (1989) Nature (London) 339:317-320
35. Swann K, Igusa Y, Miyazaki S (1989) EMBO J 8:3711-3718
36. Endo M, Tanaka M, Ogawa Y (1970) Nature (London) 228:34-36
37. Fabiato A (1985) J Gen Physiol 85:247-289
38. Smith JS, Coronado R, Meissner G (1986) J Gen Physiol 88:573-588
39. Jackson TR, Patterson SI, Thastrup O, Handley MR (1988) Biochem J 253:81-86
40. Foskett JK, Roifman CM, Wong D (1991) J Biol Chem 266:2778-2782
41. Foskett JK, Wong D (1991) J Biol Chem 266:14535-14538
42. Capiod T, Combettes L, Noel J, Claret M 1991 J Biol Chem 266:268-273
43. Rousseau E, LaDine J, Lui QY, Meissner G (1988) Arch Biochem Biophys 267:75-86
44. Neylon CB, Irvine RF (1990) FEBS Lett 275:173-176
45. Wakui M, Osipchuck YV, Petersen OH (1990) Cell 63:1025-1032
46. Malgaroli A, Fesce R, Meldolesi J (1990) J Biol Chem 265:3005-3008
47. Berridge MJ (1991) Proc R Soc London B 244:57-62
48. Schmid A, Dehlinger-Kremer M, Schulz I, Gogelein H (1990) Nature (London) 346:374-376
49. Petersen OH, Gallacher DV, Wakui M, Yule DI, Petersen CCH, Toescu EC (1991) Cell Calcium 12: 135-144
50. Diarra A, Sauvé R (1992) Pflügers Arch 422:40-47
51. Thastrup O (1990) Agents and Actions 29:9-15
52. Rooney TA, Renard DC, Sass EJ, Thomas AP (1991) J Biol Chem 266:12272-12282
53. Brown GR, Sayers LG, Kirk CJ, Michell R, Michelangeli F (1992) Biochem J 282:309-312
54. Messiaen L, Taylor CW, Berridge MJ (1992) J Physiol (London) 455:623-640
55. Toescu EC, O'Neill SC, Petersen OH, Eisner DA (1992) J Biol Chem 267:23467-23470

56. Giannini G, Clementi E, Ceci R, Marziali G, Sorrentino V (1992) Science 257:91-94
57. Hatzelmann A, Haurand M, Ullrich V (1990) Biochem Pharmacol 39:559-567
58. Hecker M, Brune B, Decker K, Ullrich V (1989) Biochem Biophys Res Commun 159:961-968
59. Thorn P, Brady P, Llopis J, Gallacher DV, Petersen OH (1992) Pflügers Arch 422:173-178
60. Bootman MD, Taylor CW, Berridge MJ (1992) J Biol Chem 267:25113-25119
61. Nunn D, Taylor CW (1992) Mol Pharmacology 41:115-119
62. Bezprozvanny I, Watras J, Ehrlich E (1991) Nature (London) 351:751-754
63. Iiono M (1990) J Gen Physiol 95:1103-1112
64. Bootman MD, Berridge MJ, Taylor CW (1992) J Physiol 450:163-178

CHAPTER 20

INOSITOL 1,4,5-TRISPHOSPHATE ANALOGS: A NEW CLASS OF THERAPEUTIC AGENTS.

Darren E. Richard, Sylvie Bernier and Gaétan Guillemette

Department of Pharmacology, Faculty of Medecine, University of Sherbrooke, Quebec, Canada

INTRODUCTION

Many hormones and neurotransmitters exert their biological effects by stimulating the hydrolysis of phosphatidylinositol 4,5-bisphosphate. This perturbation of membrane phospholipids represents a transduction mechanism, implicating receptors that initiate a signal cascade resulting in the mobilization of calcium, the activation of protein kinase C (PKC), and the release of arachidonic acid. A common characteristic of these receptors is that they are linked to a membrane mechanism of transduction that produces second messengers (1-4). The key reaction of this mechanism of transduction is the hydrolysis of a phosphoinositide by phospholipase C (PLC β) (5). This reaction is under the control of a G protein-coupled receptor (6). The products of this hydrolysis are diacylglycerol (DAG) and inositol 1,4,5-trisphosphate ($InsP_3$). DAG stimulates protein kinase C while $InsP_3$ mobilizes intracellular calcium via a specific receptor located on calcium reservoirs (Fig. 1). The $InsP_3$ receptor has been detected and characterized in a wide variety of tissues including the adrenal cortex (7-9), the liver (10-12), the anterior pituitary gland (13,14) and the brain (15,16). It has been purified from rat brain extracts (17), from

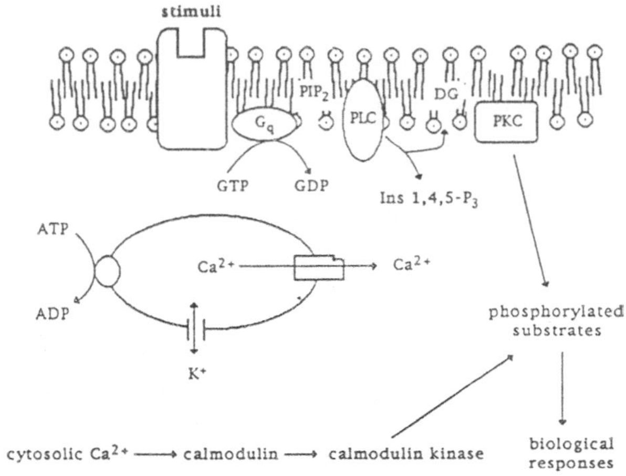

Figure 1. *Shematic representation of the diverse mechanisms implicated in intracellular calcium regulation.*

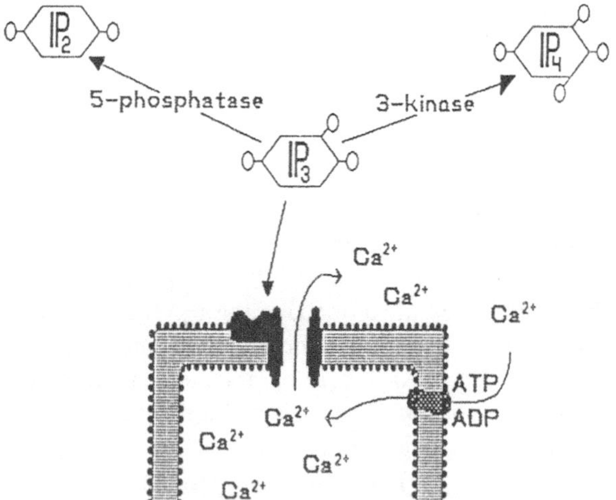

Figure 2. *The different recognition sites of the second messenger $InsP_3$ (Receptor, $InsP_3$ phosphatase and $InsP_3$ kinase).*

bovine aorta smooth muscle (18), from rat vas deferens (19) and partially purified from bovine liver extracts (20). It has been cloned and sequenced from rat and mouse cerebellum extracts (21,22). It has also been demonstrated by reconstitution in lipid vesicles that this glycoprotein is a calcium channel (23). During stimulations with Ca^{2+} mobilizing hormones, $InsP_3$ does not accumulate inside the cells, since it is rapidly metabolized by very active enzymes (Fig. 2). Inositol 1,4,5-trisphosphate phosphomonoesterase ($InsP_3$ phosphatase) removes the phosphate group at position 5 to produce inositol 1,4-bisphosphate ($InsP_2$) (24), an inactive metabolite in regards to Ca^{2+} mobilization. Inositol 1,4,5-trisphosphate kinase ($InsP_3$ kinase) is another well characterized enzyme that adds a phosphate group at position 3 to produce inositol 1,3,4,5-tetrakisphosphate ($InsP_4$) (25,26). $InsP_4$ has been suggested to be involved in the mechanism of Ca^{2+} entry into cells (27).

Pathologies Related to $InsP_3$ Signaling and Metabolism

Some efforts have been devoted to try and identify pathologies or physiological disorders that could be linked to the activity of the $InsP_3$ receptor, to Ca^{2+} mobilization, or to $InsP_3$ metabolism. In the course of a study on ataxic mutant mice, Furuichi et al. (21) found a membrane glycoprotein P_{400} that was reduced in the cerebellum of deficient mutants. This P_{400} protein was found to be identical to the $InsP_3$ receptor protein.

It has been suggested that the $InsP_3$ receptor is implicated in olivopontocerebellar atrophy (OPCA) (brain disorder causing a problem of muscular incoordination). This disorder is characterized by a severe degeneration of Purkinje cells. An important reduction (61%) of $InsP_3$ receptor binding was found in the cerebellum of patients with OPCA (28).

It has also been reported that there is a 50% to 70 % diminishment of $InsP_3$ receptor binding in certain regions of the brain (parietal cortex and hyppocampus) of Alzheimer patients (29).

Alterations of $InsP_3$ binding sites in pheochromocytomas (chromaffin cell tumors that secrete abundant amounts of catecholamines which cause hypertension) were also described. $InsP_3$ receptors of tumor cells were of higher affinity than those of non functional adenomas (Kd of 5.77 ± 1.07 versus 11.3 ± 1.98 nM) (30).

High levels of free Ca^{2+} have been found in cells of patients suffering from hypertension. These levels were correlated with elevated blood pressure (31-36). These increased levels of Ca^{2+} have been observed in a variety of tissues from hypertensive rats: caudal artery (37), mesenteric artery (38), kidney (39), thoracic aorta (40), T lymphocytes (41) and blood platelets (42). It has recently been suggested that an elevated production of $InsP_3$, resulting from an augmented hormonal activation of PLC, could cause an exagerated intracellular calcium release. This increased level of intracellar calcium could be responsible for inducing the contraction of vascular smooth muscle in hypertension (43-45). Other studies have demonstrated that hypertensive rats express an increased number of $InsP_3$ receptors

Table 1. **Interaction properties of natural $InsP_3$ analogues on $InsP_3$ recognition sites.**

Ligand	$InsP_3$ receptor	$InsP_3$ phosphatase	$InsP_3$ kinase
$Ins(1,4,5)P_3$	2-300 nM[17,19,55-63]	3-72 μM[64]	0.2-1.5 μM[64]
$Ins(1,3,4,5)P_4$	40 nM[58] 100 nM[47] 350 nM[56] 15 μM[16]	0.5->150 μM[64]	no effect[64]
$Ins(1,4)P_2$	Inactive at 1 μM[63]	no effect (400 μM)[65]	--
$Ins(cyclic1{:}2,4,5)P_3$	1.7 μM[63]	38 μM[68] 1 mM[67]	30 μM[68]
$Ins(1,3,4)P_3$	>> 7 μM[16] inactive at 1 μm[19]	32 μM[69]	1.8 mM[69]
$Ins(1,3,4,5,6)P_5$	1.2 μM[73] 6.6 μM[74] 26 μM[75]	no effect (10 μM)[79]	--
$Ins(1,2,3,4,5,6)P_6$	2 μM[47] 12 μM[74] 15 μM[75] 20 μM[73] 26 μM[16] 36 μM[76]	no effect (10 μM)[79]	--

This table presents typical examples of values reported rather than an exhaustive list.

References: (14) Guillemette et al., 1987; (16) Willcocks et al., 1987; (17) Suppatapone et al., 1988; (19) Mourey et al., 1990; (47) Bernier and Guillemette, in press; (55) Guillemette et al., 1990; (56) Chilvers et al., 1990; (57) Guillemette et al., 1989; (58) Ely et al., 1990; (59) Theibert et al., 1990; (60) Marks et al., 1990; (61) Chadwick et al., 1990; (62) Theibert et al., 1990; (63) Willcocks et al., 1989; (64) Shears, 1991; (65) Downes et al., 1982; (67) Connolly et al., 1986; (68) Hirata et al., 1989; (69) Polokoff et al., 1988; (73) Rivera et al., 1990; (74) Palmer et al., 1989; (75) Varney et al., 1990; (76) Challiss et al., 1989; (79) Hughes and Shears, 1990.

in the heart (46) and in the vascular smooth muscle (47).

Enzymatic metabolism of $InsP_3$ is also implicated in certain pathologies. For many years, lithium (Li^+) has been known to be effective in the treatment of acute mania and in the prophylaxis of manic illness (48). A theory concerning the underlying causes of this illness is that specific groups or types of neurons are considered to be hyperactive (49). Li^+ interferes with inositol phospholipid dependent signalling mechanisms by inhibiting inositol monophosphatase (50,51). Inositol monophosphatase is a crucial enzyme in the process of recycling inositol and its inhibition might have severe consequences on phosphatidyl inositol production, eventually attenuating the synthesis of the second messengers $InsP_3$ and diacylglycerol. This diminished concentration of second messengers could suppress the manic symptoms.

Lin et al. (52) have reported that focal cerebral ischemia (blockade of cerebral artery) caused the disappearance of $InsP_3$ kinase activity in the rat cerebral cortex. This change in $InsP_3$ metabolism may be associated with changes in intracellular Ca^{2+} homeostasis that underly the pathophysiology of neuronal cell death.

Recently, it was found that patients suffering from Lowe's oculocerebrorenal syndrome (a X-gene linked developmental disorder affecting the eyes, the brain and the kidneys) have a gene that encodes a protein highly homologous to $InsP_3$ phosphatase (53). It seems that this protein induces problems in inositol phosphate metabolism. This altered $InsP_3$ metabolism could be a cause of the illness.

Finally, lymphocytes infected with HIV (Human Immunodeficiency Virus) have been shown to contain lower $InsP_3$ and $InsP_4$ phosphatase activity. This decreased activity could interfere with the inositol polyphosphate signal transduction system, causing an abnormal regulation of intracellular calcium (54).

Development of $InsP_3$ Analogs

In view of the fundamental importance of the $InsP_3$ signalling system in cell biology, it is desirable to have access to pharmacological tools which facilitate intervention at $InsP_3$ receptors and at $InsP_3$ metabolic enzymes. $InsP_3$ receptor agonists or antagonists and compounds which could block or stimulate the polyphosphoinositide pathway may represent useful therapeutic drugs.

Natural $InsP_3$ Derived Analogues

The naturally occuring messenger, Ins $(1,4,5)P_3$ is the most active of all inositol phosphates. The affinity of $Ins(1,4,5)P_3$ for its receptor and metabolic enzymes varies among tissues and species (17,19,55-64). As mentioned before, $Ins(1,4,5)P_3$ is principally metabolized by two enzymes, $InsP_3$ phosphatase and $InsP_3$ kinase. The products of these enzymes, $Ins(1,4)P_2$ and $Ins(1,3,4,5)P_4$ are considerably less active than $InsP_3$ (table 1) at the receptor level (16,19,47,56,58). These findings demonstrate that the two enzymes effectively inactivate the Ins(1,4,5)P_3 calcium mobilizing effect. Ins(1,3,4,5)P_4 is a substrate for InsP_3 phosphatase, with a relatively good affinity for this enzyme (64), while Ins(1,4)P_2 produced no effect on InsP_3 phosphatase (65). The InsP_3 metabolic pathway also produces other InsP_3 metabolites. Ins(cyclic 1:2,4,5)P_3 can accumulate after prolonged stimulation (66) and has a controversial action on the InsP_3 receptor. Some studies have shown that it could potently promote Ca^{2+} mobilization (67) while other studies showed that it had a minor contribution to Ca^{2+} mobilization (63). For both metabolic enzymes, Ins(cyclic 1:2,4,5)P_3 showed a good affinity (68). Ins(1,3,4)P_3, the product of Ins(1,3,4,5)P_4 dephosphorylation, is a relatively inactive compound at the InsP_3 receptor level (16,19,69-72). Ins(1,3,4)P_3 has a high affinity for InsP_3 phosphatase and a very low affinity for InsP_3 kinase (69). Ins(1,3,4,5,6)P_5 and Ins(1,2,3,4,5,6)P_6 have weak affinities for the InsP_3 receptor (16,47,73-76) but can be present in cells at concentrations as high as 50 μM (77,78). These high concentrations may be sufficient to allow a significant interaction with the InsP_3 receptor. The interaction of InsP_5 and InsP_6 with InsP_3 kinase has not been studied. However they both are inactive on InsP_3 phosphatase at concentrations as high as 10 μM (79).

Table 2. Interaction properties of natural cellular compounds on InsP_3 recognition sites.

Ligand	InsP_3 receptor	InsP_3 phosphatase	InsP_3 kinase
ATP	373 μM[16] 1.0 mM[14] 1.3 mM[56]	5 mM[80]	460 μM[81]
GTP	315 μM[16]	--	--
2,3-Diphosphoglycerate	250 μM[82] 300 μM[83]	300 μM[82] 340 μM[83] 97 μM[84]	10.5 mM[82]
Heparin	3 μg/ml[55] 10 μg/ml[57]	no effect (>2mg/ml)[57]	0.4 μg/ml[57]

This table presents typical examples of values reported rather than an exhaustive list.

References: (14) Guillemette et al., 1987; (16) Willcocks et al., (55) Guillemette et al., 1990; (56) Chilvers et al., 1990; (57) Guillemette et al., 1989; (80) Morris et al., 1988; (81) Shears, 1990; (82) Guillemette et al., 1990; (83) Bootman et al., 1990; (84) Cooke et al., 1989.

Natural Cellular Compounds

Other non-inositide natural intracellular components also affect the action and/or metabolism of $InsP_3$ (table 2). ATP and GTP (found in large amounts in cells) have a weak effect on $InsP_3$ receptor binding (14,16,56). ATP was tested on the metabolic enzymes and demonstrated a very low affinity for the $InsP_3$ phosphatase and kinase (80,81). 2,3 diphosphoglycerate, a compound known to regulate the oxygen-binding affinity of hemoglobin has been shown to be a weak competitive inhibitor of $InsP_3$ receptor, $InsP_3$ phosphatase and $InsP_3$ kinase (82-84). Finally heparin, a sulphated polysaccharide known for its anticoagulating effects, can compete with $InsP_3$ for its specific binding sites (55,57). Figure 3 shows the effect of heparin on $InsP_3$ induced calcium release. It seems that the negatively charged sulphate groups of heparin may interact with some domains of the receptor responsible for binding phosphate groups of $InsP_3$. While having no effect on $InsP_3$ phosphatase, heparin also inhibits $InsP_3$ kinase with a good affinity. Althought having relatively non-specific or weak interactions, these naturally occuring compounds can serve as prototypes in the design and synthesis of more potent and specific $InsP_3$ analogues.

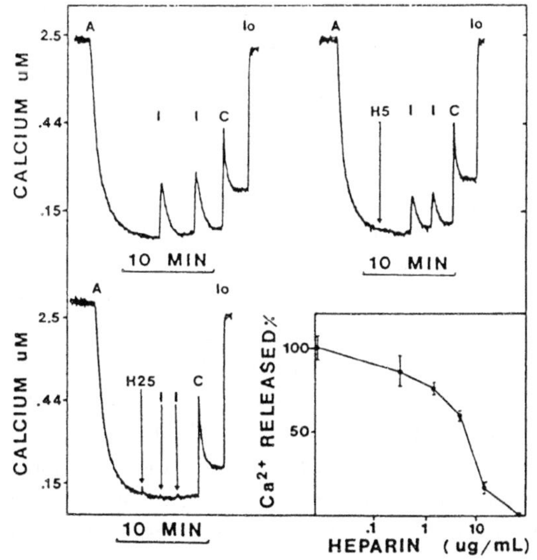

Fig. 3. Effect of heparin on the $InsP_3$ induced Ca^{2+} release. Bovine adrenal cortex microsomes were incubated at 37°C and their Ca^{2+} uptake and release activities were monitered by using fura-2. Typical traces in which the Ca^{2+} taken up by an ATP-dependent process is partially released upon $InsP_3$ addition. A, 1mM ATP; I, 1 μM $InsP_3$; C, 1 nmol of Ca^{2+}; Io, 1 μM ionomycin; H5, 5 μg/ml heparin; H25, 25 μg/ml heparin. Lower right, the dose-response relationship between heparin and $InsP_3$-induced Ca^{2+} release. 100% represents the amount of Ca^{2+} released by 1 μM $InsP_3$ in the absence of heparin.

Non Natural and Synthetic Analogues

In the last few years, a number of compounds that interact with $InsP_3$ recognition sites have been discovered. Some showed very good affinities and selectivities towards the recognition sites. A major family of compounds produced have been the synthetic $InsP_3$ analogues. These products have the same basic structure as $InsP_3$ with modifications to the functional groups (OH, PO_4^-). These compounds have helped a great deal in understanding the interaction between the $InsP_3$ molecule and its recognition sites. A major contribution has been the inositol phosphorothioates. Myo-inositol 1,4,5-trisphosphorothioate (Ins(1,4,5)PS_3) binds to the $InsP_3$ receptor with a slightly lower affinity than $InsP_3$ (85) (table 3). It is a full and potent intracellular Ca^{2+} releasing agonist, being only about 3-4 fold less potent than $InsP_3$, in a number of cells (86-89). It is resistant to dephosphorylation by $InsP_3$ phosphatase and can produce a sustained calcium transient (89,90). It can act as a competitive inhibitor of $InsP_3$ phosphatase and in that regard it is the most potent inhibitor yet reported (84). It has no significant effect on $InsP_3$ kinase (90).

Another interesting group of analogues are the fluoro-substituted $InsP_3$ molecules. To examine the importance of the hydroxyl groups on $InsP_3$, hydroxyl groups (OH^-) were replaced by fluor atoms at specific positions on the inositol ring. D-3-deoxy-3-fluoro-myo inositol 1,4,5-trisphosphate (3F-Ins(1,4,5)P_3) has a fluor atom instead of a hydroxyl group at position 3. This analogue showed a very good affinity for all three $InsP_3$ recognition sites (91). 3F-Ins(1,4,5)P_3 potently released calcium from intracellular stores with an affinity similar to that of $InsP_3$. Since 3F-Ins(1,4,5)P_3 is slowly metabolized, the release of calcium was somewhat sustained. This slow metabolism is due to the inactivation of $InsP_3$ kinase. The enzyme can bind the molecule but the fluor atom renders the phosphorylation at position 3 impossible. Other fluorinated $InsP_3$ analogues have also been synthesized and they showed interesting results (92). This approach of fluorination may produce useful compounds for inhibition of $InsP_3$ metabolizing enzymes.

D-6-deoxy-myo-inositol 1,4,5-trisphosphate (D-6 deoxy Ins(1,4,5)P_3) is a compound that permitted to highlight the importance of the hydroxyl group in position 6 of the $InsP_3$ molecule. Removal of the hydroxyl group in position 6 reduced the affinity of the resulting compound by about 70 fold (93). This hydroxyl group seems to be very important for $InsP_3$ receptor binding. It was suggested that this group may stabilize the $InsP_3$ molecule, by forming a hydrogen bond that induces a favorable conformation for interaction with the $InsP_3$ receptor (94). Ca^{2+} release studies showed that D-6-deoxy Ins(1,4,5)P_3 was a weak agonist for releasing Ca^{2+}, but that its action was prolonged since Ca^{2+} re-uptake was retarded. This is a good indication of resistance to metabolism. Further studies reveled that $InsP_3$ phosphatase exhibited a good affinity for D-6-deoxy Ins(1,4,5)P_3 (94). The compound however was not a substrate for the enzyme. The same study revealed that D-6-deoxy Ins(1,4,5)P_3 was also a good inhibitor of $InsP_3$ kinase.

Benzene 1,2,4-trisphosphate (BzP_3), a planar rigid benzene ring analogue of $InsP_3$ has also been synthesized (95). This compound contains three phosphate groups in positions 1,2 and 4 of a benzene ring. BzP_3 showed a low affinity for the $InsP_3$ receptor. However the compound had a good affinity for the two $InsP_3$ metabolizing enzymes (96). It seems that the flexibility of the inositol ring is important for $InsP_3$ receptor binding, and less important for the interaction with the enzymes.

Recently, a novel $InsP_3$ analogue, L-chiro-inositol 2,3,5-trisphosphate (L-ch-Ins(2,3,5)P_3) has been synthesized, and tested on $InsP_3$ receptor and metabolic enzymes (91). This molecule is similar to $InsP_3$ except that it has an axial hydroxyl group in position 3 instead of an equatorial hydroxyl group. A ten fold loss in receptor binding affinity and Ca^{2+} mobilizing potency was observed for L-ch-Ins(2,3,5)P_3 as compared to $InsP_3$. Ca^{2+} mobilization studies showed a very slow reuptake of the released Ca^{2+}, indicating a diminished metabolism. Indeed, low concentrations of L-ch-Ins(2,3,5)P_3 inhibited $InsP_3$ phosphatase and $InsP_3$ kinase activities and the compound was not metabolized by either enzyme. This is an interesting compound that can interact with the $InsP_3$ receptor and that is resistant to $InsP_3$ metabolizing enzymes, showing a sustained release of Ca^{2+}.

Certain chemical compounds with structures apparently different from that of $InsP_3$ have also been found to interact with $InsP_3$ recognition sites. The polyoxoanion, decavanadate (table 2) competitively blocked $InsP_3$ receptor binding and Ca^{2+} mobilization (97,98). Decavanadate also potently inhibited the two metabolic enzymes. Unfortunately, this compound exhibited a poor selectivity since it interferes with the activity of all the known recognition sites for $InsP_3$.

Polyvinyl sulphate is another highly sulphated compound that has been tested on $InsP_3$ recognition sites (99,100). It inhibited $InsP_3$ binding in a non-competitive manner. This inhibition was found not to be an allosteric interaction but rather corresponded to an irreversible syntopic antagonism. Polyvinyl sulphate also potently inhibited $InsP_3$ phosphatase and $InsP_3$ kinase in a non-competitive manner. Dextran sulphate (DS-5000), another polyanionic compound, was tested on the recognition sites for $InsP_3$ (99). While having weak effects on the $InsP_3$ receptor, Ca^{2+} release and $InsP_3$ phosphatase, it showed a good selectivity for $InsP_3$ kinase inhibition (4000 to 7000 times more potent).

Two chemical dyes of polyanionic nature, cibacron blue and patent blue, were reported to have a slight affinity for $InsP_3$ receptor binding and $InsP_3$ phosphatase (83). Recently four more dyes have been found to selectively inhibit $InsP_3$ recognition sites (101). Thymol blue was more potent to inhibit $InsP_3$ binding than $InsP_3$ metabolizing enzymes. Phenol red selectively inhibited $InsP_3$ phosphatase. Tetrabromophenol blue inhibited $InsP_3$ phosphatase and $InsP_3$ kinase, and to a lesser extent (10 times less), $InsP_3$ receptor binding. Finally tetraiodosulfonephthalein selectively inhibited $InsP_3$ phosphatase. The enzymatic inhibitions observed with most chemical dyes were of non-competitive reversible nature, indicating an allosteric type of interaction. These compounds all possess the same basic structure but differ from each other by the nature of their functional groups. Unfortunately, the affinity of these compounds for the different $InsP_3$ recognition sites is relatively low.

Table 3. Interaction properties of non-natural analogues on $InsP_3$ recognition sites.

Ligand	$InsP_3$ receptor	$InsP_3$ phosphatase	$InsP_3$ kinase
$Ins(1,4,5)PS_3$	121 nM[85]	6 µM[84]	no effect (10 µM)[90]
$3F\text{-}Ins(1,4,5)P_3$	8 nM[91]	3.9 µM[91]	8.6 µM[91]
$6\text{-}deoxy\text{-}Ins(1,4,5)P_3$	5.2 nM[93]	76 µM[94]	5.7 µM[94]
BzP_3	43 µM[95]	32 µM[96]	6.1 µM[96]
$L\text{-}ch\text{-}Ins(2,3,5)P_3$	60.4 nM[91]	7.7 µM[91]	7.1 µM[91]
Decavanadate	0.5-2.6 µM[97,98]	1.5 µM[97]	5 µM[97]
Polyvinyl sulfate	30-250 nM[99,100]	5 µM[99]	15 nM[99]
Dextran sulfate	34 µM[99]	57 µM[99]	8 nM[99]
Dyes:			
Cibacron Blue	96 µM[83]	88 µM[83]	--
Patent Blue	279 µM[83]	340 µM[83]	--
Thymol Blue	105 µM[101]	950 µM[101]	590 µM[101]
Tetrabromophenol Blue	250 µM[101]	25 µM[101]	12 µM[101]
Phenol Red			
Tetraiodophenol-	510 µM[101]	100 µM[101]	1100 µM[101]
sulfonephthalein	600 µM[101]	220 µM[101]	35 µM[101]

This table presents typical examples of values reported rather than an exhautive list.

References: (83) Bootman et al., (84) Cooke et al., 1989; (85) Nunn et al., 1990; (90) Taylor et al., 1989; (91) Safrany et al., 1992; (93) Bonis et al., 1991; (94) Safrany et al., 1991; (95) Boulay et al., 1991; (96) Poitras et al, in press; (97) Sturpish et al., 1991; (98) Föhr et al., 1991; (99) Luong et al., 1991; (100) O'Rourke and Feinstein, 1990; 1990; (101) Richard et al., unpublished results.

CONCLUDING REMARKS

Many pathologies have been associated to the $InsP_3$ signalling pathway ($InsP_3$ production, $InsP_3$ receptor, Ca^{2+} release, metabolism) and it is essential to develop products that are capable of controling the different steps of this important pathway. In the last few years, considerable progress has been made in understanding how $InsP_3$ interacts with its different recognition sites. Many compounds have been shown to share structural analogy with $InsP_3$. A few of them showed interesting properties that could be exploited for specific or selective modulation of $InsP_3$ recognition site activities. These compounds may also represent prototypes that will permit the development of new compounds exhibiting higher potencies and higher selectivities. These new compounds will certainly represent useful tools for biochemical and pharmacoligical studies aimed at the understanding of the mechanism of cellular calcium regulation. It is hoped that in the near future some of these compounds may find therapeutical applications for the treatment of previously mentioned illnesses and of other pathologies that may be caused by a defect of calcium regulation.

REFERENCES

1. Michell RH (1975) Biochem Biophys Acta 415:81-147
2. Berridge MJ (1981) Mol Cell Endocrinol 24:115-140
3. Irvine RF, Dawson RMC and Freinkel N (1982) Contemp Metab 2:302-342
4. Nishizuka Y (1983) Trends Biochem Sci 8:13-16
5. Putney JW, Takemura H, Hughes AR, Horstman DA and Thastrup O (1989) FASEB J 3:1899-1905
6. Pang IH and Sternweiss PC (1990) J Biol Chem 265:18707-18712
7. Baukal AJ, Guillemette G, Rubin R, Spat A and Catt KJ (1985) Biochem Biophys Res Commun 133:532-538
8. Guillemette G, Balla T, Baukal AJ, Spat A and Catt KJ (1987) J Biol Chem 262:1010-1015
9. Rossier MF, Capponi, AM and Vallotton MB (1989) J Biol Chem 264:14078-14084
10. Spat A, Bradford PG, McKinney JS, Rubin RP and Putney JW (1986) Nature 319:514-516
11. Guillemette G, Balla T, Baukal AJ and Catt KJ (1988) J Biol Chem 263:4541-4548
12. Mauger JP, Claret M, Pietri F and Hilly M (1989) J Biol Chem 264:8821-8826
13. Spat A, Lukacs GL, Eberhardt I, Kiesel L and Runnebaum B (1987) Biochem J 244:493-496
14. Guillemette G, Balla T, Baukal AJ and Catt KJ (1987) Proc Natl Acad Sci USA 84:8195-8199

15. Worley PF, Baraban JM, Supattapone S, Wilson VS and Snyder SH (1987) J Biol Chem 262:12132-12136
16. Willcocks AL, Cooke AM, Potter BVL and Nahorski SR (1987) Biochem Biophys Res Commun 146:1071-1078
17. Supattapone S, Worley PF, Baraban JM and Snyder SH (1988) J Biol Chem 263:1530-1534
18. Chadwick CC, Saito A and Fleischer S (1990) Proc Natl Acad Sci USA 87:2132-2136
19. Mourey RJ, Verma A, Supattapone S and Snyder SH (1990) Biochem J 272:383-389
20. Kamata H, Hirata M, Ozaki S, Kusaka I, Kagawa Y and Hirata H (1992) J Biochem 111:546-552
21. Furuichi T, Yoshikawa S, Miyawaki A, Wada K, Maeda N and Mikoshiba K (1989) Nature 342:32-38
22. Mignery GA, Sudhof TC, Takei K and DeCamilli P (1989) Nature 342:192-195
23. Ferris CD, Huganir RL, Supattapone S and Snyder SH (1989) Nature 342:87-89
24. Storey DJ, Shears SB, Kirk CJ and Michell RH (1985) Nature 312:374-376
25. Irvine RF, Letcher AJ, Heslop JP and Berridge MJ (1986) Nature 320:631-634
26. Hansen CA, Mah S and Williamson JR (1986) J Biol Chem 261:8100-8103
27. Irvine RF (1990) FEBS Lett 263:5-9
28. Kish SJ, Li PP, Robitaille Y, Currier R, Gilbert J, Schut L and Warsh JJ (1989) Brain Res 489:373-376
29. Young LT, Kish SJ, Li PP and Warsh JJ (1988) Neuroscience Letters 94:198-202
30. Yamada M, Mizuma H, Monden T, Murakami M, Iriuchijima T, Kobayaski I, Ishida T and Mori M (1992) Neuroendocrinology 56:69-73
31. Erne P, Bolli P, Burgisser E and Buhler FR (1984) N Engl J Med 310:1084-1088
32. Bruschi G, Bruschi ME, Caroppo M, Orlandini G, Spaggiari M and Cavatorta A (1985) Clinical Science 68:179-184
33. Cooper R, Lipowski J, Ford E, Shamsi N, Feinberg H and Le Breton G (1989) Hypertension 13:139-144
34. Lechi A, Lechi C, Bonadonna G, Sinigaglia D, Corradini P, Polignano R, Arosio E, Covi G and DeTogni P (1987) Hypertension 9:230-235
35. Cooper RS, Shamsi N and Katz S (1987) Hypertension 9:224-229
36. Haller H, Lindschau C, Quass P and Distler A (1992) Am J. Hypertens 5:117-124
37. Bhalla RC, Sharma RV and Bendhack LM (1988) In: Kwan CY (ed) Membrane abnormalities in hypertension, CRC Press, pp 334-358
38. Hermsmeyer K and Erne P (1989) Am J Hypertens 2:655-658
39. Dowd TL and Gupta RK (1992) J Biol Chem 267:3637-3643
40. Jelicks LA and Gupta RK (1990) J Biol Chem 265:1394-1400

41. Batlle DC, Janss G, Lapointe M, Llibre J and Saleh A (1990) Am J Hypertens 3:343-348
42. Oshima T, Young EW, Bukoski RD and McCarron DA (1991) Hypertension 18:758-762
43. Meyer P and Marche P (1988) Am J Med Sci 295:396-399
44. Uehara Y, Ishii M, Ishimitsu T and Sugimoto T (1988) Hypertension 11:28-33
45. Kato H, Fukami K, Shibasaki F, Homma Y and Takenawa T (1992) J Biol Chem 267:6483-6487
46. Fukuda K, Yoshikawa H, Yoshikawa A, Baba A, Nishio I and Masuyama Y (1990) J Hypertension 8 (suppl 3): Abstract P8.09, S119
47. Bernier S and Guillemette G (1993) Am J Hypertens in press
48. American Psychiatric Association (1975) Am J Psychiatry 132:977-1001
49. Wachtel H (1990) Pharmacopsychiatry 23:27-32
50. Hallcher LM and Sherman WR (1980) J Biol Chem 255:10896-10901
51. Berridge MJ, Downes CP and Hanley MR (1982) Biochem J 206:587-595
52. Lin TA, Lin TN, He YY, Hsu CY and Sun GY (1992) Biochem Biophys Res Commun 184:871-877
53. Attree O, Olivos IM, Okabe I, Bailey CL, Nelson DL, Lewis RA, McInnes RR and Nussbaum RL (1992) Nature 358:239-242
54. Nye KE, Riley A and Pinching AJ (1992) Clin Exp Immunol 89:89-93
55. Guillemette G, Favreau I, Boulay G and Potier M (1990) Molec Pharmac 38:841-847
56. Chilvers ER, Challiss RAJ, Willcocks AL, Potter BVL, Barnes PJ and Nahorski SR (1990) Br J Pharmac 99:297-302
57. Guillemette G, Lamontagne S, Boulay G and Mouillac B (1989) Molec Pharmac 35:339-344
58. Ely JA, Hunyady L, Baukal, A and Catt, KJ (1990) Biochem J 268:333-338
59. Theibert AB, Supattapone S, Ferris CD, Danoff SK, Evans RK and Snyder SH (1990) Biochem J 267:441-445
60. Marks AR, Tempst P, Chadwick CC, Riviere L, Fleisher S and Nadal-Ginard B (1990) J Biol Chem 265:20719-20722
61. Chadwick CC, Saito A and Fleischer S (1990) Proc Natl Acad Sci USA 87:2132-2136
62. Theibert AB, Supattapone S, Worley PF, Baraban JM, Meek JL and Snyder SH (1987) Biochem Biophys Res Commun 148:1283-1289
63. Willcocks AL, Strupish J, Irvine RF and Nahorski SR (1989) Biochem J 257:297-300
64. Shears SB (1991) Pharmac Ther 49:79-104
65. Downes CP, Mussatt MC and Michell RH (1982) Biochem J 203:169-177
66. Sekar MC, Dixon JF and Hokin LE (1987) J Biol Chem 262:340-344
67. Connolly TM, Wilson DB, Bross TE and Majerus PW (1986) J Biol Chem 261:122-126
68. Hirata M, Watanabe Y, Ishimatsu T, Ikebe T, Kimura Y, Yamaguchi K, Ozaki S and Koga T (1989) J Biol Chem 264:20303-20308

69. Polokoff MA, Bencen GH, Vacca JP, DeSolms, SJ, Young SD and Huff JR (1988) J Biol Chem 263:11922-11927
70. Strupish J, Cooke AM, Potter BVL, Gigg R and Nahorski SR (1988) Biochem J 253:901-905
71. Gawler DJ, Potter BVL and Nahorski SR (1990) Biochem J 272:519-524
72. Parker I and Ivorra I (1991) J Physiol 433:207-227
73. Rivera J, Bernal AL, Varney M and Watson SP (1990) Endocrinology 127:155-162
74. Palmer S, Hughes KT, Lee DY and Wakelam MJO (1989) Cell Signall 1:147-156.
75. Varney MA, Rivera J, Lopez-Bernal A and Watson SP (1990) Biochem J 269: 211-216.
76. Challiss RAJ, Willcocks AL, Mulloy B, Potter BVL and Nahorski SR (1991) Biochem J 274:861-867
77. Pittet D, Schlegel W, Lew DP, Monod A and Mayr GW (1989) J Biol Chem 264:18489-18493
78. Stephens LR, Hawkins PT, Stanley AF, Moore T, Poyner DR, Morris PJ, Hanley MR, Kay RR and Irvine RF (1991) Biochem J 275:485-499
79. Hughes PJ and Shears SB (1990) J Biol Chem 265:9869-9875
80. Morris AP, Gallacher DV, Irvine RF and Petersen OH (1987) Nature 330:653-655
81. Shears SB (1990) Cell Signall 2:191-195
82. Guillemette G, Favreau I, Lamontagne S and Boulay G (1990) Eur J Pharmac 188:251-260
83. Bootman MD, Pay GF, Rick CE and Tones MA (1990) Biochem Biophys Res Commun 166:1334-1339
84. Cooke AM, Nahorski SR and Potter BVL (1989) FEBS Lett 242:373-377
85. Nunn DL, Potter BVL and Taylor CW (1990) Biochem J 265:393-398
86. Taylor CW, Berridge MJ, Brown KD, Cooke, AM and Potter BVL (1988) Biochem Biophys Res Commun 150:626-632.
87. Thévenod F, Dehlinger-Kremer M, Christian AL, Potter BVL and Schulz I (1989) J Membr Biol 109:173-186
88. Valdivia L, Valdivia HH, Potter BVL and Coronado R (1990) Biophys J 57:1233-1243
89. Nahorski SR and Potter BVL (1989) Trends Pharmacol Sci 10:139-144
90. Taylor CW, Berridge MJ, Cooke AM and Potter BVL (1989) Biochem J 259:645-650
91. Safrany ST, Wilcox RA, Liu C, Potter BVL and Nahorski SR (1992) Eur J Pharmacol 226:265-272
92. Safrany ST, Sawyer D, Wojcikiewicz RJH, Nahorski SR and Potter BVL (1990) FEBS Lett 276:91-94
93. Bonis D, Sezan A, Manduit P, Dubreuil D, Cleophax J, Gero SD and Rossignol B (1991) Biochem Biophys Res Commun 175:894-900

94. Safrany ST, Wojcikiewicz RJH, Strupish J, Nahorski SR, Dubreuil D, Cleophax J, Gero SD and Potter BVL (1991) FEBS Lett 278:252-256
95. Boulay G, Poudrier R, Luong TT, Fournier A and Guillemette G (1991) Pharmacol (Life Sci Adv) 10:67-73
96. Poitras M, Bernier S, Boulay G, Fournier A and Guillemette G (1993) Eur J Pharmacol 244:203-210
97. Strupish J, Wojcikiewicz RJH, Challiss RAJ, Safrany ST, Willcocks AL, Potter BVL and Nahorski SR (1991) Biochem J 227:294
98. Föhr KJ, Wahl Y, Engling R, Kemmer TP and Gratzl M (1991) Cell Calcium 12:735-742
99. Luong TT, Poitras M, Boulay G and Guillemette G (1991) Eur J Pharmacol 208:213-221
100. O'Rourke F and Feinstein MB (1990) Biochem J 267:297-302
101. Richard DE, Bernier S, Boulay G and Guillemette G (1993) Eur J Pharmacol Submitted

CHAPTER 21

THE ROLE OF PLASMA MEMBRANE IN THE PATHOGENESIS OF OSTEOBLASTIC METASTASIS

Michael Koutsilieris and Peter A. Rogers

*Endocrine-Oncology Clinic, Research Center,
Laval University Hospital Center,
and Hormonal Bioregulation Laboratory,
Laval University, Quebec, Canada*

INTRODUCTION

Malignant tumors are the result of unregulated growth of a *"tumorigenic cell"* which proliferates without respecting; a) its own regulatory mechanisms (failure of autocrine regulation), b) cell-cell interactions with adjacent cells (failure of paracrine regulation), and c) systemic regulatory mechanisms (failure of hormonal regulation).

The proliferation of normal cells is regulated by specific intracellular signals affecting the cell cycle (1,2). In mammalian cells there are three major plasma membrane associated pathways altering intracellular signals of proliferation: a) the cyclic AMP (cAMP) pathway, b) the phosphatidylinositol (PI)-Ca^{++}-diacylglycerol (DAG) system, and c) the growth factor receptor-tyrosine kinase system. These pathways can function independently or cooperatively to induce either stimulation or inhibition of DNA replication and cell division (1).

The products of certain oncogenes *(oncoproteins)* are now known to be implicated in the uncontrolled growth of cancer cells (2). These *oncoproteins* are similar to the protein products of *proto-oncogenes* such as the growth factor family that regulate the proliferation of normal cells (3). A notable difference between these two classes of proteins however, is that production of proto-oncogene-encoded proliferation signal terminates following cessation of the inductive stimulus.

To date several *oncoproteins* have been chemically characterized and associated with the development of specific tumors in vivo (1). For example, PDGF b subunit is the *oncoprotein* encoded by simian sarcoma virus *(v-sis)* oncogene. This particular oncoprotein is secreted by *"tranformed cells"* and induces the unregulated proliferation of transformed cells by binding to its own receptor in an autocrine manner. Another oncogene *(erdB)* codes for an *oncoprotein* which is a truncated EGF receptor molecule (lacking the NH_2-terminal domain). This *oncoprotein*, which lacks the extracellular domain of EGF receptor, produces a constitutive activation of its intracellular tyrosine kinase domain, thereby stimulating the uncontrolable proliferation of the cells. Furthermore, certain oncogenes such as the *v-abl, v-src* , and *v-fps* are known to produce *oncoproteins* that are *tyrosine protein kinases*. Other oncogenes, such as the ras-family produce *oncoproteins* with *guanine nucleotide-binding protein activity (GTPases)*. In addition *v-erbA*-like oncogenes encoded for *oncoproteins* that are *nuclear transcription factors* (1-5). It is important to note that apart from the genetic mechanism of tumorigenesis, tumors also arise from abnormal cellular evolution during differentiation, as well as by epigenetic changes at the cell membrane and of cell-cell interactions (5).

INVASION AND METASTASIS

The most important clinical manifestation of tumor progression in cancer patients is the ability of the tumor cell to invade surrounding tissue structures and subsequently to metastasize to organs distant from the primary site of tumor development (5). By definition, benign tumors do not metastasize. The invasive and metastatic capacity of malignant tumors should not be considered as defining malignancy since some malignant tumors do not demonstrate these two characteristics.

The capacity of tumor cells to produce invasion and metastasis differs among the cell clones of a given malignant tumor, although all are derived from the same *"progenitor malignant cell"* (6). The events of the metastatic cascade include; *I: invasion*; a) detachment of the metastatic cells from the primary tumor, b) migration to the adjacent tissue, and *II: metastasis*; a) adhesion to the wall of lymphatic and blood vessels, b) entry into vessels, c) dissemination through the systemic circulation, d) arrest in specific tissue permitting the adhesion of metastatic cells, e) extravasation, and f) proliferation at a distant site on a organ permissive to the growth of metastatic cells (6). Defining the physical principles and biological mechanisms which permit the migration and subsequent implantation of the malignant cell assumes a particular importance with respect to future therapeutic intervention and other potential pharmacological control strategies.

The ability of a cell to move is, for the most part, mediated through the cytoplasmic actin microfilament system and its modulation by a group of associated proteins (7). Depending on the cell type and state of differentiation, actin-binding proteins (ABPs) permit specific rearrangements of the three-dimensional state of the microfilaments in the peripheral cytoplasm which in turn allows the cell to become motile. Two of these microfilament-associated proteins, gelsolin and actin-binding protein (ABP; the nonmuscle isoform of filamin) have attracted a significant amount of attention in recent years due their clear effect on the state of the microfilament system (8,9). Since differences exist with respect to motility between tumor cells, it is tempting to speculate that such differences may be due to corresponding differences in the expression and functional deployment of specific ABPs. In this respect it is interesting to note that several cell lines of human melanoma which are deficient in ABP had a significantly decreased translocation capacity as evaluated in a chemotactic chamber (10). Further direct evidence for the central role of ABP in cellular motility has emerged from experiments in which locomotor capacity was restored to cells deficicient in ABP by genetic transfection which resulted in normal levels of ABP. The relationship between actin structural proteins and cellular motility has been recently reviewed (11). Many important questions regarding the possible functional relationship between cell motiliy, metastatic potential and malignant transformation remain unanswered. More extensive research efforts in this direction should prove to be fruitful.

The interface between the cytoskeleton and the cell membrane constitutes a point of control of cell shape and cell attachment to adjacent cells. The proteins which are found in this particular domain of the cell are implicated in cellular motility and responsiveness to growth factors and other external stimuli (reviewed in references 11,12). Cytoplasmic actin filaments terminate at the plasma membrane in specialized regions referred to as focal adhesions. These supramolecular assemblies are concentrated in those areas of the membrane which are involved in adhesion to the substratum and are found associated with the transmembrane adhesion proteins, the integrins. The number of focal adhesions is inversely related to the rate of cellular translocation. The dynamics of several of the various proteins associated with the focal adhesions appear to be regulated by phosphorylation which may also be indicative of a signalling role for focal adhesions (13,14 and references therein). A 125 kD focal adhesion kinase (FAK), $pp125^{FAK}$, $pp60^{src}$ and other protein kinases are concentrated in the adhesion sites.

There is little doubt that there exists cytoskeletal involvement in the metastatic process and it is reasonable to assume at this point in time that our understanding of this involvement is in its early stages. Fidler (15), writing on the pathogenesis of metastasis, has perhaps best defined the problem by stating, "Belief that certain rules govern metastasis implies that elucidation and understanding of these rules will lead to better therapeutic intervention.". This line of thinking is directly applicable to the role of the cytoskeleton in the molecular etiology of cancer.

BONE METASTASES

Skeletal metastases are very common clinical manifestations in every advanced stage cancer patient. Some primary tumors such as prostate cancer, lung cancer, thyroid cancer, and breast cancer, are referred to as osteophylic due to their ability to metastasize preferentially onto the skeleton (16).

While succesful growth of an implanted tumor on skeleton should depend upon the biological characteristics of tumor clones and the local interactions with bone cells, the distribution of metastases between one part of the skeleton and another would appear to depend more upon local vascularity and drainage factors (6). The axial skeleton is more often involved than the appendicular skeleton and this correlates well with the distribution of persistent red bone marrow in adults. The spine, pelvis, and the ribs are the earliest sites of metastasis, while the skull, femur, humerus, scapula, and sternum are commonly involved at a later stage (17). Analysis of the images obtained from 1,355 bone scans performed in advanced cases of cancer in 16 medical institutions documented a broadly similar distribution of bone metastases for all solid tumors but prostate, cervix and bladder cancer, all detected to have a propensity to involve the bone of the pelvis (18). Within the spine, the vertebral body is more commonly involved as compared to pedicles, and the lumbar and dorsal spine are more often affected than cervical and sacrum. The higher incidence of metastases in the lumbar spine is suspected to be caused by the greater anatomical mass as well as by the vertebral venous system transporting metastatic cancer cells (19). This venous system comprises a paravertebral network of thin-valved veins which communicates freely with the pelvic and intercostal veins. The existance of the paravertebral venous system most probably explains the high frequency of the metastases into the pelvic bones by prostate, bladder and cervical cancer (20). It is important to note that arguments concerning the role of lymphatic flow versus blood flow in the development of osseous metastases are academic, as tumor cells have been documented to move freely between the lymphatic and the vascular system (21,22).

Among solid tumors metastasizing to the skeleton prostate adenocarcinoma and breast cancer are those which give rise to the osteoblastic reaction (23). The incidence of pure osteoblastic metastases is 95% for metastatic prostate cancer and 8% for metastatic breast cancer. Pure lytic lesions in prostate cancer are an extremely rare phenomenon. Consequently, metastatic prostate adenocarcinoma is a unique cancer in its ability to promote new bone formation (24).

Finally, skeletal metastases in prostate cancer patients frequently are the only sites of disease progression after an initial response to hormone therapy (25-27). Therefore, several theories have emerged describing the role of various tumor clones with apparently different patterns of metastatic potential (28) and a remarkable heterogeneity with regard to hormone sensitivity (25,26).

An effective therapeutic approach for advanced prostate cancer would require further knowledge of the biology of metastatic prostate cancer cells and detailed description of the pathogenetic interactions between metastatic cells and bone tissue (29,30).

PATHOPHYSIOLOGY OF THE OSTEOBLASTIC METASTASES

Direct evidence that human prostate cancer tissues contained mitogens for osteoblasts emerged when human prostate cancer, benign prostatic hyperplasia, normal adult, and normal pre-pubertal prostate tissues were extracted and the extracts analysed for mitogenic activity using several cell bioassay systems (31-40). Prostate extracts contained several mitogens for fibroblasts and osteoblast-like cells. The nature of this mitogenic material was proteinaceous since tryptic digestion and exposure to 100°C inhibited its activity (31-33).

Since prostate-derived mitogen(s) for osteoblast-like cells was acid stable, and retained on and eluted from C18 silica columns, it was assumed that reverse-phase high performance liquid chromatography (r-HPLC) would be a suitable approach to purify this material (31,32, 34-36). Indeed, r-HPLC was employed to purify several mitogens with broad cell type specificity from prostate cancer and benign prostatic hyperplasia extracts (32,34-36). Some preparations revealed mitogenic selectively for osteoblasts (35). The use of r-HPLC enabled the purification of several peptides with molecular weights varying from 6 kDa to 24 kDa, all exerting selective mitogenic activity for osteoblasts (31,36). It was therefore postulated that a precursor protein was the parent molecule of all these active peptides (31).

Benign prostatic hyperplasia, prostate cancer, human normal adult prostate, but not pre-pubertal prostate extracts contained selective mitogens for osteoblasts (31,32,35). Since none of the normal and malignant tissue extracts tested thus far contained selective mitogenic activity for osteoblasts, prostate tissue was hypothesized to be the site of production and/or the site of the biological processing of this material to its active form (31,35). Metastasis is a unique event in this particular situation by allowing the physical proximity of prostate cancer cells to osteoblasts and thereby enabling this selective growth factor(s) to stimulate the osteoblasts (31,35). This mechanism has been implicated in the pathogenesis of osteoblastic metastases caused by prostate cancer (31). The putative mitogen derived from human prostate tissue which appears to specifically affect osteoblasts has not yet been obtained in a form which is suitable for refined structural analysis.

A PC-3 cell line has been established from a hormone-indepedent osteoblastic tumor in a prostate cancer patient and explored as a potential in vitro system for the study of the paracrine interactions between metastatic prostate cancer cells and bone cells (41). Xenopus oocytes directed the synthesis of substances mitogenic to osteoblasts following the injection of mRNA isolated from PC-3 cells (40). Recently, similar if not identical mitogens to those described in human prostate tissues have been detected in PC-3 cell CM (conditioned medium) (41-43). One PC-3 cell CM-derived mitogen contains a NH_2-terminal sequence homologous to that of urokinase-type plasminogen activator (uPA) (42). This fragment could represent the product of a specific enzymatic conversion of uPA molecules to a lower molecular weight fragment (amino acid residues 1-134) by prostate cancer cells (42,43). Interestingly, elevated uPA levels have been detected in stage D2 prostate cancer patients (44,45) and uPA has been shown to be the predominant plasminogen activator of the prostate gland (46-48).

Our studies have used rat PA-III cells that succesfully produce the osteolytic and osteoblastic reaction onto rat skeleton (49,50). In this sytem development of PA-III cell tumors on rat bone required disruption of local periosteum by the inoculating needle (51). This was indicative of the crucial role of proliferative stimuli originated from skeletal tissue favoring the development of prostate cancer cell-induced tumor (52). The local osteoblastic reaction was histologically similar to the one produced by human prostate cancer cells (31-40,41-43,52).

Based on data indicating the significant role of growth factors in cancer biology and bone physiology (53,54), we investigated the effects of IGF-I and IGF-II on the proliferation of PA-III cells. The results of these experiments revealed; a) the presence of IGF-I receptors on PA-III cell membrane, and b) the ability of IGFs produced by osteoblast-like cells (UMR 106 cells) to stimulate PA-III cells (55,56). Therefore, osteoblast-derived IGF-I was probably the paracrine stimulus for the initial growth of PA-III cells leading to the establishment of the PA-III cell-induced tumor on rat bone.

Subsequently, we documented the ability of the PA-III cell conditioned medium (CM) to digest the IGF-binding proteins (IGFBPs) detected in UMR 106 osteoblast-like CM (56). IGFBPs in osteoblasts have been proposed to play an important role in regulating the local bioavailability of the IGFs (57). Therefore, the ability of the PA-III cell CM to hydrolyze the IGFBPs (evaluated by Western blotting assays) using the IGFBPs in UMR 10 cell CM was indicative of the possible role of protease(s) in IGF-dependent paracrine interactions between prostate cancer cells and osteoblasts (56,57). Purification and sequence analysis revealed that the IGFBPs-hydrolyzing protease of PA-III cell CM is in fact the rat urokinase-type plasminogen activator molecule (58).

It is conceivable that; a) osteoblasts stimulate the proliferation of PA-III cells via an IGF-dependent mechanism, and b) PA-III cell CM containing uPA which hydrolyzes IGFBPs produced by osteoblasts. IGF availability is therefore increased at the sites of PA-III cell-induced tumors which in turn stimulate the proliferation of; a) PA-III cell-induced tumor in a paracrine manner, and b) osteoblasts in a autocrine manner (59).

In this system, stimulation of osteoblasts could occur via other uPA-mediated mechanisms involving; a) the production of an active NH_2-terminal fragment of uPA (42,43) or, b) conversion of plasminogen to plasmin which in its turns activates the latent form of the osteoblast-derived TGF b 1 (60-64). Under physiological conditions TGF b 1 would; a) enhance osteoblastic functions, b) stimulate the production of plasminogen activator inhibitors (PAIs), and c) inhibit

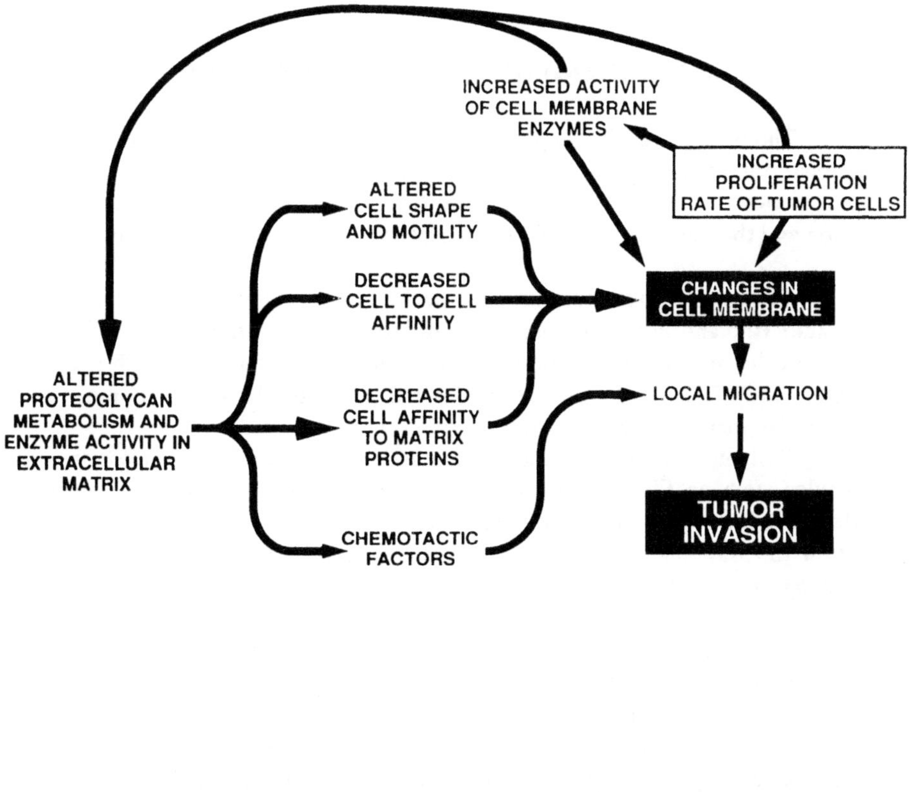

Figure 1: *Dynamic modulation of the cell membrane of cancer cells leading to local invasion and metastasis. These changes include; a) production or activation of cell membrane enzymes inducing an altered activity of the enzymes and of the proteoglycan metabolism in the extracellular matrix (ECM), b) production or activation of chemotactic factors leading to the local migration of malignant cells, c) alteration of the affinity of the tumor cell for the various extracellular matrix proteins and for other adjacent cells, and d) alterations in the shape of tumor cells resulting in higher motility of tumor cells in the extracellular matrix. These changes enable malignant cells to invade adjacent tissues and subsequently to metastasize to distant organs.*

uPA production (63,64) This latter mechanism is known to regulate extracellular matrix deposition by osteoblasts (63,64).

In the case of osteoblastic metastases, that although the increased levels of active TGF b 1 would stimulate the growth and function of osteoblasts (60-64), it would not regulate the uncontrolable production of large quantities of uPA by the adjacent metastatic prostate cancer cells (56-58). Thus a resulting vicious circle of paracrine interactions between metastatic prostate cancer cells and osteoblasts may stimulate further the proliferation of metastatic prostate cancer cells as well as bone cells locally (59).

PLASMA MEMBRANE AND OSTEOBLASTIC METASTASIS

The plasma membrane plays a key role in the establishment of structural and functional interactions between; a) extracellular matrix and cytoskeleton and b) among various cell types (7-9). In addition, the plasma membrane defines the position and the motility of membrane receptors and matrix adhesion molecules (Fig. 1).

Cell to cell interactions involve local regulators (growth factors) that interact in an autocrine and paracrine manner modulating the proliferation of cells (12-24). Control of the bioavailability and the specific activity of certain growth factors such as IGFs and TGF b 1 can be modulated by the plasma membrane and by extracellular matrix enzymatic systems (5,11,15).

Cell surface glycocoproteins such as integrins, laminin receptor and matrix molecules such as fibronectin, type I collagen, vitronectin, laminin, fibrin, von Willebrand factor and throbospondin play crucial roles in cell motility and cell function (59,65). The binding sites of adhesion proteins recognize proteins that contain specific ligand sequences adjacent to Arg-Gly-Asp sequence (RDS sequence). These sequences are found in integrins, fibronectin, uPA receptor molecule (uPA.R), IGFBP-1 and IGFBP-2. Thus it is possible that these molecules may occupy the same molecular domains on the plasma membrane (59). The plasma membrane of osteoblasts would be predicted to contain uPA.R-uPA complexes in close proximity to IGFBPs-IGFs complexes (IGFBP-1 and IGFBP-2) with their own position dictated by the RDS sequences of the integrin-like adhesion molecules and/or of the integrin-bound or fibronectin-bound proteoglycans, such as syndecans (65). Osteoblasts are known to produce the highest expression levels of IGFs and TGF b 1 after liver and platelets, respectivelly. Colocalization of these important regulators of osteoblast function on the osteoblast plasma membrane enables this cell to exert a tight control over extracellular matrix deposition. Mobilization of the plasma membrane-dependent array of regulatory substances may either regulate matrix resorption (uPa-plasmin activation of extracellular matrix enzymes) or matrix formation (activation of latent TGF b 1 that inhibits uPA production/ increase PAI

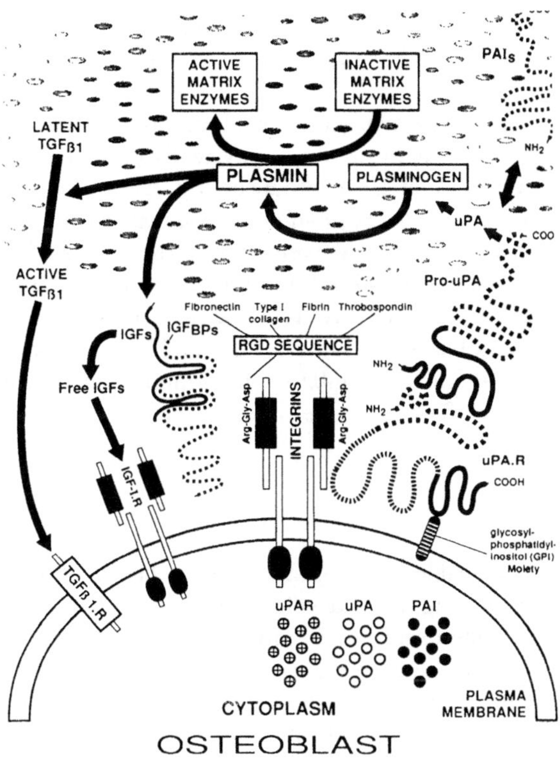

Figure 2. *Autocrine regulation of the function of osteoblasts mediated by cell membrane and extracellular matrix proteins. The Arg-Gly-Asp sequence (RGD sequence) on integrins enable these interactions by positioning uPA.R close to the RGD sequence thus directing pro-uPA closer to IGFBP-1 and IGFBP-2. IGFBP-1 and IGFBP-2 bind IGFs with an increased affinity for IGFs.R thereby creating a reservoir of IGFs around osteoblasts which can be released after hydrolysis of IGFBPs. Activation of pro-uPA to uPA transforms plasminogen to plasmin which in turn activates extracellular matrix enzymes. This cascade of events activates the latent form of TGF b 1 and increases the bioavailability of IGFs by hydrolyzing IGFBPs. TGF b 1 and IGFs can act via their receptor to stimulate the production of collagen type I and other extracellular bone matrix proteins. The action of TGF b 1 and IGFs will reduce the production of pro-uPA and increase the production of uPA inhibitor (PAI). This will inactivate extracellular uPA which then inhibits the transformation of plasminogen to plasmin thus inhibiting the activation of extracellular matrix enzymes. This mechanism regulates extracellular matrix deposition in osteoblasts. The presence of metastatic prostate cancer cells interrupts the regulation of this mechanism by secreting high amounts of uPA which stimulate osteoblasts by the continuous and high bioavailability of IGFs/TGF b 1. This can lead to the accumulation of increasing amounts of woven bone deposited at the sites of metastasis caused by metastatic prostate cancer cells.*

secretion that inactivates uPA-plasmin system enhances local osteoblast function by increasing bioavailable IGFs after hydrolysis of IGFBPs) (Fig. 2).

The development of a metastatic bone tumor of prostate cancer origin permits metastatic prostate cancer cells to simply be physically close to osteoblasts. Thus large and uncontrolable amounts of uPA produced by metastatic prostate cancer cells probably interrupt the coordinated regulation of bone matrix deposition. Local increases of bioactive IGFs and TGF b 1 should explain; a) further metastatic prostate cancer cell growth (paracrine effects of osteoblast-deriving IGFs on metastatic prostate cancer cells), and b) stimulation of the proliferation and function of osteoblasts locally (autocrine action of osteoblast-deriving TGF b 1 and IGFs, as well as other mitogens for osteoblasts) (56-59). This unregulated and continous deposition of bone martix results in the formation of woven bone observed at the sites of prostate cancer cell metastases.

Concluding remarks:

It is now reasonable to suggest that modulation of the expression of the paracrine factors participating in the development of metastatic bone tumors as well as in the reaction of the bone to metastatic prostate cancer tumors by either systemic or local administration of anticancer pharmaceutic agents may modify the evolution of PA-III cell-induced tumors on rat bone in vivo (66). This may provide certain clues and insights as to how we may treat the osteoblastic metastases in advanced prostate cancer patients.

Acknowledgements

This study was supported by a grant to Dr. M. Koutsilieris from the Cancer Research Society of Montreal (CRS Inc., Montreal). Dr. Koutsilieris is a scholar (chercheur boursier) of the 'Fonds de la Recherche en Santé du Quebec' (FRSQ).

REFERENCES

1. Druker BJ, Mamon HJ and Roberts TM (1989) N Engl J Med 321:1383-1391
2. Baserga R, Calabretta B, Travali S, Jaskulski D, Lipson KE and DeRiel JK (1988) Ann NY Acad Sci 551:283-288
3 Weinberg RA (1985) Science 230:770-776
4. Bishop JM (1985) Cell 42:23-38
5. Galeotti T and Masotti L (1992) In: Ohnishi ST, Ohnishi T (eds) Cellular Membrane: a key to disease processes, pp 314-335

6. Talmage JE and Filder IJ (1982) In: Stoll BA (ed) Prolonged Arrest of Cancer, pp 137-163
7 Way M and Weeds A (1990) Nature (News and Views) 344:292-293
8. Finidori J, Friederich E, Kwiatkowski DJ and Louvard D (1992) J. Cell Biol 116:1145-1155
9. Janmey PA, Hvidt S, Lamb J, Stossel TP (1990) Nature 345:89-92
10. Cunningham CC, Byers HR, Hartwig JH, Stossel TP and Kwiatkowski DJ (1989) J Cell Biol 109:277-282.
11. Cunningham CC (1992) Cancer Metastasis Rev 11:69-77
12. Luna EJ and Hitt AL (1992) Science 258:955-963
13. Burridge K, Petch LA and Romer LH (1992) Current Biol 10:537-539
14. Zachary I and Rozengurt E (1992) Cell 71:891-894
15. Fidler IJ (1991) BioEssays 13:551-554
16. Stoll BA (1983) In: B Stoll A, Parbhoo A (eds) Bone Metastasis: monitoring and treatment, pp 65-72
17. Beliveau RE and Spencer RP (1975) Cancer 36:359-364
18. Tofe AJ, Francis MD and Harvey WJ (1975) J Nucl Med 16:986-991
19. Krishnamurthy GT, Tubis M, Hiss J and Bland WH (1977) J A M A 237:2504-258
20. Batson OV (1940) Ann Surg 112:138-144
21. Warren S, Harris PN and Graves RC (1936) Arch Pathol 22:139-144
22. Fisher B and Fisher ER (1966) Surg Gynecol Obstet 122:791-802
23. Sharpe WS and McDonald JR (1942) Arch Pathol 17:312-322
24. Galasko CSB (1982) Clin Orthopaed Rel Res 12:20-25
25. Koutsilieris M and Tolis G (1985) Prostate 7:31-36
26. Koutsilieris M, Faure N, Tolis G, Laroche B, Robert G and Ackman CFD (1986) Urology, 27:221-228
27. Tolis G, Koutsilieris M, Herrera R, Stellos A, Martinez, A and Dufresne MD (1984) Med Oncol Tumor Pharmachother 1:129-134
28. Nicolson GL (1988) Cancer and Metastasis Rev 7:143-158
29. Killion JJ and Fidler IJ (1989) Seminars in Oncology 16:106-114
30. Galasko CSB (1976) Nature 263:507-508
31. Koutsilieris M (1987) Ph.D. Thesis, McGill University
32. Koutsilieris M, Rabbani SA and Goltzman D (1986) Prostate 9:109
33. Koutsilieris M (1988) In Vivo 2:377-182
34. Koutsilieris M, Rabbani SA and Goltzman D (1987) J Endocrinol 115:447-453
35. Koutsilieris M., Rabbani SA, Bennett HPJ and Goltzman D (1987) J Clin Invest 80:941-9452
36. Koutsilieris M (1988) Arch Hell Med Soc 5:218-226
37. Jacobs SS and Lawson RK (1980) Urology 16:488-494
38. Jacobs SC, Pinka D and Lawson RK (1979) Invest Urol 17:195-203
39. Nishi N, Matuo Y and Kunitomi K (1988) Prostate 13:39-44
40. Simpson E, Harrod J, Eilon G, Jacobs JW and Mundy GR (1985) Endocrinology117:1615-1619

41. Perkel VS, Mohan S, Herring SJ, Baylink DJ and Linkhart TA (1990) Cancer Res 50:6902-6908
42. Rabbani S, Desjardins J, Bell AW, Danville D, Mazar A, Henkin J and Goltzman D (1990) Biochem Biophys Res Commun 173:1058-1064
43. Rabbani SA, Mazar AP, Bernier SM, Haq M, Bolivar I, Henkin J and Goltzman D (1992) J Biol Chem. 267:1-5
44. Hienert G, Kirchheimer JC, Pflüger H and Binder BR (1988) J Urol 140:1466-1472
45. Hienert G, Kirchheimer JC, Christ G, Fflüger H and Binder BR (1988) Eur Urol 15:256-262
46. Camiolo SM, Markus G, Englander LS, Siuta, MR, Hobika GH and Kohga S (1984) Cancer Res 44:311-316
47. Rennie, PS, Bowden JF, Bruchovsky N and Cheng H (1988) Biochem J 252:759-764
48. Andreasen PA, Kristensen P, Lund LR and Dano K (1990) Endocrinology 126:2567-2662
49. Pollard M and Luckert PH (1985) J Natl Cancer Inst 75:949-953
50. Pollard M, Luckert PH and Scheu J (1988) Cancer 61:2027-2036
51. Pollard M and Luckert PH (1984) Prostate 5:661-668
52. Koutsilieris M (1992) In Vivo 6:199-204
53. Langeland N, Guttormsen AB and Holmsen H (1986) Eur J Surg Oncol 12:221-224
54. Canalis E, McCarthy M and Centrella M (1988) Endocrinology 122:22-28
55. Polychronakos C, Janthly U, Lehoux JG and Koutsilieris M (1991) Prostate 19:313-318
56. Koutsilieris M and Polychronakos C (1992) Anticancer Res 12:905-912
57. Binoux M, Roghani M, Hossenlopp P, Hardouin S and Goumelen M (1991) Acta Endocrinol 124:241-258
58. Koutsilieris M, Frenette G, Lazure C, Govindan MV and Polychronakos C (1993) Anticancer Res, in press
59. Koutsilieris M (1993) Anticancer Res, in press
60. Lyons RM, Keski-Oja J and Moses HL (1988) J Cell Biol 106:1656-1666
61. Laiho M, Saksela O, Andreasen PA and Keski-Oja J (1986) J Cell Biol 103:2403-2412
62. Laiho M, Saksela O and Keski-Oja J (1987) J Biol Chem 262:17467-17472
63. Pfeilschifter J, Erdmann J, Schmidt W, Naumann A, Minne HW and Ziegler R (1990) Endocrinology 126:703-710
64. Gehron-Robey P, Young MF, Flanders K, Roche NS, Kondaiah P, Reddi AH, Termine JD, Sporn MB and Roberts AB.(1987) J Cell Biol 105:457-464
65. Ellis V, Pyke C, Eriksen J, Solberg H and Dano K (1992) Ann NY Acad Sci 667:13-31
66. Koutsilieris M, Grondin F and Lehoux JG (1992) Anticancer Res 12:899-904

CHAPTER 22

RECEPTORS FOR VASOACTIVE PEPTIDES AND THEIR IMPLICATIONS IN PHYSIOPATHOLOGY

Domenico Regoli, Audrey Claing and Pedro D'Orléans-Juste

Department of Pharmacology
University of Sherbrooke, Sherbrooke, CANADA

INTRODUCTION

Among the numerous peptides that modulate the functions of the cardiovascular system, angiotensin, the kinins, substance P and its homologues, and the recently identified new group of agents, the endothelins, have received much attention (1,2) because of their implication in physiology and physiopathology. The vasoconstrictors, angiotensin and endothelins, as well as the vasodilators, bradykinin and substance P, act through specific receptors that are widespread in organs and tissues to serve various functions. Receptors for vasoactive peptides are particularly abundant in the cardiovascular system. In peripheral vessels, receptors for peptides have been found in endothelia, smooth muscle fibers, nerve terminals and are presumably present in other cells, such as fibroblasts, macrophages, mastocytes and others (Fig. 1).

Naturally-occurring agents and peptides in particular do not cross cell membranes or similar barriers and act on the cell surface through specific receptors. Naturally-occurring agents and their specific cellular targets, the receptors, constitute the basic components of functional systems and biological actions. These peptides and their receptors have therefore to be considered together, since they are complementary and

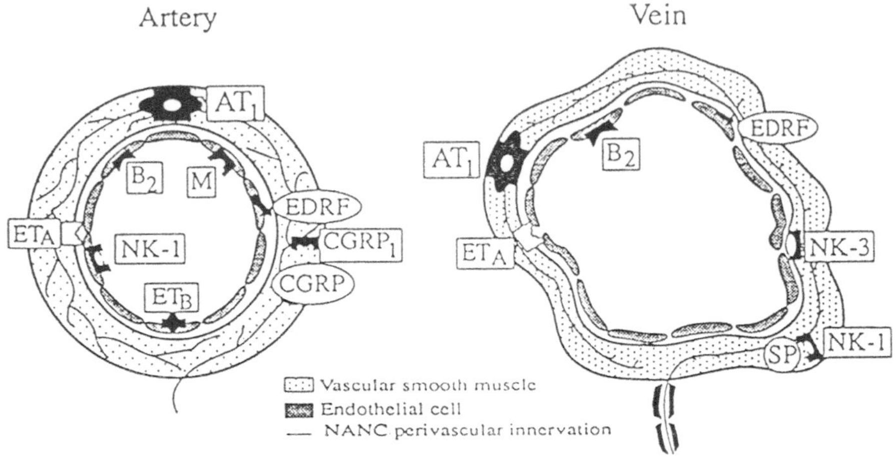

Figure 1. Cross sections of an arterial and a venous vessel. Receptors for the vasoconstrictors, angiotensin and endothelin-1 are located primarily in the smooth muscle fibers, while those of the vasodilators, bradykinin and substance P are in the arterial endothelium, from which they promote the release of EDRF (NO). In some vessels, kinin B_1 and neurokinin NK-1 receptors are found in the arterial endothelium and especially the venous smooth muscles. Vasoactive peptides modulate the release of neurotransmitters (especially noradrenaline) from the sympathetic nerves.

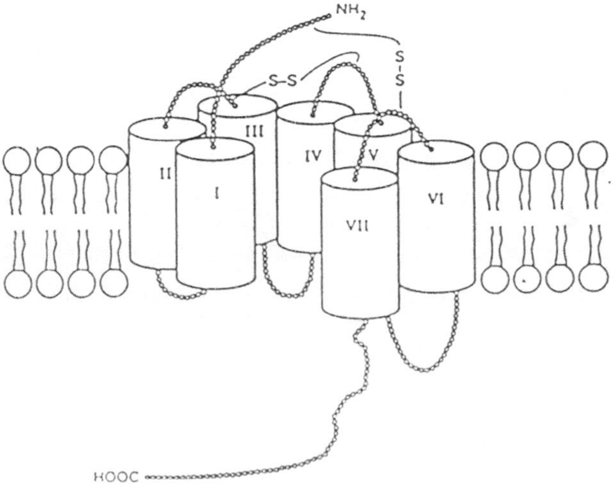

Figure 2. Schematic representation of the angiotensin II receptors. In several respects, this receptor is consistent with the convserved nature of the seven-transmembrane domains among G-protein-coupled receptors in general.

essential for the system's biological activities. They are also potential therapeutic targets for new drugs[1].

In recent years, the membrane proteins (generally small proteins whose molecular weight varies between 40,000 and 50,000) (see Table 1) have been cloned and shown to belong to a large family, the rhodopsin-like molecules, which are linked to G-proteins and to intracellular second messengers. An example of such a protein is given in Figure 2 by the structure of the angiotensin II receptor (3). Indeed, the peptide receptors are complex membraneous structures composed of seven hydrophobic domains, which are implanted in the membrane and connected through extra and intracellular loops. Extracellular loops and the superficial and middle layers of the transmembrane domains are the site of attack of the peptides, while the intracellular loops, especially the third and the initial part of the fourth loop, make contact with G-proteins, probably through cationic residues, such as Lys (K) or Arg (R) and through aromatic residues (Phe (F)). The receptor in its resting state appears to be coupled to the G-protein and therefore ready for activation of the enzymatic processes that lead to production of the second messengers in its active state.

Each system can be described in molecular, pharmacological or biochemical terms and can be characterized using biological and binding assays with agonists and antagonists that should however be as specific and as selective as possible for each receptor system in order to obtain reliable and comparable data. The major features of each system analyzed in the present chapter are presented in Tables 1A and 1B. Thus, angiotensin acts on two receptors, AT_1 and AT_2. The first has been cloned and shown to be a small protein ($\approx$41,000) composed of 359 residues (4,5). Apparent affinities observed in biological assays for angiotensin II and for DuP 753 are in the order of 8.4 to 8.9 and are very similar to those measured with the binding on membranes of rabbit adrenal cortex or of COS cells that have been made to express the AT_1 receptor by priming with the cDNA. Lower values of apparent affinities are observed in the biological tests of the xenopus oocyte sites in which changes of peak current are recorded.

As to AT_2, this is still a poorly defined entity devoid until now of biological activities, identified only by the binding of compounds such as [pN.Phe^6]Ang II, PD-123177, which have not been characterized as either agonists or antagonists. This binding site has recently been cloned and found to be a putative seven-transmembrane domain receptor with 32% to 34% homology with AT_1 receptors (6,7). Photoaffinity studies in human myometrium suggest that the AT_2 receptor is significantly larger (58,000) than the AT_1 receptor (58,000) (8). Apparent affinities of agonists evaluated in the rabbit aorta (contraction, pD_2 8.8) are very similar to those measured with binding assays (K_i 20) on the rat adrenal gland or with binding assays to COS cells (K_i 23) that were made to express AT_1 receptors. These values are similar to the

[1] Agonist-receptor system: Functional unit that subserves physiological functions and pathological events and is a potential target of therapeutic agents.

Table 1A

Major features of receptors for vasoactive peptides

		ANGIOTENSIN		ENDOTHELINS	
		AT_1	AT_2	ET_A	ET_B
RECEPTOR MOLECULAR WEIGHT AND N° OF RESIDUES		41093 359	41303 3363	48516 427	49643 441
RECEPTOR CATEGORY					
2ND MESSENGER		IP_3/Ca^{2+}	--	IP_3/Ca^{2+}	IP_3/Ca^{2+}
PHARMACOLOGICAL EFFECT	PREPARATION	Rabbit aorta	--	Rat aorta	Rat aorta
	EFFECT	Contraction	--	Contraction	Relaxation
SELECTIVE AGONISTS	STRUCTURE	Angiotensin II	--	Endothelin-1	Endothelin-3
	pD_2	8.86	--	8.35	≈ 8.0
SELECTIVE ANTAGONISTS	STRUCTURE	Dup 753	--	BQ-123	IRL-1038
	pA_2	8.4	--	6.90	6.0
BINDING ASSAY SYSTEMS		Rat adrenal gland	Human myometrium	Porcine aortic VSM	Porcine cerebellum
SELECTIVE LIGANDS	STRUCTURE	Angiotensin II (DuP 753)	Angiotensin II (CGP 42112A)	Endothelin-1 (BQ-123)	Endothelin-1 (BQ-3020)
	IC_{50} (nM)	20	0.2	7.3	0.2
MOLECULAR BIOLOGY SYSTEMS		COS-7	COS-7	CHO	CHO
BINDING	STRUCTURE	Dup 753	PD 123719	BQ-123	Sarafotoxin 6C
	K_i (nM)	23	1.7	17	0.29
BIOLOGICAL ACTIVITY	STRUCTURE	Angiotensin II	Angiotensin II	Endothelin-1	Sarafotoxin 6C
	SYSTEM	Xenopus oocyte	PC12W	CHO	CHO
	EFFECT	↑ Peak	Phosphotyrosine ↓ Phosphatase	↑ IP_3	↑ IP_3
	K_i (nM)	4.39	0.1	0.2-0.3	0.2

value (pD_2 8.0) estimated for the peak current curve obtained on the xenopus oocyte, which was made to express the AT_1 receptor. Of particular significance is the correspondence, as illustrated in Table 1, between pD_2 values reflecting the contractile effect (on the rabbit aorta), the IC_{50} values reflecting the binding affinity in the rabbit adrenal cortex, and the curve of IP_3 activation (obtained in the rat adrenal gland) reflecting the production of the second messenger. These data can be interpreted as if the interaction of angiotensin II with its AT_1 receptor follows the same pattern as the production of the second messenger and the contraction of the vascular smooth muscle. This suggests that each agonist receptor interaction contributes a quantum of biological effect, as it can be expected from the classical occupation theory. These data also suggest that the system has no spare receptors.

Receptors for vasoactive peptides

The similarity of values obtained on various parameters with agonists and with selective antagonists, such as DUP 753, provides the best evidence that the same functional entity is involved in all the processes concerning the AT_1 receptor and analyzed in Table 1A.

The other vasoconstrictors, endothelins, interact also with at least two receptor types, ET_A (9) and ET_B (10), which are proteins of similar molecular weight and structure as the angiotensin receptor sites, and belong to the rhodopsin category. Activation of these proteins has been shown to induce accumulation of phosphoinositides and cytoplasmic calcium. The two receptors have been studied in various pharmacological preparations of which the rat aorta is among the most sensitive and interesting, since it contains both ET_A and ET_B receptors which mediate respectively contraction (ET_A) and relaxation (ET_B) of this tissue. The comparison of pharmacological, biochemical (to plasma membranes) and molecular biological (to intact cells) affinities for endothelin-1 by ET_A show very similar values comprised between 8.3 and 10.0. In general, the values of IC_{50} measured with binding to membrane preparations or to cells in culture are higher than those obtained in biological assays (see also ET_B system), probably because of the inertia of the multicellular tissue and of the recording systems.

ET_A and ET_B antagonists are still in the early stage of development and the compounds available at present (e.g. BQ-123) show affinities much lower than those of the naturally-occuring agonists. Worthy of notice is the existence of a very good selective agonist for the ET_B receptor (11), BQ-3020, which shows a pD_2 of 9.7 (12).

Data obtained on the two kinin (B_1 and B_2) and the three neurokinin (NK-1, NK-2 and NK-3) receptors are presented in Table 1B. Molecular biological data for B_1 receptors are not yet available and therefore this entity cannot be fully described. It has however been expressed in the same system as the B_2 receptor (CCB-16LU) and appears to be contained in a gene that is different from that of the B_2 (13). This receptor is found in vascular smooth muscles (rabbit aorta) as well as in other cells (intestinal smooth muscle) (14), human lung fibroblasts (15,16) (see review by Marceau and Regoli, 1991). B_1 receptors show very similar affinities for agonists, such as [Lys]desArg^{9}BK and antagonists, such as [Leu^9]desArg^{10}KD, and for desArg^{10}KD in binding assays (16). B_1 receptors are induced by *in vitro* incubation of tissues, possibly because of trace amounts of cytokinins in the organ baths and solutions (17) and *in vivo* via the administration of lipopolysacharides or other similar toxins (18). B_1 receptor induction appears to be due to new receptor synthesis since it is inhibited by cycloheximide (19). Up and down regulation of kinin and angiotensin receptors and their implication in physiopathology will be described at the end of this chapter.

Table 1B
Major features of receptors for vasoactive peptides

		KININS		NEUROKININS		
		B_1	B_2	NK-1	NK-2	NK-3
RECEPTOR MOL WEIGHT AND N° OF RESIDUES			41442 364	46385 407	44425 398	51104 452
RECEPTOR CATEGORY						
2ND MESSENGER		IP_3/Ca^{2+}	IP_3/Ca^{2+}	IP_3/Ca^{2+}	IP_3/Ca^{2+}	IP_3/Ca^{2+}
PHARMACOLOGICAL EFFECT	PREPARATION	Rabbit Aorta	Rabbit jugular vein	Rabbit vena cava	Rabbit pulmonary artery	Rat portal vein
	EFFECT	Contraction	Contraction	Contraction	Contraction	Contraction
SELECTIVE ANTAGONISTS	STRUCTURE	Lys, des Arg^9 BK	Lys-BK	[Sar^9,Met $(O_2)^{11}$]-SP	[ß-Ala^8] NKA (4-10)	Senktide
	pD_2	8.60	8.63	8.66	8.60	8.30
SELECTIVE ANTAGONIST	STRUCTURE	[Leu^8] des Arg^{10} KD	HOE-140	CP-96 345	SR48968	[Trp^7,ß-Ala^8]NKA (4-10)
	pA_2	8.37	9.17	9.50	9.60	7.46
BINDING ASSAY SYSTEMS		Rabbit aorta SMC	Guinea pig epithelium	Rat brain	Rat duodenum	Guinea pig brain
SELECTIVE LIGANDS	STRUCTURE	desArg^{10}KD	Bradykinin	[Sar^9,Met$(O_2)^{11}$]-SP	Neurokinin A	Senktide
	K_i	0.97	0.96	0.93	0.92	0.93
MOLECULAR BIOLOGY SYSTEMS			COS-7	COS-CELL	CHO	COS-1
BINDING	STRUCTURE		[Hyp^3]Tyr$(Me)^8$BK	Substance P	Neurokinin A	Neurokinin B
	K_i (nM)		1.0	0.72	0.93	0.42
BIOLOGICAL ACTIVITY	STRUCTURE		Bradykinin	Substance P	Neurokinin A	Neurokinin B
	SYSTEM		CHO	COS-7	COS-7	COS-7
	EFFECT		↑ $[Ca^{2+}]$	↑ IP_3	↑ IP_3	IP_3
	K_i (nM)		1.02	0.99	0.98	0.98

The B_2 receptor is considered a stable component of the plasma membrane of many cells and has been implicated in a variety of physiological and physiopathological conditions (20). It has been cloned, even the human receptor (21,22), and shown to be a rhodopsin-like structure of 41,442 acting through the activation of phosphoinositides and the accumulation of intracellular calcium. This mechanism has been shown to be effective also for the B_1 receptor. B_2 receptors show high affinity for bradykinin and related kinins, as indicated by the pD_2 values of agonists (pD_2 8.63 in the rabbit jugular vein) and antagonists (HOE, pA_2 9.17), affinity of bradykinin in the guinea pig epithelium (K_i 1), affinity of bradykinin in COH cells in culture, as measured by the calcium permeability 1.02, as well as for the antagonists whose affinity varies between 9.2 and 10.2 (see Table 1B). The B_2 receptor has been implicated in a variety of pathological conditions, such as inflammation, asthma and allergic diseases (23,24,25). Recently, it has been implicated in chronic inflammation and pain (26).

Neurokinin receptors of three types, NK-1, NK-2 and NK-3 (27) have been cloned (even in humans, see Table 1B) and shown to be, like those of the other agents, namely rhodopsin-like structures which all work through the activation of phosphoinositide and accumulation of intracellular calcium. NK-1 receptors are found in arteries and mediate vasodilatation, leading to hypotension, and in the veins where they subserve venoconstriction and plasma extravasation. NK-1 receptors are present in large numbers in the brain and in the spinal cord to subserve numerous central effects of substance P. Numerical values shown in Table 1B indicate striking similarities within affinity values measured with agonists and highly potent antagonists in pharmacological, biochemical and molecular biological assays. Much of the same can be said about the NK-2 receptor, which was among the first to be cloned and expressed in xenopus oocytes (28). This functional site is present in the smooth muscle of pulmonary arteries and, in general, in the respiratory tract, trachea and bronchial smooth muscles, as well as in the gastrointestinal tract, especially in man, where NK-2 is the peripheral receptor most represented and most important functionally. It sustains bronchoconstriction, intestinal peristalsis and urine excretion. A good similarity is shown in Table 1B between affinities of NKA ([β-Ala^8]NKA(4-10)) and the newly discovered antagonist, SR 48968, for the NK-2 site.

NK-3 receptors belong also to rhodopsin-like compounds and work through activation of phosphoinositides and calcium (29). Unlike NK-2, NK-3 binding sites are present in various parts of the brain and are quite abundant throughout the cortex (30). They have been shown to be present at the spinal cord level and reduce pain, possibly through the release of opioids (31). They have also been found in the periphery, in some veins and in nerve terminals, both parasympathetic (guinea pig ileum) and sensory terminals, to facilitate the micturition reflex (32). A very good correspondence has been found between data obtained with pharmacological and biochemical assays, while the available antagonists are rather weak. Data reported in Table 1B shows that for each neurokinin receptor, very selective agonists have been identified and found to be highly potent. These compounds have been analyzed by Regoli et al. (31).

Table 2

Physiological roles of peptide receptors and their implication in physiopathology

AT_1	Vasoconstriction, sympathetic hypertone, aldosterone, vasopressin secretion, neurosecretion, cell proliferation, cell differentiation, HYPERTENSION, HEART FAILURE
AT_2	No role
B_1	Vasodilatation, venoconstriction, cell proliferation, collagen secretion, ENDOTOXIC SHOCK, CHRONIC PAIN
B_2	Vasodilatation, intestinal peristalsis, bronchoconstriction, plasma extravasation, INFLAMMATION, ACUTE PAIN, ASTHMA
NK-1	Vasodilatation, plasma extravasation, intestinal or bladder contraction, NEUROGENIC INFLAMMATION, PAIN, HEADACHE
NK-2	Visceral motility, bronchoconstriction, sympathetic hypertone, ASTHMA, COUGH, IRRITABLE BOWEL OR BLADDER
NK-3	Venoconstriction, visceral motility (parasympathetic tone), bradycardia, IRRITABLE BOWEL SYNDROME, ANALGESIA? HYPERTENSION?
ET_A	Vasoconstriction, stimulation of aldosterone and prostacyclin secretion, bronchoconstriction (human), ACUTE NEPHROTOXICITY
ET_B	Vasodilatation, *ex vivo* platelet aggregation, stimulation of thromboxane A_2 (guinea pig)

A comparison of affinities mesured in biological, biochemical and molecular biological assays (CHO or COS cells) is presented in table 1B for the B_2 and neurokinin receptors: a good similarity of pD_2, Ki (binding) and Ki (COS or CHO cells) has been found for all receptors analysed in table 1B.

The most relevant physiological actions of each peptide-receptor system as well as the implications in physiopathology are briefly underlined in Table 2. Angiotensin II is a potent vasoconstrictor that contributes to the maintenance of blood pressure level, both by its peripheral action on arterial vascular smooth muscles and by increasing the volume of fluid through retention of sodium (aldosterone secretion) and water (vasopressin secretion). Angiotensin also increases sympathetic tone through central and peripheral actions. The various effects of angiotensin concur to increase blood pressure and hypersecretion of renin lead to hyperproduction of angiotensin and hypertension. The functional activity of the renin-angiotensin system is also increased in heart failure, a syndrome in which converting enzyme inhibitors (that prevent the release of angiotensin II) have been found to be of therapeutic value. Accumulating evidence indicates that angiotensin II also participates in neurosecretion, cell proliferation and cell differentiation (3). No functional role has yet been identified for AT_2, which remains a binding site.

Receptors for bradykinin and related kinins have been divided into B_1 and B_2 (19). B_1 may be generated in pathological conditions and, following the administration of toxins or the increased generation of interleukins (33). B_1 funtional sites have been shown to be present in vascular smooth muscles (both arterial and venous) and visceral smooth muscles (intestine, urinary bladder, etc.) and even in endothelia. B_1 receptors mediate both vasodilatation (in peripheral vessels) and vasoconstriction in large vessels (e.g. the rabbit aorta) and stimulation of visceral smooth muscles. B_1 receptors are generated *de novo* and show large variations between normal conditions (few receptors) and pathological conditions (increased number of receptors) (19). B_1 receptors have been implicated in endotoxic shock and the irreversible hypotension that characterizes the syndrome, as well as in chronic pain (34).

B_2 receptors are widespread in endothelial and vascular smooth muscles as well as in a variety of visceral smooth muscles. They are responsible for the hypotensive effects of the kinins and for their ability to induce plasma extravasation, stimulation of intestinal motility, bronchoconstriction and activation of sensory nerve fibers to produce pain. In physiopathology, B_2 receptors and their kinins have been implicated in the inflammatory processes, particularly the early phase of inflammation in acute pain, the kinins being the most active stimulants of sensory (pain) afferents. Kinins may play a very important role in asthma through direct bronchoconstriction by promoting the release of sensory neuropeptides (35).

Neurokinins are a group of neuropeptides which are contained in sensory fibers and are released by direct activation of these fibers or by antidromic stimulation (36,37). Release of substance P and neurokinin A as well as some N-terminal extensions of neurokinin A (NPK and NPγ) lead to the activation of neurokinin receptors which have been classifed into NK-1, NK-2 and NK-3 (27) and sustain physiological and pathological roles, as indicated in Table 2.

NK-1 receptors are similar to the B_2 and have been implicated in vasodilatation and venoconstriction that lead to plasma extravasation and to the stimulation of a variety of smooth muscles, the intestine, the bladder and other organs. NK-1 receptors are widespread in the central nervous system and sustain various central effects of which the most relevant appear to be transmission of pain stimulation and regulation of cardiovascular functions (38,39). In physiopathology, NK-1 receptors have been shown to play a role in neurogenic inflammation, pain and headache (31). NK-1 receptor antagonists are, at present, among the most promising agents for the development of new anti-inflammatory and analgesic drugs. Recent data point to the possible utilization of NK-1 receptor antagonists as antiemetics (40).

NK-2 receptors are present in the smooth muscle of various viscera, especially in man (41). They mediate intestinal motility, bronchoconstriction, contraction of the bladder and the cyliary muscle. The most important physiopathological role of this functional site appears to be in asthma and irritable diseases of the intestine or the bladder (42). NK-2 receptor antagonists appear to act as an antitussive, at least in the guinea pig (43).

NK-3 receptors have been firstly demonstrated in the central nervous system where they may subserve various central actions, including the stimulation of vasopressin and oxytocin and the resulting vasoconstriction and increase in blood pressure (44). NK-3 receptors have been found in some veins (for instance, the rat portal vein), in the parasympathetic terminal and the mesenteric plexus, as well as the dorsal horn of the spinal cord. Two major roles have been suggested for these receptors, namely the stimulation of intestinal motility (45) and the promotion of opioid secretion at the spinal level with the resulting analgesia (31,46). Recent data point to the involvement of NK-3 receptors in the regulation of cardiovascular functions (47). NK-3 receptors in the central nervous system may contribute to high blood pressure by indirect effects (through promoting vasopressin release (44)) or other still unknown mechanisms.

As previously mentioned, the presence of ET_A receptors in the vascular smooth muscle has been predominantly associated with vasoconstriction. In most ET_A-containing vascular tissues, endothelin-1 has be found to be 10 to 100 times more potent than ET-3. Some of the typical preparations where ET_A receptors have been found are the rabbit aorta, the porcine coronary artery, the rabbit carotid artery, the rat uterus, the dog saphenous vein (49,50,51,52), as well as the arterial and venous mesenteric vasculature of the rat (53). It is worthy of mention that the contractile properties of human vascular tissues have been predominantly associated with the activation of ET_A receptors (54), thus justifying the development of selective ET_A receptor antagonists (BQ-123, BQ-610) as potential therapeutic agents which may be used to investigate the role of endogenous endothelin as a modulator of vascular tone in man.

The other receptor type which has been described by Sakurai *et al.* (1990) (10) is clearly associated with the hypotensive effects of endothelin-1. Indeed, systemically administered endothelin-1 in conscious and anesthetized rats and rabbits induces a transient hypotensive response followed by a protracted increase of blood pressure in these animal models. The initial hypotensive effect of ET-1 was initially associated with the activation of ET_B receptors on which ET-1 and ET-3 are equipotent. Warner *et al.* (1989) (55) were the first to suggest that distinct receptors could mediate the vasoconstrictive properties of endothelins as opposed to their vasodilatory properties, the former being mediated by ET_A receptors and the later by ET_B receptors. As illustrated in Table 2, it is also of interest to point out that ET_B receptor activation has been associated with some of the vasoconstrictive properties of endothelins, in particular in blood vessels of venous origin (human saphenous vein, canine, femoral and mesenteric veins). In addition, a typical preparation into which ET_B receptor activation has been associated with the response of endothelin is the rabbit pulmonary artery (51).

Secondly, intravascular administration of endothelin-1 in conscious and unanesthetized animals has revealed that this potent factor induces inhibition of *ex vivo* platelet aggregation, suggesting that endothelins, when administered systemically, triggers the release of prostacyclin which is also a potent modulator of the vasoconstrictive properties of those peptides. Like EDRF, the release of prostacyclin in some *in vivo* models has been associated with ET_B receptor activation which is mainly located on the vascular endothelium (51). In contrast, endothelin also triggers an ET_A mediated release of prostacyclin from the rat pulmonary and rabbit renal vasculatures (56,57). These particular prostanoid-releasing properties of endothelin-1 were abolished when these organs were pretreated with a selective ET_A receptor antagonist, BQ-123. This later observation clearly illustrated that endothelin-1 may trigger the release of prostacyclin via the activation of ET_A receptors in perfused organs in contrast to the anti-aggregatory properties of the same peptide in the rabbit, which appears to be an ET_B-related phenomenon. This discrepancy illustrates that endothelin-1 may enhance the release of endogeneous prostacyclin by different receptor types which are located on different target cells.

Since, as mentioned previously in this chapter, the endothelin antagonists are in an early phase of development, the future use of selective ET_B and ET_A agonists and antagonists may further our knowledge on the physiological contribution of the two types of receptors in physiological and pathological conditions. Nonetheless, recent reviews have suggested an important role for endothelin in the cardiovascular pathologies, such as congestive heart failure without systemic hypertension (58) and heart failure (59).

As illustrated in another recent review (60), endothelin-1 may be used as a plasmatic marker of cardiovascular injuries. Whether the use of selective antagonists will be beneficial in the prevention or treatment of these pathologies remains to be fully investigated. Nonetheless, the selective ET_A receptor antagonist, BQ-123, has been successfully used in experimental models of cyclosporine (61) and ischemia (62) induced nephrotoxicity in the rat. In these two models, BQ-123 has been shown to reverse the injury-induced reduction of glomerular filtration rate as well as tissue necrosis. It is also of interest that the use of selective ET_B receptor antagonists which interfere with the release of endothelin-1-induced release of thromboxane from the guinea pig pulmonary vasculature may have potential uses in the prevention of bronchospasm induced by this potent vasoactive peptide.

In conclusion:

- Receptors for the four peptide categories have been identified: they are genetic entities which have been characterized by pharmacological assays, binding and molecular biology.

- The angiotensin-AT_1 receptor system is definitely involved in the production and maintenance of high blood pressure and AT_1 antagonists appear to be promising new antihypertensive agents. Similar to ACE inhibitors, AT_1 antagonists may exert beneficial effects in heart failure and other cardiovascular diseases.

- The kinin-B_2 receptor system is important for peripheral vasodilatation and may contribute to the antihypertensive effect of ACE inhibitors. It appears to be involved in acute pain, inflammation and possible allergic reactions. The kinin-B_1 receptor system appears to sustain chronic pain and inflammatory reactions: it is modulated by interleukins.

- The neurokinin (SP)-NK-1 system has been implicated in pain transmission of the spinal cord and higher levels as well as in neurogenic inflammation resulting from the peripheral release of sensory neuropeptides. The neurokinin (NKA, NPK, NPγ)-NK-2 receptor system is playing a role in the smooth mucle tone of various organs and may contribute to bronchoconstriction, intestinal peristaltism and bladder motility. The neurokinin (NKB)-NK-3 receptor system is primarily active on the central nervous system and may play a role in the regulation of pain transmission as well as in cardiovascular homeostasis.

- ET_A receptor activation is predominantly associated with contraction and release of modulatory prostacyclin from the arterial vascular smooth muscle. In contrast, there is a growing body of evidence showing that the venous smooth muscle may contain both ET_A and ET_B receptors to be involved in the constriction of those particular vessels. Also, ET_B receptors located on the vascular endothelium are responsible for the release of endothelium-derived nitric oxide (EDNO) which plays a major role as a physiological antagonist of the potent constrictive and pressor properties of endothelins. The role of ET_B receptor activation and release of EDNO is particulary important in the renal vasculature which is exquisitely sensitive to endothelin-1.

Acknowledgements

The authors acknowledge the secretarial assistance of H. Morin. D.R. is a career investigator of the Medical Research Council of Canada. A.C. and P.D.J. are in receipt of a studentship and scholarship, respectively, from the Heart and Stroke Foundation of Canada.

REFERENCES

1. Rocha e Silva M, Beraldo WT and Rosenfeld G (1949) Am. J. Physiol. 156: 261-273.
2. Yanagisawa M, Kurihara H, Kimura S, Tomobe Y, Kobayashi M, Mitsui Y, Yazaki Y, Goto K, Masaki T (1988) Nature 332: 411-415.
3. Sandberg K (1994) Trends in Endocrinol. Met. 5: 28-35.

4. Murphy TJ, Alexander RW, Griendling KK, Runge MS and Bernstein KE (1991) Nature 351: 233-236.
5. Sasaki K, Yamano Y, Bardhan S, Iwai N, Murray JJ, Hasegawa M, Matsuda Y and Inagami T (1991) Nature 351: 230-233.
6. Kambayashi Y, Bardhan S, Takahashi K, Tsuzuki S, Inui H, Hamakubo T and Inagami T (1993) J. Biol Chem. 268: 24543-24546.
7. Mukoyama M, Nakajima M, Horiuchi M, Sasamura H, Pratt RE and Dzau VJ (1993) J. Biol. Chem. 268: 24539-24542.
8. Servant G, Boulay G, Bossé R, Escher E and Guillemette G (1993) Mol. Pharmacol. 43: 677-683.
9. Arai H, Hori S, Aramori I, Ohkubo H and Nakanishi S (1990) Nature 348: 730-732.
10. Sakurai T, Yanagisawa M, Takuwa Y, Miyazaki H, Kimura S, Goto K and Masaki T (1990) Nature 348: 732-735.
11. Ihara M, Noguchi K, Saeki T, Fukuroda T, Tsuchida S, Kimura S, Fukami T, Ishikawa K, Nishikibe M and Yano M (1992) Life Sci. 50: 247-255.
12. Saeki T, Ihara M, Fukuroda T, Yamagiwa M and Yano M (1991) Biochem. Biophys. Res. Commun. 179: 286-292.
13. Watling KJ (1992) Trends Pharm. Sc. 13: 266-269.
14. Gaginella TS, Kachur JF (1989) Am. J. Physiol. 256: G1-15.
15. Baenziger NL, Jong YJ, Yocum SA, Dalemar LR, Wilhelm B, Vavrek R, Stewart JM (1992) Eur. J. Cell Biol. 58: 71-80.
16. Marceau F and Regoli D (1991) In: Bradykinin Antagonists, ed. R.M. Burch, Marcel Dekker Inc. New York, USA. p. 33-49.
17. Marceau F, Barabé J, St-Pierre S and Regoli D. (1980) Can. J. Physiol. Pharmacol. 58: 536-542.
18. Regoli D, Marceau F and Lavigne J (1981) Eur. J. Pharmacol. 71: 105-115.
19. Regoli D and Barabé J (1980) Pharmacol. Rev. 32: 1-46.
20. Regoli D, Rhaleb N-E, Dion S and Drapeau G (1990) Trends Pharmacol. Sci. 11: 156-161.
21. Eggerickx D, Raspe E, Bertrand D, Vassart G and Parmentier M (1992) Biochem. Biophys. Res. Commun. 187: 1306- 1313.
22. Hess JF, Borkowski JA, Young GS, Strader CD and Ransom RW (1992) Biochem. Biophys. Res. Commun. 184: 260-268.
23. Simonsson BG, Skoogh BE, Bergh NP, Anderson R and Svedmyr N (1973) Respiration 30: 378-388.
24. Fuller RW, Dixon CM, Cuss FM and Barnes PJ (1987) Am. R. Respir. Dis. 135: 176-180.
25. Barnes PJ, Chung KF and Page CP (1988) Pharmacol. Rev. 40: 49-84.
26. Dray A and Perkins M (1993) Trends in Neurosci. 16: 99-104.
27. Henry JL (1987) In: Substance P and Related Neurokinins, eds. S.L. Henry, R. Couture, A.C. Cuello, G. Pelletier, R. Quirion and D. Regoli, Springer-Verlag, New York, p. 125.
28. Nakanishi S (1991) Ann. Rev. Neurosci. 14: 123-136.

29. Guard S and Watson SP (1991) Neurochem. Int. 18: 149-165.
30. Dam TV, Escher E and Quirion R (1990) Brain Res. 506: 175-179.
31. Regoli D, Drapeau G, Dion S and Couture R (1988) Trends Pharmacol. Sci. 9: 290-295.
32. Maggi CA, Giuliani S, Santicioli P, Abelli L, Regoli D and Meli A (1987) Eur. J. Pharmacol. 136: 189-205.
33. White MV and Kaliner MA (1992) J. Allergy and Clin. Immuno. 90: 699-704.
34. Watling KJ and Krause JE (1993) Trends Pharmacol. Sci. 14: 81-84.
35. Barnes PJ (1992) Agent Action 38: 432-438.
36. Cuello AC (1987) Neuropharmacol. 26: 971-979.
37. Marley P and Livett BG (1985) Neurobiol. 1: 201-283.
38. Jessel SJ and Iversen LL (1977) Nature 268: 433-434.
39. Yashpal K, Gauthier S and Henry JL (1987) J. Autonomic Nervous System 18: 93-103.
40. Tattersall FD, Rycroft W, Hargreaves RJ and Hill RG (1993) Eur. J. Pharmacol. 250: R5-R6.
41. Giuliani S, Barbanti G, Turini D, Quartara L, Rovero P, Giachetti A and Maggi A (1991) Eur. J. Pharmacol. 203: 365-370.
42. Maggi CA, Patacchini R, Rovero P and Giachetti A (1993) J. Autonom. Pharmacol. 13, 23-93.
43. Advenier C, Girard V, Naline E, Vilain P and Emonds-Alt X (1993) Eur. J. Pharmacol. 250: 169-171.
44. Takano Y and Kamiya Y (1991) Asian Pacif. J. Pharmacol. 6: 341-348.
45. Fosbraey P, Featherstone RL and Morton IK (1984) N.S. Arch. Pharmacol. 326: 111-115.
46. Helke CJ, Krause JE, Mantyh PW, Couture R and Bannon MJ (1990) Faseb J. 4: 1606-1615.
47. Tschöpe C, Picard P, Culmann J, Prat A, Itoi K, Regoli D, Unger T and Couture R (1992) Br. J. Pharmacol. 107: 750-755.
48. Vane JR, Anggard EE and Botting RM (1990) New Eng. J. Med. 323: 27-36.
49. Rae GA, Calixto JB and D'Orléans-Juste P (1993) Eur. J. Pharmacol. 240: 113-119.
50. Sogabe K, Nirei H, Shoubo M, Nomoto A, Ao S, Notsu Y and Ono T (1993) J. Pharmacol. Exp. Ther. 264: 1040-1046.
51. Ihara M, Saeki T, Fukuroda T, Kimura S, Ozaki S, Patel AC and Yano M (1992) Life Sci. 51: 47-52.
52. Moreland S, McMullen DM, Delaney CL, Lee VG and Hunt JT (1993) Biochem. Biophys. Res. Commun. 184: 100-106.
53. D'Orléans-Juste P, Claing A, Warner TD, Yano M and Télémaque S (1993) Br. J. Pharmacol. 110: 687-692.
54. Davenport AP, O'Reilly G, Molensar P, Maguire JJ, Kuc RE, Sharkey A, Bacon CR, Ferro A (1993) J Cardiovasc Pharmacol 22: 522-525.
55. Warner TD, Mitchell JA, de Nucci G and Vane JR (1989) J. Cardiovasc. Pharmacol. 13: S85-S88.

56. D'Orléans-Juste P, Télémaque S, Claing A, Ihara M and Yano M (1992) Br. J. Pharmacol. 105: 773-775.
57. Télémaque S, Gratton J-P, Claing A and D'Orléans-Juste P (1993) Eur. J. Pharmacol. 237: 275-281.
58. Pacher R, Bergler-Klein J, Globits S, Teufelsbauer H, Schuller M, Krauter A, Ogris E, Rodler S, Wutte M and Hartter E (1993) Am. J. Cardio. 71: 1293-1299.
59. Stewart DJ, Cernacek P, Costello KB and Rouleau JL (1992) Circ. 85: 510-517.
60. Battistini B, D'Orléans-Juste P and Sirois P (1993) Lab. Invest. 68: 600-28.
61. Fogo A, Hellings SE, Inagami T and Kon V (1992) Kidney International 42: 770-774.
62. Grover GJ, Sleph PG, Fox M and Trippodo NC (1992) J. Pharmacol. Exp. Ther. 263: 1074-1082.

INDEX

GPSR Compliance

The European Union's (EU) General Product Safety Regulation (GPSR) is a set of rules that requires consumer products to be safe and our obligations to ensure this.

If you have any concerns about our products, you can contact us on ProductSafety@springernature.com

In case Publisher is established outside the EU, the EU authorized representative is:

Springer Nature Customer Service Center GmbH
Europaplatz 3
69115 Heidelberg, Germany

Batch number: 10799746

Printed by Printforce, the Netherlands